SiC MATERIALS AND DEVICES

Volume 1

SELECTED TOPICS IN ELECTRONICS AND SYSTEMS

Editor-in-Chief: **M. S. Shur**

Published

Vol. 23: Oxide Reliability: A Summary of Silicon Oxide Wearout, Breakdown, and Reliability
ed. *D. J. Dumin*

Vol. 24: CMOS RF Modeling, Characterization and Applications
eds. *M. J. Deen and T. A. Fjeldly*

Vol. 25: Quantum Dots
eds. *E. Borovitskaya and M. S. Shur*

Vol. 26: Frontiers in Electronics: Future Chips
eds. *Y. S. Park, M. S. Shur and W. Tang*

Vol. 27: Intersubband Infrared Photodetectors
ed. *V. Ryzhii*

Vol. 28: Advanced Semiconductor Heterostructures: Novel Devices, Potential Device Applications and Basic Properties
eds. *M. Dutta and M. A. Stroscio*

Vol. 29: Compound Semiconductor Integrated Circuits
ed. *Tho T. Vu*

Vol. 30: Terahertz Sensing Technology — Vol. 1
Electronic Devices and Advanced Systems Technology
eds. *D. L. Woolard, W. R. Loerop and M. S. Shur*

Vol. 31: Advanced Device Modeling and Simulation
ed. *T. Grasser*

Vol. 32: Terahertz Sensing Technology — Vol. 2
Emerging Scientific Applications and Novel Device Concepts
eds. *D. L. Woolard, W. R. Loerop and M. S. Shur*

Vol. 33: GaN-Based Materials and Devices
eds. *M. S. Shur and R. F. Davis*

Vol. 34: Radiation Effects and Soft Errors in Integrated Circuits and Electronic Devices
eds. *R. D. Schrimpf and D. M. Fleetwood*

Vol. 35: Proceedings of the 2004 IEEE Lester Eastman Conference on High Performance Devices
ed. Robert E. Leoni III

Vol. 36: Breakdown Phenomena in Semiconductors and Semiconductor Devices
M. Levinshtein, J. Kostamovaara and S. Vainshtein

Vol. 37: Radiation Defect Engineering
Kozlovski V. and Abrosimova V.

Vol. 38: Design of High-Speed Communication Circuits
ed. R. Harjani

Vol. 39: High-Speed Optical Transceivers
eds. *Y. Liu and H. Yang*

Selected Topics in Electronics and Systems – Vol. 40

SiC MATERIALS AND DEVICES

Volume 1

edited by

Michael Shur
Rensselaer Polytechnic Institute, USA

Sergey Rumyantsev
Rensselaer Polytechnic Institute, USA
Ioffe Institute of the Russian Academy of Sciences, Russia

Michael Levinshtein
Ioffe Institute of the Russian Academy of Sciences, Russia

World Scientific

NEW JERSEY • LONDON • SINGAPORE • BEIJING • SHANGHAI • HONG KONG • TAIPEI • CHENNAI

Published by

World Scientific Publishing Co. Pte. Ltd.

5 Toh Tuck Link, Singapore 596224

USA office: 27 Warren Street, Suite 401-402, Hackensack, NJ 07601

UK office: 57 Shelton Street, Covent Garden, London WC2H 9HE

British Library Cataloguing-in-Publication Data
A catalogue record for this book is available from the British Library.

SiC MATERIALS AND DEVICES
Volume 1

ISBN 981-256-835-2

Editor: Tjan Kwang Wei

Printed in Singapore by Mainland Press

PREFACE

Silicon carbide has been known investigated since 1907 when Captain H. J. Round demonstrated yellow and blue emission by applying bias between a metal needle and a SiC crystal. In 1923, a Russian scientist, Oleg Losev, discovered two types of light emission from SiC – the emission that we would now call "pre-breakdown" light and the electroluminescent emission. The potential of using SiC in semiconductor electronics was already recognized about a half of century ago. The most remarkable SiC properties are:

- Wide band gap (3 to 3.3 eV for different polytypes
- Very large avalanche breakdown field (2.5 – 5 MV/cm)
- High thermal conductivity (3 – 4.9 W/cm K)
- High maximum operating temperature (up to 1,000 °C)
- Chemical inertness and radiation hardness

However, some of these potential advantages also cause exceptional technological difficulties in getting silicon carbide material to reach the device quality, and it has taken a few decades to travel the road from basic research to commercialization.

The breakthrough was reached in early 1990s, when nearly all basic semiconductor devices – p-n diodes, Schottky diodes, Metal Oxide Semiconductor Field Effect Transistor (MOSFETs), Metal Semiconductor Field Effect Transistor (MESFETs), bipolar junction transistors, thyristors, IMPATT diodes, and solar blind photodetectors – have been successfully demonstrated.

The first commercial SiC devices – power switching Schottky diodes and high temperature MESFETs – are now on the market.

For nitride based devices, silicon carbide has become a substrate of choice, because of its excellent thermal conductivity and decent lattice match. This includes both electronic and blue light emitting diodes and lasers, and, together with nitride based semiconductors, silicon carbide is now in the forefront of the semiconductor research.

The contributors to this two-volume book are recognized leaders in SiC technology and materials and device research. They provide complete and up-to-date review of the state-of-the-art.

This first volume has chapters on SiC material properties (Chapter 1), SiC homo- and hetero-epitaxy (Chapter 2), Ohmic contacts to SiC (Chapters 3), Schottky barrier diodes (Chapter 4), high power p-i-n rectifiers (Chapter 5), microwave SiC diodes (Chapter 6), SiC thyristors (Chapter 7), and SiC static induction transistors (Chapter 8).

This book will be useful for technologists, scientists, engineers, and graduate students who are working on silicon carbide or other wide band gap materials and devices. The book can also be used as a supplementary textbook for graduate courses on silicon carbide and wide band gap semiconductor technology.

Michael Shur
Sergey Rumyantsev
Michael Levinshtein

CONTENTS

International Journal of High Speed Electronics and Systems
Vol. 15, No. 4 (2005) 705–745

SiC MATERIAL PROPERTIES

GERHARD PENSL†, FLORIN CIOBANU, THOMAS FRANK, MICHAEL KRIEGER, SERGEY RESHANOV, FRANK SCHMID, MICHAEL WEIDNER

Institute of Applied Physics, University of Erlangen-Nürnberg, Staudtstrasse 7
D-91058 Erlangen, Germany
† *gerhard.pensl@physik.uni-erlangen.de*

This chapter briefly summarizes device-relevant material properties of the wide bandgap semiconductor silicon carbide. The polytypes 4H-, 6H- and 3C-SiC are predominantly considered. These SiC polytypes can reproducibly be grown as single crystals; they have superior electronic and thermal material properties. The conductivity type can be adjusted by shallow donors and acceptors. Special sections are related to the diffusion of dopants, to the impurity conduction, to the minority carrier lifetime and to the different types of traps generated at the interface of thermally grown SiC/SiO_2 structures.

1. Introduction

Silicon carbide (SiC) has a long and famous history. The first observation of SiC or at least of a Si-C-bond containing compound goes back to Jöns Jakob Berzelius, Professor in chemistry at the Karolinska Institute in Stockholm.[1] In 1823, Berzelius burnt an unknown compound and observed an equal number of silicon (Si) and carbon (C) atoms. Nowadays, SiC has developed into one of the leading contenders among the wide bandgap semiconductors. This leading role is due to the fact that

(a) large SiC substrates are commercially available (3 inch in diameter),
(b) SiC can homoepitaxially be grown avoiding lattice mismatch,
(c) n- and p-type conductivity can be achieved either by doping during crystal growth or afterwards by ion implantation and
(d) oxide films can thermally be grown on both the Si- and C-face.

SiC has a series of superior physical properties like the wide bandgap ranging from 2.3 eV to 3.3 eV, an electric field strength at breakdown of $E_B = (0.8\text{-}3)\times10^6$ V/cm, a saturation drift velocity of $v_S = 2\times10^7$ cm/s and a thermal conductivity of $\kappa = 4.9$ $\mathrm{Wcm^{-1}K^{-1}}$ (see Table 1 and Ref. 2, which provides detailed information on SiC properties). These material properties favor SiC for electronic devices, which are operated at high temperature, high power and/or high frequency.

In the following sections, we will briefly report and describe SiC material properties; this data may serve as a basis for subsequent chapters. We especially put emphasis on device-relevant properties like the band structure and effective masses, thermal properties, doping by shallow donors/acceptors and free charge carriers, diffusion of dopants, impurity conduction, minority carrier lifetime and the properties of SiC/SiO_2-interfaces, which are important for metal-oxide-semiconductor transistors.

Table 1. Important material parameters for the 3C-, 6H- and 4H-SiC polytype.

material parameters	3C	6H	4H
hexagonality H: *	0	0.33	0.5
indirect bandgap E_G at 300 K in eV:	2.35	3.08	3.28
lattice constant at 300 K in Å (a-direction):	4.349	3.081	3.076
(c-direction):	---	15.079	10.05
density ρ in g/cm^3:	3.21	3.21	3.21
breakdown field E_B in MV/cm at a doping level of 2.0×10^{15} cm^{-3}:	0.8	1	3
static permittivity ε_S (300K) perpendicular c:	9.72	9.66	9.76
parallel c:	—	10.03	10.32
saturated electron drift velocity v_S in cm/s at a doping level of 4.8×10^{16} cm^{-3}:	2.7×10^7	2×10^7	2×10^7
thermal conductivity κ in $Wcm^{-1}K^{-1}$:	4.9	4.9	4.9

* hexagonality H = (number of hexagonal lattice sites)/(total number of non-equivalent lattice sites)

2. Polytypism

The exciting feature of SiC is its appearance in many different polytypes.[3,4] SiC is composed of Si-C-bilayers perpendicular to the c-axis (see Fig. 1). Depending on the stacking sequence (steps to the right and to the left), a lone cubic polytype denoted 3C- or β-SiC, a great number of hexagonal polytypes denoted 2H-, 4H-, 6H-SiC etc., and a great number of rhombohedral polytypes denoted 15R-, 21R-, 27R-SiC etc. are obtained (see Fig. 2). Altogether more than 200 different SiC polytypes are identified by Laue patterns. The hexagonal and rhombohedral polytypes are collectively referred to as α-SiC. Depending on the particular polytype, a different number of non-equivalent lattice sites exists either with cubic (see hatched areas in Fig. 2) or with hexagonal (see cross-hatched areas in Fig. 2) environment (3C: one cubic site (k), 4H: one cubic site (k) and one hexa-

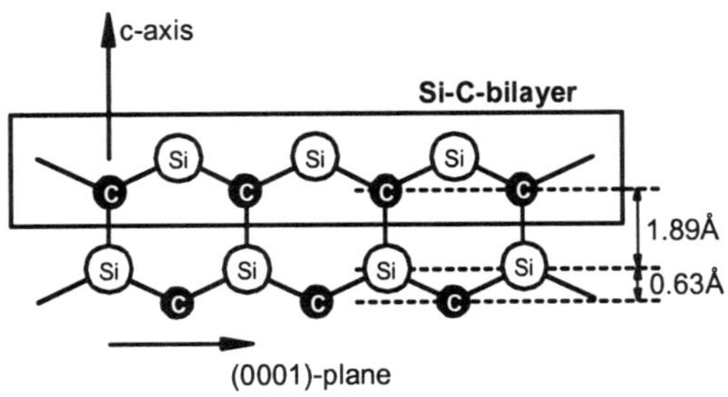

Fig. 1. Si-C-bilayer: the basic component for the formation of polytypes.

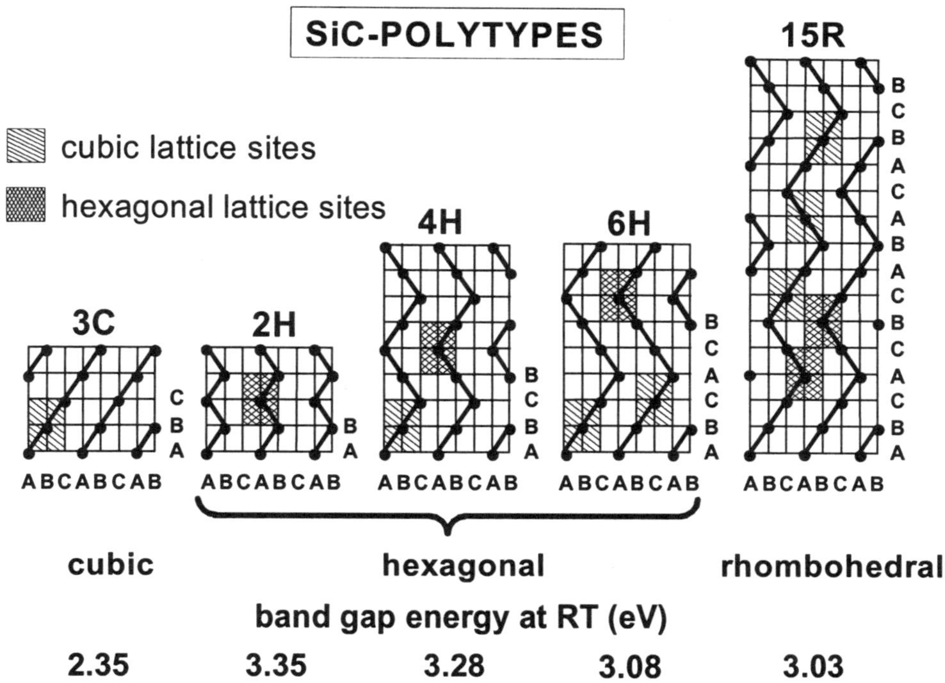

Fig. 2. SiC polytypes; the non-equivalent lattice sites are schematically indicated by the hatched (k) and cross-hatched (h) areas.

gonal site (h), 6H: two cubic sites (k_1, k_2) and one hexagonal site (h)). Identical donor species residing on non-equivalent lattice sites (see section 5.1) have different ionization energies caused by the Kohn-Luttinger effect.[5] The pure wurtzite structure (2H-SiC) has only been observed in form of whiskers. The most common polytypes for electronic devices are 4H-, 6H- and 3C-SiC.

3. Band Structure and Effective Masses

Although SiC is a tetrahedrally coordinated compound, the real-space distribution of the electronic charge density n(r) differs considerably from that of a typical covalent semiconductor. Churcher et al. pointed out that the valence charge is predominantly accumulated around the smaller C atoms (see Fig. 3).[6] This "ionicity" of the bonds does not arise from differences in the valence electrons of the two constituents (Si, C), but from the differences in their cores.

All the SiC polytypes are indirect semiconductors and the maximum of the upper valence bands is located at the center (Γ-point) of the Brillouin zone (BZ). Early theoretical work on band structures of SiC has been conducted on the cubic polytype.[7, 8] Because of the weak dispersion of the conduction bands in 4H- and 6H-SiC close to the boundary of the BZ, it was first not clear whether the conduction band minima in these

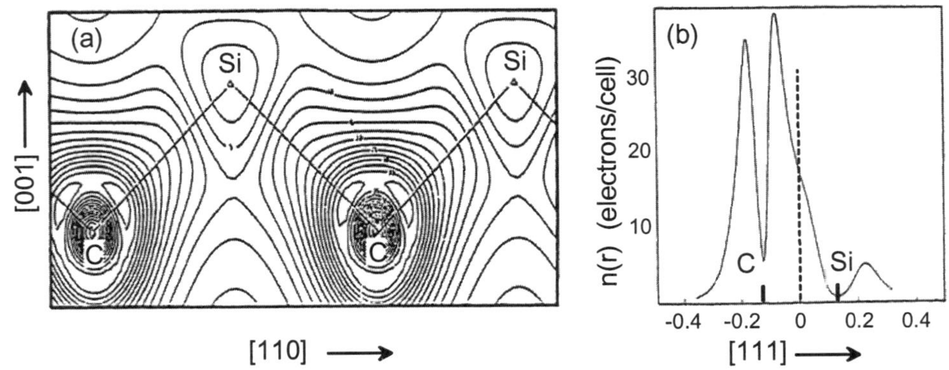

Fig. 3. Pseudo-charge density n(r) calculated for 3C-SiC. (a) n(r)-contours in the (110) plane, (b) n(r) plotted along the [111] direction. The C and Si sites are marked (after Churcher et al.).[6]

polytypes are located directly at the boundary of the BZ or somewhat inside. Fig. 4 shows the BZ of 4H-SiC and the possible positions of the conduction band minima (F or M); as a consequence the number of conduction band minima M_C strongly differs as indicated in the figure (M_C = 12 or 3). Ab-initio calculations of the ground state properties of hexagonal SiC based on the concept of density functional theory (DFT) in the local density approximation (LDA), which have recently been performed by different authors, agree well in details of the internal bands.[9, 10] As an example, the energy band structure of 4H-SiC is revealed in Fig. 5. It turns out that the conduction band minima of the three SiC polytypes under consideration are directly located at the BZ boundary (see Table 2). A peculiarity is the "camel's back" (6 meV in height) of the lowest conduction band of 6H-SiC, shifting the minima to about halfway between M and L (U-point). The puzzling feature of DFT-LDA calculations is the fact that the energy separation between conduction and valence bands is in general too small by approximately 1 eV. In order to obtain correct band gaps (see Table 1, third line) quasi-particle corrections within the GW

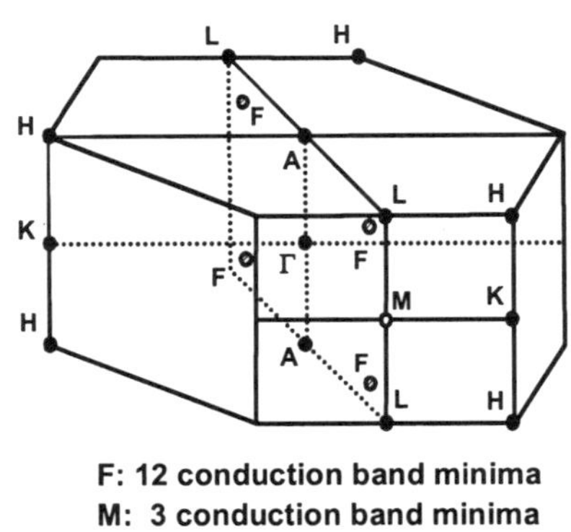

Fig. 4. Brillouin zone of 4H-SiC; possible positions of the conduction band minima in k-space (M or F) and the corresponding number of minima (3 or 12) are indicated in the inset.

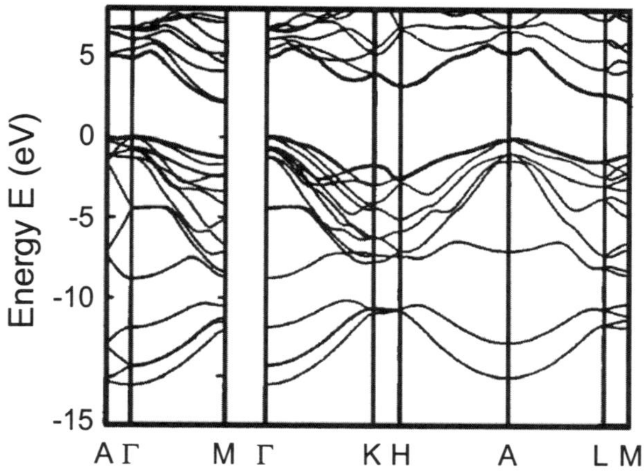

Fig. 5. Global band structure of 4H-SiC (after Wellenhofer and Rössler).[9]

approximation are required; this approach results in a rigid shift of the conduction band into the opposite direction from the valence bands, however, it does not strongly affect the internal band structure. Van Haeringen et al. have discussed the physical reasons for the strong variation of the band gap ranging from 2.3 eV to 3.3 eV with regard to the different SiC polytypes.[11] Details deduced from theoretical energy band structures are in good agreement with corresponding experimental results.[9, 10, 12]

Table 2. Parameters of the Brillouin zones and of the energy bands of three SiC polytypes.

polytype	atoms per unit cell	space group	non-equivalent lattice sites	equivalent conduction band minima position	number
3C-SiC	2	Γ_d^2	1	X	3
6H-SiC	12	C_{6v}^4	3	U on (L-M)-line	6
4H-SiC	8	C_{6v}^4	2	M	3

The temperature dependence of the band gap of 6H-SiC in the temperature range from 295 K to 673 K has been determined from the long-wavelength cutoff of the spectral quantum efficiency of UV-diodes; it is given by

$$\Delta E_G/\Delta T = -3.8\times10^{-4}\ \text{eV/K}.$$ [13, 14]

Electrons, which are thermally or optically excited from impurity states into the conduction band, dominate the electrical properties of electronic devices. These electrons occupy states close to the conduction band edges; their effective masses can be calculated from the band structure. In Table 3, theoretical values of the effective electron mass are compared with experimental results.[9, 15-20]

Table 3. Effective electron masses of 3C-, 6H- and 4H-SiC.

polytype	theory eff. electron mass (m_0^{-1})		polytype	experiment eff. electron mass (m_0^{-1})	
3C-SiC:			3C-SiC:		
$m_{\parallel X\text{-}\Gamma}$	0.68[9]	0.70[15]	$m_{\parallel X\text{-}\Gamma}$	0.667[16]	0.67[17]
$m_{\perp X\text{-}\Gamma}$	0.23	0.23	$m_{\perp X\text{-}\Gamma}$	0.247	0.22
6H-SiC:			6H-SiC:		
$m_{\parallel M\text{-}\Gamma}$	0.75	0.78	$m_{\perp}$	0.35[18]	0.42[19]
$m_{\parallel M\text{-}K}$	0.24	0.23	$m_{\parallel}$	1.4	2.0
$m_{M\text{-}L}$	1.83	1.2-2.0			
4H-SiC:			4H-SiC:		
$m_{M\text{-}\Gamma}$	0.57	0.66	$m_{M\text{-}\Gamma}$	0.58[20]	
$m_{M\text{-}K}$	0.28	0.31	$m_{M\text{-}K}$	0.31	
$m_{M\text{-}L}$	0.31	0.30	$m_{M\text{-}L}$	0.33	

[16] determined by electron cyclotron resonance, [17] determined by IR absorption, [18] determined by Raman scattering, [19, 20] determined by optically detected cyclotron resonance.

The following effective hole masses are experimentally determined:

3C-SiC: $m_h = 0.45\ m_0$,[21]

6H-SiC: $m_{h\parallel} = 1.85\ m_0$, $m_{h\perp} = 0.66\ m_0$,[22]

4H-SiC: $m_{h\parallel} = 1.75\ m_0$, $m_{h\perp} = 0.66\ m_0$.[23]

[21] determined by high-field cyclotron resonance, [22, 23] determined by optically detected cyclotron resonance.

The knowledge on the density-of-states (DOS) effective mass of the conduction (m_e^d) and valence band (m_h^d) is necessary for least-squares fits of the neutrality equation (see Eqs. (1) and (2) in section 5) to the temperature dependence of the free electron/hole concentration, which is obtained from Hall effect measurements, and also for the simulation of electronic devices. In 3C-SiC, the dispersion of the lowest conduction band is parabolic up to (150-200) meV above the minimum and consequently the DOS effective mass is independent of the temperature in this energy range. With the effective mass values listed in Table 3, m_e^d for 3C-SiC is given by:

$$m_e^d = \left(m_\perp^2 m_\parallel\right)^{1/3} = 0.35\ m_0 .$$

m_e^d and m_h^d as a function of the temperature have been calculated for the 4H and 6H SiC polytype by Wellenhofer and Rössler (see Figs. 6 and 7).[9] In 4H-SiC, the band edge value (T = 0) starts at m_e^d = 0.394 m_0 and increases slightly with occupation of the second conduction band (Fig. 6(a)). The peak observed for 6H-SiC in Fig. 6(b) is due to the camel's back structure. The temperature dependence of m_h^d in Figs. 7(a) and 7(b) is similar. At low temperatures, m_h^d is determined by the band-edge curvature of the top-most valence band at the Γ-point and saturates with increasing temperature due to the flat dispersion of this band for larger k-vectors.

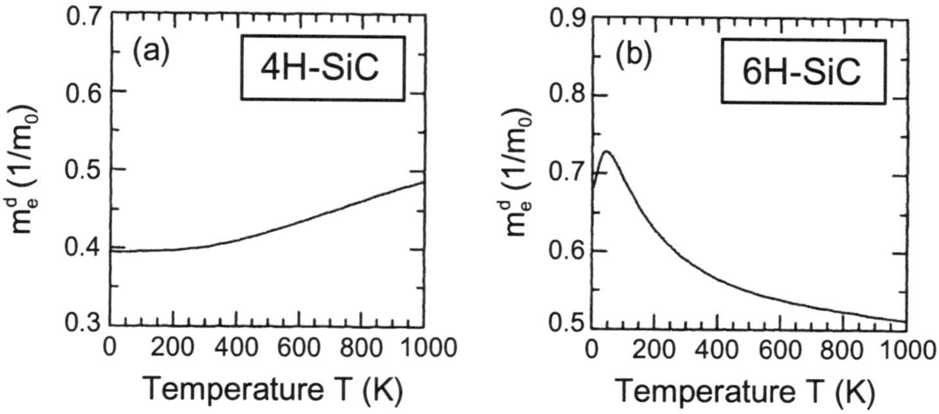

Fig. 6. Electron DOS effective mass as a function of the temperature for (a) 4H-SiC and (b) 6H-SiC (after Wellenhofer and Rössler).[9]

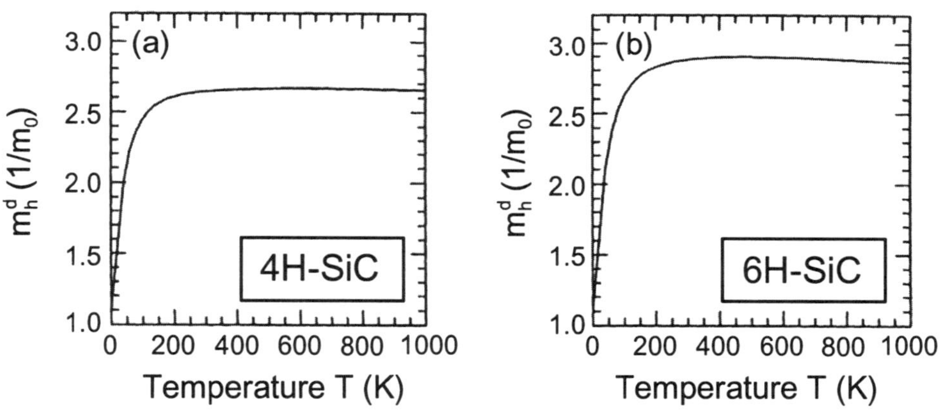

Fig. 7. Hole DOS effective mass as a function of the temperature for (a) 4H-SiC and (b) 6H-SiC (after Wellenhofer and Rössler).[9]

4. Thermal Properties

Heat capacity measurements for 6H-SiC up to 2300 K are reported by Nilsson et al. (see Fig. 8).[24] Especially thermal expansion and thermal conductivity are essential physical quantities for device fabrication. The knowledge on the thermal expansion plays an important role for cycling processes between high and low temperatures of material systems using SiC as a substrate or heteroepitaxial film. Based on its superior thermal conductivity (about 70% higher than copper) SiC may also be used as a heat sink or heat distributing material.

The axial thermal expansion coefficients of SiC perpendicular (α_{11}) and parallel (α_{33}) to the c-axis have been measured by different authors.[25-27] In Fig. 9, the thermal expansion coefficients α_{11} and α_{33} for 3C-, 4H- and 6H-SiC are displayed in the tempera-

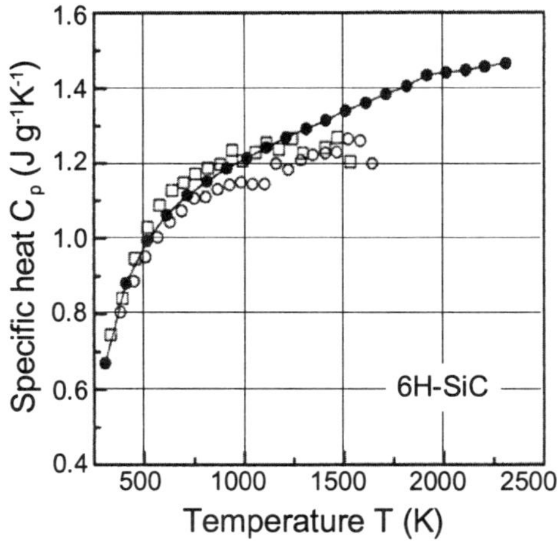

Fig. 8. Temperature dependence of the specific heat for monocrystalline 6H-SiC.[24]

ture range from 20°C to 1000°C (see symbols). Based on a least-squares fit to the experimental points, the temperature dependence of the thermal expansion coefficients can be described as second-order polynomials in temperature (see dashed and solid lines in Fig. 9).[25] For 4H-SiC, the resulting second-order polynomials are

$$\alpha_{11} = 3.21 \times 10^{-6} + 3.56 \times 10^{-9}\,T - 1.62 \times 10^{-12}\,T^2\ (1/°C),$$
$$\alpha_{33} = 3.09 \times 10^{-6} + 2.63 \times 10^{-9}\,T - 1.08 \times 10^{-12}\,T^2\ (1/°C).$$

The thermal expansion coefficients of the three investigated SiC polytypes do not significantly deviate from each other as shown in Fig. 9; however, the anisotropy factor of the thermal expansion coefficients $A=(\alpha_{11}/\alpha_{33})-1$ increases with increasing temperature and assumes a value of 0.11 at 1000°C.

The temperature dependence of the thermal conductivity κ for three SiC samples of different polytype is displayed in Fig. 10.[28] Curve 1 (full triangles) and curve 3 (full

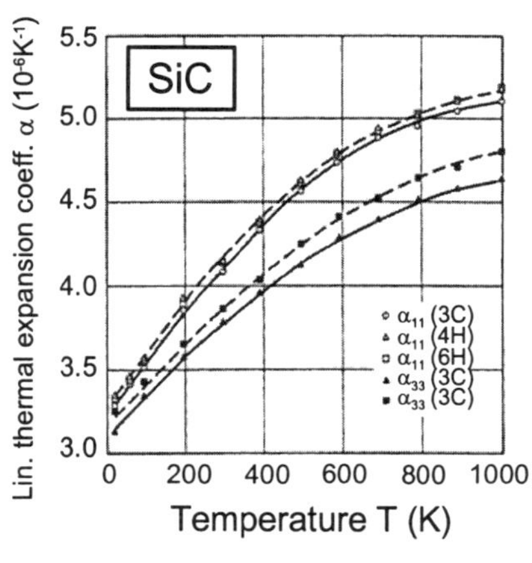

Fig. 9. Principal axial coefficients of thermal expansion for the 3C-, 4H- and 6H-SiC polytype

(after Li and Bradt).[27]

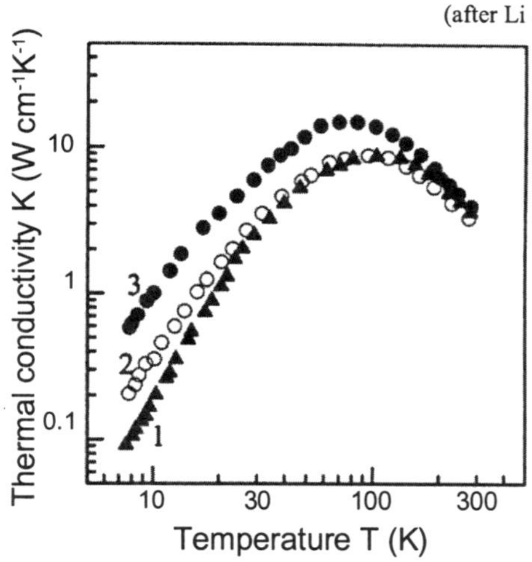

Fig. 10. Temperature dependence of the thermal conductivity κ taken perpendicular to the c-axis on single crystals of the 4H (curve 1) and 6H polytype (curve 3); curve 2 is taken on polycrystalline 3C-SiC (after Morelli et al.).[28]

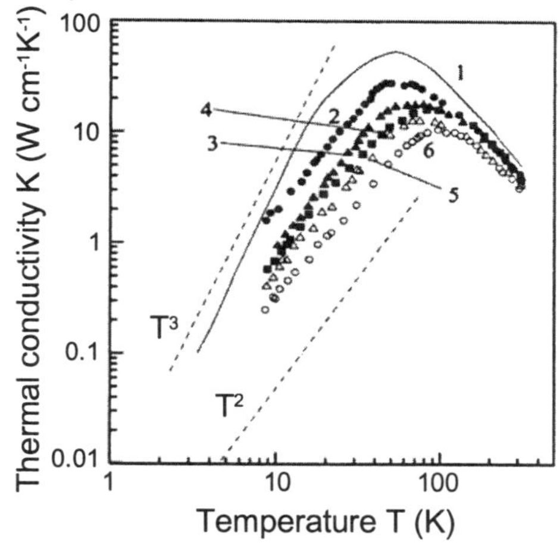

Fig. 11. Temperature dependence of the thermal conductivity κ taken perpendicular to the c-axis of 6H-SiC samples providing different electron concentrations n (in cm^{-3}); 1: very pure or highly compensated, 2: 3.5×10^{16}, 3: 2.5×10^{16}, 4: 8×10^{17}, 5: 2×10^{17} and 6: 3×10^{18} (after Morelli et al.).[28]

circles) correspond to measurements taken perpendicular to the c-axis of single crystals of the 4H- and 6H-SiC polytype, respectively, while curve 2 (open circles) is obtained from polycrystalline 3C-SiC. The temperature dependences of curves 1 and 2 tend to approach a T^3-law at low temperatures, which is characteristic for scattering of phonons by grain boundaries. Curve 3 can better be described by a T^2-dependence, which is attributed to scattering of phonons by electrons in an impurity band.[29] In Fig. 11, it is demonstrated that the thermal conductivity κ in 6H-SiC single crystals decreases with increasing electron concentration and that the slope at low temperatures can be represented by a T^2-law, which confirms the dominance of electron scattering.

5. Dopants and free charge carriers

Nominally undoped SiC crystals are n-type caused by residual nitrogen (N), which is present in the growth chamber of a Lely furnace or in the reactor of a chemical vapor deposition (CVD) system. n- and p-type conductivity of SiC can be adjusted in a wide range by shallow donors and acceptors, which are incorporated either during crystal growth or subsequently by ion implantation. Hall effect is predominantly used to determine the free carrier concentration as a function of temperature n(1/T) or p(1/T) according to

$$n(1/T) = \frac{r_{H,e}(T)}{|e|\cdot|R_{H,e}|}, \qquad p(1/T) = \frac{r_{H,h}(T)}{|e|\cdot|R_{H,h}|}.$$

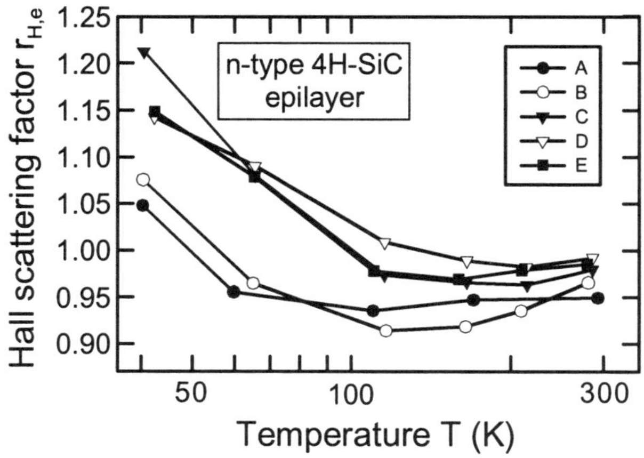

Fig. 12. Hall scattering factor $r_{H,e}$ versus temperature for five lightly N-doped 4H-SiC epilayers. Samples A to E are doped at N concentrations between 1×10^{16} cm^{-3} and 8×10^{16} cm^{-3} (see Schmid et al.).[34]

The Hall coefficients $R_{H,e}$ and $R_{H,h}$ are purely experimental quantitities and are determined by the Hall effect measurement. In order to determine the free carrier concentrations n(1/T) and p(1/T), the Hall scattering factors $r_{H,e}(T)$ and $r_{H,h}(T)$ have to be known. The Hall scattering factors in lightly N-doped 4H- and 6H-SiC epitaxial layers have been determined by Rutsch et al. (Figs. 12 and 13).[30, 31] The observed Hall scattering factors vary between 0.85 and 1.21 in the investigated temperature range. They can therefore be taken as $r_{H,e} \approx 1$ (for both SiC polytypes) without causing a large error bar. The low field Hall scattering factor $r_{H,h}(T,B)$ for holes in Al-doped 4H- and 6H-SiC is reported by Pensl et al. (Fig. 14).[32, 33] In this case, $r_{H,h}(T,B)$ strongly deviates from 1; it assumes values between 1.4 at 100 K and 0.5 at 800 K and has to be taken into account for the evaluation of p(1/T). A detailed discussion of the Hall scattering factors in SiC is given in the review article of Schmid et al..[34]

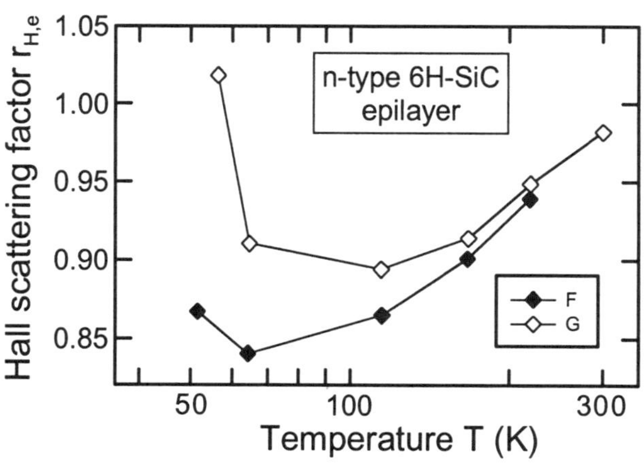

Fig. 13. Hall scattering factor $r_{H,e}$ versus temperature for two lightly N-doped 6H-SiC epilayers; sample F : 1×10^{16} cm^{-3}, sample G: 2×10^{15} cm^{-3} (see Schmid et al.).[34]

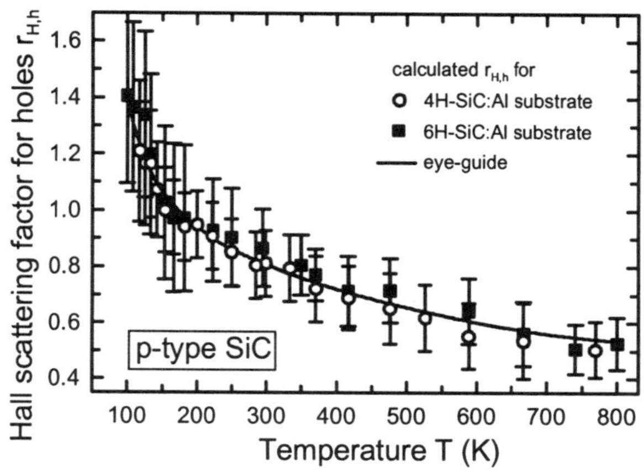

Fig. 14. Low field Hall scattering factor $r_{H,h}$ (T,B) for holes determined for 4H- and 6H-SiC as a function of the temperature; the solid line represents an eye guide indicating averaged values of $r_{H,h}$ (T,B) (see Schmid et al.).[34]

The concentration and ionization energy of the dopants as well as the concentration of the compensation are determined by the fit of the neutrality Eq. (1) or Eq. (2) to the temperature variation of the free carrier concentration n(1/T) or p(1/T). Based on the Boltzmann approximation, the neutrality equation for a semiconductor containing one donor (residing on i inequivalent lattice sites) or one acceptor species is given by

$$n + N_{comp} = \sum_i \frac{N_i}{1 + g(i) \cdot \frac{n}{N_C} \exp\left(\frac{\Delta E_i}{k_B T}\right)}, \tag{1}$$

$$p + N_{comp} = \frac{N_{acc}}{1 + g(acc) \cdot \frac{p}{N_V} \exp\left(\frac{\Delta E_{acc}}{k_B T}\right)}, \tag{2}$$

where index i = k for 3C-SiC and i = k, h for 4H- and 6H-SiC.

n	= free electron concentration,
p	= free hole concentration,
N_i	= concentration of the ith donor (residing at k or h lattice site),
ΔE_i	= ionization energy of the ith donor,
N_{acc}	= acceptor concentration,
ΔE_{acc}	= acceptor ionization energy,
N_{comp}	= concentration of the compensation,
k_B	= Boltzmann constant,
N_C	$= 2 \cdot M_C \left(2\pi \cdot m_e^d k_B T / h^2\right)^{3/2}$ = effective density-of-states (DOS) of the conduction band,
N_V	$= 2 \cdot M_V \left(2\pi \cdot m_h^d k_B T / h^2\right)^{3/2}$ = DOS of the valence band,
M_C	= number of conduction band minima,
m_e^d	= DOS effective mass of electrons,
m_h^d	= DOS effective mass of holes.

The DOS effective mass of electrons m_e^d and holes m_h^d in 4H- and 6H-SiC is calculated by Wellenhofer and Rössler (see Figs. 6 and 7).[9] The degeneracy factors g(i) and g(acc) for donors and acceptors, respectively, are given by[35]

$$g(i) = g_i^0 + \sum_j g_i^j \exp\left(-\Delta E_i^j / k_B T\right), \quad (3)$$

$$g(acc) = g^0 + \sum_j g^j \exp\left(-\Delta E_j / k_B T\right), \quad (4)$$

where g_i^0, g^0 and g_i^j, g^j are the degeneracy factors of the ground and the jth excited state and ΔE_i^j, ΔE_j are the energy separations between these two states.

5.1. *Shallow donors and electrons*

N atoms residing on C lattice sites and phosphorus (P) atoms residing on Si lattice sites serve as the shallow donors in SiC.[36, 37] Figs. 15 (a) and 15 (b) show Hall effect results taken on a high-quality unintentionally doped, n-type 4H-SiC epilayer grown by CVD. The parameters of N-donors determined by the fit of neutrality equation (1) to the experimental points (see solid curve in Fig. 15 (a)) are listed in the figure caption. The electron Hall mobility reaches a maximum value of 2×10^4 cm^2/Vs at 50 K (Fig. 15 (b)). The electron Hall mobility in general is smaller by approximately a factor two in 6H-SiC. It strongly depends on the concentration of charged impurities (donors and compensa-

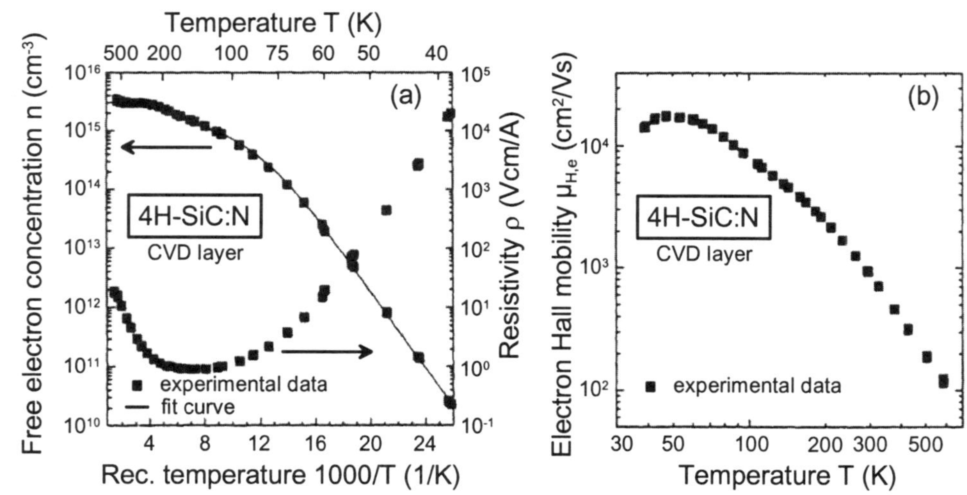

Fig. 15 (a) Hall effect and conductivity measurements from an unintentionally doped 4H-SiC epitaxial layer as a function of the temperature. The ionization energies of the cubic and hexagonal N donor are: $\Delta E_N(k) = 102$ meV and $\Delta E_N(h) = 59$ meV, respectively; $N_{total} = 3\times10^{15}$ cm^{-3}. (b) Electron Hall mobility versus temperature.

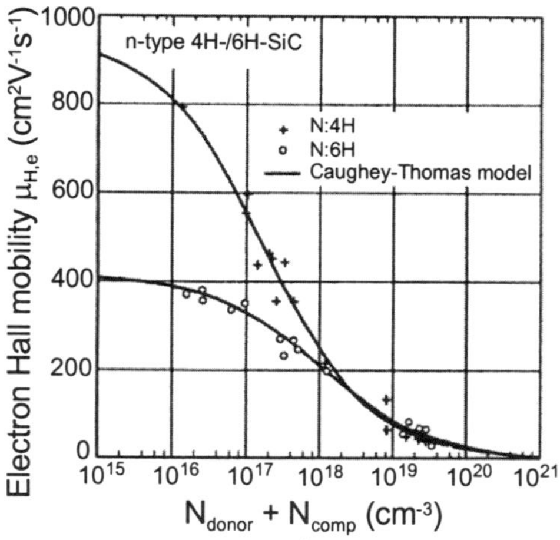

Fig. 16. Electron Hall mobility as a function of the sum of dopant and compensation concentration taken at room temperature for 4H-/6H-SiC samples.[38]

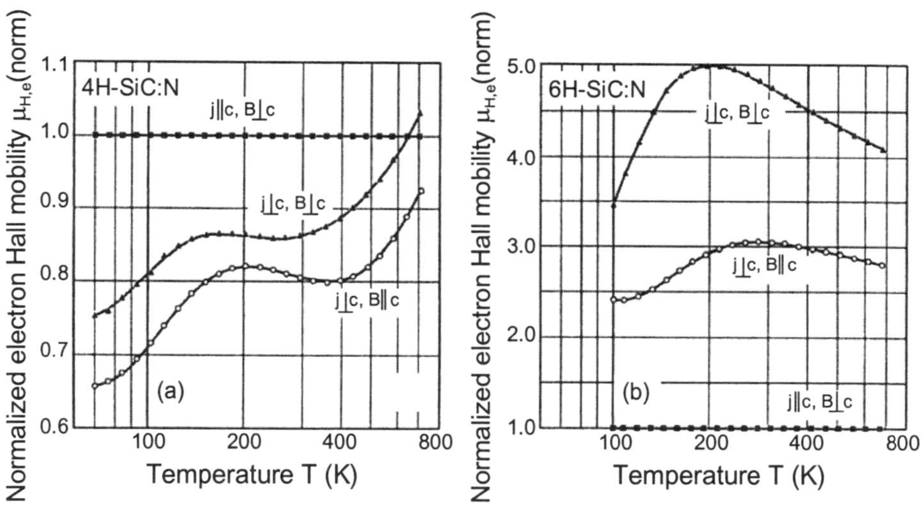

Fig. 17. Normalized electron Hall mobility parallel ($\mu_{H,e}$(norm) = μ(j||c, B⊥c)/μ(j||c, B⊥c)) and perpendicular ($\mu_{H,e}$(norm) = μ(j⊥c, B⊥c)/μ(j||c, B⊥c), $\mu_{H,e}$(norm) = μ(j⊥c, B||c)/μ(j||c, B⊥c)) to the c-axis for (a) 4H-SiC and (b) 6H-SiC. The concentrations of nitrogen donors N_D and of the compensation N_{comp} are $N_D = 2.6 \times 10^{18}$ cm^{-3} / $N_{comp} = 9.0 \times 10^{16}$ cm^{-3} in the investigated 4H-SiC bars and $N_D = 4.5 \times 10^{16}$ cm^{-3} / $N_{comp} = 1.6 \times 10^{16}$ cm^{-3} in the investigated 6H-SiC bars (after Schadt et al.).[39]

tion) as well as on the crystal quality. The dependence of $\mu_{H,e}$ on the donor and compensation concentration is displayed in Fig. 16.[38] The electron Hall mobility in 4H-SiC shows a much smaller anisotropy parallel and perpendicular to the c-axis than in 6H-SiC as is demonstrated in Fig. 17.[39]

The ionization energy of N and P donors decreases with increasing donor concentration. The plot in Fig. 18 shows experimental data for cubic and hexagonal N donors in 6H-SiC (full and open circles) and the corresponding calculated dependences based on a model, which has been proposed by Éfros et al. (solid and dashed curves).[40, 41] Experimentally determined ionization energies of cubic and hexagonal N donors in 3C-, 4Hand 6H-SiC for N donor concentrations ranging approximately from 5×10^{14} cm^{-3} to 1×10^{19} cm^{-3} are summarized in Table 4.[42-45] Ionization energies of P donors assume values, which are close to the ionization energies of N donors.[46-48]

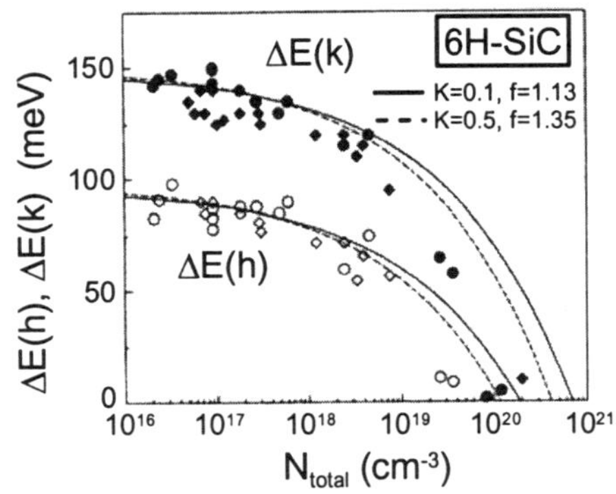

Fig. 18. Ionization energy of cubic (full circles) and hexagonal (open circles) N donors in 6H-SiC as a function of the total N donor concentration.[40] The dashed and solid curves are calculated according to Éfros et al..[41]

Sufficiently low N donor concentrations of a few 10^{14} cm^{-3} can be reached in high-quality SiC epilayers by CVD.[49] The crucial point, however, is highly N-doped SiC. The electrical activity of N atoms incorporated into SiC either during the Lely growth process or afterwards by ion implantation saturates at concentrations of (3-5) $\times10^{19}$ cm^{-3}.[50, 51] This observation is demonstrated by the Hall effect results given in Fig. 19. Three N-box profiles of 0.8 μm in depth and mean concentrations of N1 = 2.5×10^{18} cm^{-3}, N2 = 5×10^{19} cm^{-3} and N3 = 3×10^{20} cm^{-3} have been formed in p-type 4H-SiC epilayers by multiple implantation at 500°C and subsequent annealing at 1700°C for 30 min. The temperature dependence of the free electron concentration in Fig. 19 (a) clearly indicates that sample N3 with the highest implanted N-box profile (open circles) results in a smaller free electron concentration than sample N2, which has been implanted at a mean N concentration of only 5×10^{19} cm^{-3} (black circles). This result is confirmed by the electron Hall mobility in Fig. 19 (b); the three mobility curves show a continuous shift to lower values with increasing concentration of electrically active N donors. N donors at concentrations above 5×10^{19} cm^{-3} are deactivated either by forming precipitates or electrically inactive complexes, which are thermally stable up to high temperatures.[52-55]

Table 4. Electrically and optically determined ionization energies of 3C-, 4H- and 6H-SiC doped with N donors at concentrations ranging from 5×10^{14} cm^{-3} to 1×10^{19} cm^{-3}.[42-45]

polytype	cubic lattice site (k)		hexagonal lattice site (h)	
	Hall effect	optical method	Hall effect	optical method
3C-SiC	18-48	53-56	–	–
4H-SiC	96-105	92	41-59	52-55
6H-SiC	125-150	137-142	84-100	81

Figs. 20 (a) and 20 (b) reveal Hall effect results taken on three P-implanted 4H-SiC layers (P1, P2 and P3), which have been implanted and annealed under identical conditions like samples (N1-N3). The free electron concentration measured in sample P3 un-

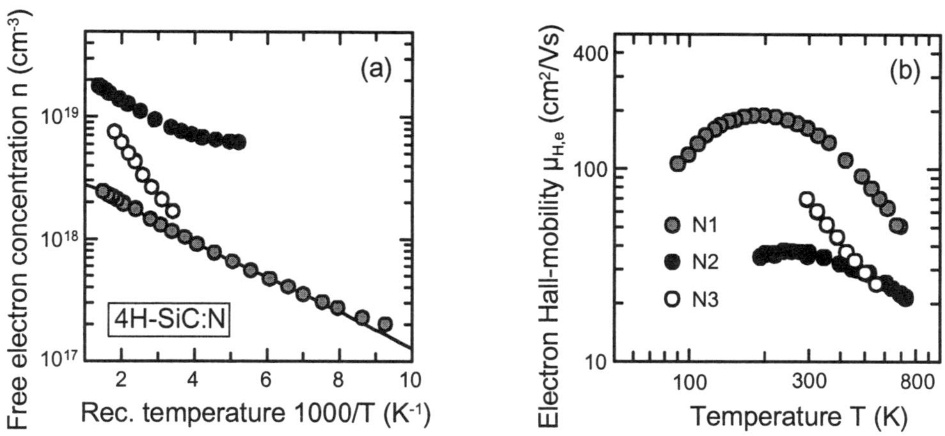

Fig. 19. Hall effect and conductivity results obtained from N-implanted samples (N1 = 2.5×10^{18} cm^{-3} (grey circles), N2 = 5×10^{19} cm^{-3} (black circles) and N3 = 3×10^{20} cm^{-3} (open circles)). (a) Free electron concentration n as a function of the reciprocal temperature, (b) electron Hall mobility $\mu_{H,e}$ vs. temperature (after Laube et al.).[51]

ambiguously demonstrates that P atoms (opposite to N atoms) can electrically be activated at concentrations above 10^{20} cm^{-3}. Schmid et al. reported that the incorporation of P donors is independent of the orientation of the sample surface ((0001)- or (11-20)-face).[56] P-donors have the disadvantage that they cannot be incorporated into SiC in-situ during bulk or CVD growth at concentrations higher than approximately $\approx10^{18}$ cm^{-3}.[57-60] The reason is not clear yet, it is argued that the sticking coefficient of P atoms at the growth front is small at temperatures above 1500°C because of the high vapor pressure of P.

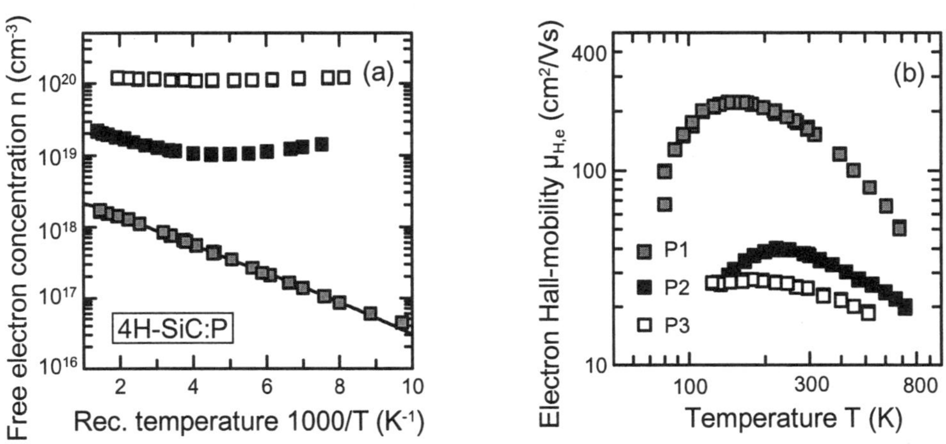

Fig. 20. Hall effect and conductivity results obtained from P-implanted samples (P1 = 2.5×10^{18} cm^{-3} (grey squares), P2 = 5×10^{19} cm^{-3} (black squares) and P3 = 3×10^{20} cm^{-3} (open squares)) (a) Free electron concentration n as a function of the reciprocal temperature, (b) electron Hall mobility $\mu_{H,e}$ vs. temperature (after Laube et al.).[51]

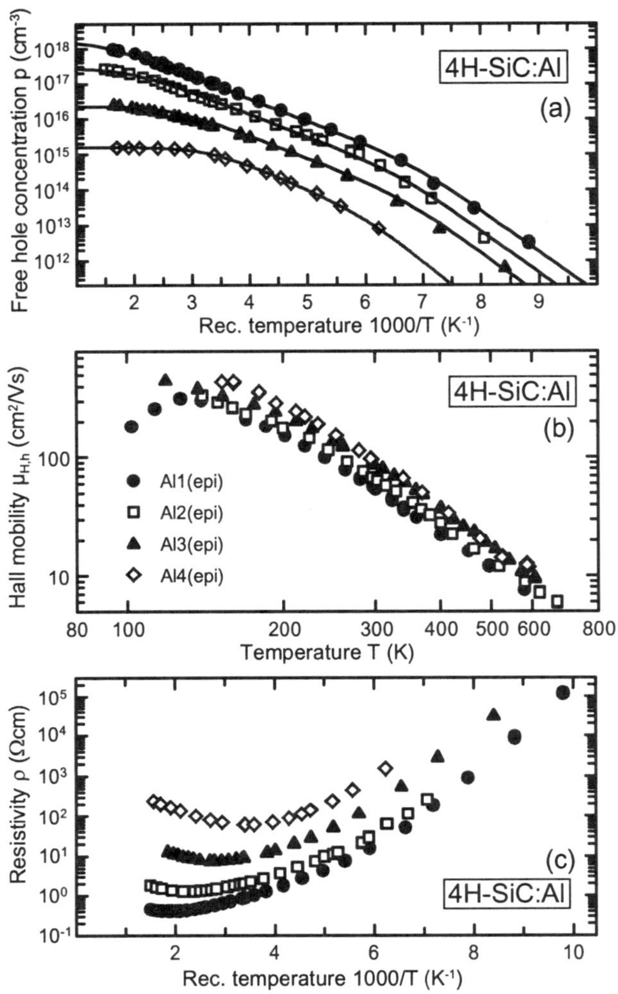

Fig. 21. Hall effect and resistivity measurements taken on four p-type 4H-SiC epilayers with different Al acceptor concentration; the symbols correspond to experimental points. (a) Free hole concentration vs. reciprocal temperature, (b) Hall mobility of holes $\mu_{H,h}$ as a function of temperature, and (c) resistivity ρ as function of temperature. The following Al acceptor concentrations are obtained from least-square-fits of Eq. (1) to the experimental data (see solid curves in Fig. 21 (a)). Al1: 1.6×10^{18} cm^{-3}, Al2: 2.7×10^{17} cm^{-3}, Al3: 2.8×10^{16} cm^{-3}, and Al4: 1.8×10^{15} cm^{-3}; the ionization energy of Al acceptors varies between 216 meV and 240 meV depending on the acceptor concentration, and the compensation is in the range of 3×10^{13} cm^{-3} to 7×10^{14} cm^{-3} (after Schmid et al.).[34]

P donors can also be incorporated into SiC via neutron transmutation; it is based on the capture of thermal neutrons by ^{30}Si according to the reactions ${}^{30}_{14}\mathrm{Si}(n,\gamma)\,{}^{31}_{14}\mathrm{S}$ and the subsequent β^- decay resulting in the ${}^{31}_{15}\mathrm{P}$ isotope. Because of the large recoil energies of the γ-quantum and the β^--particle (780 eV and 32.3 eV, respectively), this doping process requires annealing steps at elevated temperatures ($T \geq 1700°C$) to electrically activate the P donors and to reduce the lattice damage.[61, 62] The generated defect centers (e. g. D_I-

center) are thermally extremely stable. In low temperature photoluminescence (LTPL) spectra, the corresponding defect lines (L_V-lines) can be observed even after an anneal at 2000°C.[61]

5.2. *Shallow acceptors and holes*

p-type conductivity in SiC can be achieved by overcompensation of the residual N donors with shallow acceptors. Aluminum (Al) and boron (B) residing on Si lattice sites are the prevailing acceptors in SiC.[63-65] These acceptors can be incorporated into the SiC lattice either during crystal growth or by subsequent ion implantation. In Figs. 21 (a) to (c), Hall effect and resistivity measurements taken on four p-type, Al-doped high-quality 4H-SiC epilayers are displayed. The free hole concentration p(1/T) in Fig. 21 (a) has been determined by taking into account the temperature-dependent Hall scattering factor $r_{H,h}$ as given in Fig. 14. The parameters of Al-acceptors determined from the least-squares-fits of neutrality equation (2) to the experimental points (see solid curves in Fig. 21 (a)) are summarized in the figure caption. For low-doped p-type SiC, the following ionization energies are reported:[66, 45]

$$\Delta E_{Al} = (200 \text{ to } 250)\text{ meV}, \quad \Delta E_B = (300 \text{ to } 400)\text{ meV}.$$

Because of the fact that the valence band structure is similar for the different SiC polytypes and the maximum of the valence band is located at the Γ-point, the ionization energy of acceptors is largely independent of the SiC polytype and the lattice site. The ionization energy depends on the acceptor concentration and the degree of compensation K as is shown in Fig. 22 for the Al acceptor in 6H-SiC.[67] For compensations in the range of $0.01 < K < 0.5$, the ionization energy strongly decreases with increasing acceptor concentration.

The Hall mobility of holes in SiC is smaller by approximately a factor six than the corresponding Hall mobility of electrons due to their heavier mass. Fig. 23 shows the Hall mobility of holes as a function of the concentration of acceptors and of the compensation in 4H- and 6H-SiC.[38]

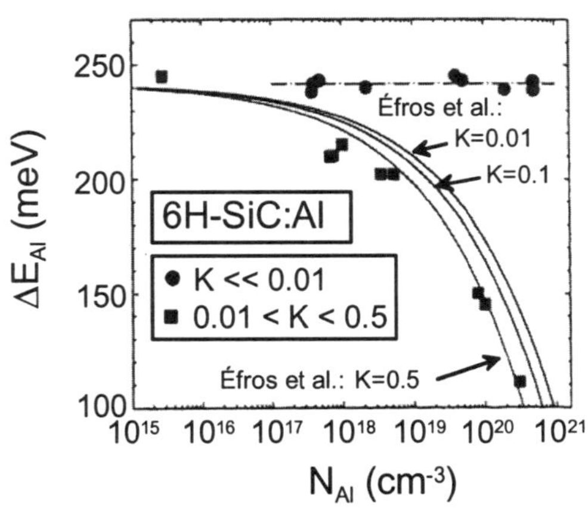

Fig. 22. Al ionization energy ΔE_{Al} as a function of the Al concentration and degree of compensation K; symbols are experimental points, dashed and solid curves are calculated after Éfros et al..[41, 67]

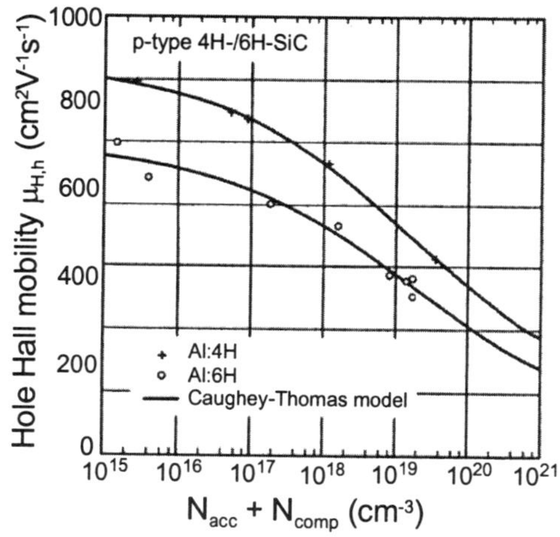

Fig. 23. Hole Hall mobility as a function of the sum of dopant and compensation concentration taken at room temperature for 4H-/6H-SiC samples.[38]

Although it is reported in the literature that the solubility of Al in SiC is extremely high ($c^{eq} \approx 10^{21}$ cm^{-3}, for T ≈ 2000 ... 2400°C), there is still a lack of 4H-SiC bulk material providing a sufficiently low p-type resistivity (ρ ≤ 0.1 Ωcm), which would be required for SiC-based high power switches to reduce the internal losses.[68]

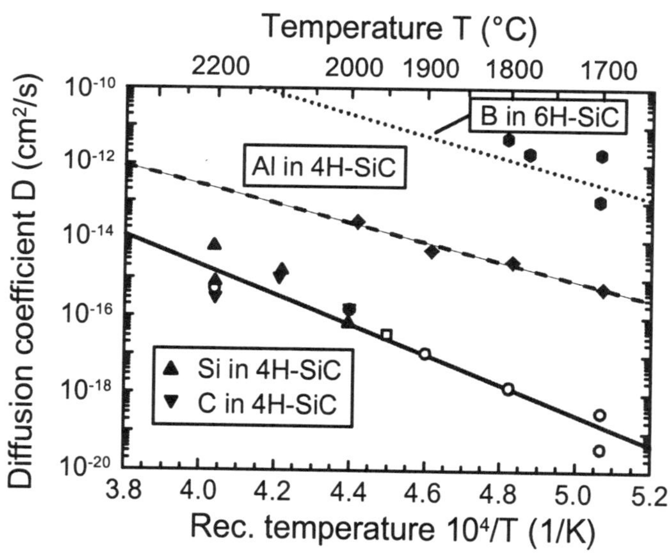

Fig. 24. Temperature dependence of different diffusion coefficients; B: dotted line (Ref. 71), full hexagons (Ref. 76), Al: dashed line and full diamonds (Ref. 70), Si: full triangles (Ref. 79), C: reversed full triangles (Ref. 79), native point defect: open symbols (Ref. 79).

6. Diffusion of Dopants

Among the frequently used dopant atoms in SiC like N, P, Al, and B, strong diffusion has been observed for B and only a weak diffusion for Al.[69] Linnarsson et al. conducted Al diffusion experiments in highly Al-doped 4H-SiC epilayers ($N_{Al} > 10^{20}$ cm^{-3}) grown by CVD and estimated the temperature dependence of an effective diffusion coefficient D*(Al) in the temperature range between 1700°C and 2000°C by the following expression (see dashed straight line in Fig. 24):

$$D^*(Al) = 1.4\times10^{-3} \times \exp(-4.8\text{eV}/kT).^{70}$

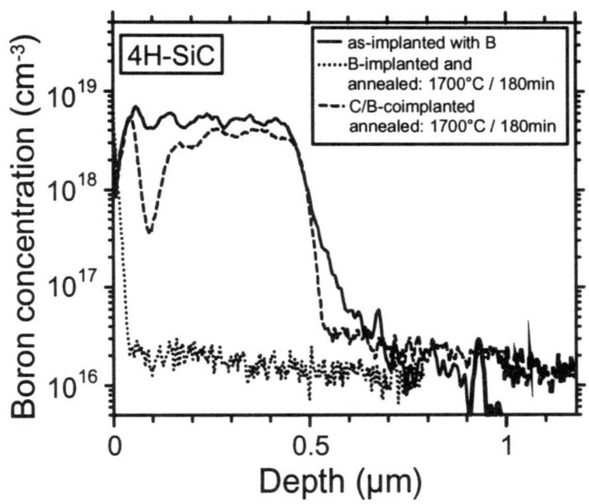

Fig. 25. B profiles in 4H-SiC as determined by SIMS subsequent to different processing steps; the processes applied are described in the legend (after Laube et Pensl).[73]

Early experiments on the B diffusion in p-type 6H-SiC have been performed by Mokhov et al. with the track autoradiography method, which is based on the nuclear reaction $^{10}B(n,\alpha)^7Li$.[71] The investigated 6H-SiC samples were Al-doped and, in addition, B-diffused from the vapor phase; the dotted straight line in Fig. 24 represents the temperature dependence of the diffusion coefficient D(B). Recent diffusion experiments on implanted B-profiles in 4H-/6H-SiC provide a transient-enhanced diffusion (TED) to the surface due to a non-equilibrium concentration of native defects, which are generated by the implantation, and a moderate B diffusion of the trailing edge to the undamaged bulk region.[72-74] The dotted B profile in Fig. 25, which is determined by SIMS subsequent to an anneal at 1700°C for 180 min, indicates that the implanted B atoms are completely out-diffused. The TED of B can effectively be suppressed either by co-implantation of C (see dashed curve in Fig. 25) or by a preanneal at lower temperatures (e. g. at 900°C), which leads to a recombination of native defects (see dashed and dotted B profiles in Fig. 26). Native defects are apparently necessary for the diffusion of B. Based on detailed simulations of diffused B profiles, Bracht et al. have demonstrated that the B diffusion is controlled by the kick-out mechanism.[75,76] B atoms residing on Si-lattice sites are kicked-out by Si self-interstitials (Si_i) and become a mobile species on interstitial positions:

$$B_i \leftrightarrow B_{Si} + Si_i$$

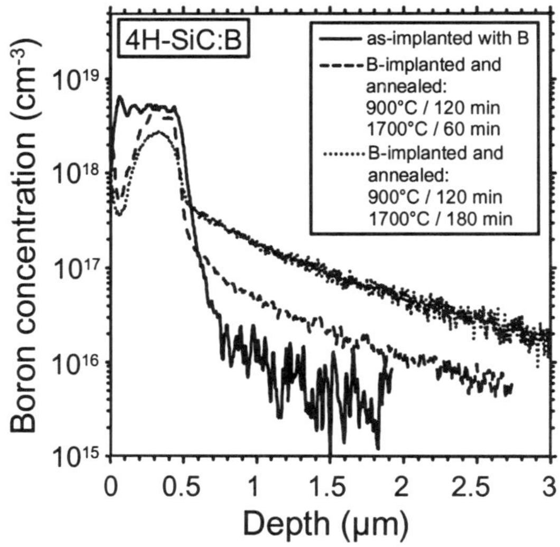

Fig. 26. B profiles in 4H-SiC as determined by SIMS subsequent to different two-step anneals; the annealing steps are described in the legend. With increasing annealing time at 1700°C a Fickian B diffusion develops into the undamaged bulk (dashed and dotted curves) (after Laube et al).[72]

The full hexagons in Fig. 24 are obtained from fitting the experimental B profiles shown in Fig. 26; they agree well with the B diffusion coefficients determined by Mokhov et al..[71] Ab initio calculations confirm that Si interstitials mediate the B diffusion and the migration path of B interstitials strongly depends on the position of the Fermi level.[77, 78] The formation of stable B-C pairs at hexagonal interstitial sites may explain the effective reduction of the TED of B in B-/C- co-implanted SiC samples.

Self-diffusion experiments in isotopically enriched 4H-SiC result in the Si and C diffusion coefficients, which are marked in Fig. 24 as full and reversed full triangles, respectively.[79] The open symbols along the solid straight line in Fig. 24 represent the contribution of a native point defect to the self-diffusion; this data was deduced from B diffusion in SiC.[79, 80]

7. Impurity Conduction

The conductivity of doped semiconductors is governed by different transport mechanisms. At sufficiently low temperatures, the charge transport is no longer due to the propagation of free charge carriers in an energy band but occurs between impurity states by phonon assisted tunneling. This charge transport is termed impurity or hopping conduction, which was at first observed in SiC by Busch and Labhart.[81] If the doping concentration is below the Mott-Anderson transition (n-type SiC $\approx 10^{19}$ cm^{-3}; p-type SiC

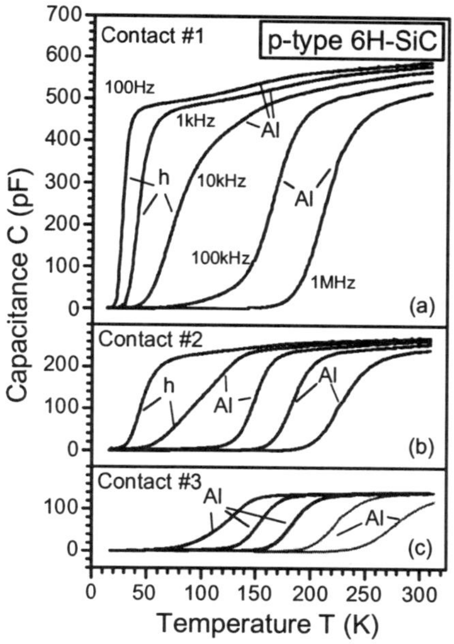

Fig. 27. Capacitance as a function of the temperature taken at different frequencies;
(a) contact #1: [Al] = 3.4×10^{18} cm^{-3},
(b) contact #2: [Al] = 7.2×10^{17} cm^{-3},
(c) contact #3: [Al] = 1.8×10^{17} cm^{-3}
(after Krieger et al.).[89]

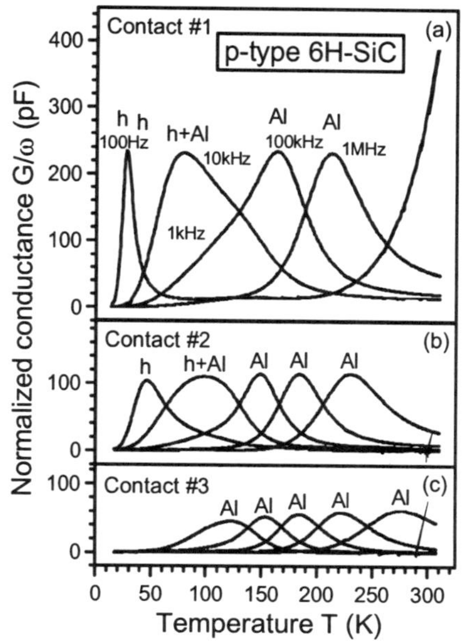

Fig. 28. Normalized conductance as a function of the temperature taken at different frequencies;
(a) contact #1: [Al] = 3.4×10^{18} cm^{-3},
(b) contact #2: [Al] = 7.2×10^{17} cm^{-3},
(c) contact #3: [Al] = 1.8×10^{17} cm^{-3}
(after Krieger et al.).[89]

$\approx 10^{20}$ cm^{-3}), the overlap of the impurity wave function is only weak and the impurity states are localized.[82] In this case, conduction takes place by hopping of electrons from occupied to unoccupied localized states, which requires a non-vanishing concentration of compensation. The ionization energy of each impurity state is modulated by the Coulomb interaction with all the neighboring charged impurities. As a consequence, the charge carrier transport is thermally activated. The required activation energy ε_3 is constant in a wide range of temperatures. At very low temperatures, when only a few states in a narrow band around the Fermi level can contribute to the conduction, the activation energy ε_3 decreases with decreasing temperature. This regime is termed variable range hopping (VRH).[83] For SiC variable range hopping becomes important for $T \ll 20$ K. The activation energy ε_3 has in detail been studied for germanium.[84] A theoretical modeling of ε_3 is sophisticated and several different approaches have been made in the past.[85-87] A detailed overview on impurity conduction is given in Refs. 82 and 88.

For p-type 6H-SiC, the activation energy ε_3 has been determined by admittance spectroscopy (AS) as well as by temperature dependent resistivity and Hall effect investigations.[89] Figs. 27 (a)-(c) and Figs. 28 (a)-(c) reveal capacitance C(T)- and normalized conductance G/ω(T)-curves, respectively, which are taken on three different Schottky contacts prepared on Al-doped 6H-SiC samples (Figs. 27/28 (a): [Al] = 3.4×10^{18} cm^{-3} (contact #1), Figs. 27/28 (b): [Al] = 7.2×10^{17} cm^{-3} (contact #2), Figs. 27/28 (c): [Al]=1.8×10^{17} cm^{-3} (contact #3)). The C(T)-curves taken on contact #1 and #2 show two series of steps each (marked with h, Al in the figures) and the corresponding G/ω(T)-curves show two series of peaks (also denoted by h, Al) for the applied probe frequencies

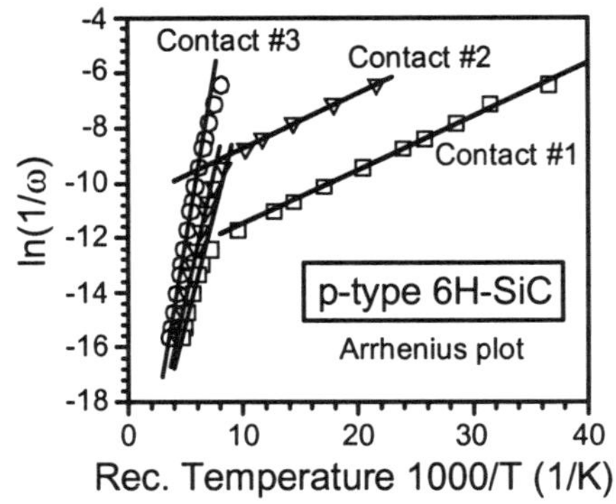

Fig. 29. Arrhenius plots obtained from the G/ω(T)-peak maxima for three contacts with different Al-concentrations. Contacts #1 and #2 reveal two different slopes corresponding to the Al-acceptor and hopping conduction, respectively, whereas contact #3 shows only the steep slope for the Al acceptor.[89]

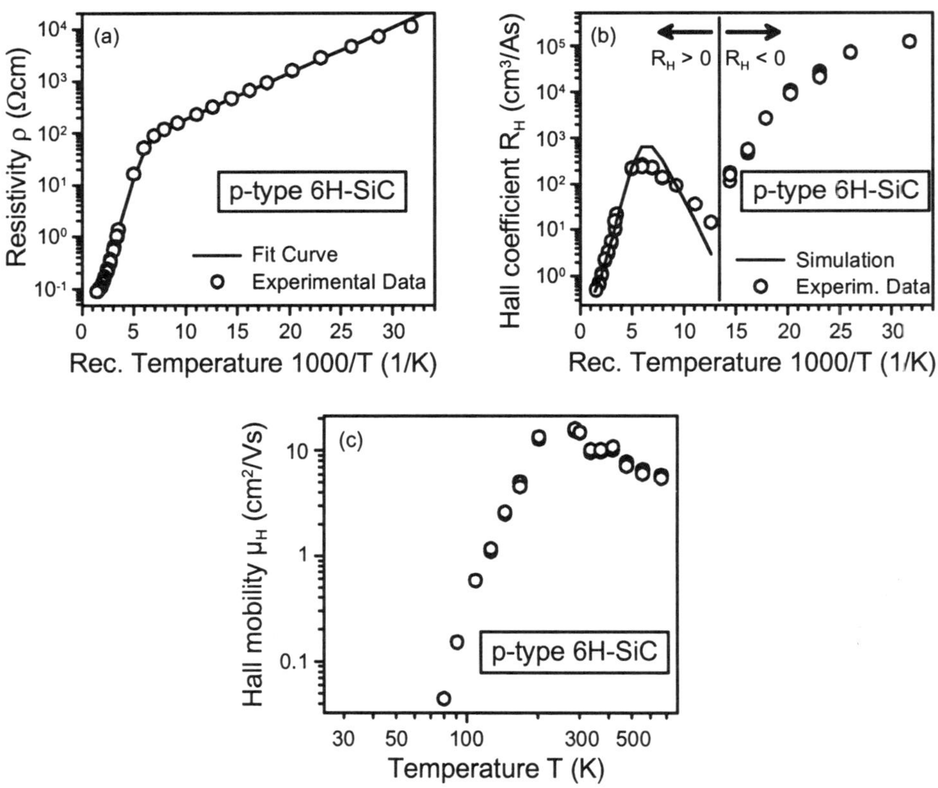

Fig. 30. (a) Resistivity ρ(1/T), (b) Hall coefficient R_H(1/T) and (c) Hall mobility μ_H(T) determined in van der Pauw arrangement (sample size = 1.1×1.1 mm^2); the sample is cut from the same wafer area used for the AS investigations. Open circles correspond to experimental points, the solid curves in (a) and (b) are a least-squares-fit to the experimental data and a simulation using the two-band-model, respectively.[89]

(100 Hz - 1 MHz). The capacitance steps and conductance peaks marked with h, which are observed at low temperatures in Figs. 27 (a) / (b) and Figs. 28 (a) / (b), shift moderately with temperature variation and lead to a small activation energy of $\Delta E(h) = (17 \pm 2)$ meV in the Arrhenius plot of Fig. 29. This value is due to the thermal activation energy ε_3 for hopping conduction. The capacitance steps and conductance peaks marked with Al result in the steep increase of the Arrhenius plot (Fig. 29) at elevated temperatures and correspond to the activation energy of the Al acceptor ($\Delta E(Al) = (141 - 211)$ meV).

ε_3 is also obtained from the temperature dependent resistivity measurement shown in Fig. 30 (a), which is taken on a sample with [Al] = 1.6×10^{19} cm^{-3}. At high temperatures, the steep slope of the $\rho(1/T)$-curve is dominated by the Al acceptor. At $T \approx 160$ K, there is a knee point and with decreasing temperatures the slope becomes flatter. From this slope, an activation energy of $\varepsilon_3 = 17.5$ meV is obtained, which agrees well with the AS result. The Hall coefficient R_H (Fig. 30 (b)) has a maximum at $T \approx 160$ K and decreases with decreasing temperatures indicating that the conductivity mechanism has changed. The maximum of R_H can be described in the framework of the two-band-model after Shklovskii and Éfros.[82] At $T \approx 75$ K, the sign of the Hall coefficient R_H changes from $R_H > 0$ to $R_H < 0$ indicating that the Hall effect is now governed by hopping conduction. However, the sign of R_H is not related to a particular type of charge carriers as in the case of band conduction. Because of the random distribution and localization of acceptor states in crystalline SiC the situation at low temperatures is quite similar to amorphous semiconductors. Under the influence of a magnetic field, charge carriers can take different paths to hop from occupied to unoccupied sites, which may lead to a sign reversal of the Hall coefficient. Theoretical models confirm the appearance of this sign anomaly of the Hall effect, but there exists no fundamental theory for this effect.[90, 91]

Similar investigations were conducted by Evwaraye et al. on n-type 4H-SiC samples containing nitrogen concentrations of [N] = $(1.5 - 3) \times 10^{18}$ cm^{-3}.[92] These authors obtained an activation energy for hopping conduction of $\varepsilon_3 = (4 - 5)$ meV from temperature-dependent resistivity measurements and of $\varepsilon_3 = (2.3 - 3.0)$ meV from admittance spectroscopy spectra. Although the observed activation energies ε_3 are very low, the conductivity in this regime of impurity conduction also strongly decreases caused by the extremely reduced mobility of charge carriers (see Fig 30 (c)).

8. Minority Carrier Lifetime

The minority carrier lifetime is one of the important parameters for the performance of bipolar devices. It controls e. g. the level of base modulation and the voltage drop across the device as well as the recovery time and switching losses. Both Si and SiC are indirect semiconductors and high-quality crystals made of these semiconductors should provide lifetimes in the ms-range. This expectation is correct for Si, however in case of SiC, the highest observed values of the lifetime just reach the µs-range. The main recombination channels, which determine the total lifetime τ_r in the bulk of a semiconductor, are the Shockley-Read-Hall (SRH) recombination (τ_{SRH}) via defect centers, the radiative recombination (τ_{rad}) and the Auger recombination (τ_{Auger}):[93, 94]

$$1/\tau_r = 1/\tau_{SRH} + 1/\tau_{rad} + 1/\tau_{Auger}\,.$$

In SiC, which is an indirect semiconductor, the radiative recombination is negligible small; its contribution to the total recombination process is 10^{-4}. Auger recombination reduces τ_r at high excess carrier densities. Galeckas et al. have demonstrated that the Auger recombination dominates τ_r in 4H-SiC at excess concentrations above 10^{18} cm^{-3}.[95] Below this value, electron-hole pairs recombine predominantly via defect centers (SRH recombination). Efficient recombination centers are frequently defect centers energetically located in the middle of the bandgap like gold in Si.[96] The recombination centers, which lead to the strong reduction of the minority carrier lifetime in SiC, are not identified yet.

The minority carrier lifetime determined in SiC samples by electrical or optical measurements is in general not a well defined quantity. It can be influenced by surface recombination, by the injection level of minorities or by processing steps, which may introduce additional lifetime killers. As a consequence, analysis techniques, which determine the lifetime in different parts of the device structure, e. g. close to the space charge region of an implanted p-n junction or in the base of a diode, may result in widely differing values of the lifetime. These discrepancies are based on physical reasons and are not due to deficiencies of the measurement. The lifetime, therefore, more likely describes the property of an electron or hole under particular environmental conditions than a property of the semiconductor itself. A detailed theoretical treatment of the carrier lifetime, which is governed by recombination processes, is given in Ref. 94. In the following, typical hole lifetimes τ_h obtained from different electrical and optical measurement techniques are reported; all the investigations are conducted on one and the same commercially purchased n-type 4H-SiC epilayer.

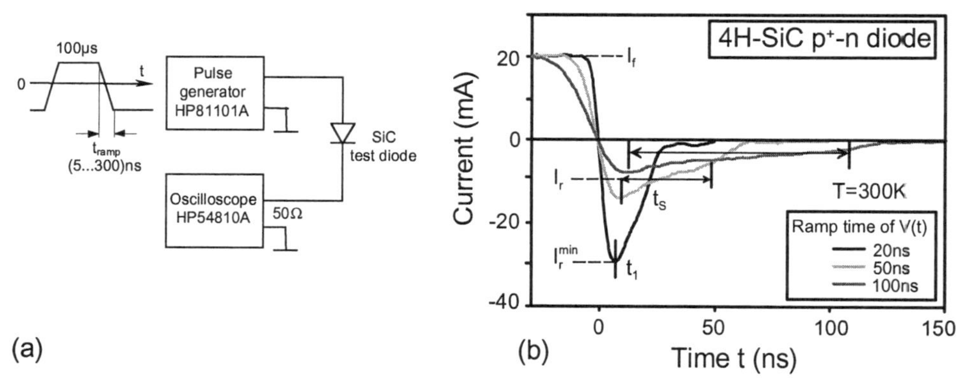

Fig. 30. (a) Electrical circuit for CRT measurements: (b) CRT switched current transients I(t) of a 4H-SiC p^+-n mesa-diode taken with different ramp-down times of the voltage pulse. The current transients are recorded with the oscilloscope.

For the electrical lifetime measurements, p^+-n mesa-diodes were used, which were fabricated by implantation of an Al box profile ($N_{Al} = 3\times10^{18}$ cm^{-3}, depth = 0.4 µm) into the 4H-SiC epilayer ($N_N \approx 10^{16}$ cm^{-3}).[97] Mesa-diodes were formed by reactive ion etching to a depth of 1.2 μm (contact diameter = 0.6 mm). In order to determine device-relevant hole lifetimes, two different electrical analysis techniques are applied, the current recovery time (CRT) and the open circuit voltage decay (OCVD) method. [98-102] The CRT method determines the lifetime of holes at the edge of the space charge region, while the OCVD method considers the decay of holes in the n-base.

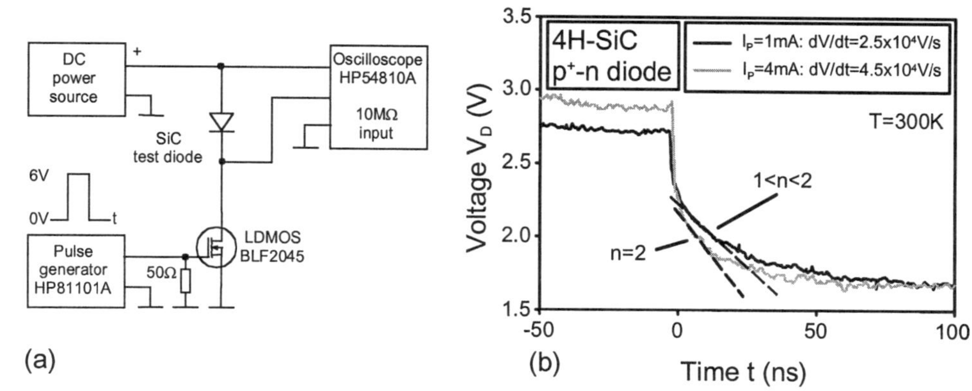

Fig. 31. (a) Electrical circuit for OCVD measurements; (b) OCVD transients $V_D(t)$ for different injection levels taken on a 4H-SiC p^+-n mesa-diode. The voltage transients are measured over the 10 MΩ input resistor of the oscilloscope.

In Fig. 30 (a), the electrical circuit for CRT measurements is shown. Forward voltage pulses (pulse length = 100 μs) supplied by a pulse generator can be applied to the reverse-biased p^+-n mesa-diode. The 50 Ω input of the oscilloscope, connected in series with the mesa-diode, serves as the load. The ramp-down time of the voltage can be adjusted by the pulse generator (≥ 5 ns). Fig. 30 (b) displays the current transients for three different ramp-down times of the applied voltage. For fast ramp-down ($t_{ramp} \leq 20$ ns), the current transients have a pronounced minimum (I_r^{min}) at t_1 and provide no plateau. With increasing t_{ramp}, the current minimum decreases and a plateau (width t_s) is formed. The calculation of τ_h according to Ref. 100 is based on the quantities I_f, I_r^{min} and t_1 (see Fig. 30 (b)). It results in $\tau_h \approx 50$ ns for short ramp-down times and decreases down to 25 ns, when t_{ramp} increases up to 200 ns.

The calculation of τ_h based on the width t_s of the reverse current plateau is given by Ref. 98:

$$\mathrm{erf}\sqrt{\frac{t_s}{\tau_h}} = \left(1+\frac{I_r}{I_f}\right)^{-1}.$$

In this case, the hole lifetime increases proportional with the ramp-down time; for $t_{ramp} \approx$ 100 ns, a value of 100 ns is obtained. The extrapolation to short ramp-down times (5 ns) yields $\tau_h \approx 50$ ns in agreement with the value given above. Levinshtein et al. explained the dependence of τ_h on the ramp-down time t_{ramp} by a thin defect-rich layer (thickness ≈ 0.1 μm) at the interface between the p^+-n junction and the n-base of the diode, in which the hole lifetime is much smaller than in the n-base.[101] This explanation is supported by deep level transient spectroscopy (DLTS) investigations taken on the p^+-n 4H-SiC mesa-structures, which reveal in the considered region several minority traps at concentrations of a few 10^{14} cm^{-3}.

For the OCVD measurements, a dc voltage source is used (see Fig. 31. (a)). The forward current through the p^+-n test diode is switched off by a fast MOSFET, which is triggered by a pulse generator. The transient of the voltage drop over the p^+-n diode $V_D(t)$

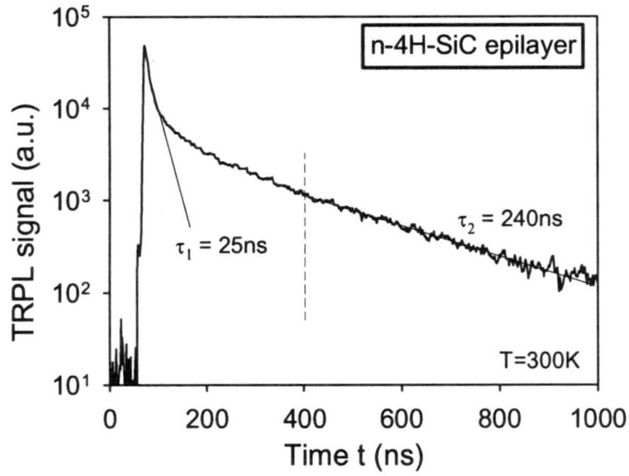

Fig. 32. TRPL decay curve of the total photoluminescence taken on the n-type 4H-SiC epilayer, which was employed for the p^+-n mesa-diodes. The PL measurement was conducted at room temperature with a laser excitation wavelength $\lambda = 266$ nm.

is measured by an oscilloscope with 10 MΩ input. The duty cycle (ratio of current pulse length to repetition rate) is 10^{-4} to keep the p^+-n diode at constant temperature. OCVD transients $V_D(t)$ for different injection levels are displayed in Fig. 31 (b). The injected excess carriers recombine in the n-base and lead to the measured decay curves. The hole lifetime is obtained from the linear part of the decay curves according to the following equation:[102]

$$\tau_h = -\frac{nkT/q}{dV_D(t)/dt},$$

where n is 1 or 2 depending on whether the injection level is low or high. With the numerical values as determined in Fig. 31. (b), the hole lifetime in the n-base results in

$$\tau_h \approx 1\ \mu s.$$

As expected these value exceeds the hole lifetimes determined at the edge of the space charge region.

Levinshtein et al. measured the minority carrier lifetime in n-p-n-p 4H-SiC thyristors at high current densities and compared the experimental data with calculations based on the decrease of the emitter injection coefficient.[103] These authors determined electron lifetimes of $\tau_n = 80$ ns at T = 300 K and $\tau_n = 280$ ns at 500 K.

In order to be able to compare the electrically determined hole lifetimes with corresponding optical data, time-resolved photoluminescence (TRPL) measurements are conducted directly on the n-type 4H-SiC epilayer in areas between the p^+-n mesa-diodes. A survey on the TRPL technique is given in Ref. 104. As mentioned above, the radiative recombination channel is extremely weak in SiC and is sensitively interconnected with all the other non-radiative recombination processes. The advantage of the optical method is that it is non-destructive. It is assumed that the decay of the photoluminescence can be approximated by an exponential dependence. The time constant of this exponential function is considered as a measure for the minority carrier lifetime. The crucial question is to what extent optically determined lifetimes are affected by surface recombination.

The TRPL measurements were conducted at room temperature.[97] A flash lamp pumped, frequency-quadrupled pulsed Nd:YAG-laser ($\lambda = 266$ nm) with a pulse width of 5 ns and a repetition rate of 30 Hz served as the light source. The decay of the total PL signal was detected with a photomultiplier. In order to suppress surface recombination, we considered only the section of the decay curves after the TRPL signal was decreased below 50 % of its maximum value. The decay curve in Fig. 32 can be approximated by two exponential dependencies:

$$\tau_{h,1} = 25 \text{ ns and } \tau_{h,2} = 240 \text{ ns}.$$

The reason for the steep decrease of the PL intensity during the first 200 ns is not clarified yet. Due to the short wavelength of $\lambda = 266$ nm, most of the excess carriers are confined in a 0.3 μm thick surface layer. Shishkin et al. argue that during the first 400 ns non-radiative surface recombination makes a more significant contribution and diffusion removes very quickly excess carriers.[105] Table 5 summarizes the hole lifetimes, which are determined by electrical and optical methods on the identical 4H-SiC epilayer.

Table 5. Comparison of hole lifetimes determined by electrical and optical methods on one and the same 4H-SiC epilayer.[97]

method	hole lifetime (ns)	remark
CRT (after Ref. 100)	50	for short t_{ramp} (≈ 5 ns)
	25	for $t_{ramp} \geq 100$ ns (method no longer valid)
CRT (after Ref. 98)	100	for $t_{ramp} = 100$ ns
	linear increase with t_{ramp}	explanation by Ref. 101
	50	extrapolation to $t_{ramp} \approx 5$ ns
OCVD (after Ref. 102)	1000	obtained from linear part of decay curve
TRPL	25 240	decay of PL intensity during first 200 ns probably affected by surface recombination

9. Properties of SiC/SiO_2 Interfaces

SiC can thermally be oxidized and is, therefore, like Si a suitable candidate for metal-oxide-semiconductor field effect transistors (MOSFETs). Since the early nineties intensive research work has been conducted on the thermal oxidation of SiC; for a review, see Ref. 106. Because of the nearly isotropic electrical properties and the high electron Hall mobility in the bulk the 4H- and 3C-SiC polytype are theoretically superior for high power MOSFET applications.[49, 107] Regarding the breakdown electric field, the quality of

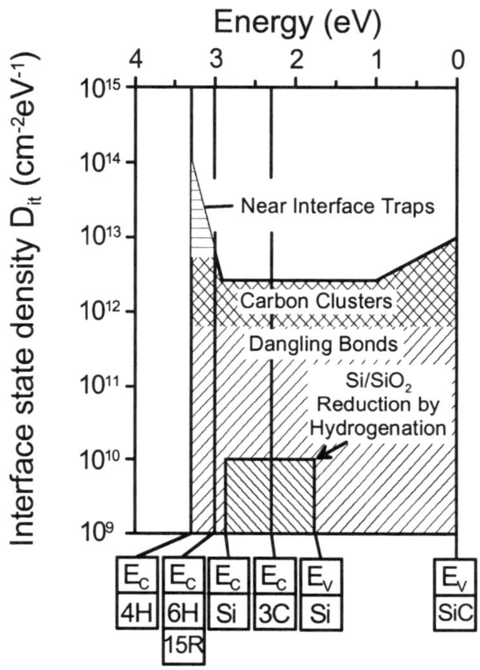

Fig. 33. Schematic representation of the density of states D_{it} at the SiC/SiO_2 interface of different polytypes; the band edges (E_V, E_C) of the SiC polytypes are marked on the x-axis. D_{it} is composed of dangling bond centers, carbon clusters and near interface traps (NITs). For comparison, also D_{it} of the Si/SiO_2 interface is schematically shown.

oxides thermally grown on SiC is comparable to that of SiO_2 layers grown on Si, however, the density of charged interface states at SiC/SiO_2 interfaces is about two orders of magnitude higher than at Si/SiO_2 interfaces.[106, 108] Especially 4H-SiC/SiO_2 interfaces suffer from an extremely high density of interface states (D_{it}) close to the conduction band edge, which leads to small effective electron drift mobilities in the channel of MOSFETs and reduces in this way the performance of the devices.[109, 110]

The density of states D_{it} at SiC/SiO_2 interfaces is composed of three different types of traps (see Fig. 33):

(a) Dangling bond centers

So-called P_b-centers, which are caused by structural misfit between the semiconductor and the oxide, are expected at SiC/SiO_2 interfaces in analogy to Si/SiO_2 interfaces. These

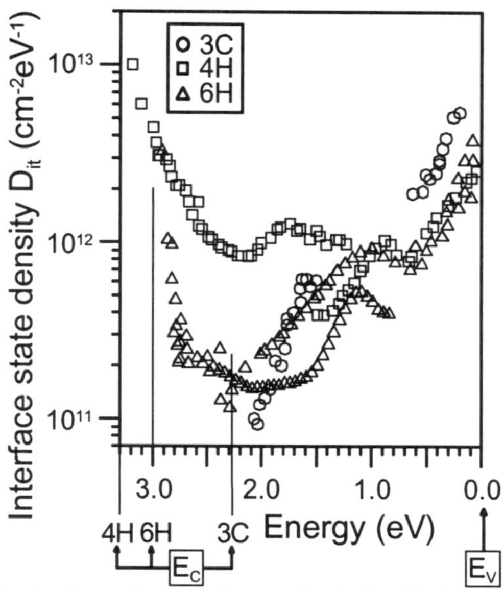

Fig. 34. Interface state density D_{it} as a function of energy for the 3C- (circles), 4H- (squares) and 6H-SiC (triangles) as determined by ac conductance spectroscopy and deep level transient spectroscopy (DLTS). The zero point of the energy scale corresponds to the top of the SiC valence bands.

dangling bond centers are in detail discussed in the literature; their density is of the same order like at Si/SiO_2 interfaces and plays only a marginal role compared to the total density of interface traps at SiC/SiO_2 interfaces.[109, 111] These centers are not subject of the research activity of this project.

(b) Carbon clusters

Excess carbon at SiC/SiO_2 interfaces has been observed by angle-resolved X-ray photoelectron spectroscopy.[112] Atomic-force microscopy and combined transmission electron microscopy/energy loss spectroscopy investigations revealed carbon particles at SiC/SiO_2 interfaces.[113, 114] ESR investigations discovered paramagnetic centers, which are related to dangling bonds of C atoms in a surrounding of amorphous carbon.[115] Internal electron photoemission (IPE) spectra show a significant similarity between spectra of electron states in the SiC band gap of SiC/SiO_2 structures and spectra taken at interfaces of hydrogenated amorphous carbon (a-C:H) films deposited on SiO_2.[106] In a-C:H, the highest occupied states correspond to π-bonds of sp^2-hybridized carbon atoms arranged in clusters of different sizes. In addition, it has been observed that exposure of SiC surfaces to a special UV-ozone clean (anti-carbon clean) prior to the oxidation leads to a reduction of D_{it}.[116]

Fig. 34 shows D_{it} at SiC/SiO_2 interfaces across the band gap of the 3C-, 4H- and 6H-SiC polytype as obtained from capacitance- and ac conductance - voltage measurements.[106, 117] D_{it} has a broad distribution across the entire band gap; the density of states is higher in the lower part of the band gap, however, it also steeply increases close to the conduction band edge of 4H-SiC. With exception of this particular increase, the distribution of D_{it} can be explained in the framework of the "carbon-cluster-model" assuming a combination of π-bonded carbon clusters of different sizes at the SiC/SiO_2 interface.[106]

(c) Near interface traps (NITs)

According to Hall effect investigations, D_{it} approaches 10^{14} $cm^{-2}eV^{-1}$ close to the conduction band edge of $4H\text{-}SiC/SiO_2$ MOS capacitors; these defect states are not correlated with the π-bonds of sp^2-hybridized carbon atoms.[118] It is, therefore, concluded that they belong to a chemically different defect species. Comparative photon stimulated electron tunneling (PST) investigations conducted on both SiC/SiO_2 and Si/SiO_2 interfaces result in acceptor states in the oxide close to the interface, which are located in a narrow energy range at around 2.8 eV below the conduction band edge of SiO_2.[106] This energy range coincides with the conduction band edge of 4H-SiC, while it is resonant in the conduction band of Si and 3C-SiC.[119] In 4H-SiC, these states can trap electrons from the 4H-SiC conduction band, which leads to a reduction of the free electron concentration and to an additional scattering by charged impurities in the channel of MOSFETs. In this way, the electron Hall mobility $\mu_{H,e}$ determined in the bulk of SiC can be strongly reduced. We have termed these states: "Near Interface Traps" (NITs).[106] Although the atomic structure of NITs is unknown, their density correlates with the presence of excess Si in the oxide that means they appear to be associated with an oxygen deficiency.[120]

Many experiments are described in the literature to improve the electrical properties of SiC/SiO_2 interfaces and of SiC-based MOSFETs. In the following a selection of proposals is summarized:

(i) Influence of SiC polytypes

The top of the valence band of SiC polytypes has a fixed position with respect to the lower conduction band edge of SiO_2. It is observed that D_{it} near the SiC conduction band edge is strongly reduced changing the SiC polytype from 4H- over 6H-/15R- to 3C-SiC that means with shrinking band gap.[106, 110]

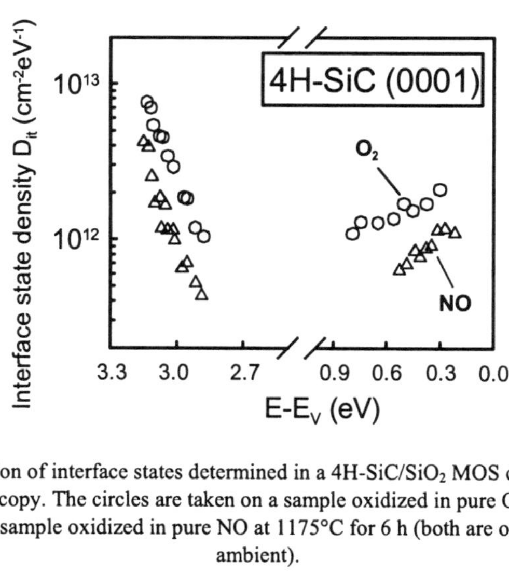

Fig. 35. Energy distribution of interface states determined in a $4H\text{-}SiC/SiO_2$ MOS capacitor as obtained from ac conductance spectroscopy. The circles are taken on a sample oxidized in pure O_2 at 1120°C for 24 h, the triangles are taken on a sample oxidized in pure NO at 1175°C for 6 h (both are oxidized in nominally dry ambient).

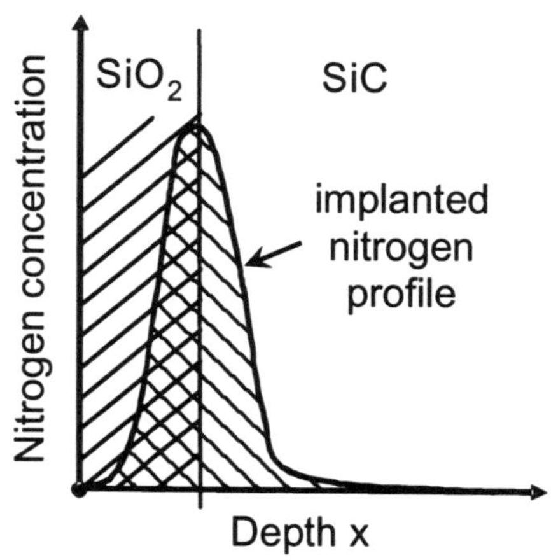

Fig. 36. Scheme of N-implanted (Gaussian profile), n-type 4H-SiC MOS structure. The dashed vertical straight line indicates the position of the SiC/SiO_2 interface.

(ii) Orientation-dependent effects
Oxidized Si (0001)-surfaces of hexagonal SiC polytypes result in lower D_{it} values than C (000-1)-surfaces. This observation is attributed to a higher density of carbon clusters related to the higher availability of carbon at the (000-1)-surface.[106, 121, 122] This suggestion gained additional support from experiments performed with tilted surfaces and experiments conducted on a-planes with (11-20)- and (1-100)-faces.[121-124] Reduced D_{it} values close to the conduction band edge of 4H-SiC and improved performance of the mobility was observed by the Kyoto group at (11-20)- and (03-38)-planes.[124-128] This group also pointed out the different behavior of energetically shallow and deep traps in 4H-SiC MOS capacitors with regard to the (0001)- and (11-20)-orientation concluding that the chemical origin of these traps is different.[124, 128]

(iii) Nitridation-related effects
One of the most important developments in improving the SiC/SiO_2 interface properties concern nitridation of the oxide; first experiments have been reported by Li et al..[129] Nitridation has been carried out in NO, N_2O or NH_3 ambient either directly during the oxidation process or by post-oxidation annealing of the SiC/SiO_2 structures.[129-136] Conductance measurements reveal that a significant reduction of D_{it} can be achieved by nitridation compared to the standard oxidation in the lower part of the band gap as well as close to the conduction band edge (see Fig. 35).[111, 130]

An effective reduction of D_{it} is reached in the complete upper half of the band gap of 4H-SiC by implantation of a surface-near Gaussian N profile and subsequent oxidation in pure O_2 ambient (see Fig. 36).[137] By optimizing the implanted N concentration ($N_{max} \approx 3\times10^{18}$ cm^{-3}) and positioning the SiC/SiO_2 interface into the maximum of the implanted Gaussian profile, the density of interface traps could be reduced to approximately 10^{10} $cm^{-2}eV^{-1}$ corresponding to the detection limit of the employed analysis system (see Fig. 37).

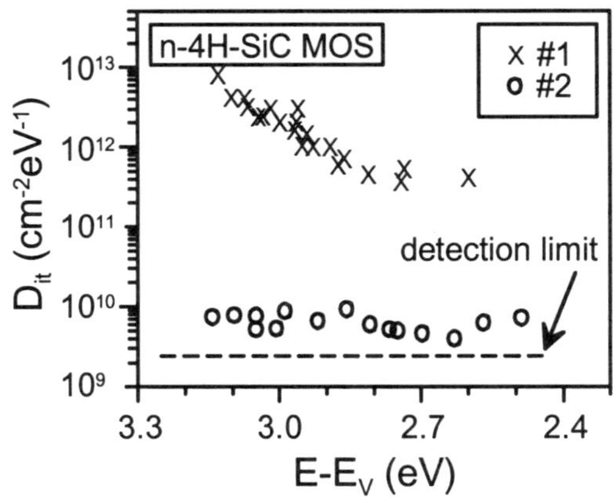

Fig. 37. D_{it} as a function of energy for an n-type 4H-SiC MOS capacitor oxidized in pure O_2 (crosses) and a sample processed under identical conditions, which has been implanted with a Gaussian N profile prior to the oxidation (circles).

The effect of nitridation on the effective electron drift mobility $\mu_{eff,e}$ in the inversion channel of p-type SiC MOSFETs seems to considerably depend on the experimental conditions (e.g. composition of nitrogen ambient, contaminations introduced during the oxidation, temperature ramps). A large scatter of values for $\mu_{eff,e}$ ranging from 165 cm^2/Vs to 260 cm^2/Vs for 3C-SiC and from 10 cm^2/Vs to 150 cm^2/Vs for 4H-SiC MOSFETs is reported in the literature.[138-141] The role of nitrogen in reducing traps at the SiC/SiO_2 interface is likely a complex one, because nitrogen affects traps in the band gap of SiC of different chemical origin at the same time. McDonald et al. observed a saturation of the passivation of interface states at an nitrogen content of 2.5×10^{14} cm^{-2}.[142] These authors suggest that large clusters composed of excess interfacial carbon or silicon are step by step dissolved by forming strong C≡N or Si≡N bonds. X-ray photoelectron spectroscopy experiments have detected Si≡N bonds after incorporation of nitrogen; such strong bonds may relax the strain at the interface leading to a decrease of NITs and remove open Si and/or C bonds at the oxidizing surface.[131] All the proposed mechanisms for the passivation of traps at SiC/SiO_2 interfaces by the nitridation process are presently speculative and have to be theoretically and experimentally clarified.

Acknowledgments

The support of this work by the German Science Foundation (SiC-Forschergruppe) is gratefully acknowledged.

References

1. J. J. Berzelius, "Untersuchungen über die Flußspatsäure und deren merkwürdigsten Verbindungen", Annal. d. Physik B **77** (1824) 1-48 and 169-230.

2. M. E. Levinshtein, S. L. Rumyantsev, M. S. Shur, *Properties of Advanced Semiconductor Materials: GaN, AlN, InN, BN, SiC, SiGe*, Wiley & Sons, Inc., New York, 2001.
3. W. F. Knippenberg, "Growth Phenomena in Silicon Carbide", Philips Res. Repts. **18** (1963) 161-274.
4. D. Pandey, P. Krishna, "The Origin of Polytype Structures", in *Progress in Crystal Growth and Characterization* Vol. **7**, *Crystal Growth and Characterization of Polytype Structures*, ed. P. Krishna, Pergamon Press, Oxford, 1983, pp. 213-258.
5. L. Patrick, "Kohn-Luttinger Interference Effect and Location of the Conduction Band Minima in 6H-SiC", Phys. Rev. B **5** (1972) 2198-2206.
6. N. Churcher, K. Kunc, V. Heine, "Calculated ground-state properties of silicon carbide", J. Phys. C: Solid State Phys. **19** (1986) 4413-4426.
7. F. Herman, J. P. van Dyke, R. L. Kortum, "Electronic Structure and Optical Spectrum of Silicon Carbide", Mater. Res. Bull. **178** (1968) S167-S178.
8. P. E. van Camp, V. E. van Doren, J. T. Devreese, "Ground State and Electronic Properties of Silicon Carbide and Boron Nitride", phys. stat. sol. (b) **146** (1988) 573-587.
9. G. Wellenhofer, U. Rössler, "Global Band Structure and Near-Band-Edge States", phys. stat. sol. (b) **202** (1997) 107-124.
10. F. Bechstedt, P. Käckell, A. Zywietz, K. Karch, B. Adolph, K. Tenelsen, J. Furthmüller, "Polytypism and Properties of Silicon Carbide", phys. stat. sol. (b) **202** (1997) 35-62.
11. W. van Haeringen, P. A. Bobbert, W. H. Backes, "On the Band Gap Variation in SiC Polytypes", phys. stat. sol. (b) **202** (1997) 63-80.
12. W. R. L. Lambrecht, S. Limpijumnong, S. N. Rashkeev, B. Segall, "Electronic Band Structure of SiC Polytypes: A Discussion of Theory and Experiment", phys. stat. sol. (b) **202** (1997) 5-33.
13. W. J. Choyke, L. Patrick, "Absorption of Light in Alpha Silicon Carbide Near the Band Edge", in *Silicon Carbide*, eds. J. P. O'Connor, J. Smiltens, Pergamon Press, Boston, 1960, pp. 306-311.
14. G. Pensl, W. Suttrop, H. Zhang, R. Helbig, P. Glasow, P. Lanig, G. Ziegler, "Fabrication of Implanted 6H-SiC UV-Photodiodes", in *Advances in Physics of Materials*, eds. Z. H. Zaidi, M. Husain, Today & Tomorrow's Printers & Publishers, New Delhi, 1989, pp. 159-173.
15. C. Persson, U. Lindefelt, "Detailed band structure for 3C-, 2H-, 4H-, 6H-SiC, and Si around the fundamental band gap", Phys. Rev. B 54 (1996) 10257-10260.
16. K. Kaplan, R. J. Wagner, H. J. Kim, R. F. Davis, "Electron cyclotron resonance in cubic SiC", Solid State Commun. **55** (1985) 67-69.
17. G. Pensl, Th. Troffer, "Silicon Carbide - A Promising Wide-Band-Gap Semiconductor for Electronic Devices", Solid State Phenomina **47-48** (1996) 115-126.
18. H. Harima, S. Nakashima, T. Uemura, "Raman scattering from anisotropic LO-phonon-plasmon-coupled mode in n-type- and 6H-SiC", J. Appl. Phys. **78** (1995) 1996-2005.
19. N. T. Son, O. Kordina, A. O. Konstantinov, W. M. Chen, E. Sörman, B. Monemar,

E. Janzén, "Electron effective masses and mobilities in high-purity 6H-SiC chemical vapor deposition layers", Appl. Phys. Lett. **65** (1994) 3209-3211.
20. D. Volm, B. K. Meyer, D. M. Hofmann, W. M. Chen, N. T. Son, C. Persson, U. Lindefelt, O. Kordina, E. Sörman, A. O. Konstantinov, B. Monemar, E. Janzén, "Determination of the electron effective-mass tensor in 4H-SiC", Phys. Rev. B **53** (1996) 15409-15412.
21. J. Kono, S. Takeyama, H. Yokoi, N. Miura, M. Yamanaka, M. Shinohara, K. Ikoma, "High-field cyclotron resonance and impurity transition in n-type and p-type 3C-SiC at magnetic fields up to 175 T", Phys. Rev. B **48** (1993) 10909-10916.
22. N. T. Son, C. Hallin, E. Janzén, "Hole effective masses in 6H-SiC from optically detected cyclotron resonance", Phys. Rev. B **66** (2002) 045304.
23. N. T. Son, P. N. Hai, W. M. Chen, C. Hallin, B. Monemar, E. Janzén, "Hole effective masses in 4H-SiC", Phys. Rev. B **61** (2000) R10544-R10546.
24. O. Nilsson, H. Mehling, R. Horn, J. Fricke, R. Hofmann, S. G. Müller, R. Eckstein, D. Hofmann, "Determination of the thermal diffusivity and conductivity of mono crystalline silicon carbide (300-2300 K)", High Temperatures - High Pressures **29** (1997) 73-79.
25. A. Taylor, R. M. Jones, "The Crystal Structure and Thermal Expansion of Cubic and Hexagonal Silicon Carbide", in *Silicon Carbide*, eds. J. P. O'Connor, J. Smiltens, Pergamon Press, Boston, 1960, pp. 147-161.
26. E. L. Kern, D. W. Hamill, H. W. Deem, H. D. Sheets, "Thermal Properties of β-Silicon Carbide from 20 to 2000°C", Mat. Res. Bull. Vol **4** (1969) S25-S32.
27. Z. Li, R. C. Bradt, "Thermal expansion of the hexagonal (4H) polytype of SiC", J. Appl. Phys. **60** (1986) 612-614.
28. D. Morelli, J. Heremans, C. Beetz, W. S. Woo, G. Harris, C. Taylor, "Carrier concentration dependence of the thermal conductivity of silicon carbide", Inst. Phys. Conf. Ser. No. **137** (1993) 313-315.
29. D. T. Morelli, J. P. Heremans, C. P. Beetz, W. Sik Yoo, H. Matsunami, "Phonon-electron scattering in single crystal silicon carbide", Appl. Phys. Lett. **63** (1993) 3143-3145.
30. G. Rutsch, R. P. Devaty, W. J. Choyke, D. W. Langer, L. B. Rowland, E. Niemann, F. Wischmeyer, "Hall Scattering Factor and Electron Mobility of 4H-SiC: Measurements and Numerical Simulation", Mater. Sci. Forum **338-342** (2000) 733-736.
31. G. Rutsch, R. P. Devaty, W. J. Choyke, D. W. Langer, L. B. Rowland, "Measurement of the Hall scattering factor in 4H and 6H SiC epilayers from 40 to 290 K and in magnetic fields up to 9 T", J. Appl. Phys. **84** (1998) 2062-2064.
32. G. Pensl, F. Schmid, F. Ciobanu, M. Laube, S. A. Reshanov, N. Schulze, K. Semmelroth, H. Nagasawa, A. Schöner, G. Wagner, "Electrical and Optical Characterization of SiC", Mater. Sci. Forum **433-436** (2003) 365-370.
33. G. Pensl, F. Ciobanu, M. Krieger, M. Laube, S. Reshanov, F. Schmid, G. Wagner, H. Nagasawa, A. Schöner, "Hall Scattering Factor of Holes and Shallow Defect Centers in Aluminum-doped SiC", Mat. Res. Soc. Symp. Proc. **742** (2003) 163-174.
34. F. Schmid, M. Krieger, M. Laube, G. Pensl, G. Wagner, "Hall Scattering Factor for Electrons and Holes in SiC", in *Silicon Carbide - Recent Major Advances*, eds.

W. J. Choyke, H. Matsunami, G. Pensl, Springer, Berlin, 2004, pp. 517-536.
35. J. S. Blakemore, *Semiconductor Statistics*, Dover Publications, Inc., New York, 1962, pp. 140-148.
36. H. H. Woodbury, G. W. Ludwig, "Electron Spin Resonance Studies in SiC", Phys. Rev. **124** (1961) 1083-1089.
37. S. Greulich-Weber, "EPR and ENDOR Investigations of Shallow Impurities in SiC Polytypes", phys. stat. sol. (a) **162** (1997) 95-151.
38. W. J. Schaffer, G. H. Negley, K. G. Irvine, J. W. Palmour, "Conductivity Anisotropy in Epitaxial 6H and 4H SiC", Mater. Res. Soc. Symp. Proc. **337** (1994) 595-600.
39. M. Schadt, G. Pensl, R. P. Devaty, W. J. Choyke, R. Stein, D. Stephani, "Anisotropy of the electron Hall mobility in 4H, 6H, and 15R silicon carbide", Appl. Phys. Lett. **65** (1994) 3120-3122.
40. A. Schöner, "Elektrische Charakterisierung von flachen und tiefen Störstellen in 4H- 6H- und 15R-Siliciumkarbid", Ph. D. Thesis, University of Erlangen-Nürnberg, 1994.
41. A. L. Éfros, Nguyen Van Lien, B. I. Shklovskii, "Impurity band structure in lightly doped semiconductors", J. Phys. C: Solid State Physics **12** (1979) 1869-1881.
42. L. S. Aĭvazova, S. N. Gorin, V. G. Sidyakin, I. M. Shvarts, "Electrical properties of cubic silicon carbide", Sov. Phys. Semicond. **11** (1997) 1069-1070.
43. W. Götz, A. Schöner, G. Pensl, W. J. Choyke, R. Stein, S. Leibenzeder, "Nitrogen donors in 4H-silicon carbide", J. Appl. Phys. **73** (1993) 3332-3338.
44. W. Suttrop, G. Pensl, W. J. Choyke, R. Stein, S. Leibenzeder, "Hall effect and infrared absorption measurements on nitrogen donors in 6H-silicon carbide", J. Appl. Phys. **72** (1992) 3708-3713.
45. G. Pensl, T. Dalibor, "Impurity and defect levels" in *Landolt-Börnstein* **41**, Subvolume A2, Part β, ed. M. Schulz, Springer, Berlin, 2003, pp. 8-33.
46. A. I. Veĭnger, A. G. Zabrodskiĭ, G. A. Lomakina, E. N. Mokhov, "Paramagnetic and electrical properties of transmutation-generated phosphorus impurity in 6H-SiC", Sov. Phys. Solid State **28** (1986) 917-919.
47. T. Troffer, C. Peppermüller, G. Pensl, K. Rottner, A. Schöner, "Phosphorus-related donors in 6H-SiC generated by ion implantation", J. Appl. Phys. **80** (1996) 3739-3743.
48. K. Abe, T. Ohshima, H. Itoh, Y. Aoki, M. Yoshikawa, I. Nashiyama, M. Iwami, "Hot-Implantation of Phosphorus Ions into 6H-SiC", Mater. Sci. Forum **264-268** (1998) 721-724.
49. T. Kimoto, S. Nakazawa, K. Hashimoto, H. Matsunami, "Reduction of doping and trap concentrations in 4H-SiC epitaxial layers grown by chemical vapor deposition", Appl. Phys. Lett. **78** (2001) 2761-2763.
50. H.-J. Rost, D. Schulz, D. Siche, "High Nitrogen Doping During Bulk Growth of SiC", in *Silicon Carbide - Recent Major Advances*, eds. W. J. Choyke, H. Matsunami, G. Pensl, Springer, Berlin, 2004, pp. 163-178.
51. M. Laube, F. Schmid, G. Pensl, G. Wagner, M. Linnarsson, M. Maier, "Electrical activation of high concentrations of N^+ and P^+ ions implanted into 4H-SiC", J. Appl. Phys. **92** (2002) 549-554.
52. D. Åberg, A. Hallén, P. Pellegrino, B. Svensson, "Nitrogen deactivation by

implantation-induced defects in 4H-SiC epitaxial layers", Appl. Phys. Lett. **78** (2001) 2908-2910.

53. F. Schmid, M. Laube, G. Pensl, G. Wagner, M. Maier, "Electrical activation of implanted phosphorus ions in [0001]- and [11-20]-oriented 4H-SiC", J. Appl. Phys. **91** (2002) 9182-9186.
54. F. Schmid, G. Pensl, "Comparison of the electrical activation of P^+ and N^+ ions co-implanted along with Si^+ or C^+ ions into 4H-SiC", Appl. Phys. Lett. **84** (2004) 3064-3066.
55. M. Bockstedte, A. Mattausch, O. Pankratov, "Solubility of nitrogen and phosphorus in 4H-SiC: A theoretical study", Appl. Phys. Lett. **85** (2004) 58-60.
56. F. Schmid, T. Frank, G. Pensl, "Experimental evidence for an electrically neutral (N-Si)-complex formed during the annealing process of Si^+-/N^+-co-implanted 4H-SiC", Mater. Sci. Forum **483-485** (2005) 641-644.
57. K. Semmelroth, F. Schmid, D. Karg, G. Pensl, M. Maier, S. Greulich-Weber, J.-M. Spaeth, "Growth of Phosphorus-Doped 6H-SiC Single Crystals by the Modified Lely Method", Mater. Sci. Forum **433-436** (2003) 63-66.
58. P. Desperrier, R. Müller, A. Winnacker, P. J. Wellmann, "Growth of phosphorus-doped n-type 6H-SiC crystals using a modified PVT technique and phosphine as source", Mater. Sci. Forum **457-460** (2004) 727-730.
59. R. Wang, I. B. Bhat, T. P. Chow, "Epitaxial growth of n-type SiC using phosphine and nitrogen as the precursors", J. Appl. Phys. **92** (2002) 7587-7592.
60. A. Henry, E. Janzén, "Epitaxial Growth and Characterization of Phosphorus Doped SiC Using TBP as Precursor", Mater. Sci. Forum **483-485** (2005) 101-104.
61. S. Tamura, T. Kimoto, H. Matsunami, M. Okada, S. Kanatawa, I. Kimura, "Nuclear Transmutation Doping of Phosphorus into 6H-SiC", Mater. Sci. Forum **338-342** (2000) 849-853.
62. F. H. C. Carlsson, L. Storasta, B. Magnusson, J. P. Bergman, K. Sköld, E. Janzén, "Neutron Irradiation of 4H SiC", Mater. Sci. Forum **353-356** (2001) 555-559.
63. Le Si Dang, K. M. Lee, G. D. Watkins, W. J. Choyke, "Optical Detection of Magnetic Resonance for an Effective-Mass-like Acceptor in 6H-SiC", Phys. Rev. Lett. **45** (1980) 390-394.
64. A. G. Zubatov, V. G. Stepanov, Yu. A. Vodakov, E. N. Mokhov, "ESR in certain boron-doped SiC polytypes", Sov. Tech. Phys. Lett. **8** (1982) 120-121.
65. T. Matsumoto, O. G. Poluektov, J. Schmidt, E. N. Mokhov, P. G. Baranov, "Electronic structure of the shallow boron acceptor in 6H-SiC; mA pulsed EPR/ENDOR study at 95 GHz", Phys. Rev. B **55** (1997) 2219-2229.
66. T. Troffer, M. Schadt, T. Frank, H. Itoh, G. Pensl, J. Heindl, H. P. Strunk, M. Maier, "Doping of SiC by Implantation of Boron and Aluminum", phys. stat. sol. (a) **162** (1997) 277-298.
67. A. Schöner, N. Nordell, K. Rottner, R. Helbig, G. Pensl, "Dependence of the aluminum ionization energy on doping concentration and compensation in 6H-SiC", Institute of Physics Conf. Series No. **142** (1996) 493-496.
68. Yu. A. Vodakov, E. N. Mokhov, "Diffusion and Solubility of Impurities in Silicon Carbide", in *Silicon Carbide*, eds. R. C. Marshall, J. W. Faust, JR., C. E. Ryan, University of South Carolina Press, Columbia, 1974, pp. 508-519.

69 S. Seshadri, G. W. Eldridge, A. K. Agarwal, "Comparison of the annealing behavior of high-dose nitrogen-, aluminum-, and boron-implanted 4H-SiC", Appl. Phys. Lett. **72** (1998) 2026-2028.
70. M. K. Linnarsson, M. S. Janson, A. Schöner, B. G. Svensson, "Aluminum and boron diffusion in 4H-SiC", Mat. Res. Soc. Symp. Proc. **742** (2003) 291-301.
71. E. N. Mokhov, E. E. Goncharov. G. G. Ryabova, "Diffusion of boron in p-type silicon carbide", Sov. Phys. Semicond. **18** (1984) 27-30.
72. M. Laube, G. Pensl, H. Itoh, "Suppressed diffusion of implanted boron in 4H-SiC", Appl. Phys. Lett. **74** (1999) 2292-2294.
73. M. Laube, G. Pensl, "Transient-Enhanced Diffusion of Boron in SiC", Mater. Sci. Forum **338-342** (2000) 941-944.
74. G. Pensl, V. V. Afanas'ev, M. Bassler, T. Frank, M. Laube, M. Weidner, "Physics of SiC Processing", Mater. Sci. Forum **338-342** (2000) 831-836.
75. H. Bracht, N. A. Stolwijk, M. Laube, G. Pensl, "Diffusion of boron in silicon carbide: evidence for the kick-out mechanism", Appl. Phys. Lett. **77** (2000) 3188-3190.
76. K. Rüschenschmidt, H. Bracht, M. Laube, N. A. Stolwijk, G. Pensl, "Diffusion of boron in silicon carbide", Physica B **308-310** (2001) 734-737.
77. M. Bockstedte, A. Mattausch, O. Pankratov, "Different roles of carbon and silicon interstitials in the interstitial-mediated boron diffusion in SiC", Phys. Rev. B **70** (2004) 115203-1152203-13.
78. R. Rurali, E. Hernández, P. Godignon, J. Rebollo, P. Ordejón, "First-principles studies of the diffusion of B impurities and vacancies in SiC", Phys. Rev. B **69** (2004) 125203-125203-11.
79. K. Rüschenschmidt, H. Bracht, N. A. Stolwijk, M. Laube, G. Pensl, G. R. Brandes, "Self-diffusion in isotopically enriched silicon carbide and its correlation with dopant diffusion", J. Appl. Phys. **96** (2004) 1458-1463.
80. A. O. Konstantinov, "Nature of point defects generated during diffusion of acceptor impurities in silicon carbide", Sov. Phys. Semicond. 26 (1992) 151-156.
81. G. Busch, H. Labhart, "Über den Mechanismus der elektrischen Leitfähigkeit des Siliciumcarbids", Helv. phys. acta **14** (1946) 463-492.
82. B. I. Shklovskii, A. L. Éfros, *Electronic Properties of Doped Semiconductors*, Springer Series in Solid-State Sciences **45**, Berlin, 1984, pp. 35-39.
83. N. F. Mott, "Conduction in glasses containing transition metal ions", J. Non-Crystal. Solids **1** (1968) 1-17.
84. H. Fritzsche, M. Cuevas, "Impurity conduction in transmutation-doped p-type germanium", Phys. Rev. **119** (1960) 1238-1245.
85. N. F. Mott, "On the transition to metallic conduction in semiconductors", Canad. J. Phys. **34** (1956) 1356-1368.
86. A. Miller, E. Abrahams, "Impurity conduction at low concentrations", Phys. Rev. **120** (1960) 745-755.
87. A. L. Éfros, B. I. Shklovskii, I. Y. Yanchev, "Impurity conductivity in low compensated semiconductors", phys. stat. sol. (b) **50** (1972) 45-52.
88. N. F. Mott, W. D. Twose, "The Theory of Impurity Conduction", Adv. Physics **10** (1961), 107-163.
89. M. Krieger, K. Semmelroth, G. Pensl, "Impurity conduction observed in Al-doped

6H-SiC", Mater. Sci. Forum **457-460** (2004) 685-688.
90. D. Emin, "The sign of the Hall effect in hopping conduction", Phil. Mag. **35** (1977), 1189-1198.
91. M. Grünewald, P. Thomas, D. Würtz, "The sign anomaly of the Hall effect in amorphous tetrahedrally bonded semiconductors: a chemical-bond orbital approach", J. Phys. C **14** (1981), 4083–4093.
92. A. O. Evwaraye, S. R. Smith, W. C. Mitchel, M. D. Roth, "Determination of the activation energy ε_3 for impurity conduction in n-type 4H-SiC", Appl. Phys. Lett. **68** (1996) 3159-3161.
93. R. N. Hall, "Electron-Hole Recombination in Germanium", Phys. Rev. **87** (1952) 387 and W. Shockley, W. T. Read, "Statistics of the Recombination of Holes and Electrons", Phys. Rev. **87** (1952) 835-842.
94. D. K. Schroder, *Semiconductor Material and Device Characterization*, John Wiley & Sons, INC., New York, 1998, pp. 420-508.
95. A. Galeckas, J. Linnros, V. Grivckas, U. Lindefelt, C. Hallin, "Evaluation of Auger Recombination Rate in 4H-SiC", Mater. Sci. Forum **264- 268** (1998) 533-536.
96. W. M. Bullis, "Properties of Gold in Silicon", Solid State Electron. **9** (1966) 143-168.
97. S. A. Reshanov, G. Pensl, "Comparison of Electrically and Optically Determined Minority Carrier Lifetimes in 6H-SiC", Mater. Sci. Forum **483-485** (2005) 417-420.
98. B. Lax, T. Neustadter, "Transient Response of p-n Junctions", J. Appl. Phys. **25** (1954) 1148-1154.
99. B. Tien, C. Hu, "Determination of Carrier Lifetime from Rectifier Ramp Recovery Waveform", IEEE Electron. Device Lett. **9** (1988) 553-555.
100. S. R. Dhariwal, R. C. Sharma, "Determination of Carrier Lifetime in p-i-n Diodes by Ramp Recovery", IEEE Electron. Device Lett. **13** (1992) 98-101.
101. M. E. Levinshtein, T. T. Mnatskanov, P. Ivanov, J. W. Palmour, S. L. Rumyantsev, R. Singh, S. N. Yurkov, "'Paradoxes' of Carrier Lifetime Measurements in High-Voltage SiC Diodes", IEEE Trans. Electron. Devices **48** (2001) 1703-1709.
102. B. R. Gossik, "On the Transient Behavior of Semiconductor Rectifiers", J. Appl. Phys. **27** (1956) 905-911.
103. M. E. Levinshtein, J. W. Palmour, S. L. Rumyantsev, R. Sing, "Forward current-voltage characteristics of silicon carbide thyristors and diodes at high current densities", Semicond. Sci. Technol. **13** (1998) 1006-1010.
104. J. P. Bergman, O. Kordina, E. Janzén, "Time Resolved Spectroscopy of Defects in SiC", phys. stat. sol. (a) **162** (1997) 65-78.
105. Y. Shishkin, R. P. Devaty, W. J. Choyke, "Optical Lifetime Measurements in 4H-SiC", Mater. Sci. Forum **338-342** (2000) 679-682.
106. V. V. Afanas'ev, M. Bassler, G. Pensl, M. Schulz, "Intrinsic SiC/SiO_2 Interface States", phys. stat. sol. (a) **162** (1997) 321-337.
107. T. Tachibana, H. S. Kong, Y. C. Wang, R. F. Davis, "Hall measurements as a function of temperature on monocrystalline SiC thin films", J. Appl. Phys. **67** (1990) 6375-6381.
108. D. Alok, P. K. McLarty, B. J. Baliga, "Electrical properties of thermal oxide grown on n-type 6H-silicon carbide", Appl. Phys. Lett. **64** (1994) 2845-2846.

109. G. Pensl, M. Bassler, F. Ciobanu, V. Afanas'ev, H. Yano, T. Kimoto, H. Matsunami, "Traps at the SiC/SiO_2-Interface", Mat. Res. Soc. Symp. Proc. **640** (2001) H3.2.1-H3.2.11.
110. R. Schörner, F. Friedrichs, D. Peters, D. Stephani, "Significantly Improved Performance of MOSFET's on Silicon Carbide Using the 15R-SiC Polytype", IEEE Electron Device Lett. **20** (1999) 241-244.
111. V. V. Afanas'ev, F. Ciobanu, G. Pensl, A. Stesmans, "Contributions to the Density of Interface States in SiC MOS Structures", in *Silicon Carbide - Recent Major Advances*, eds. W. J. Choyke, H. Matsunami, G. Pensl, Springer, Berlin 2003, pp. 343-371.
112. V. K. Vathuya, D. N. Wang, M. H. White, "On the correlation betwen the carbon content and the electrical quality of thermally grown oxides on p-type 6H-Silicon Carbide", Appl. Phys. Lett. **73** (1998) 2161-2163.
113. A. Koh, L. Kubler, C. Wright, S. P. Wilks, P. A. Mawby, W. R. Bowen, "Comparative surface studies on wet and dry sacrificial thermal oxidation on silicon carbide", Appl. Surf. Sci. **174** (2001) 210-216.
114. K. C. Chang, T. Nuhfer, L. M. Porter, Q. Wahab, "High-carbon concentrations at the silicon dioxide-silicon carbide interface identified by electron energy loss spectroscopy", Appl. Phys. Lett. **77** (2000) 2186-2188.
115. P. J. McFarlane, M. E. Zvanut, "Characterization of paramagnetic defect centers in three polytypes of dry heat treated, oxidized SiC", J. Appl. Phys. **88** (2000) 4122-4127.
116. V. V. Afanas'ev, A. Stesmans, M. Bassler, G. Pensl, M. Schulz, C. I. Harris, "Elimination of SiC/SiO_2 interface states by preoxidation ultraviolet-ozone cleaning", Appl. Phys. Lett. **68** (1996) 2141-2143.
117. F. Ciobanu, G. Pensl, H. Nagasawa, A. Schöner, S. Dimitrijev, K.-Y. Cheong, V. V. Afanas'ev, G. Wagner, "Traps at the Interface of $3C\text{-}SiC/SiO_2$-MOS-Structures", Mater. Sci. Forum **443-446** (2003) 551-554.
118. N. S. Saks, M. G. Ancona, E. W. Rendell, "Using the Hall effect to measure interface trap densities in silicon carbide and silicon metal-oxide-semiconductor devices", Appl. Phys. Lett. **80** (2002) 3219-3221.
119. V. V. Afanas'ev, A. Stesmans, "Photon-stimulated tunneling of electrons in SiO_2: evidence for a defect-assisted process", J. Phys.: Condens. Matter **9** (1997) L55-L60.
120. V. V. Afanas'ev, A. Stesmans, "Analysis of near-interfacial SiO_2 traps using photon stimulated electron tunneling", Microelectron. Eng. **36** (1997) 149-152.
121. J. N. Shenoy, M. K. Das, G. L. Chindalore, J. A. Cooper, Jr., M. R. Melloch, J. W. Palmour, K. G. Irwine, "Recent advances in silicon carbide MOS technology", Inst. Phys. Conf. Ser. **142** (1996) 745-748.
122. J. N. Shenoy, M. K. Das, J. A. Cooper, Jr., M. R. Melloch, J. W. Palmour, "Effect of substrate orientation and crystal anisotropy on the thermally oxidized SiO_2/SiC interface", J. Appl. Phys. **79** (1996) 3042-3045.
123. F. Lanois, D. Planson, P. Lassagne, C. Raynaud, E. Bano, "6H-SiC MOS Capacitors on Sloped Surfaces: Realisation, Characterisation and Electrical Results", Mater. Sci. Forum **264-268** (1998) 1029-1032.
124. H. Yano, T. Kimoto, H. Matsunami, "Shallow states at $SiO_2/4H\text{-}SiC$ interface on

(11-20) and (0001) faces", Appl. Phys. Lett. **81** (2002) 301-303.
125. H. Yano, T. Kimoto, H. Matsunami, K. Asano, Y. Sugawara, "High Channel Mobility in Inversion Layers of 4H-SiC MOSFET's by Utilizing (11-20) Face", IEEE Electron Dev. Lett. **20** (1999) 611-613.
126. H. Yano, T. Hirao, T. Kimoto, H. Matsunami, "A cause for highly improved channel mobility of 4H-SiC metal-oxide-semiconductor field-effect transistors on the (11-20) face", Appl. Phys. Lett. **78** (2001) 374-376.
127. T. Hirao, T. Kimoto, H. Matsunami, "4H-SiC MOSFETs on (03-38) Face", Mater. Sci. Forum **389-393** (2002) 1065-1068.
128. H. Matsunami, "Silicon Carbide Technology in New Era", Mater. Sci. Forum **389-393** (2002) 3-8.
129. H. F. Li, S. Dimitrijev, H. B. Harrison, D. Sweatman, "Interfacial characteristics of N_2O and NO nitrited SiO_2 grown on SiC by rapid thermal processing", Appl. Phys. Lett. **70** (1997) 2028-2030.
130. G. Y. Chung, J. R. Williams, K. McDonald, L. C. Feldman, "4H-SiC oxynitridation for generation of insulating layers", J. Phys.: Condens. Matter **16**, (2004) S1857-S1871.
131. P. Jamet, S. Dimitrijev, "Physical properties of N_2O and NO-nitrited gate oxides grown on 4H SiC", Appl. Phys. Lett. **79** (2001) 323-325.
132. R. Schörner, P. Friedrichs, D. Peters, D. Stephani, S. Dimitrijev, P. Jamet, "Enhanced channel mobility of 4H-SiC metal-oxide-semiconductor transistors fabricated with standard polycrystalline silicon technology and gate-oxide nitridation", Appl. Phys. Lett. **80** (2002) 4253-4255.
133. P. Jamet, S. Dimitrijev, P. Tanner, "Effects of nitridation in gate oxides grown on 4H-SiC", J. Appl. Phys. **90** (2001) 5058-5063.
134. P. T. Lai, S. Chakraborty, C. L. Chan, Y. C. Cheng, "Effects of nitridation and annealing on interface properties of thermally oxidized SiO_2/SiC metal-oxide-semiconductor system", Appl. Phys. Lett. **76** (2000) 3744-3746.
135. H.-F. Li, S. Dimitrijev, D. Sweatman, H. B. Harrison, "Effect of NO Annealing Conditions on Electrical Characteristics of n-type 4H-SiC MOS Capacitors", J. Electron. Mater. **29** (2000) 1027-1031.
136. G. Y. Chung, C. C. Tin, J. R. Williams, K. McDonald, M. Di Ventra, S. T. Pantelides, L. C. Feldman, R. A. Weller, "Effect of nitric oxide annealing on the interface trap densities near the band edges in the 4H polytype of silicon carbide", Appl. Phys. Lett. **76** (2000) 1713-1715.
137. F. Ciobanu, G. Pensl, V. Afanas'ev, A. Schöner, "Low Density of Interface States in n-type 4H-SiC MOS Capacitors Achieved by Nitrogen Implantation", Mater. Sci. Forum **483-485** (2005) 693-696.
138. J. Wan, M. A. Capano, M. R. Melloch, J. A. Cooper, Jr., "N-Channel 3C-SiC MOSFETs on Silicon Substrate", IEEE Electron. Dev. Lett. **23** (2002) 482-484.
139. T. Ohshima, K. Kiong Lee, Y. Ishida, K. Kojima, Y. Tanaka, T. Takahashi, M. Yoshikawa, H. Okumura, K. Arai, T. Kamiya, "The Electrical Characteristics of Metal-Oxide-Semiconductor Field Effect Transistors Fabricated on Cubic Silicon Carbide", Jpn. J. Appl. Phys. **42** (2003) L625-L627.
140. G. Gudjønsson, H. Ö. Ǿlafsson, E. Ö. Sveinbjörnsson, "Enhancement of inversion

channel mobility in 4H-SiC MOSFETs using a gate oxide grown in nitrous oxide (N_2O)", Mater. Sci. Forum **457-460** (2004) 1425-1428.
141. H. Ö. Ǿlafsson, "Detection and removal of traps at the SiO_2/SiC interface", Ph. D. Thesis, Chalmers University of Technology, Göteborg, 2004.
142. K. McDonald, R. W. Weller, S. T. Pantelides, L. C. Feldman, G. Y. Chung, C. C. Tin, J. R. Williams, "Characterization and modeling of the nitrogen passivation of interface traps in SiO_2/4H-SiC", J. Appl. Phys. **93** (2003) 2719-2722.

International Journal of High Speed Electronics and Systems
Vol. 15, No. 4 (2005) 747–780

SiC HOMOEPITAXY AND HETEROEPITAXY

ANDREY S. BAKIN

Institute of Semiconductor Technology
Braunschweig Technical University,
Hans-Sommer-Straße 66, D-38106, Germany

SiC bulk material quality and surface preparation do not satisfy all the requirements for direct device production. It is necessary to have high quality thick epitaxial layers with low background doping concentration for the fabrication of SiC high power, high voltage, high frequency devices. Different aspects of SiC homo- and heteroepitaxial growth are discussed in this chapter. The wafer surface has a large impact on epitaxial layers, heterostructures and finally on device properties. Thus wafer processing before epitaxial growth is discussed in detail.

1. Introduction

Most semiconductor optoelectronic and many microelectronic devices are fabricated on heterostructures which consist of layers of different materials but the same crystal structure. In this case advanced epitaxial growth techniques are needed. But even in the case of well-developed epitaxial growth technologies device parameters can be degraded if the substrate material quality or processing is poor. Therefore wafer fabrication technology is equally important for state-of-the-art semiconductor device manufacturing. A comparison of liquid and gas phase growth techniques shows that liquid phase methods provide low cost fabrication of large single crystals and are normally the method of choice for large scale production of crystals for further wafer manufacturing. Unfortunately it is hard to employ this method for certain semiconductors since due to specific phase diagrams extremely high pressures are needed to prevent the decomposition of the source material at temperatures lower than the melting point. This problem exists for instance in the case of GaN, SiC, or ZnO. The implementation of gas phase growth techniques provide the fabrication of the best quality SiC bulk crystals and epitaxial layers as well as device heterostructures which cannot be realized by liquid phase growth.

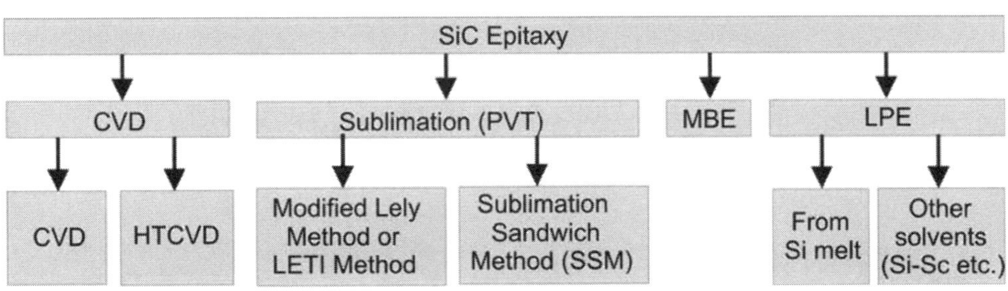

Fig. 1. Methods of SiC epitaxial growth

Currently SiC bulk material quality and surface preparation are not high enough for direct device production, therefore SiC epitaxial growth plays a key role in SiC device technology. The following methods are employed for SiC epitaxial growth: chemical vapor deposition (CVD), sublimation growth or physical vapor transport (PVT), molecular beam epitaxy (MBE) and liquid phase epitaxy (LPE) (cf. Fig. 1). CVD approach is now one of the standard industrial techniques for compound semiconductor epitaxial growth. A special technique called high-temperature CVD (HTCVD) was developed for manufacturing of thick SiC epitaxial layers and even crystal growth. Sublimation or PVT method was initially developed as seeded bulk growth but also used for epitaxial layer growth. Further advancements in sublimation epitaxy were made by development of sublimation sandwich method or closed-spaced technique. Liquid phase epitaxy was initially carried out employing growth from Si-melt. An alternative approach used growth from solutions of rare-earth metals , for instance Sc. Different aspects of SiC homo- and heteroepitaxial growth will be discussed in this chapter.

2. SiC homoepitaxial growth

Silicon carbide is a typical material with polytypism. More than 170 polytypes of this material have been reported. These polytypes show different properties. For instance the band gap energies of different SiC polytypes range from 2.42 eV to 3.33 eV. SiC homoepitaxial growth is the epitaxial layer growth of the same polytype as of the SiC substrate.

Thick epitaxial layers (at least more than 50 μm depending on the voltage) with low background doping concentration (lower than 10^{15} - $5x10^{14}$ cm^{-3}) are a requirement for the fabrication of SiC high power, high voltage devices.

Now let us discuss briefly the available methods of SiC homoepitaxial growth.

2.1. Sublimation growth technology

SiC sublimation epitaxy is a comparatively simple approach and is performed similar to the sublimation crystal growth but the growth duration is shorter and the growth temperature is typically lower than for bulk crystal growth.

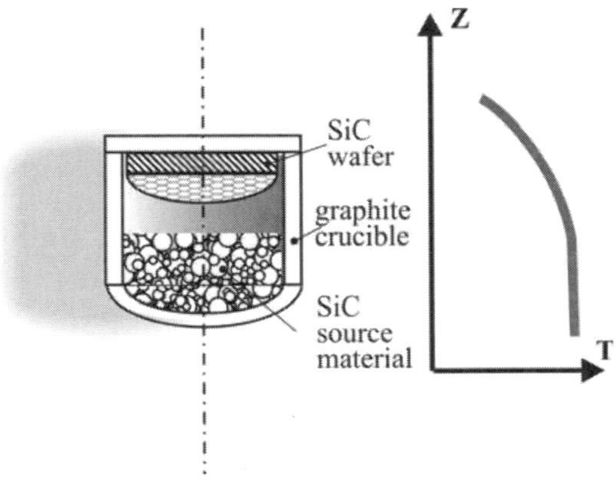

Fig. 2 Schematic view of sublimation approach and temperature distribution along the crucible.

Seeded sublimation SiC growth was first elaborated in the late seventies by Tairov and Tsvetkov [1-5] at the former LETI - at present St.-Petersburg Electrotechnical University. Sometimes it is called LETI-method or Modified-Lely (M-Lely). Since it is a seeded crystal growth it is at the same time a SiC sublimation epitaxial growth. The crucible resembles that one presented in Fig. 2. SiC powder, small crystallites or a SiC polycrystalline plate placed inside a cylindrical graphite crucible are used as a source. In the case of epitaxial growth the distance between the source material and substrate is significantly shorter than for bulk growth. Thus epitaxial growth occurs in conditions closer to thermodynamic equilibrium. Additional sources can be placed into the crucible or provided by additional gas flow in order to control native point defect concentrations or doping over the gas phase composition. The crucible is often rotated in order to gain a higher homogeneity of growth. The growth temperature is from 1700 °C to 2500 °C for bulk growth and from 1700 °C to 2100 °C for epitaxial growth. Inductive radio-frequency (RF) or resistive heating are typically employed in sublimation growth reactors. The temperature gradient ΔT is maintained in such a way that the SiC source is heated to a higher temperature than the substrate (cf. Fig. 2). Typical values of ΔT are from 20 to 60 °C/cm. The SiC source material sublimes and then the vapor condenses on the coldest part of the growth cell - on the substrate and on the substrate holder. The layers are grown in vacuum or argon environment. Some experiments were made in hydrogen ambience. The method is comparatively inexpensive and provides the growth of high quality epitaxial layers. Typical growth rates of epitaxial layers provided by this method are from 0.1 to 5 μm/min. Epitaxial growth occurs in a more equilibrium conditions and crystalline quality of sublimation grown epitaxial layers is typically higher than of the bulk material used as a substrate.

The composition of equilibrium vapor phase over the surface of SiC differs significantly from the solid phase composition, and in addition this vapor phase has nonstoichiometric composition with respect to atoms forming a solid binary compounds. The main species in the equilibrium vapor phase over SiC are Si, Si_2C, SiC_2 and SiC. The incongruent evaporation of silicon carbide makes it difficult to controllably grow high quality SiC layers and crystals employing the sublimation method.

An advanced approach to sublimation epitaxy is presented by sublimation sandwich method [6-9]. The source material (in this case typically polycrystalline plate or ingot with a flat surface) is placed parallel and in close proximity to the single crystalline wafer. The growth therefore occurs in the near-equilibrium conditions.

The main purpose of sublimation growth is to improve crystal quality, and reduce defect density [10-32]. At present commercially available SiC wafers are typically 50 mm and 75 mm in diameter, and 100 mm SiC wafers are available in limited amounts.

The growth parameters are of great importance for crystal growth optimization [25, 33, 34, 35, 36, 37]. The control of the rate-determining step of crystal and epitaxial growth is an important issue for high quality material growth in the case of sublimation. Investigations of the rate-determining step of SiC sublimation crystal growth were carried out for the temperature in the range from 2300 ºC to 2500 ºC and growth in an Ar atmosphere [25]. The total vapor pressure for most of the experiments was 50 hPa, however, some of the experiments were carried out under pressures from 5 to 1000 hPa. The ranges of temperature and source-crystal distance values where gas phase diffusion or surface processes predominates the growth rate for a total vapor pressure 50 hPa were estimated. Further investigations show that SiC crystals grown in the diffusion-controlled area, i.e. high temperature and/or large Δl, are of higher quality. In the case of sublimation epitaxy the distances Δl are very small and the growth conditions can be changed by variation of the total vapor pressure and temperature.

As mentioned above, silicon carbide exhibits polytypism. Different polytypes provide the possibility of manufacturing multi-polytype heterostructures of one material in the same device. Uncontrolled formation of different polytypes (inclusions, lamellae etc.) on the other hand is an additional type of defect. Specified polytype structures can be obtained employing a certain polytype SiC wafer and by controlling the growth kinetics [5, 13, 24]. Formation of certain polytypes are influenced by the crystallization kinetics of the initial stages of crystal growth [24, 32]. After outgasing of the inner graphite parts of the furnace and growth crucible an argon pressure is maintained at ~400 hPa in order to suppress the mass transfer in the vapor phase and consequently to suppress an unintended crystallization before the growth temperature is reached. Then, at the growth temperature the argon pressure is reduced. In the simplest case the decrease of the argon pressure follows an exponential law with a time constant τ [24]. The pumping rate at the initial stages of growth defines the polytype formation at this stage. The initial Ar pressure employed for the subsequent growth varied from 50 hPa to below 10^{-3} hPa. The influence of the pump down time (which determines the deposition kinetics at the initial stage of growth) on the SiC crystal yield of 3C, 4H, 6H polytypes is presented in [24]. High pumping rates lead to a short duration of the initial stage of growth (with low growth rate limited by diffusion mass transfer in the gas phase) and due to this to an increase of dislocation and pinhole density.

There is also additional specifics of certain polytype growth. In the case of 4H-SiC sublimation growth it is necessary to choose C-(0001)-face of the substrate in order to avoid formation of 6H-SiC layer.

Control of the growth kinetics influences the formation of certain polytypes at the initial stages of SiC growth and then the growth of the single-crystalline material of this polytype is continued. However, different polytype inclusions can be formed during further epitaxial growth due to instabilities of the growth parameters such as temperature or oversaturation, and defects at certain areas of the growth front. Formation of a certain polytype is also influenced by the vapor phase composition [38]. The visual observation of the photoluminescence of the samples under UV radiation at temperatures from 300 to 80 K can be used as a rapid method of polytype determination and of monitoring different polytype inclusions. These images provide the possibility to control the yield of the required SiC polytype as well as to reveal the distribution of both doping and structural peculiarities [20].

Recently a lot of work has been done to improve sublimation epitaxy as an effective technique for the growth of thick epitaxial layers [39-41]. Thickness and doping inhomogeneity provided by this method for the growth on 1.5" wafer is about 5%. Low-doped n-type 6H-SiC layers with (N_d-N_a) of about (7-9)$\cdot 10^{14}$ cm^{-3} and hole diffusion length of about 2.5-3 mm were obtained [42]. Layers grown by sublimation epitaxy were also employed for fabrication of nuclear particle detectors [43, 44]. An additional advantage of the sublimation approach is the possibility to grow heteropolytype structures, for instance 3C-SiC/6H-SiC for advanced devices applications [45-48]. The development of multi-wafer SiC sublimation epitaxy equipment for industrial applications would be important for cost-effective production of thick epitaxial layers.

2.2. Molecular beam epitaxy (MBE)

The next approach to epitaxial growth of SiC from the gas phase is the molecular-beam epitaxy (MBE). This technique employs growth in extremely thermodynamically nonequilibrium conditions under ultra high vacuum conditions. The source materials evaporate and are transported as molecular beams (since in ultra high vacuum there are no collisions with other atoms within the source-substrate distance). Finally they reach the heated and rotating substrate. Both, growth rate and composition, are very well controlled. MBE provides epitaxial layers of the highest purity, with extremely precise thicknesses (down to monoatomic layers), at low temperatures, typically 600-1200 °C (much lower than other SiC growth technologies). Typical growth rate provided by this method is from 0.1 to 2 nm/min.

Solid source MBE [49] was successfully employed to grow SiC layers on 2 to 5 off (0001)SiC wafers at the temperatures from 800 °C to 1000 °C and the growth rate of about 0.03 nm/min [50]. In this case pyrolitic carbon and polycrystalline Si were evaporated using electron beams. Gas source MBE implementing alternating beams of Si_2H_6 as a Si source and of C_2H_4 as a carbon source was also employed to grow 3C-SiC on (0001) and (0114) planes of SiC wafers at the temperatures in the range from 700 °C to 1150 °C [51, 52]. Studies of gas-source MBE carried out on 3.5° and 5° off-axis (towards [1120]) (0001) 6H-SiC wafers at the growth temperatures 1000 °C and 1150 °C showed that the grown layers were 3C-SiC with a high density of twins [52]. 3C-SiC layers grown on (0114) 6H-SiC wafers were free of twins. Actually (0114) 6H-SiC wafer can be also qualified as a 54.7° off-axis (0001) 6H-SiC wafer. Detailed investigation of the homoepitaxial 6H-SiC growth by the gas source MBE is described in [53]. The advantage of the MBE method is the low growth temperature and possibility of growing different SiC polytypes, so realizing heterostructure consisting of layers of different polytypes with different band gaps, for instance such structures as 4H/3C/4H-SiC(0001) and 6H/3C/6H-SiC(0001) [54]. Formation of certain polytype under MBE growth can be observed in situ employing RHEED investigations. The control of different polytype growth is realized by variation of growth temperature or providing Si- or C-rich growth conditions. The disadvantage of the method is the necessity of ultra-high vacuum in the growth chamber which reduces the productivity of this technology and makes it rather expensive with respect to mass production of epitaxial structures. Growth rates are extremely low. These considerations make MBE an interesting method particularly for research applications.

2.3. Liquid phase epitaxy (LPE)

Liquid Phase Epitaxy (LPE) is a comparatively simple and low cost technique where growth occurs on the three-phase equilibrium line of the corresponding phase diagram (providing the epitaxial growth in equilibrium with the liquid phase). SiC LPE was initially realized employing a supersaturated solution of Si and C in Si melt. But maximal growth rate even at 1700ºC did not exceed 7 µm/hr. Other elements, such as rear-earth Sc etc., were added to the solution in order to increase C solubility and growth rate was increased up to 150 µm/hr if Si-Sc solvent was used [99-103]. The difficulty of the epitaxial layer surface morphology control causes limited use of this technology. LPE growth reactors for SiC epitaxial growth are commercially available from Epigress [55].

2.4. CVD growth technology

Chemical vapor deposition (CVD) is one more method for manufacturing advanced epitaxial structures. The pressure in the reactor can be atmospheric and in this case it is an atmospheric pressure CVD (APCVD). A better control of the growth process is provided by low-pressure CVD (LPCVD) employing pressures from 10 mbar to 1000 mbar. The use of low pressures reduces gas phase nucleation. Low-pressure chemical vapour deposition (LPCVD) is widely used in silicon carbide growth. Gas flows are precisely controlled by mass-flow controllers. The substrate is placed on a rotating and heated graphite susceptor. Molecules from the gas phase diffuse to the substrate surface, decompose there and are adsorbed by the surface and proceed with surface reactions and formation of the epitaxial layer. CVD provides precise control of the gas phase, of growth rates with the possibility to reproducibly grow layers with thicknesses of several monoatomic layers but at the same time possibility to grow thick epitaxial layers. The productivity of this method is high and at present it is the main epitaxial technology for SiC device production.

In the case of SiC Chemical Vapor Deposition (CVD) gases like SiH_4 (as a Si source), C_3H_8 (as a carbon source) are transported to the heated wafer, are decomposed there and take part in a chemical reaction, providing the deposition of SiC. Typical temperatures of SiC CVD homoepitaxy are from 1400 ºC to 1700 ºC at pressures from 100 to 1000 mbar. Inductive radio-frequency or resistive heating of a graphite susceptor is typically used to obtained the desired growth temperature. SiC coated susceptors are typically used in order to prevent the hydrogen etching of the graphite susceptor. Growth rates provided by CVD approach are typically up to 10 µm per hour [56-58]. N-type SiC is typically realized by nitrogen doping and p-type – by aluminium. Vanadium doping is used to provide semi-insulating SiC material. High quality SiC epitaxial layers were realized employing CVD technique [59, 60].

Existing reactors providing a good uniformity, high growth rates and wide range of doping, both n-type and p-type. To improve the the uniformity of the SiC layers on large SiC wafers and/or of a set of SiC wafers in one run, vertical reactors have been developed. To obtain even higher growth rates, high-temperature CVD reactor have been developed [75, 77, 78].

There are different types of CVD reactors: hot-wall and cold-wall. Each of the approaches has certain advantages and disadvantages. The most pure SiC layers were grown employing hot-wall reactor. This type of reactor is typical for the CVD setups manufactured by Epigress [55]. Schematic view of a simple hot-wall reactor is presented in Fig. 3. The substrate is placed inside a heated graphite susceptor and the gases passing through the reactor are decomposed more effectively than in cold wall reactor. In the example shown in Fig. 3 the susceptor is heated employing induction heating. The walls around the substrate are coated with growing material, providing stable growth conditions during the run and from run to run. Epigress manufactures reactors for 1×2", 1×3", 1×4", 3×2", 7×2", 5×3" and 5×4" wafers.

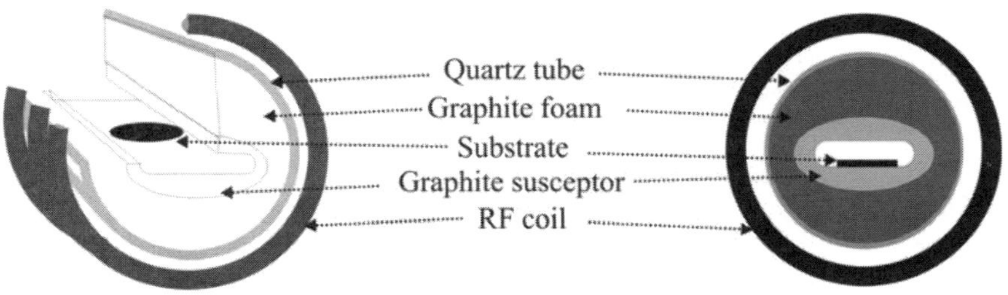

Fig. 3 Schematic view of a simple hot-wall CVD reactor: a –"3D" view and b – cross-section.

Another example of an industrial hot-wall horizontal reactor manufactured by EMF [61]. Two resistive heaters – upper and lower - are employed in this system. According to EMF this reactor low inertia and provides rapid temperature ramps if necessary. EMF manufactures reactors for 1×2", 1×3", 1×4" and 3×3" wafers.

An example of a vertical cold wall reactor is a rotating disc system manufactured by Veeco (former Emcore). In the case of a cold wall reactor there is no parasitic deposition on the walls over the substrate and correspondingly no particles fall down on the growing layer as in the case of a hot wall reactor. Additionally the thermal conditions during the growth and from run to run are constant since there is no deposition on the walls near the substrate.

Epitaxial layer uniformity in a multi-wafer reactor is one of the key issues for industrial manufacturing of epitaxial structures and devices based on them. Scientists from Cree recently reported 4H-SiC epitaxial layer uniformity in a hot-wall planetary multiwafer reactor [62]. For this reactor design hot-wall configuration provided minimization of silicon supersaturation and reduced parasitic deposits. Epitaxial growth on 4° and 8° off-axis SiC substrates with growth rates up to 10 μm/hr was investigated. The total average intrawafer thickness and doping inhomogeneity for 7x3" reactor on 8° off-axis wafers are about ±1% and ±5% respectively. In the case of the 6x4" reactor this values are 3 and 9%, respectively.

The Si to C ratio in the gas phase over the substrate or layer during the epitaxy influences the growth rate, layer quality and dopant incorporation. It is essential for compound semiconductors growth that the deviation from stoichiometry in the crystalline phase (caused in the case of SiC by the Si to C ratio) influence the incorporation of dopant atoms. This fact was employed to improve the SiC layer doping control and this approach was called "site competition mechanism" [63]. It is known, for instance; that nitrogen atoms, which are typically used for SiC n-type doping, occupy carbon lattice places. Hence, the nitrogen incorporation into the growing material will be reduced if growth occurs in carbon-rich conditions in the gas phase. Implementation of Si-rich conditions will improve the incorporation of nitrogen and provide manufacturing of highly-doped SiC material. An additional problem concerning the Si to C ratio control during CVD epitaxy can arise due to the unintentional influence of carbon from the graphite parts of the reactor transported by hydrogen (which is typically used as a carrier gas). This carbon changes the real Si to C ratio in the gas phase over the substrate.

One of the main areas of SiC application are high-power, high-frequency, and high-temperature electronics. An advantage of the CVD technique is that the gas phase Si/C ratio can be controlled during the growth process. The improvements in the epitaxial growth techniques [64-68] as well as the improvement in substrate quality [69, 70] and availability have made it possible to produce high-power devices demonstrating the advantages of SiC [71]. Thick active layers of high quality are required for high-voltage, high power devices. The CVD growth rates of the epitaxial layers at typical growth temperatures from 1500 to 1600 °C are still too low to grow an active layer of 50 μm in a time reasonable for commercial production. Sublimation growth on the other hand has the advantage of high growth rates but does not provide a good control of the gas phase Si/C ratio over the growing crystal which is equivalent with the concentration of intrinsic point defects. This advanced control and high growth rates can be realized in high-temperature chemical vapour deposition (HTCVD) [77, 73].

The high temperature chemical vapor deposition (HTCVD) epitaxial growth process was developed in order to grow epitaxial layers at high growth rates as well as even to grow high quality SiC boules [77]. This growth approach was developed on the base of the CVD technology in a vertical quartz reactor fitted to the operation at the temperatures up to 2300 °C. The graphite cylinder crucible with a hole in the bottom center (gas inlet), and with small holes placed close to the perimeter of the top lid (gas outlet) is thermally isolated by graphite foam (cf. Fig. 4). The crucible is inductively heated. He, H_2, Ar or a mixture of H_2 and Ar are used as carrier gas. Concentrated or diluted silane is the Si-source and concentrated propane is used as an addition to the carbon pressure from the graphite walls if necessary.

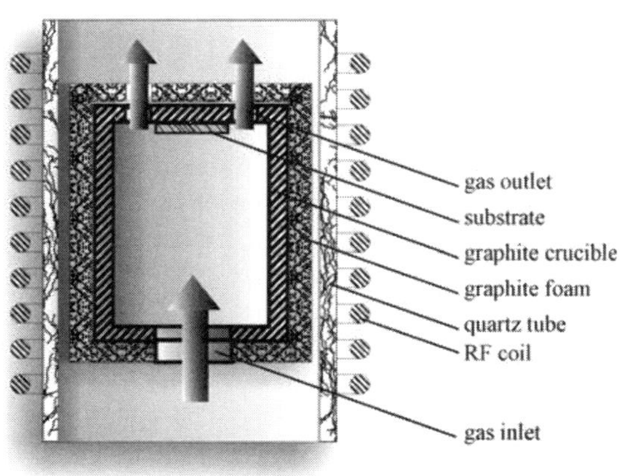

Fig. 4 Schematic drawing of the graphite crucible used for HTCVD

Due to the high temperatures of the process implementation of the hydrogen as the carrier gas makes the growth control complicated since the etch rates of the substrate and of the graphite crucible are too high and a large amount of hydrocarbons are formed (caused by etching of the graphite walls by hydrogen). A different crucible material can be considered for instance tungsten, vanadium or tantalum [72]. However, at such high temperatures the metals will be also attacked by the H_2 or form metal carbides reacting with the hydrocarbons. A good alternative in this case is implementation of He as the carrier gas.

The surface morphology of the grown crystals is very sensitive to the growth conditions. The carrier gas flow has to be carefully chosen with respect to the temperature, the carrier gas mixture, and the geometry of the system. Especially a homogeneous temperature of the substrate is difficult to maintain, since the high growth rates require high precursor feed rates which permanently cool the growth surface. This effect was revealed by an optical microscopy of the surface of a 100 μm thick SiC layer grown in unoptimized conditions at 2200°C at a growth rate of 200 μm/h on a 6H-SiC off-axis substrate [78, 77]. As in other cases of compound semiconductors gas phase growth a careful control of the gas-phase composition is necessary to provide reproducible growth of high-quality SiC layers especially in the case of growth of semiinsulating material. Similar to sublimation growth it is necessary to maintain a controlled supersaturation not only during the growth but also during the temperature ramp up. The highest growth rates obtained for this HTCVD method are comparable to typical sublimation growth rates. In principle their upper value is limited by the surface mobility at the growth temperature.

The described HTCVD is of great importance for the SiC growth especially for the production of high-purity semiinsulating material. In contrast to sublimation growth this method provides full control of the gas phase over the growing crystal during the growth process. It is possible to control native point defects and to realize excellent purity of the grown crystals. As a result it is possible to obtain semi-insulating material without additional compensation with dopants [73]. Recent advancements of the HTCVD technology led to the growth of large SiC boules and to the fabrication of SiC wafers from them including semiinsulating ones [73, 74, 75, 76, 77, 78, 79, 80]. Recently commercial SiC wafer production employing HTCVD is announced by Okmetic AB, Sweden [73].

For epitaxial layer growth HTCVD is too expensive and it would be an advantage to proceed using less costly standard epitaxial CVD technique but with significantly higher growth rates providing growth of thick epitaxial layers for high power devices in a reasonable growth durations. Recently a new approach to increase the SiC growth rate at standard for CVD growth temperatures (about 1500 °C) were made employing HCl in situ etching during the growth in a hot-wall CVD reactor employing silane and propane precursors [81]. Growth rate of about 28 μm/hr was reported in this case leading to growth of high quality epitaxial layers. Increase in growth rates is typically achieved by increase of the silane flow but this typically cause formation of silicon clusters and leads to 3D island growth. Presence of HCl prevents formation of silicon clusters and allows implementation of high silane flows. Further increase of growth rate up to 55 μm/hr resulted in a polycrystalline growth [81]. So optimization of growth conditions is obviously needed in order to obtain single-crystalline layers at such high growth rates. It is expected that this approach will provide growth rates up to 100 μm/hr [81].

At present SiC epitaxial layers still have too high a density of various defects. These defects could affect the properties of the device structures realized on the base of such layers and further optimisation of the SiC epitaxial growth process is needed to improve the layer quality.

2.5. Defects in the epitaxial layers

Different surface defects can be formed on the SiC epitaxial layers. Initially a serious problem was the formation of triangular inclusions of 3C-SiC polytype on the surface during the 6H- or 4H-SiC epitaxial growth. It was found out that if the growth occurs on 6H- or 4H-SiC wafers cut from the boules exactly along the (0001) basal plane (±≤ 0.5°) or on wafers with the off-axis angle less than 2° large flat terraces between the atomic steps lead to the nucleation in the middle of the terraces instead of at the edges of the steps resulting in the formation of 3C-SiC inclusions. As the CVD epitaxial growth temperature is comparatively low there exists a probability of nucleation of both hexagonal and cubic polytype. The so called "step controlled epitaxy" approach is employed to improve the quality of homoepitaxially grown 4H- and 6H-SiC layers [82, 83, 84, 85]. SiC wafers cut from the boules not exactly along the (0001) basal plane but with an off-axis angle from 3° to 8° towards [1120]. Such an off-axis cut leads to the formation of a substrate surface with a greater density of atomic steps and short flat terraces between them. The bigger the off-axis – the smaller the terraces length and it is easier for the species coming from the gas phase during the growth to reach the proper places at the steps on the surface and so to reproduce the polytype of the substrate instead of building a new island eventually of another polytype. This approach provides good control of the reproduction of the polytype of the growing layer from the substrate and reduces density of defects formed during the epitaxy. For this reason homoepitaxial growth is at present carried out on the 6H-SiC wafers with the off-axis angle of about 4° and on the 4H-SiC wafers with the off-axis angle of about 8° due to reported higher polytype stability. On the other hand the 4-off-axis substrates are cheaper than 8° off-axis ones. But stronger step-bunching occurs if 4° off-axis are used. Recently investigations were carried out to develop 4H-SiC epitaxial process on 4° off-axis 3" substrates by optimization of growth parameters [86].

One of the most severe problems concerning device fabrication (short-cutting of p-n-junctions etc.) on SiC-basis is caused by micropores (three-dimensional micropores with characteristic dimensions of about 30 – 40 µm) and micropipes or pinholes (micropores with dimensions $x >> y, z$). This defect is sometimes called "device-killer". The most common micropipes or pinholes are small holes approximately 0.1 - 5 µm in diameter which penetrate the layers and the subsequent substrates. The density of the micropipes is in the order of 10 - 10^3 cm^{-2}. Micropipes formation was firstly theoretically described as a process of elastic stress compensation around dislocation line by Frank [87]. Most of the micropipes are caused by several screw dislocations bunched together forming a so called super-screw dislocation [88]. According to the Frank-theory [89] the resulting large Burgers vector of a super-screw dislocation could lead to the formation of a hollow core in the center of the dislocation and the resulting micropipe is one of the critical defects in SiC-based device fabrication [90].

An example of a system of pores and micropipes is shown in a set of images in Fig. 5 depicting the propagation of the whole system in the depth of a 6H-SiC commercial wafer which is observed in the transparent crystal by varying the focus depth of an optical microscope. A weak contrast (slightly brighter hexagonal area, marked with a white dotted line) is visible on the enlarged photos representing a hexagonal growth defect area. A system of pores and micropipes associated with this defect is also seen on the photos.

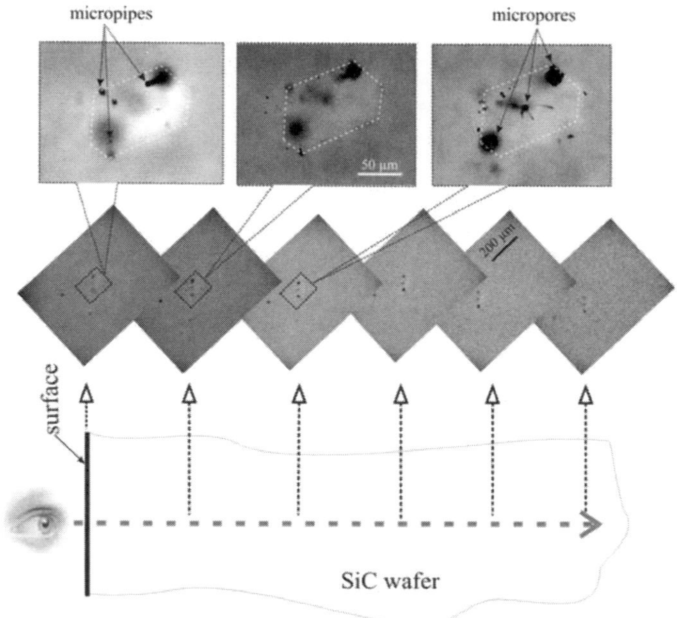

Fig. 5 Optical microscopy images of a system of pores and micropipes on different depths inside a commercial 6H-SiC wafer (observed in the transparent crystal by varying the focus depth over approximately 500 μm)

Pores and micropipe defects in substrates serve as sources for the formation of other defects (pores, inclusions of different polytypes, twins etc.) in epitaxial layers subsequently grown on these substrates. Hence one of the most important problems of substrate material fabrication for high temperature and high power devices is to reduce the formation of micropipe defects in bulk SiC crystals [91, 92, 93, 94] as well as to prevent their reproduction into the epitaxial layer grown on the substrates.

The main reason for the formation of micropipes and dislocations is nonoptimized conditions of nucleation and growth. Thus, the initial stages of SiC growth at the growth temperature as well as the nucleation of initial layers at lower temperatures during the heating stage have been studied [13]. Even at high Ar pressures up to 100 hPa the nucleation of initial layers below the growth temperature was observed. Also the replication of micropipe defects from the substrate to the growing crystal must be reduced [15].

The process of closing a micropipe during the initial stages of sublimation growth was investigated, observing SEM images of the initial micropipe in the substrate and the same place after further stages of growth and continuous closing of the micropipe defect after several μm of sublimation growth was demonstrated [15]. Finally the micropipe was completely closed during further growth. This was achieved using growth conditions under which the lateral growth rate is much higher than the growth rate in the direction perpendicular to the growing surface. The total thickness of the grown layer with demonstrated closed micropipe was about 10 μm [15].

Thus, reproduction from the substrate into the growing layer can be reduced during sublimation growth of layer or crystal and a low micropipe density material (below 80 cm^{-2}) was grown on substrates with micropipe densities as high as 300-600 cm^{-2} [15]. The densities of micropipe defects and dislocations were revealed using selective defect etching of the (0001)Si side in molten KOH. The results of the KOH etching were verified by a direct SEM observation of micropipes on both SiC (0001) sides.

Micropipes filling process by CVD was first reported in [94]. The growth of 6H-SiC layers occurred in a hot-wall reactor using silane-propane-hydrogen system on 6H-SiC (0001) Si- and C-face substrates at the growth temperature from 1500 °C to 1600 °C and C to Si ratio from 1.5 to 3. The off-axis substrates used in this study were 3°-4° towards the <1120> direction. Growth rates from 1 to 2.5 µm/hr were employed and layers with thickness from 15 to 40 µm were obtained and investigated. It was observed that disturbances of the surface like micropipes stop lateral growth on the off-axis substrate and give rise to the formation of pyramidal defect due to the anisotropy of growth in different directions (cf. Fig. 6) [94].

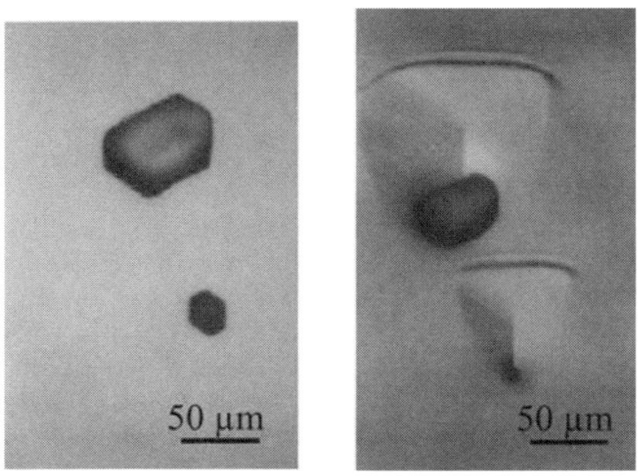

a b

Fig. 6 Formation of a pyramidal defect: a - surface of the 6H-SiC substrate with micropipes before epitaxial growth; b – the same place after epitaxial growth of 6H-SiC layer

Closing of micropipes by epitaxial CVD growth was observed in the case of smaller micropipes (cf. Fig. 7) [94]. Also bigger micropipes could be presumably closed in this case if the growth of thicker layers would be realized.

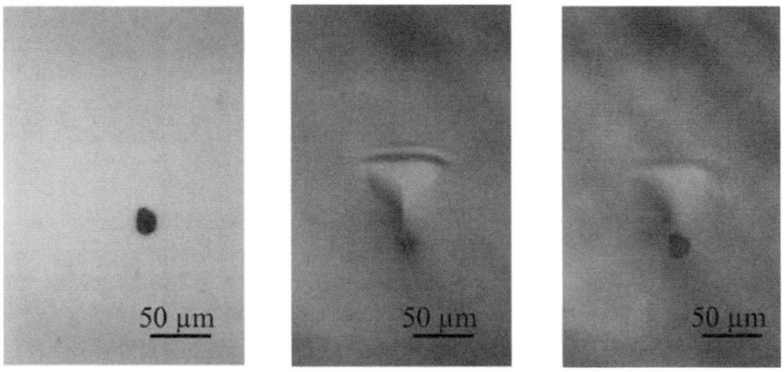

Fig. 7 Closing of a micropipe by CVD epitaxial growth: a - surface of the 6H-SiC substrate with micropipes before epitaxial growth; b – the same place after epitaxial growth of 6H-SiC layer; c – surface of the substrate focused through the epitaxial layer.

It was shown that a micropipe can be closed completely but if it was caused by a screw or superscrew dislocation – the dislocation will be replicated to the layer. Formation of the growth spirals in such places have been observed.

The presence of dopants, such as Ga and Al enhances step-bunching around the micropipes at the areas where a micropipe comes to the substrate surface making it difficult to close a micropipe (cf. Fig. 8) [94].

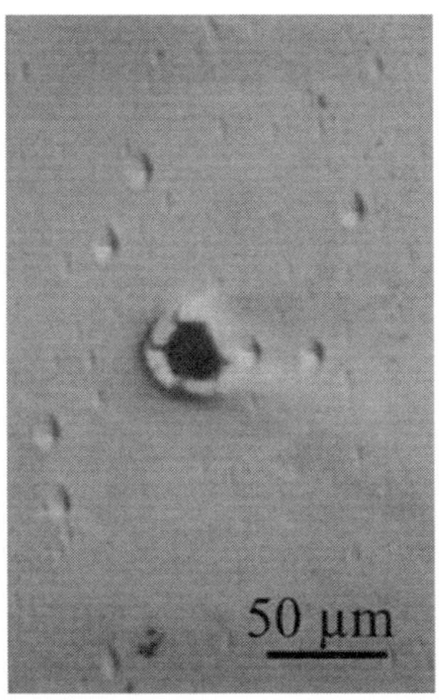

Fig. 8 Pronounced step-bunching around micropipe in the presence of Ga dopant during the SiC epitaxy

Micropipe filling process under 4H-SiC CVD epitaxy have been reported in [95]. 5 times lower micropipe density was observed in CVD grown SiC epitaxial layers comparing to initial 4H-SiC SiC wafers. Closing of micropipes under the SiC CVD epitaxial growth was recently further investigated in [96, 97, 98]. In [97] micropipe filling was demonstrated for the CVD growth of 4H-SiC layer on 2" on-axis C-face substrate. A hillock with the step spiral structure was observed on the layer surface at the filling point of the micropipe giving evidence of the spiral growth as the filling mechanism.

Beyond sublimation epitaxy and CVD micropipe closing was also demonstrated employing liquid phase epitaxy [99, 100, 101, 102, 103].

Special annealing experiments have been carried out in order to investigate the micropipes behaviour in SiC at high temperatures, to understand micropipe healing processes and to reduce further the micropipe density [13, 32]. The annealing temperatures are 2000 °C and 2300 °C. The partial pressures are the same during annealing and growth. For optical investigations of pinholes and micropores in the samples the annealing was periodically interrupted. Typical micropores with a characteristic dimension of the cross-section from 1 to 80 μm are taken into consideration. Pinholes breaking and healing forming a filled micropipe due to high temperature annealing have been observed. The effect of pinhole breaking and transformation into filled pipes can be related to the dissociation of super-screw dislocations into multiple unit screw dislocations. This approach was further investigated by Toyota Central Labs and Denso Corp.[104]. A TEM cross-sectional investigation of a healed micropipe formed in the core of a super-screw dislocation due to annealing at 2230 °C was carried out [104]. It was shown that the super-screw dislocation (micropipe is formed in its core) was dissociated into several unit screw dislocations. Additional formation of a large amount of stacking faults, edge and partial dislocations was observed in the healed micropipe area. The latter defects are introduced to compensate the strain which was associated with the super-screw dislocation. Stacking faults were not observed in the areas where the micropipe still exists. The time dependence of the length of 5 different pinholes during the annealing step are were investigated for two different temperatures. The length of pinholes obviously can be reduced by 2 times employing 10 hours annealing at 2000 °C or alternatively by not more than 1 hour annealing at 2300 °C. Certainly the screw dislocations which were in the core of the pinhole are still in the crystal volume.

These results demonstrate the effectiveness of high-temperature annealing for healing micropipes in SiC crystals. This approach originally reported in [13, 32] was further investigated, developed and successfully employed for an improvement of SiC crystals by Toyota Central Labs and Denso Corp.[104]. The "healed" substrates can be used as micropipe-free substrates for further bulk crystal growth or device fabrication.

The micropipes or hollow cores are closed but the screw dislocations still present in the epitaxial layer. These defects still influence the device performance. Correlation between screw dislocation density and breakdown voltage of diodes fabricated on this material have been reported in [105]. For certain devices there is an insufficient improvement observed in the device parameters manufactured on filled micropipe material. The density of screw dislocations in epitaxial layers must be also reduced.

Screw dislocations propagate typically along the c-axis. Therefore, also efforts were made to avoid screw dislocation propagation employing the epitaxial layer growth on a-axis SiC wafers (SiC substrate surface parallel to c-axis). Screw dislocations were eliminated but the formation of other defects was observed [106, 107] and it is hard to eliminate these defects.

One more effective approach to reduction of defect density in epitaxial layers was reported [108]. Porous SiC buffers obtained by anodization was employed in this case. The

SiC layers grown on such a buffer were investigated employing both x-ray diffraction and x-ray topography and nearly 10 times lower defect density is reported [108]. Recently further efforts were made in order to develop this approach [109].

Substrate material and surface preparation play an important role in the characteristics of layer-substrate structures. The surface preparation of the substrate prior to epitaxial growth is an important aspect of the whole growth process. Sawing of the crystal into wafers and further grinding and polishing of the wafers cause the formation of a damaged layer and strain, introduce scratches, adsorbed layers, impurity aggregates, mounds into the substrate surface. Surface damage influences the nucleation process and causes the formation of defects in the growing crystalline phase. A chemical treatment of the substrate surface can also leave residues giving rise to nucleation of dislocations and stacking faults. An improved substrate surface processing is an extremely important issue to be accomplished for the optimization of the epitaxial growth in each substrate-layer system.

The thickness of the damaged layer after mechanical treatment of the substrate depends on the type of processing and on the material itself. Typically after mechanical polishing the upper layer of the crystal contains a lot of cracks, whereas the deeper areas of the damaged layer contain dislocations. The damaged layer can be investigated by different methods: etching, x-ray diffractometry (XRD), reflection high-energy electron diffraction (RHEED), scanning electron microscopy (SEM), transmission electron microscopy (TEM), atomic force microscopy (AFM), X-Ray reflection on glancing angle, indirect electrophysical and optical methods etc. It should be noted that the sensitivity of the epitaxial layer or the final device to the quality of the wafer processing is often higher than the sensitivity of some of the above mentioned investigation methods.

Commercially available 6H- and 4H-SiC substrates also contain different types of defects formed during the substrate fabrication and processing steps. Due to extremely high chemical inertness it is difficult to remove mechanically damaged layers near the substrate surface. Typical methods of SiC substrate surface processing are: wet chemical etching, sublimation etching, oxidation, ion plasma etching, pre-growth in situ hydrogen etching.

Wet chemical etching of the SiC substrate in molten KOH leads to a significant roughness of the surface. Already in 1979 it was suggested [110] to introduce an "in situ" pre-growth thermal etching of the substrate in a negative temperature gradient (source is colder than substrate). This technique was developed for the sublimation sandwich method (SSM) of SiC epitaxial layer growth [110] and is also used in the M-Lely method. Unfortunately this approach can lead to some instabilities during the important initial stage of the growth and thus to the formation of additional defects (it is difficult to change the temperature gradient and at the same time maintain a constant substrate temperature).

Dry etching is a standard approach to surface processing and patterning of semiconductors. Etching rates for SiC are found to be as high as 0.45 μm/min for reactive ion etching and 1.35 μm/min for plasma etching [111, 112, 113]. RHEED measurements show that the mechanical treatment of the SiC surface generates a thick damaged layer including an amorphous surface layer with a thickness of about 5 nm [14]. The quality of the samples surface is significantly better after the ion plasma etching of an 8 μm thick layer. A "faceted" microstructure on the surface is revealed in this case. In the presence of defects in the substrate like micropipes the formation of large area surface defects under ion plasma etching is observed [14].

A simple method of substrate surface improvement after mechanical treatment is oxidation [14]. The method is simple but it is necessary to use several cycles of oxidation and further etching in 10% HF for removal of enough material. An advanced oxidation approach involves the final removal of the oxidized layer directly in the furnace during heating up of the substrate to the growth temperature. This approach however does not provide a low roughness of the resulting SiC substrate surface.

One more approach to improve the SiC substrate surface is hydrogen etching directly in-situ in a CVD reactor carried out in the temperature range of normal growth [114, 115]. Three different etching temperatures 1500 °C, 1550 °C and 1600 °C were studied. The Si-face 6H- and 4H-SiC (0001) substrates were used to investigate this process. Both on-axis and 3°-4° off-axis substrates were investigated. Hydrogen etching durations were from 10 to 30 minutes employing typical epitaxial process hydrogen carrier gas flows 10 or 14 l/min [115].

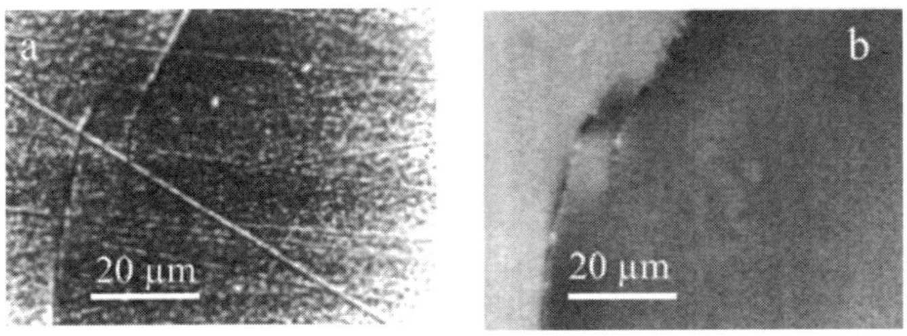

Fig. 9 Optical microscopy images with Nomarski contrast of the surface morphology of a 6H SiC on-axis substrate: a- before etching; b- after hydrogen etching at 1500 °C for 30 min.

Optical microscopy and AFM investigations of the surface of on- and off-axis substrates before the treatment revealed a large number of scratches propagating in different directions, grain boundaries, small pits (up to 10 μm in diameter), cracks, micropipes [115]. The scratches are uniformly distributed over the substrates (cf. Fig. 9 a). The density of other defects was higher in the near the edge parts of the substrates. The scratches were typically up to 20 nm deep and up to several hundred nm broad. Although it is known that etching of SiC in hydrogen at temperatures below 1600 °C is a very slow process [116, 117], the effectiveness of this etching approach is demonstrated by Fig. 9. The scratches were very efficiently removed by hydrogen etching (cf. Fig. 9 a and Fig. 9 b)[115]. Defects of the substrate like micropipes and grains or misoriented area boundaries become more pronounced (cf. Fig. 9 a and Fig. 9 b). A distinct difference of the etching results of off- and on-axis substrates was observed [115]. Optical microscopy and AFM investigations revealed that the scratches are completely removed from the surface of the on-axis substrate after etching for 30 min at 1500 °C or 1550 °C. In the case of the off-axis substrate some scratches were observed on the surface after the same etching process. The AFM investigations of the etched on-axis substrate revealed that the surface is well covered with a regular straight terraces grid with 250 nm step separation. The estimated value of the step height reflecting the size of the unit cell and off-angle was found to be 1.5 nm. The evaluated off-angle was found to be 0.3° for the investigated on-axis SiC substrate. Some triangular features with tie sides in high symmetry direction are observed on the terraces. The height of the features do not exceed 1.5 nm. Due to damages the terraces are of different size and not well-ordered prior to the etching [115].

The same hydrogen etching also decrease density of scratches on the off-axis substrate but the well-ordered surface is not achieved. As expected the terraces are narrower due to the higher off-angle and they are not so straight as in the case of the on-axis substrate.

So, in-situ hydrogen etching of the SiC substrates greatly improves the surface quality. On the other hand silicon-droplets were observed on the surface after treating SiC in hydrogen above 1400 °C in [118]. Addition of small amount of HCl during etching of SiC in hydrogen efficiently removes the Si from the SiC surface preventing the formation of dislocation related defects in the layer. Layers grown without an optimised in-situ etching were often decorated with pit defects originating from screw dislocations [119]. Si droplets were not observed in the investigations described above [114] supposingly due to a slight carbon-pressure from the system (SiC-coated graphite susceptor and graphite felt).

Hydrogen etching at higher temperatures (from 1600 °C to 1800 °C) is also successfully employed as a pre-growth in-situ substrate treatment for further 4H-SiC epitaxial layers growth [120, 121]. This method is at present typically used for in-situ SiC substrate etching prior to epitaxial layer growth [122, 119, 120, 121, 123]. By this very effective approach the micropipe density in SiC boules (grown on SiC substrate etched in hydrogen) is reduced [124].

Further problems concerning substrates for epitaxial process are strain and the presence of misoriented areas (domains) or the so-called mosaic structure [13, 20, 125, 126]. The presence of domains was first revealed by HRXRD [125]. The domains can be also observed in transient polarized light, by X-ray topography, as well as by optical microscopy after selective etching in molten KOH. For most cases the misorientation did not exceed 1' for 28-32 mm in diameter wafers and several angle minutes for 32-40 mm in diameter wafers. The borders of misoriented areas are low-angle boundaries which are formed by dislocations. XRD rocking curves taken from Lely substrates (even employing large x-ray beam sizes up to 10 mm) typically show one sharp (0006) or (0004) peak (for 6H and 4H polytypes, respectively) with a FWHM of about 12 arcsec. Rocking curves from M-Lely wafers are typically wider (20 – 80 arcsec, depending on the x-ray beam size and size of domains) and consist of several peaks from different slightly misoriented domains. The angle of misorientation can be estimated from these measurements. An effective approach to investigation of the slightly misoriented areas in SiC wafers is mapping. This can be performed using x-ray topography (Berg-Barrett and Lang methods), cathodoluminescence (CL), or photoluminescence (PL). Additionally etching in molten KOH is employed to reveal dislocations including those that are forming low angle boundaries of the misoriented areas [20]. As it is shown in [20] it is possible to use PL or CL mapping at 80 K for a first estimation of the defect structure (which also causes the nonuniformity of doping). Simple optical observation provides information mostly on polytype inclusions. PL/CL images of the whole wafer or epitaxial layer reveal the distribution of both doping and structural peculiarities and they correlate well with results obtained by IR absorption [20]. It is easy to find even small different polytype inclusions. Partial correlation between the CL/PL mapping and the x-ray topography images was observed. But only x-ray topography provides a detailed information on defect structure of the wafer or epitaxial layer including misoriented areas, dislocations, micropipes, strain.

The main reasons for the formation of misoriented areas in SiC epitaxial layers are the reproduction of such areas from the substrate. In-situ hydrogen etching before epitaxy can be also employed to reduce the mosaicity in SiC layers grown by CVD on the commercially available 6H- and 4H-SiC substrates which typically contain slightly misoriented regions or mosaicity [127, 128]. 6H-SiC layers for these investigations were grown in a hot-wall reactor [129]. The temperature and time of hydrogen etching were varied in the ranges 1500-1600 °C and 10-30 minutes, respectively. Rocking curves of the basal plane (0006) were taken from the same area of a wafer before growth and after a layer has been grown [130]. An FWHM of about 10 arcsec for the layer fabricated under optimal etch and growth conditions is obtained even though the FWHM of the substrate is about 75 arcsec. The best results are presumably obtained if the same etch and growth temperatures are used. Without hydrogen etching the FWHM of the layer may even increase compared to that measured initially for the same place of the substrate before epitaxy. In such cases the layer has a similar domain structure as the substrate [130]. Similar results with an optimal temperature of the pre-growth etching for the polytype control in the layers are also found for etching with HCl containing gas mixtures [131].

The improvement of the crystalline structure of the layers compared to the substrates due to hydrogen etching was typically observed for layers thinner than 30 μm [130]. Domain structure similar to those of the substrates appear in thicker layers. Presumably strain is formed during the growth of single crystalline layers on substrates with domain structure and it increases with the layer thickness. And dislocations forming low-angle boundaries are generated after certain critical value of strain.

Doping of the layers also influence misoriented areas reproduction. It was observed that the Ga-doped layer has practically the same domain structure as the substrate [130]. This might be due to decoration of the boundaries of the slightly misoriented regions with the dopant. A difference in lattice parameter between substrate and layer is also observable since the "center of gravity" of the curves differ from each other.

The above presented results show that the in-situ hydrogen etching of the SiC substrates prior to growth significantly improves the crystalline quality of the subsequently grown CVD layers. High-quality SiC epitaxial layers showing XRD FWHM of 13.5 arcsec were demonstrated on these etched substrates with high growth rates of up to 70 μm/h [120]. Also significant improvement of the SiC bulk crystal quality can be achieved if the SiC wafer was etched in hydrogen prior to growth [124].

3. SiC heteroepitaxial growth

Presently compound-semiconductor heteroepitaxial structures offer enhanced possibilities in particular in combination with substrates of different material. Heteroepitaxial devices on Si are one of the most promising approaches providing all the benefits of both materials, low cost manufacturing and monolithic integration with the Si-based microelectronics. Large mismatch in lattice parameter and thermal expansion coefficient between SiC layer and Si substrate causes formation of lattice defects and residual stress.

Various approaches were employed to improve the growth of high quality compound semiconductor layer on Si with different lattice mismatch. These are use of special buffer layers (thick buffer, graded buffer, misfit grainlets), liquid interface, area-selective growth , thermal annealing of the layer in intermediate stages of growth and after growth, growth on patterned substrates up to thin freestanding structures, defect binding layer (for instance strained layer superlattices (SLS), precipitates of dopants or intrinsic components). For the cases of growth with large mismatch (for instance GaN on sapphire etc.) selective nucleation followed by lateral epitaxial overgrowth (LEO) [132], or pendeoepitaxy [133] has proven to be successful in decreasing the defect density by localizing the disturbed area to certain regions of the wafer which are not used in further device fabrication.

Off-oriented substrates where the surface serves as a source of steps for nucleation of the layer are often employed for advanced heteroepitaxial growth of compound semiconductors. For applications of heteroepitaxial structures on Si monolithically integrated with silicon microelectronics it is necessary to develop the growth of these structures on exactly oriented (001) substrates since these are the basis of all industrial processes of involved silicon microelectronics production. For manufacturing of high quality device structures gas phase growth is used , which is dominated by the CVD approach.

Cubic silicon carbide (3C-SiC) on silicon is at present a topic of considerable interest due to its potential in electronic device applications, micromechanical systems, etc. [134, 135, 136, 137, 138, 139]. Large area 3C-SiC wafers were not available and 3C-SiC epitaxy is mostly developed as a heteroepitaxial process employing predominantly Si wafers, and sometimes sapphire or hexagonal SiC wafers. 3C-SiC and Si lattice mismatch of about 20% makes the 3C-SiC epitaxial growth on Si complicated and leading to defect formation in epitaxial layers.

CVD and MBE methods were used for SiC on Si growth. Actually only CVD is at present employed for SiC epitaxial growth on Si. Significant improvements in this field have been demonstrated in the last years. Si surface carbonization process is typically used prior to SiC growth to improve the 3C-SiC layer quality. Carbonization process is realized providing the flows of carbon-containing gases (CH_4, C_3H_8, CBr_4 etc.) over the Si wafer surface at the temperatures of about 1000 °C – 1400 °C. Implementation of low growth temperatures is an advantage and growth of monocrystalline 3C-SiC layers have been realized at temperatures as low as 750°C [140] or 1000 °C [141]. In order to obtain such low growth temperature methylsilane is used as a source for both Si and C. But in this case there is no possibility to vary Si to C ratio in the gas phase and so to optimize further the layer growth.

Ge can be employed as a surfactant to promote two-dimensional SiC CVD growth at 1000 °C [141]. Both 3C-SiC and 6H-SiC were obtained on Si(111) wafer.

The investigations of InP on Si growth showed that for the growth of heteroepitaxial lattice-mismatched layers on exactly oriented (001)Si substrates it is necessary to provide an optimized roughness of the surface of the substrates (carry out nano-scale patterning) in order to reduce formation of antiphase domains (APD) and layer strain [142]. This conclusion was supposed to be transferred to defect elimination in other heteroepitaxial systems on exactly oriented substrates. InP growth experiments were carried out on (001)Si substrates with different roughness. It was shown that there is an optimum range of Si wafer roughness which provides APD-free InP layer growth. The influence of the Si substrate roughness is related to double-atomic or monoatomic steps formation on the surface of Si substrate. Moreover if nano-scale patterns are formed on the Si surface, the crystalline defects will merge during epitaxy and cancel out each other more rapidly. On the other hand too high values of surface roughness leads to APD formation. First ever reported III/V on Si RTD structures were demonstrated employing this approach to heteroepitaxial growth [143]. Several methods can be employed for the maskless processing of the Si substrates. A nm-scale patterning of Si surface can be carried out by boiling Si substrates in deionized water for 10-40 minutes [144].

Another approach to nano-scale patterning is oxidation (thickness of oxide about 0.5-1 μm) and subsequent removal of the oxide layer can be employed followed by a final standard pre-epi cleaning [142]. A significant roughening of the substrate surface is observed after both procedures with clear differences for the different wafer suppliers.

It is important to investigate the surface of Si substrate at every technological step prior to the epitaxial growth in order to understand better the processes of defect formation and reduction and to optimize wafer surface processing. AFM investigations of Si surface after the high temperature steps (Si annealing at T = 940 °C and further AsH_3 preflow at 760 °C) show that the surface after the AsH_3 preflow is well covered with regular straight terraces with a step width of 54 nm and [1 10] oriented (cf. Fig. 1). The estimated step height is 0.3 nm. This means that the nominally on-axis Si substrate is actually an off-axis one (in [$\bar{1}$10] direction with the angle of misorientation of about 0.32°. The edges of the terraces are typically rough on the nm-scale. It can be presumed that the Si surface is etched by atomic hydrogen produced of decomposing AsH_3 in the reactor. Steps on the surface of nominally on-axis wafers were also observed for SiC wafers after hydrogen etching as described above.

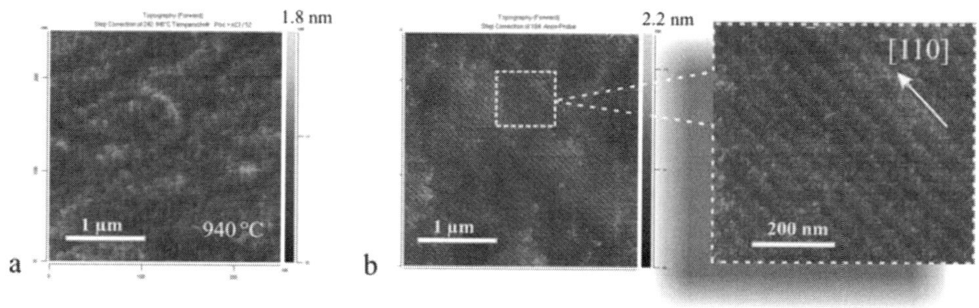

Fig. 10 AFM topographs of Si substrate surfaces: a - after in-situ annealing at T = 940 °C for 18 min; b - after further AsH_3 preflow at 760 °C

So this is one of the possible approaches to Si wafer in situ treatment before epitaxial process. Influence of Si substrate roughness on the formation of defects in 3C-SiC grown by CVD method was recently reported [145] and suppression mechanism of double positioning growth in 3C-SiC(111) crystal by using an off-axis Si(110) substrate is recently discovered [146]. Si(110) substrate was used in order to decrease lattice mismatch at the interface. Implementation of the off-axis substrates prepared by mechanical treatment and having nano-scale roughness on the surface including scratches, bumps. It was shown that nano-scale morphology of the substrate plays an extremely important role in defect formation at the SiC and Si interface. So these results agree with the general conclusions made in [142] that for high quality compound semiconductor heteroepitaxial growth it is necessary to control wafer surface and to provide an optimized roughness of the surface of the substrates in order to reduce defects, APD formation and to reduce layer strain. Proper nano-scale patterning provides possibility to employ even on-axis substrates instead of off-axis substrates for high-quality heteroepitaxial growth.

Recently a defect-annihilation mechanism for planar defects within the structure of 3C-SiC grown on Si was found by Hoya Corporation [147]. This mechanism drastically reduce the defect-density level in 3C-SiC. 3C-SiC monocrystalline material was epitaxially grown on a 150 mm in diameter "undulant-Si" substrate – a substrate with a wavy surface. It is an intentially damaged substrates surface (by polishing with diamond paste in [1 10] direction) and than the damaged layer is removed by oxidation and oxide etching leaving "undulant" or wavy surface. Implementation of the "undulant" Si substrate suppressed APD formation in a thin near interface SiC layer and that is why twins of only one orientation were observed in the final layer. This approach seems to be similar to the described above InP/Si approach. Simply the misfit for InP/Si is 8 % and in the case of SiC/Si it is 20%. Thus, presumably higher surface roughness is needed.

The investigations with SiC on Si as well as the investigations of InP on Si growth show that for the growth of heteroepitaxial lattice-mismatched layers on exactly oriented (001)Si substrates it is necessary to provide an optimized roughness of the surface of the substrates in order to reduce APD formation and to reduce layer strain. This conclusion can be further transferred to defect elimination in other heteroepitaxial systems on exactly oriented substrates.

Recently influence of ultra-violet (UV) stimulation on SiC epitaxial growth have been reported. Low temperature growth of SiC on Si(100) using low-pressure chemical vapour deposition (LPCVD) offering the possibility to control the Si/C ratio in the gas phase was developed employing a horizontal infrared-heated MOCVD reactor originally designed for III/V epitaxial growth [148]. The precursors carbon tetrabromide (CBr_4) and monosilane (SiH_4, 2% in H_2) were used. The growth temperature for most experiments was 940°C with a total pressure of 20 to 100 hPa. Exactly oriented Si (100) substrates were employed. Experiments with varied CBr_4/SiH_4 flow ratios were carried out. A short Si substrate carbonization step [134] was added in the experiments at the beginning of the SiC growth process. The growth rate of SiC was about 0.25 μm/hr. The samples were grown with or without UV stimulation in order to estimate the influence of the UV stimulation. A 150 W xenon lamp as an ultra-violet radiation source has been added to the set-up [148]. It was observed that the UV stimulation increased the uniformity of the grown surfaces and in most cases decreased their roughness.

Lateral epitaxial overgrowth of SiC on Si was demonstrated already many years ago [149] and was recently further developed [150] in order to reduce lattice mismatch caused defect density in 3C-SiC on Si. APCVD technique employing hydrogen as a carrier gas, hexachloro-disilane and propane as Si and C precursors, respectively, was used in this work. Growth occurred on (001)Si wafers with SiO_2 as a mask for LEO. It is known that high SiC growth temperatures are incompatible with such masking material as SiO_2 leading to damage of silicon dioxide layer. This challenge was solved in the work employing optimized temperature, heating prozess and optimized SiO_2 pattern. The SiC nucleation on silicon dioxide was avoided employing narrow SiO_2 pattern smaller than 2 μm so that all reaction atoms initially adhered on the SiO_2 can reach the Si surface [150]. Implementation of HCl etch provides formation of LEO optimized process pattern of SiO_2/Si.

Compound semiconductor layers grown on Si are also of special interest for the fabrication of micromechanical devices. It is possible to integrate micromechanical devices and to fabricate micro-electro-mechanical systems (MEMS) on Si. SiC is a promising material for sensors such as pressure, temperature or UV sensors. SiC is mechanically and chemically stable and replaces silicon in MEMS devices and gas-sensors for use in extreme environments (high-temperature or corrosive environments, for example). On the other hand this advantage makes it difficult to employ standard patterning approaches - wet and even dry etching - because of low etching rates and poor mask selectivity [151, 152]. SiC etching rates up to 1.35 μm/min are possible when employing ion etching or plasma etching. Wet chemical etching is only possible in molten potassium hydroxide or with photo-electrochemical processes [153].

Selective deposition of cubic silicon carbide on silicon-on-insulator (SOI) substrates was employed for the fabrication of a high temperature pressure sensor [154]. A novel alternative processing approach to bulk micromachining of polycrystalline SiC using Si molds was recently reported [155]. But this is a rather complicated approach. A new approach to the low-temperature low-pressure chemical vapour deposition (LPCVD) growth of SiC on Si(001) patterned with a SiO_2 or Si_xN_y mask for selective growth of SiC on Si was developed and by employing a lift-off process for the first time SiC on Si was patterned without etching. The 40-190 nm thick SiC layers were patterned employing this novel approach [156, 157] with SiO_2 or Si_xN_y as sacrificial layer. The SiO_2 mask was produced employing thermal oxidation of the silicon substrate at 1200°C up to a thickness of about 1 μm and patterned in HF (Fig. 11 a). The stripes of the masks are [001] and [011] oriented and serve as sacrificial layer for the following lift-off process (widely used in silicon technology and described schematically in Fig. 11) which is carried out in buffered HF (to lift-off SiC on SiO_2 or Si_xN_y) (cf. Fig. 11 b). Finally, the Si is patterned employing etching in 30% KOH using the SiC as mask (cf. Fig. 11 d) [156, 158]. Deposition of SiC was observed both on the free and on the masked areas of the surface. In the case of growth employing SiO_2 masks the lift-off yield approached 100%.

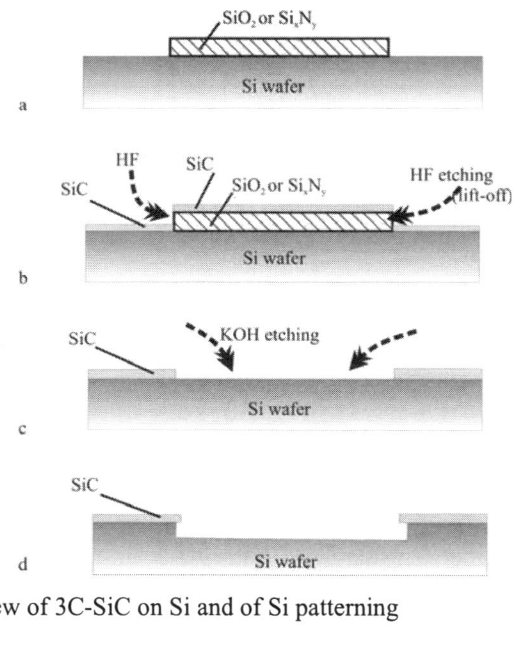

Fig. 11 Schematic view of 3C-SiC on Si and of Si patterning

After etching the Si through the SiC mask undercut is found for structures oriented along <100> directions (Fig. 12 a, b)[156, 158]. Cracking or fracturing of the overhanging SiC layers is not observed even in the case of a very small radius of bending (Fig. 12 a). Investigations of a second SiC layer deposition on an already patterned SiC on Si structure show that for the [011]-oriented stripes further growth occurs everywhere except the boundaries between the SiC layer on the stripe and the sidewalls of the patterned Si and also except the narrow area of the planar Si wafer near the lower part of the Si sidewalls (Fig. 12 d)[156, 158]. For the [001]-oriented stripes the SiC growth occurs on the first SiC layer including the bended part of the layer covering partly the Si sidewalls and on the Si wafer except an approximately 20 μm wide area near the sidewalls (Fig. 12 c)[156, 158]. These features are revealed employing further Si etching. The presence of the 20 μm area near the sidewalls free of SiC gives a further possibility to promote SiC and Si patterning employing smaller spacings between the stripes.

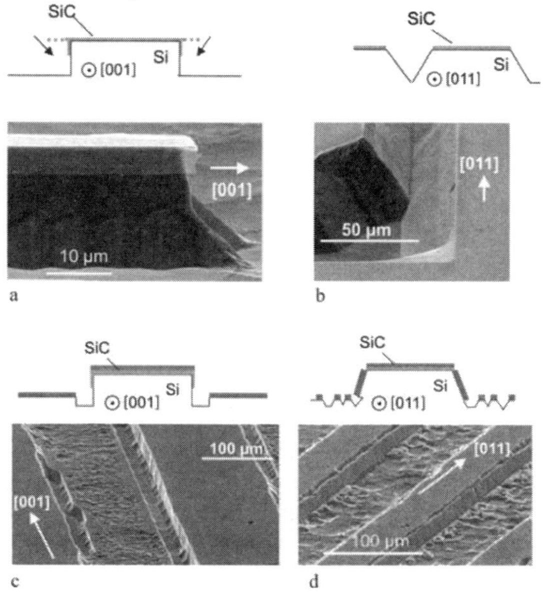

Fig. 12 SEM images of a patterned SiC on Si structure: a - [001]-oriented stripe after first SiC layer growth (~ 40 nm), lift-off and Si patterning; b – [011]-oriented stripe after the same process; c - [001]-oriented stripe after second SiC growth (~ 180 nm) and Si patterning; d - [011]-oriented stripe after the same process as c.

The etching selectivity between Si and SiC is excellent. The approach can be used for fabrication of micromechanical devices, gas sensors, biomedical applications etc. or for further pendeo-epitaxial growth [159], or for the fabrication of micromechanical SiC structures on Si [156–158]. By employing the lift-off process described above for the first time SiC on Si could be patterned without SiC etching and a SEM image of a micro-patterned SiC resonant structure is shown in Fig. 13 [156–158]. The presented mesa structures are defined by SiC lift-off and subsequent Si etching in KOH. As clearly seen the edges of the fabricated resonant structure are sharp. Further Si-etching from the backside is needed for the final formation of the resonant structure.

Fig. 13 SEM image of a micro-patterned SiC resonant structure on Si (001) substrate manufactured employing a lift-off technique

4. Summary

Significant progress in SiC homo- and heteroepitaxial technology have been observed in recent years, epitaxial layer quality and SiC devices performance have been improved. Thick epitaxial layers with low background doping levels which are necessary for further high power device development have been reproducibly realized. The main challenge in the development of the wide range of SiC high power devices is the presence of defects in epitaxial layers (some of them caused by the defects in SiC wafers). Further improvement of SiC epitaxial layers quality is necessary to realize reliable SiC-based devices. Therefore reduction of critical defects in epitaxial layers, reduction of lifetime limiting defects are still necessary in order to push further development of SiC based devices. CVD and sublimation epitaxy are the most promising techniques for SiC high-power devices production. Several commercial reactors for SiC CVD epitaxy providing good layer thickness and doping uniformity are currently available.

References

[1] Yu.M. Tairov, V.F. Tsvetkov, Proc. 1st European Conference of Crystal Growth, Zurich (1976) p.188.

[2] Yu.M. Tairov, V.F. Tsvetkov, J.Crystal Growth, V. 43 (1978) 209

[3] Yu.M. Tairov. J. of Crystal Growth, V. 52 (1981) pp.146.

[4] Yu.M. Tairov and V.F. Tsvetkov, in: Progress in Crystal Growth and Characterization, Vol. 7, Pergamon, London (1983) pp.111-162.

[5] Yu.M. Tairov. Materials Science and Engineering B29 (1995) pp.83-89.

[6] E.N.Mokhov, Yu.A.Vodakov, G.A.Lomakina in: Problems of Physics and Technology of Wide Bandgap Semiconductors, LIYaF, Leningrad (1979) p.136-149 (in Russian).

[7] Yu.A. Vodakov, E. N. Mokhov, M. G. Ramm and A. D. Roenkov Doping Peculiarities of SiC Epitaxial Layers Grown by Sublimation Sandwich-Method (Springer Proc. in Phys.) V. 56 p 329

[8] M. Kunzer, K. F. Dombrowski, F. Fuchs, U. Kaufman, J. Schneider, P. G. Baranov and E. N. Mokhov 1995 6th Int. Conf. SiC (ICSCRM) (Kyoto, Japan); Inst. Phys. Conf. Ser. p 257

[9] S. Yu. Karpov, Yu. N. Makarov, E. N. Mokhov, M. G. Ramm, M. S. Ramm, A. D. Roenkov, R. A. Talalaev, and Yu. A. Vodakov. Analysis of silicon carbide growth by sublimation sandwich method. Journal of Crystal Growth, 173 (1997) pp. 408-416.

[10] D.D. Avrov, A.S. Bakin, S.I. Dorozhkin, V.P. Rastegaev, Yu.M. Tairov. In: book of abstracts 1st European Conference on Silicon Carbide and Related Materials, Fodele Beach Hotel, Heraklion, Crete, Greece, October 6-9 (1996) p. 5.

[11] S.I. Dorozhkin, A.S. Bakin, D.D. Avrov, V.P. Rastegaev, Yu.M. Tairov. In: book of abstracts 1st European Conference on Silicon Carbide and Related Materials, Fodele Beach Hotel, Heraklion, Crete, Greece, October 6-9 (1996) p. 30.

[12] B.A.Kirillov, A.S.Bakin, Yu.M.Tairov, S.N.Solnyshkin. Simulation of heat and mass transfer during growth of silicon carbide single crystals. Semiconductors, V. 31, 7 (1997) pp. 672-676.

[13] A.S. Bakin, S.I. Dorozhkin. Improvements in SiC Defect Structure Control. Trans.Second Int. High Temperature Electronics Conf. Charlotte - NC. 5-10 June. (1994) V. 2, P.169.

[14] A.S. Bakin A.S., S.I. Dorozhkin, S.S. Bogachov, Kazak-Kazakevich, I.G. Lyutetskaja, A.P. Sazanov. Materials Science and Engineering (1996) 370-373.

[15] A.S. Bakin, S.I. Dorozhkin, Yu.M. Tairov, Yu. Dyomin. In: High Temperature Electronics Conference (HiTEC'96) v.1, Albuquerque, USA, June 10-14 (1996) pp. II-21-II-24.

[16] A.S. Bakin, B.A. Kirillov, Yu.M. Tairov. In: High Temperature Electronics Conference (HiTEC-III), v.2, Albuquerque, USA, June 10-14 (1996) pp. P207-P210.

[17] D.D. Avrov D.D., A.S. Bakin, S.I. Dorozhkin. Inst. Phys. Conf. Ser. No 142: Chapter 1 Paper pres. at the 6th International Conf. on Silicon carbide and Related Materials-1995 (ICSCRM '95), Kyoto, Japan, pp. 73-76.

[18] A.S. Bakin, S.I. Dorozhkin, Yu.M. Tairov. Semiconductors, V. 28, n.10 (1994) pp.1021-1022

[19] M. Sasaki, Y. Nishio, S. Nishino, S. Nakashima, H. Harima. Book of abstracts of ICSCIII-N'97, Stockholm, September 1997, pp. 200-201.

[20] A.S. Bakin, S.I. Dorozhkin, A.S. Zubrilov, N.D. Kuznetsov, Yu. M. Trans. Tech. Public., Switzerland, Mater. Science Forum, V. 264-268 (1998) pp. 53-56

[21] Yu.M. Tairov, V.F. Tsvetkov, J. Cryst. Growth, V.43, 2 (1978) p. 209

[22] Y.M. Tairov, V.F. Tsvetkov. Progress in controlling the growth of polytypic crystals.-In: Progr. in Crystal Growth and Characterization, Vol. 7, Crystal Growth and Characterization of Polytype Structures. Pergamon Press (1983) pp.111-162.

[23] Y.M. Tairov, V.F. Tsvetkov. Growth of Polytypic Crystals.-in: Crystals. Growth, Properties, and Applications, Springer-Verlag Berlin Heidelberg New York Tokyo (1984) pp. 3-35.

[24] Y.M.Tairov, V.F.Tsvetkov. Controlled Growth of Polytypes and their Importance for Science and Technology.- In: X-ray and Neutron Structure Analysis in Materials Science, Plenum Press, New York and London (1987) pp.331-351.

[25] M. Tuominen., R. Yakimova, A.S. Bakin, I. Ivanov, A. Henry, A. Vehanen, E. Janzen. Inst. Phys. Conf. Ser. No 142: Chapter 1 Paper pres. at the 6th International Conf. on Silicon carbide and Related Materials-1995 (ICSCRM '95), Kyoto, Japan, pp. 45-48.

[26] R.C. Glass, D. Henshall, V.F. Tsvetkov and C.H. Carter. SiC-Seeded Crystal Growth. MRS Bulletin, March 1997, pp.30-35.

[27] J. Takahashi, N. Ohtani, M. Kanaya. J. of Crystal growth, 167 (1996) pp. 596-606.

[28] M.M. Anikin, R. Madar, A. Rouault, I. Garcon, L.Di. Cioccio, J.L. Robert, J. Camassel, J.M. Bluet. In "Silicon Carbide and Related Materials 1995", ed. by S.Nakashima, H.Matsunami, S.Yoshida, and H.Harima (Inst. Phys. Conf. Ser. No 142; Inst. Phys. Publishing, Bristol and Philadelphia), p. 33-36, (1996).

[29] G. Ziegler, P. Lanig, D. Theis, C. Weyrich. IEEE Trans. Electron. Devices (USA) vol. ED-30, 1983, pp. 277-288.

[30] R.A. Stein, Physica B, 185 (1993) p.211.

[31] D.L. Barrett, J.P. McHugh, H.M. Hobgood, R.H. Hopkins, P.G. McMullin, and R.C.Clarke, J.Cryst. Growth v. 128 (1993) p.358

[32] A.S. Bakin, S.I. Dorozhkin. State-of-the-Art in Defect Control of Bulk SiC. (Invited talk). In: Conf. Proceedings of 1998 High Temperature Electronic Materials, Devices and Sensors Conference (IEEE Catalog Number 98EX132), February 22-27, 1998, San Diego, California USA, pp. 2-13

[33] K. Koga and T. Yamaguchi. Prog. Crystal Growth and Charact. 23 (1991) pp. 127-151

[34] T. Nakata, K. Koga, Y. Matsushita, Y. Ueda and T. Nima Springer Proc. in Physics, Amorphous and Cryst. Silicon carbide and Related Materials. Springer Verlag Berlin, Heidelberg 43 (1989) pp. 26-29.

[35] D.L. Barrett, I.P. McHugh, H.M. Hobgood, R.H. Hopkins, P.G. Mullin, R.C. Clarke and W.I.Choyke. J. Cryst. Growth 128 (1993) pp. 358-362.

[36] D.L. Barrett, R.G. Seidensticker, W. Gaida, R.H. Hopkins and W.I. Choyke. J.Cryst. Growth, V. 109 (1991) pp. 17-23

[37] S. Nishino, T. Higashino, T. Tanaka, S. Saraie. J. of Crystal Growth V. 147 (1995) pp. 339-342

[38] Yu. Vodakov Et al. Fiz. Tverd. Tela (Leningrad) V. 24, 1982, p. 1377 (Sov. Phays. Solid State V. 24 (1982) p. 780)

[39] N.S. Savkina, A. A. Lebedev, D. V. Davydov, A. M. Strel'chuk, A. S. Tregubova, M. A. Yagovkina. "New results in sublimation growth of SiC epilayers". Mat. Sci. and Eng. B 61-62, (1999) 165.

[40] A.A. Lebedev, D.V. Davydov, N.S. Savkina, A.S. Tregubova, M.P.Scheglov, R.Yakimova, M.Syvajarvi,E. Janzen, "Structural defects and Deep-levels in 4H-SiC epilayers Grown by sublimation epitaxy in Vacuum", Semiconductors , 34, p. (2000) 1133-1136.

[41] S.Y.Davydov, N.S.Savkina, A.A.Lebedev, M.Syvajarvi and R.Yakimova Calculation of SiC Epitaxial Growth rate in Vacuum . Mat.Science Forum V. 457-460 (2004) p. 249.

[42] N.S. Savkina, A. A. Lebedev, D. V. Davydov, A. M. Strel'chuk, A. S. Tregubova, C. Raynaud, et al. Low-doped n-type epilayers grown by sublimation epitaxy, Mat.Science & Eng, B77 (2000) p. 50-54.

[43] A.M.Ivanov, N.B.Strokan, A.A.Lebedev, D.V.Davydov, N.S.Savkina, E.V.Bogdanova. N- and p-type 6H-SiC films for the Creation Diode and Triode Structure of Nuclear Particle Detectors. Material Science forum, 389-393, (2002) pp. 1439-1442.

[44] N. B. Strokan, A. M. Ivanov, N. S. Savkina, A. M. Strelchuk, A. A. Lebedev, M. Syväjärvi, and R. Yakimova ." Detection of strongly and weakly ionizing radiation by triode structure based on SiC films" J. Appl. Phys. 93 (2003) pp. 5714-5719.

[45] A.A.Lebedev, A.M.Strel'chuk,D.V.Davydov, N.S.Savkina, A.S.Tregubova, A.N.Kuznetsov, V.A.Soloviev, N.K.Poletaev. Electrical characteristics of p-3C-SiC/n-6H-SiC heterojunctions grown by sublimation epitaxy on 6H-SiC substrates, Applied Surface Science, V. 184, N1-4, p. 421-426.

[46] A. A. Lebedev, A. M. Strel'chuk, N. S. Savkina, E. V. Bogdanova, A. S. Tregubova, A. N. Kuznetsov, and L. M. Sorokin "Investigation of the p-3C-SiC/n+-6H-SiC Heterostructures with Modulated Doping" Technical Physics Letters, 28, N12, p. 1011-1014 (2002).

[47] N.S.Savkina, A.M.Strel'chuk, L.M.Sorokin, G.N.Mosina, A.S.Tregubova, V.V.Soloviev, A.A.Lebedev "Characterization of 3C-SiC/6H-SiC Heterostructures grown by Vacuum sublimation" " Material Science Forum V. 433-466 (2003) p 293-296.

[48] A.A. Lebedev, A.M.Strel'chuk, N.S.savkina, E.V.Bogdanova, A.S.Tregubova, A.N.Kuznetsov, D.V.Davydov. Investigation of p-3C-SiC/n+6H-SiC heterojunctions with low Doped p-3C-Sic region. Material Science Forum V. 433-466, (2003) p 427-430.

[49] Y. Onum. Silicon carbide films evaporated in vacuum on synthetic micas. Jpn. J. Appl. Physics 8 (1969) p. 401

[50] Fissel A., Schroter B. and Richter W. Low-temperature growth of SiC thin films on Si and 6H-SiC by solid-source molecular beam epitaxy. Appl. Phys. Lett., V. 66, (1995) pp. 3182-3184.

[51] H. Matsunami. Progress in epitaxial growth of SiC, Physica B, V. 185 (1993) p. 65.

[52] T. Yoshinobu, N. Nakayama, H. Shiomi, T. Fuyuki, H. Matsunami. Lattice-matched epitaxial growth of single crystalline 3C-SiC on 6-SiC substrates by gas source molecular beam epitaxy. Appl. Physics Letters, 60 (1992) p. 824.

[53] R.S. Kern, R.F. Davis,. Homoepitaxial SiC growth by molecular beam epitaxy. Phys. Stat. Soidi, (b) 202 (1997) p. 379.

[54] Fissel. Artificially layered heteropolytypic structures based on SiC polytypes: molecular beam epitaxy, characterization and properties. Physics Reports, V. 379, N. 3-4 (2003) pp. 149-255.

[55] http://www.epigress.com

[56] R. Rupp, Yu. N. Makarov, H. Behner, and A. Wiedenhofer. Silicon Carbide Epitaxy in a Vertical CVD Reactor: Experimental Results and Numerical Process Simulation, phys. stat.sol. (b), 202, 281, 1997.

[57] A.A. Burk, Jr. and L.B. Rowland, Homoepitaxial VPE Growth of SiC Active Layers, Physica Status Solidi (b), 202 (1997) p. 263.

[58] T. Kimoto, A. Itoh, A. and H. Matsunami. Step-Controlled Epitaxial Growth of High-Quality SiC Layers, Physica Status Solidi (b), 202, (1997) p. 247.

[59] D.J. Larkin. An overview of SiC Epitaxial Growth. MRS Bulletin, March 1997, pp.3641.

[60] S. Nishino, J.A. Powell, H.A. Will. Appl. Phys. Lett., V.42, N.5 (1983) p. 460.

[61] http://www.emf.co.uk/

[62] A.A. Burk, M.J. O'Loughlin, M.J.Paisley, A.R. Powell, M.F. Brady, R.L. Leonard. Large area SiC epitaxial layer growth in a warm-wall planetary VPE reactor. Proc. 5th European Conference on Silicon carbide and Related Materials (ECSCRM2004), Aug. 31 – Sept. 4, Bologna, Italy (2004), Material Science Forum, V. 483-485, pp. 137-140.

[63] D. J. Larkin, P. G. Neudeck, J. A. Powell, and L. G. Matus, "Site-Competition Epitaxy for Superior Silicon Carbide Electronics," Applied Physics Letters, vol. 65 (1994) pp. 1659-1661.

[64] A. Itob and H. Matsunami. IEEE Electron Device Lett. 16 (1995) p. 280.

[65] P.J. Neudeck. D. J. Larkin. J. A. Powell. L. G. Mains. and C. S. Salupo. AppI. Phys. Lett. 64 (1994) p. 1386.

[66] S. Sriram. R. C. Clarke. A. A. Burk. H. M. Hobgood. P G. McMullin. P.A. Orphanos. R. R. Siergie. T. J. Smith. C. D. Brandi. M. C. Driver. And R. H. Hopkins. IEEE Electron Device Lett. 15, 1994, p. 458.

[67] R. Rupp, P. Lanig, J. Volkl, and D. Stefani. J. Cryst. Growth 146 (1995) p. 37.

[68] O. Kordina. A. Henry. J. P. Bergman. N. T. Son. w. M. Chen. C. Hallin, and E. Janzen. App. Phys. Let. 66 (1995) p. 1373.

[69] V. F. Tsvetkov. S. T. Allen, H. S. Kong. and C. H. Carter. Jr.. Presented at the International

Conference on SiC and Related Materials, 18-21 September 1995, Kyoto. Japan.

[70] H. M. Hobgood. D. L. Barrett. 3. P. McHugh, R. C. Clarke. S. Sriram. A. A. Burk. J. Greggi. C. D. Brandi. R. H. Hopkins. and W.J. Choyke. J. Cryst. Growth 137 (1995) p. 181.

[71] 0. Kordina. 3. P. Bergman. A. Henry. E. Janzen. S. Savage. 3. Andre. L. P. Ramberg. U. Lindefelt, w. Hermansson. and K. Bergman. Appl. Phys. Let. 67 (1995) p. 1561.

[72] A. 0. Konstantinov and P. A. Ivanov. Silicon Curbide onal Related Moterials. edited by M. G. Spencer. R. P. Devaty. 3. A. Edmond, M. Asif Khan. R. Kaplan. and M. Rabman. Inst. of Physics at Conf. Ser., No. 137 IOP. Bristol (1994) pp. 37-40.

[73] http://www.okmetic.com/

[74] A. Ellison, B. Magnusson, C. Hemmingsson, W. Magnusson, T. Iakimov, L. Storasta, A. Henry, N. Henelius, and E. Janzén; Proc. of the MRS 2000 Fall Meeting (Boston, Massachusetts, USA; Nov 27 - Dec 1, 2000) Proc. of MRS fall meeting (2000) V.640 (2001) p. H1.2.

[75] E. Janzén, A. Ellison, B. Magnusson, and G. Pozina; Proc. of the Workshop on Challenges in Semi-Insulating Nitrides and SiC (Laugarvatn, Iceland; July 14 – 18, 2002)

[76] B. Magnusson, A. Ellison, and N.T. Son; Proc. of the GAAS2002, Wide bandgap semiconductors workshop (Milan, Italy; September 25, 2002)

[77] O. Kordina, C. Hallin, A. Ellison, A.S. Bakin, I. Ivanov, A. Henry, R. Yakimova, M. Tuominen, A. Vehanen, E. Janzen. J. of Applied Physics Letters., V. 69, n.11 (1996) pp1456-1460.

[78] O. Kordina, C. Hallin, A.S. Bakin, I. Ivanov, A. Henry, M. Tuominen, R. Yakimova, A. Vehanen, E. Janzen. Digest of the 6th International Conf. on Silicon carbide and Related Materials-1995 (ICSCRM '95), Kyoto, Japan, p. 609.

[79] A. Ellison, J. Zhang, W. Magnusson, A. Henry, Q. Wahab, J.P. Bergman, C. Hemmingson, N.T. Son, E. Janzen. Materials Science Forum, V. 338-342 (2000) pp. 131-136.

[80] U. Zimmermann, J. Österman, J. Zhang, A. Henry, A. Hallen. Materials Science Forum, V. 389-393 (2002) pp. 1285-1288.

[81] R.L. Myers, O. Kordina, S. Rao, R. Everly, S.E. Saddow. Increased growth rate in SiC CVD reactor using HCl as an additive. Proc. 5th European Conference on Silicon carbide and Related Materials (ECSCRM2004), Aug. 31 – Sept. 4, Bologna, Italy (2004), Material Science Forum, V. 483-485, pp. 73-76.

[82] J. A. Powell, P. Pirouz, W.J. Choyke. Growth and Characterization of Silicon Carbide Polytypes for Electronic Applications, Semiconductor Interfaces, Microstructures, and Devices: Properties and Applications, Z.C. Feng, Eds., Institute of Physics Publishing, Bristol, United Kingdom (1993) 257.

[83] J.A. Powell, and D.J. Larkin. Processed-Induced Morphological Defects in Epitaxial CVD Silicon Carbide, Physica Status Solidi (b), 202 (1997) p. 529.

[84] H.S. Kong, Glass, and B. Davis, R. F., Chemical Vapor Deposition and Characterization of 6H-SiC Thin Films on Off-Axis 6H-SiC Substrates, Journal of Applied Physics, 64 (1988) p. 2672.

[85] Kimoto, T., Itoh, A. and Matsunami, H., Step-Controlled Epitaxial Growth of High-Quality SiC Layers, Physica Status Solidi (b), 202 (1997) 247.

[86] B. Thomas, C. Hecht. Epitaxial growth of n-type 4H-SiC on 3" wafers for power devices. Proc. 5th European Conference on Silicon carbide and Related Materials (ECSCRM2004), Aug. 31 – Sept. 4, Bologna, Italy (2004), Material Science Forum, V. 483-485, pp. 141-146.

[87] F.C. Frank, Acta Crystallogr. 4 (1951) p. 497.

[88] W.M. Vetter and M. Dudley. In "Defects in Electronic Materials II", Mat. Res. Soc. Symp. Proc., 442 (1997) pp. 661-665

[89] X.R. Huang, M. Dudley, W.M. Vetter, W. Huang, S. Wang, C.H. Carter Jr. Appl. Phys. Lett. V. 74 (1999) p. 353

[90] P.G. Neudeck and J.A. Powell. IEEE Electron Device Letters, V. 15, No. 2 (1994) pp. 63-65.

[91] S. Nishino, Y. Kojima and J. Saraie. Springer Proc. Phys.(Germany) V. 56 (1992) p.15-21.

[92] S. Nishino, T. Higashino, T. Tanaka, and J. Saraie. J. Crystal Growth V. 147 (1995) pp.339-342.

[93] J.A. Powell, P.G. Neudeck, D.L. Larkin, J.W. Yang, P. Pirouz. Inst. Phys. Conf. Ser. (UK) no.137, (1994) p.161-4.

[94] A.S. Bakin, C. Hallin, O. Kordina, E. Janzen. Defect Formation on the Surface of the 6H-SiC CVD Layers. Techn. Digest of the 6th International Conf. on Silicon carbide and Related Materials-1995 (ICSCRM '95), Kyoto, Japan, p. 355-356.

[95] S.E. Saddow, M.S. Mazzola, S.V. Rendakova, V.A. Dmitriev. Silicon carbide CVD homoepitaxy on wafers with reduced micropipe density. Materials Science and Engineering B61-62 (1999) pp. 158-160.

[96] I. Kamata, H. Tsuchida, T. Jikimoto, and K. Izumi, Jpn. J. Appl. Phys., Part 1 39 (2000) p. 6496.

[97] G. Wagner, D Schulz, J. Doerschel. Reduction of the micropipe density of 4H-SiC (0001) wafers by hot-wal-CVD to improve the seed quality for 4H-Sic bulk growth. Proc. 5th European Conference on Silicon carbide and Related Materials (ECSCRM2004), Aug. 31 – Sept. 4, Bologna, Italy (2004), pp. 60-61.

[98] K. Kojima, H. Okumura, S. Kuroda, K. Arai. 2-inch 4H-SiC homoepitaxial layer grown on on-axis C-face substrate by CVD methd. Proc. 5th European Conference on Silicon carbide and Related Materials (ECSCRM2004), Aug. 31 – Sept. 4, Bologna, Italy (2004), Material Science Forum, V. 483-485, pp. 93-96.

[99] R. Yakimova, M. Tuominen, A.S. Bakin, J.-O. Fornell, A. Vehanen, E. Janzén. Inst. Phys. Conf. Ser. No 142: Chapter 1 Paper pres. at the 6th International Conf. on Silicon carbide and Related Materials-1995 (ICSCRM '95), Kyoto, Japan, pp. 101-104.

[100] S.V. Rendakova, I.P. Nikitina, A.S. Tregubova, V.A. Dmitriev, 1997 Electronic Materials Conference, June 25-27, 1997, Fort Collins, Colorado, Technical Program with Abstracts, p. 42.

[101] M. Syväjärvi, R. Yakimova, and E. Janzén. Growth of SiC from Liquid Phase: Wetting and Dissolution of SiC. Proc. of the 1st European Conference on SiC and Related Materials; Crete, Greece; 6-9 Oct., 1996; Diam. Relat. Mater. 6, (1997) p. 1266.

[102] S. Rendakova, N. Kuznetsov, N. Savkina, M. Rastegaeva, A. Andreev, M. Minbaeva, A. Morozov, and V. Dmitriev, Mater. Res. Soc. Symp. Proc. 512 (1998) p. 131.

[103] S.V. Rendakova, I.P. Nikitina, A.S. Tregubova, and V.A. Dmitriev. Micropipe and Dislocation Density Reduction in 6H-SiC and 4H-SiC Structures Grown by Liquid Phase Epitaxy, Journal of Electronic Materials, 27, (1998) p. 292.

[104] A. Okamoto, Y.Seno, N. Sugiyama, F. Hirose, K. Hara, T. Tani, D. Nakamura, N. Kamiya and S. Onda. Materials Science Forum V. 389-393 (2002) pp. 103-106.

[105] Q. Wahab, A. Ellison, C. Hallin, A. Henry and E. Janzen. Study of breakdown limiting defects in 4H-SiC substrates and epilayers on the performance of high voltage Schottky diodes. Abstracts of the Int. Conf. On Silicon Carbide and Related Materials NC, USA, October 1999, (1999) p. 298.

[106] J. Takahashi, and N. Ohtani. Modified-Lely SiC Crystals Grown in [1100] and [1120] Directions, Physica Status Solidi (b), 202, (1997) p. 163.

[107] G.W. Eldridge, D.L. Barrett, A.A. Burk, H.M. Hobgood, R.R. Siergiej, C.D. Brandt, M.A. Tischler, G.L. Bilbro, R.J. Trew, W.H. Clark, and R.W. Gedridge, Jr. High Power Silicon Carbide IMPATT Diode Development. 2nd Annual AIAA SDIO Interceptor Technology Conference, Albuquerque, NM, American Institute of Aeronautics and Astronautics, Washington DC, (1993) Rep. 93-2703.

[108] M. Mynbaeva, N. Savkina, D. Davidov, A. Zubrilov, A. Strel'chuk, A. Tregubova, A. Lebedev, I. Kotousova, N. Seredova, V. Dmitriev. Porous SiC substrate materials for high quality epitaxial and bulk growth. Abstracts of the MRS Fall Meeting, November 29-December 3, 1999, (1999) p. 251.

[109] N.S.Savkina, V.V.Ratnikov, V.B.Shuman, A.A.Lebedev. Effect of sublimation Growth on the structure of Porous SiC:SEM and X-ray Diffraction Investigation. Materials Science Forum, V. 353-356, (2001) p. 119-122.

[110] E.N. Mokhov, Yu.A. Vodakov, G.A. Lomakina in: Problems of Physics and Technology of Wide Bandgap Semiconductors, LIYaF, Leningrad, 1979, p.136-149 (in Russian).

[111] P.H. Yih and A.J. Steckl, J. Elecrochem. Soc., V. 142 nr. 8 (1995) pp. 2853-2860.

[112] P. Chabert, Deep etching of silicon carbide for micromachining applications: etch rates and etch mechanisms, J. Vac. Sci. Technol., B 19 (2001) pp. 1339-1345

[113] S. Tanaka, K. Rajana, T. Abe and M. Esashi, Deep reactive ion etching of silicon carbide, J. Vac. Sci. Technol. B 19 (2001) pp. 2173-2176

[114] O. Kordina, C. Hallin, R.C. Glass, A. Henry and E. Janzen IOP Conf. Ser. 137 (IOP, Bristol) (1994) p. 41-44.

[115] C. Hallin, A.S. Bakin, F. Owman, P. Martensson, 0. Kordina and E. Janzen, Inst. Phys. Conf., Ser. N.142, Chap. 3 (1996) pp.613-616.

[116] I.M. Harris, P.C. Gatos and A.F. Witi. J. Elecirochem. Soc.: Solid State Science 116 (1969) p. 3110.

[117] T.L. Chu and R.B. Campbell. J. Electrochem. Soc., V.112 (1965) p. 955.

[118] I. Powell, D.J. Larkin and P.B. Abel. J. of Electronic Mat. 24 (1995) p. 295.

[119] J. Zhang, O. Kodina, A. Ellison and E. Janzen. Materials Science Forum V. 389-393 (2002) pp. 239-242

[120] K. Masahara, T. Takahashi, M. Kushibe, T. Ohno, J. Nishio, K. Kojima, Y. Ishida, T. Suzuki, T. Tanaka, S. Yoshida and K. Arai. Materials Science Forum V. 389-393 (2002) pp. 227-230

[121] M. Hasegawa, A. Miyauchi, K. Masahara, Y. Ishida, T. Takahashi, T. Ohno, J. Nishio, T. Suzuki, T. Tanaka, S. Yoshida and K. Arai. Materials Science Forum V. 389-393 (2002) pp. 227-230

[122] K. Nakayama, Y. Miyanagi, K. Maruyama, Y. Okamoto, H. Shiomi and S. Nishino. Materials Science Forum V. 389-393 (2002) pp. 235-238

[123] K. Takahashi, M. Uchida, T. Yokogawa, O. Kusumoto, K. Yamashita, R. Miyanaga and M. Kitabatake. Materials Science Forum V. 389-393 (2002) pp. 243-246

[124] S.E. Saddow, T. Elkington and M.C.D. Smith. Materials Science Forum V. 389-393 (2002) pp. 399-402

[125] R.C. Glass, L.O. Kjelberg, V.F. Tsvetkov, J.E. Sundgren, E. Janzen. J. of Crystal Growth, V. 132 (1993) p. 504

[126] A.S. Bakin, S.I. Dorozhkin, A.O. Lebedev, B.A. Kirillov, A.A. Ivanov, Yu.M. Tairov. J. of Crystal Growth, V. 198/199 (1999) pp. 1015-1018

[127] R.C. Glass, L. Kjellberg, V.F. Tsvetkov, J.E. Sundgren and E. Janzen. J.Cryslal Growth 132 (1993) pp. 504-512

[128] M. Tuominen, R. Yakimova, R.C. Glass, T. Tuomi and E. Janzen. J.Crystal Growth 144 (1994) pp. 267-276

[129] O. Kordina, C. Hallin, R.C. Glass, A. Henry and E. Janzen, 1994, OP Conf. Series 137 (1OP, Bristol) pp. 41-44

[130] A. Bakin, C. Hallin, O. Kordina, E. Janzen. High Resolution XRD Study of the Silicon Carbide CVD Growth. Inst. Phys. Conf. Ser. No 142: Chapter 2 (1996) pp. 433-436.

[131] S.A. Powell, J.B. Petit, J.H. Edgar, I.G. Jenkins, L.G. Matus, J.W. Yang, P. Pirouz, W. Choyke, Clemen and M. Yoganathan. Appl. Phys. Lett. 59 (1991) pp. 333-335

[132] B. Gerard, X. Marcadet, P. Etienne, D. Pribat, J. Nagle, F. Carlin, M. Ilegems, D. Friedrich, J. Eichholz, H. Bernt. MRS Symposium Proceedings Series, 533 (1998) FF4.2

[133] K. Linthicum, T. Gehrke, D. Thomson, E. Carlson, P. Rajoagopal, T. Smith, D. Batchelor, R. Davis. Appl. Phys. Lett., 75 (1999) pp. 196-198.

[134] S. Nishino, J.A. Powell, H.A. Will. Appl. Phys. Lett., v.42, n.5 (1983) p. 460.

[135] S. Nishino, S. Suhara, H. Ono, and H. Matsunami, Appl. Phys. 61 (1987), p. 4889

[136] S. Stefanescu, A.A. Yasseen, C.A. Zorman, and M. Mehregany, in Technical Digest, The 10th Int. Conf. Solid-State Sensors and Actuators (Sendai, Japan, June 1999), p. 194

[137] G. H. Krötz, M. H. Eickhoff, H. Möller, Sensors and Actuators 74 (1999) p. 182

[138] J. Pezoldt, P. Masri, M. Rouhani Laridjani, M. Averous, T. Wöhner, J.A. Schaefer, Th. Stauden, G. Ecke, R. Pieterwas, L. Spieß, International Conference of SiC and Related Materials 1999.

[139] C. Jacob, J. Chung, M.-H. Hong, P. Pirouz, and S. Nishino, Mat. Res. Soc. Symp. 622 (2000) p. T4.1.1-6.

[140] I. Golecki, F. Reidinger, and J. Marti, Appl. Phys. Lett. 60 (1992) p. 1703.

[141] W. L. Sarney, L. Salamanca-Riba, P. Zhou, M. G. Spencer, C. Taylor, R. P. Sharma, and K. A. Jones, J. Electron. Mater. 29 (2000) p. 359.

[142] A. Bakin, D. Piester, I. Behrens, H.-H. Wehmann, E. Peiner, A. Ivanov, D. Fehly, A.Schlachetzki. Growth of InP Layers on Nanometer-Scale Patterned Si Substrates. Journal of Crystal Growth & Design, American Chemical Society, Vol. 3, nr. 1 (2003) pp. 89-93

[143] S.Neumann, A. Bakin, P.Velling, W.Prost, H.-H.Wehmann, A.Schlachetzki, F.-J.Tegude. Growth of III/V Resonant Tunneling Diode on Si Substrate with LP-MOVPE , Journal of Crystal Growth V. 248 (2003) pp. 380-383

[144] http://www.bell-labs.com/org/physicalsciencees/projects/spt/spt_3.html

[145] T. Isshiki, M. Nakamua, T. Nishiguchi, K. Nishio, S. Ohshima, S. Nishino. Influence of substrate roughness on the formation of defects in 3C-SiC grown on Si (110) substrate by hetero-epitaxial CVD method. Proc. 5th European Conference on Silicon carbide and Related Materials (ECSCRM2004), Aug. 31 – Sept. 4, Bologna, Italy (2004), Material Science Forum, V. 483-485, pp. 185-188.

[146] M. Nakamura, T Isshiki, T. Nishiguchi, K. Nishio, S. Ohshima, S. Nishino. Suppression mechanism of double positioning growth in 3C-SiC(111) crystal by using an off-axis Si(110) substrate. Proc. 5th European Conference on Silicon carbide and Related Materials (ECSCRM2004), Aug. 31 – Sept. 4, Bologna, Italy (2004), Material Science Forum, V. 483-485, pp. 181-184.

[147] http://www.hoya.co.jp/english/index.cfm

[148] A. S. Bakin, A. A. Ivanov, D. Piester, T. Riedl, F. Hitzel, H.-H. Wehmann, and A. Schlachetzki, Materials Science Forum, vols. 353-356 (2001) pp. 163-166.

[149] T. Nishigawa, T. Nakao, J. Saraie. Jpn. J.. Applied Physics, 27 (1988) p. 964.

[150] S. Sugishita, A. Shoij, Y. Mukai, T. Nishiguchi, T. Isshiki, S. Ohshima, S. Nishino. Lateral Epitaxial overgrowth of 3C-SiC grown on Si substrate by CVD method. Proc. 5th European Conference on Silicon carbide and Related Materials (ECSCRM2004), Aug. 31 – Sept. 4, Bologna, Italy (2004), Material Science Forum, V. 483-485, pp. 177-180.

[151] Schmid U, Eickhoff M, Richter Ch, Kroetz G and D. Schmitt-Landsiedel, Etching characteristics and mechanical properties of a-SiC:H thin films Sensors Actuators, A 94 (2001) pp. 87–94

[152] H. Morkoc, S. Strite, G.B. Gao, M.E. Lin and M. Burns. Large-band-gap SiC, III–V nitride, and II–VI ZnSe-based semiconductor device technologies J. Appl. Phys. V. 76 (1994) pp. 1363–1398

[153] C.E. Stutz. J. Electron. Mater.,V. 27 (1998) pp. 81–83

[154] M. M. Eickhoff, H. Koller, G. Kroetz, J.v. Berg and R. Ziermann. Sensors Actuators, A 74 (1999) pp. 56–59

[155] X. Song, S. Guo, C. A. Zorman, C. H. Wu, A. A. Yasseen, and M. Mehregany. Mater. Sci. Forum, V. 338-342 (2000) pp. 1141-1144.

[156]A. Bakin, I. Behrens, A. Ivanov, E. Peiner, D. Piester, H.-H. Wehmann and A. Schlachetzki, Growth of 3C-SiC on Si(100) by LPCVD and patterning of the grown layers. Materials Science Forum, V. 389-393 (2002) pp. 327-330.

[157] I. Behrens, E. Peiner, A. Bakin, and A. Schlachetzki, Micromachining of silicon carbide on silicon fabricated by low-pressure chemical vapour deposition, J. Micromech. Microeng., V. 12, (2002) pp. 380-384.

[158] I. Behrens, E. Peiner, A. Bakin, and A. Schlachetzki. Proc. 12th Micromechanics Europe Workshop (MME2001), September 16-18, 2001, Cork, Ireland, pp. 43-45.

[159] Y. Okui, C. Jacob, S. Ohshima and S. Nishino. Materials Science Forum, V. 389-393 (2002) pp. 331-334.

International Journal of High Speed Electronics and Systems
Vol. 15, No. 4 (2005) 781–820

World Scientific
www.worldscientific.com

OHMIC CONTACTS TO SIC

FABRIZIO ROCCAFORTE, FRANCESCO LA VIA, and VITO RAINERI

CNR-IMM, sezione di Catania, Stradale Primosole 50, I-95121, Catania, Italy

In this chapter, the most significant results obtained in the last decade in the field of ohmic contacts to SiC are reviewed.
First, the basic concepts related to the physics of ohmic contacts and to the contact resistance measurement techniques are briefly reported.
Then, some aspects concerning the formation of low resistance (10^{-5}-10^{-6} Ωcm^2) ohmic contacts on n-type and for p-type SiC are discussed, focusing on Ni-based and Al/Ti-based contacts.
Examples of innovative applications on practical devices are also reported, as the simultaneous formation of ohmic contacts on n- and p-type SiC for vertical power MOS devices, obtained adding Al to a standard Ni contact, and a single-metal technology for ohmic and rectifying contacts in MESFETs, using Ti or Ni without post-deposition annealing.

1. Introduction

Ohmic contacts play a very important role in the signal transfer to and from the semiconductor and the external circuitry. In particular, the resistance of the contact must be negligible with respect to that of the bulk device (the device on-resistance), since a voltage drop at the contact adds an undesired contribution to the dissipated power, thus decreasing the efficiency of the system.[1]
Because of its excellent properties such as a wide band gap, a high critical electric field, a high thermal conductivity and a high electron saturation velocity, silicon carbide (SiC) is one of the most prominent candidates for the next generation semiconductor devices.[2,3]
However, although most of the common processes of silicon technology, like ion implantation, oxidation, self aligned silicides formation, etc., have been applied to SiC, and a variety of high-power and high-frequency SiC devices (Schottky diodes, MESFETs, VJFETs, SITs, etc.) have been demonstrated in the last years,[3,4,5,6] the best expression of the potentialities of the material remains strictly related to the improvement

of some technological concerns. Among them, one of the key technology issues which may limit the performances of SiC electronic devices is that of ohmic contacts.

The most important figure of merit of ohmic contacts is the specific contact resistance ρ_c, an intrinsic interfacial property, which is independent of the contact geometry. It has been demonstrated that high-frequency and high-power SiC electronics devices, particularly those in which a high current density flows horizontally (MESFETs), need to have ohmic contacts with values of ρ_c in the range 10^{-5}-10^{-6} Ωcm^2.[7,8]

Now, it is known that in order to form a good ohmic contact to a semiconductor, a metal resulting into a low Schottky barrier with the semiconductor material must be selected. In these terms, the wide band gap of SiC, about 3 times higher than that of silicon, makes almost all metals on 6H- and 4H-SiC to have a Schottky barrier height larger than ~1 eV, which presents a serious challenge to obtain a good ohmic contact on SiC with a low specific contact resistance.

For all these reasons, a great deal of effort has been spent in the last decades to study the physics of metal contacts on SiC and to achieve low resistivity reliable ohmic contacts. The interest of the scientific community towards the SiC material and device related issues resulted in a good number of review papers dealing with ohmic contacts.[2,7,9,10,11,12,13]

Metal/SiC contacts are generally non-ohmic after metal deposition, due to the high Schottky barriers at the interface, determining the rectifying properties. Hence, in order to overcome this problem, beyond using heavily doped material for creating tunnel ohmic contacts, the most common method to form an ohmic contact to SiC is the deposition of a metal layer followed by an annealing process. In fact, the post-deposition annealing may result into a reaction of the metal with the SiC (i.e. with the formation of silicides, carbides, or ternary phases), with a consequent reduction of the barrier height or thickness.

A large number of ohmic contacts materials have been investigated in the last decades, both in terms of structural characterization and of electrical performances.

For ohmic contacts on the n-type SiC material, the most promising metal is Ni. Since the early works of Crofton et al.,[14] it has been demonstrated that Ni films annealed at temperatures in the range 900-1000°C can form good ohmic contacts on n-type SiC with specific contact resistance values as low as $\sim 1\times10^{-6}$ Ωcm^2.[14] The physical problems related to the formation of nickel silicides ohmic contacts were widely investigated. As an example, many works report about the redistribution of carbon inside the silicide layer and its role on the electrical properties of the contacts (focusing on aspects like the Schottky to ohmic transition, the thermal stability, etc.). Moreover, the control of the thermal budget required for ohmic contacts formation is an important factor to achieve a good process compatibility for practical SiC devices.

For the p-type material, due to the higher values of the Schottky barrier heights, ohmic contact formation is even more difficult than in the n-type material.

Much research activity has been focused on Al/Ti contacts, which give a specific contact resistance in the $\sim 10^{-5}$ Ωcm^2 range.[15,16] Big attention has been given to the possible mechanisms of ohmic contact formation of Al/Ti alloys on SiC; however this topic remains still under discussion.

In addition to the low resistance, the thermal stability of ohmic contacts is required for SiC devices like sensors operating at high temperatures and in harsh environments, for the aerospace applications, satellites, nuclear power instruments, etc. Recent works reported on long term stability of Ni- and Ti-based contacts[17,18,19] up to 600°C in oxidizing atmosphere and up to 1000°C in inert ambient.

This paper reviews the most significant work done in the last decade on ohmic contacts on silicon carbide. In particular, with respect to the previous review articles, this work tries to cover a larger number of aspects, going from the basic physics of the metal/semiconductor contact and measurements techniques, to the physical mechanisms of ohmic contact formation, the thermal stability, the applications to devices, etc..

First, some basic concepts related to the Schottky barrier formation, the physics of ohmic contacts and the specific contact resistance measurements techniques are briefly reported. Thereafter, the major results on ohmic contacts to n-type and p-type SiC will be discussed, with particular attention to some scientific aspects concerning the physics of contact formation (i.e. in particular for the cases of annealed Ni and Al/Ti contacts). Examples of innovative applications on practical devices are also reported, focusing on new technological processes such as the simultaneous formation of ohmic contacts on n- and p-type SiC for vertical power MOS devices proposed by Kiritani et al.[20] and the single metal contact technology developed by Tanimoto et al.[21] for ohmic and rectifying contacts in MESFETs.

2. Metal-Semiconductor Contacts

Metal-semiconductor contacts fall into two basic categories, the *ohmic* and the *rectifying* (or *Schottky*) contacts.

An ohmic contact has a linear and symmetric current-voltage characteristic for positive and negative applied voltages and a negligible resistance compared with that of the bulk of the device.[1] Conversely, a rectifying Schottky contact is characterized by the current flow for only one voltage sign.

In order to introduce the important physical parameters which affect the contact performance, the classical description of the Schottky barrier formation is briefly reported.

Fig. 1a shows the energy band diagram of a metal and an n-type semiconductor before they are brought into contact. The metal work function $q\Phi_m$ and the semiconductor work function $q\Phi_s$ are the energies required to bring an electron from the Fermi level of the material to the vacuum. The energy difference between the vacuum level and the bottom of the semiconductor conduction band E_c is defined as semiconductor electron affinity $q\chi_s$.

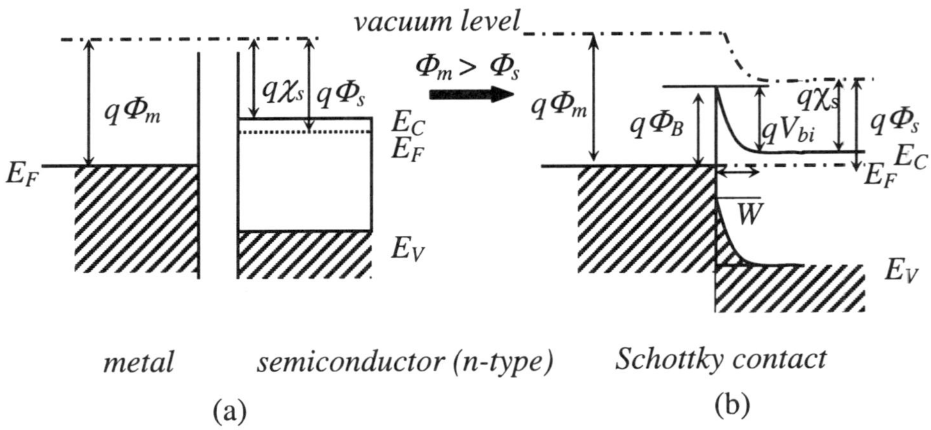

Fig. 1: Energy band diagram for a metal-semiconductor (n-type) contact, in the case $\Phi_m > \Phi_s$ before (a) and after (b) they are brought into contact, showing the formation of a rectifying contact with a Schottky barrier height $q\Phi_B$.

When the metal and the semiconductor are brought into contact, provided the semiconductor work function $q\Phi_s$ is lower than the metal work function $q\Phi_m$, electrons will flow from the n-type semiconductor to the metal, leaving behind a positively charged donors region over the depletion width W. This charge flow continues until the thermodynamic equilibrium is reached and the two Fermi levels line up. In this way the energy level of the electrons in the semiconductor will be raised near the contact by an amount qV_{bi}, as shown in Fig. 1b.

The difference between the metal work function $q\Phi_m$ and the semiconductor electron affinity $q\chi_s$ is defined as the *Schottky barrier height* $q\Phi_B$:

$$q\Phi_B = q(\Phi_m - \chi_s) \tag{1}$$

The Schottky barrier height is the most important parameter in a metal-semiconductor contact and it determines the electrical behavior of both an ohmic or a Schottky contact. The Schottky barrier can be seen as the energy necessary for electrons in the metal to penetrate into the semiconductor. On the other hand, the build-in potential V_{bi} is the barrier for electrons on the semiconductor side.

It is worth noting that the Schottky barrier height $q\Phi_B$ is almost independent of the semiconductor doping concentration N_D. Actually, there exists a weak dependence of $q\Phi_B$ on N_D through the image force lowering of the barrier $q\Delta\Phi_B$ ($\Delta\Phi_B \propto N_D^{1/4}V^{1/4}$).[22] On the other hand, the barrier width W depends on the doping level, being $W \propto N_D^{-1/2}$.

In general, if $\Phi_m > \Phi_s$, a rectifying contact is formed. For lightly doped semiconductors ($N < 1\times10^{17}$ cm^{-3}) the main conduction mechanism is the *thermoionic emission*, i.e. the carriers having sufficient thermal energy to surmount the Schottky barrier can pass from a material to the other. In this case, the application of a voltage V across a metal-semiconductor junction, according to the thermoionic emission theory,[22] leads to a current density J through the contact given by:

$$J = A^{**}T^2 e^{-\frac{q\Phi_B}{kT}}\left(e^{\frac{qV}{nkT}} - 1\right) \quad (2)$$

where A^{**} is the effective Richardson constant, q is the electron charge, k is the Boltzmann constant, T the absolute temperature and n is the ideality factor.

If the condition $\Phi_m < \Phi_s$ is satisfied, the formation of an ohmic contact can occur by band alignment, as shown in the energy band diagram reported in Fig. 2. Here, the transfer of

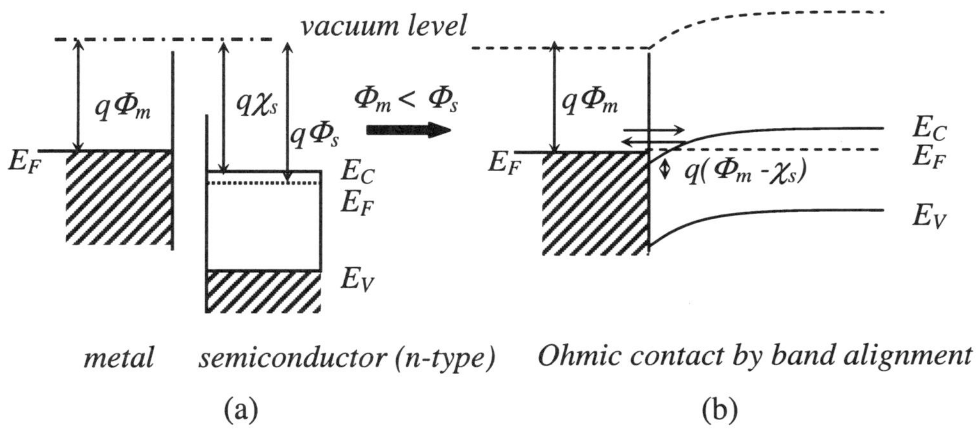

Fig. 2: Energy band diagram for a metal-semiconductor (n-type) contact, in the case $\Phi_m < \Phi_s$ before (a) and after (b) they are brought into contact. In this case ohmic contact formation occurs by band alignment.

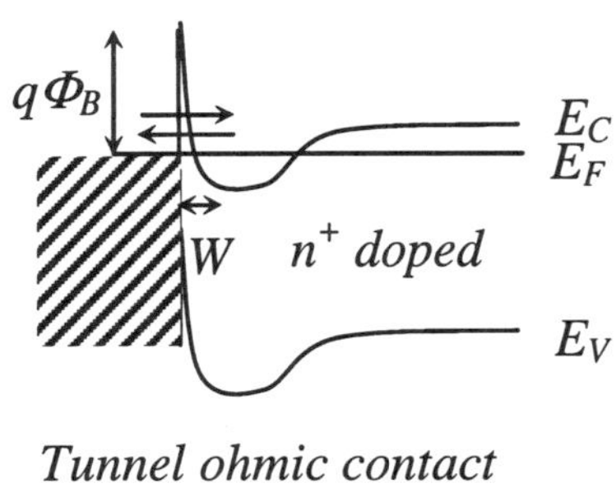

Fig. 3: Energy band diagram for a metal-semiconductor (n-type) contact, in the case of heavy doping. The high doping leads to a reduction of the depletion width W, thus allowing the electrons to tunnel through the thin barrier.

electrons from the metal to the semiconductor lowers the energy levels at the interface. In this case, when a voltage V is applied, the electrons will flow in either direction without overcome any barrier.

However, the work function's value for almost all metals is about 5-6 eV, while the electron affinity for semiconductors is around 4 eV. Therefore, it is difficult to find metals with a low work function, satisfying the condition $\Phi_m < \Phi_s$, particularly for wide band gap semiconductors like SiC or GaN. For that reason, almost all metals after deposition form rectifying contacts on moderately doped SiC.

In order to overcome this problem, the common

method to form ohmic contact in wide band gap semiconductors like SiC is using a heavily doped material, or inducing the formation of a high carrier concentration at the interface, for example by ion-implantation doping and post-annealing techniques. In this way, when a metal is deposited onto a heavily doped ($N>1\times10^{19}cm^{-3}$) semiconductor layer, the Schottky barrier will have the same height as in a low doped material but a lower thickness W, as schematically illustrated in Fig. 3. In this condition, the electrons can tunnel through the thin barrier, with the consequent formation of a tunnel ohmic contact. In this case, the carriers transport is ruled by the *field emission* mechanism.

An intermediate mechanism of the current transport through the barrier is the *thermoionic field emission*, which is generally dominant for intermediate semiconductor doping levels N, in the range 10^{17}-$10^{19}cm^{-3}$. This conduction mechanism involves the carriers that do not have sufficient thermal energy to surmount the barrier as in the case of the classic thermoionic emission but their thermal energy is sufficient to tunnel at an energy higher than the Fermi level, i.e. where the barrier is thinner.

The parameter which determines the dominant conduction mechanism is:

$$E_{00} = \frac{h}{4\pi}\left(\frac{N}{m\varepsilon}\right)^{\frac{1}{2}} \tag{3}$$

where h is the Plank's constant, m the effective mass and ε the dielectric constant.

E_{00} gives the relationship between the temperature T and the semiconductor doping N, i.e. kT/qE_{00} gives the ratio between the thermoionic emission current to the two other conduction mechanisms. For lightly doped semiconductors ($N<1\times10^{17}$ cm^{-3}), $kT/qE_{00}>>1$ and the thermoionic emission is dominant. For heavily doped semiconductors ($N>1\times10^{19}$ cm^{-3}), $kT/qE_{00}<<1$ and the tunneling dominates. For intermediate doping levels ($1\times10^{17}<N<1\times10^{19}$ cm^{-3}), i.e. $kT/qE_{00}\approx1$, the thermonic field emission process is dominant.

3. Specific Contact Resistance

The *contact resistance* R_c, measured in Ω, is the physical parameter which characterizes the resistance of a metal-semiconductor interface. Since the contact resistance is the total resistance of a metal-semiconductor contact, it obviously depends on the area and on the geometry of the contact. However, the most useful quantity used to define the performance of an ohmic contact is the *specific contact resistance* ρ_c. In fact, a metal-semiconductor interface has specific properties, which depend on several factors like the barrier height of the metal, the surface preparation method, the interface roughness, etc.. All these factors are in principle independent of the contact geometry but they strongly influence the value of the contact resistance. Hence, it is convenient to introduce the specific contact resistance ρ_c, which is independent of the contact area (the unit is $\Omega\, cm^2$) and is particularly useful when comparing contacts of different sizes. The meaning of the specific contact resistance and its difference from the total contact resistance becomes clear by making an analogy with the resistance R and the resistivity ρ in a resistor.

In principle, the specific contact resistance should be defined as the product of the total resistance R_c times the contact area A_c ($\rho_c = R_c\, A_c$). However, in a "real" contact, like a syntered metallic contact, it may be not true that the entire contact interface takes part to the conduction.

Hence, the general definition of the specific contact resistance ρ_c expressed as a function of the current density J is given by:[11,23]

$$\rho_c = \left(\frac{\partial J}{\partial V}\right)_{V=0}^{-1} \tag{4}$$

As pointed out in section 1, depending on the semiconductor doping level, a different mechanism will rule the carrier transport mechanism through the barrier. Hence, the specific contact resistance ρ_c will have a different dependence on the barrier height $q\Phi_B$.

As an example, for lightly doped semiconductors ($N<1\times10^{17}$ cm^{-3}), where the thermoionic emission dominates the current transport, an expression for the specific contact resistance can be derived by applying Eq. (4) to the thermoionic current density J (Eq. (2)):

$$\rho_c = \frac{k}{qA^{**}T}\exp\left(\frac{q\Phi_B}{kT}\right) \tag{5}$$

Then, in this condition ρ_c depends on the Schottky barrier $q\Phi_B$ but is independent of the semiconductor doping level N.

On the other hand, for heavily doped semiconductors ($N>1\times10^{19}$ cm^{-3}), i.e. where the tunnelling mechanism is dominant, it can be shown[24,25] that the specific contact resistance ρ_c is proportional to an exponential of Φ_B divided the square root of the doping:

$$\rho_c \propto \exp\left(\frac{\Phi_B}{\sqrt{N}}\right) \tag{6}$$

Hence, the specific resistance of tunnel ohmic contacts is strongly dependent both on the Schottky barrier $q\Phi_B$ and on the semiconductor doping N.

In the intermediate doping concentration range ($1\times10^{17}<N<1\times10^{19}$ cm^{-3}), where the thermoionic field emission transport mechanism is dominant, the specific contact resistance ρ_c has a dependence of the type:[23,24]

$$\rho_c \propto \exp\frac{\Phi_B}{E_{00}\coth\left(\frac{E_{00}}{kT}\right)} \tag{7}$$

In general, because of the inhomogeneous distribution of the current density and of the voltage below a real contact, the specific contact resistance ρ_c cannot be directly measured or simply determined using Eqs. (5-7). Therefore, indirect measurement techniques must be used.

The method of choice for the determination of the specific contact resistance in SiC, as well as in many semiconductors, is the Transmission Line Model (TLM).[26,27,28]

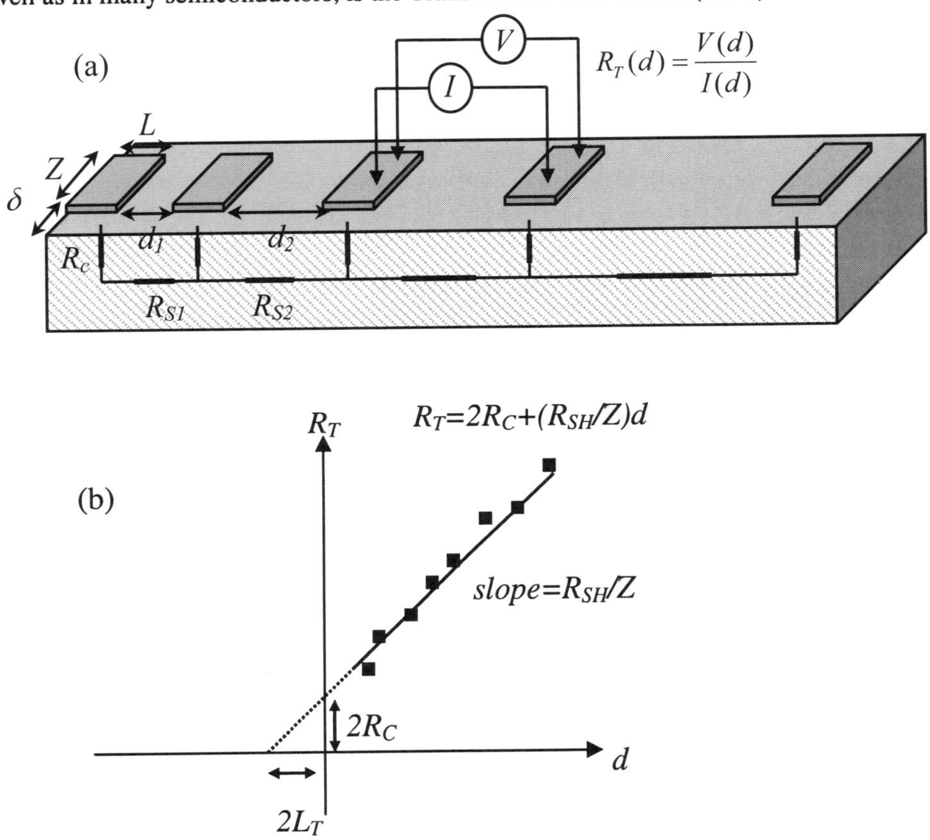

Fig. 4: (a) Scheme of a typical TLM structure; the total resistance between two adjacent pads can be determined by a current voltage (I-V) measurement. (b) Plot of the total resistance R_T as a function of the distance between the pads d, from which the contact resistance R_C, the sheet resistance R_{SH} and the transfer length L_T can be determined.

The test structures for the TLM measurements consist of an array of identical rectangular metallic pads of length L and width Z, formed on an semiconductor mesa of width $Z+2\delta$, at different spacing d, as schematically illustrated in Fig. 4a
Assuming that $\delta << Z$, the total resistance R_T between two adjacent pads placed at a distance d depends on both the contact resistance R_C and on the semiconductor sheet resistance R_{SH} and is given by:

$$R_T = 2R_C + \left(\frac{R_{SH}}{Z}\right)d \tag{8}$$

Therefore, by current-voltage (I-V) measurements between the pads, the total contact resistance R_T at various pad spacing d can be determined. By reporting R_T as a function of d, a linear plot is obtained, as that shown in Fig. 4b, in which the intercept with the y-axis is $2R_c$ while that with the x-axis is $2L_T$. L_T is the so called *transfer length* defined as the distance from the contact edge at which the current density drops to 1/e of its original value.
The effective contact area, i.e. the fraction of the geometric area which takes part to the current conduction, is given by $A_c = ZL_T$. The semiconductor sheet resistance R_{SH} can be extracted by the slope of the linear plot.
Once L_T has been determined, by the R_T vs d plot, the specific contact resistance can be obtained using the general expression:

$$\rho_c = R_c Z L_T tgh\left(\frac{L}{L_T}\right) \tag{9}$$

In some special cases, the specific contact resistance ρ_c can be approximated with a simple form.
For example, for electrically long contacts ($L>>L_T$), the specific contact resistance ρ_c can be determined by the product of the contact resistance times the effective contact area as:

$$\rho_c = R_c Z L_T \tag{10}$$

On the other hand, in the short contact limit ($L<<L_T$), the effective contact area is coincident with the geometric area, and ρ_c is simply given by:

$$\rho_c = R_c Z L \tag{11}$$

In the general case of alloyed contacts, in which the semiconductor sheet resistance immediately below the contact can be different from to its bulk value R_{SH}, the correct value of the transfer length L_T is not simply given by the x-axis intercept of the R_T vs d plot but it must be obtained by an additional measurement of *end-contact resistance*, as described in Refs. [23,28].

The previously described procedure is based on the assumption of identical contact pad and diffusion layer widths. However, in the practical cases the contact width Z is smaller than the mesa width, thus giving rise to lateral current crowding effects that reduce the accuracy of the analysis. In fact, when the diffusion layer width is wider than the contact width Z, at given R_{SH} and ρ_c, lateral current crowding produces a decrease in the measured total resistance R_T, caused by the outer conductive paths, acting like a parallel conductance. Hence, applying the simple one dimensional TLM analysis leads to a lower slope (i.e. a lower extracted R_{SH}) in the plot in Fig. 4b, thus resulting into a higher y-intercept and an overestimation of the value of ρ_c. To take into account these parasitic effects, corrective factors can be evaluated using complicated two-dimensional models.[29]

Practically, lateral current crowding effects can be avoided using the Circular Transmission Line Model (c-TLM), an implementation of the TLM in which circular structures of different radius are used for the determination of the specific contact resistance. The fabrication of this kind of test structures is easier than the linear TLM structure. In fact, in the c-TLM structures only one lithography masking level is required and no dry etching processes is necessary to define the diffusion layer, as instead occurs for the rectangular TLM patterns. However, in the c-TLM an extremely low metal sheet resistance is required to obtain a good accuracy of the measurement. A detailed description of this technique can be found in Refs. [30,31].

Another method for measuring the specific contact resistance is the four terminal *Kelvin method*.[32] In a diffused or implanted test structure, a current I is forced between two pads (1 and 2) while the voltage is measured between two other pads (3 and 4) (for a schematic of a typical Kelvin test structure see Ref. [23]). If a high impedance voltmeter is used, a very small current flows between the pads 3 and 4. Then, the potential at the pad 4 is the same as that in the semiconductor contact under the contact 3 and the specific contact resistance ρ_c can be determined by the voltage drop between the pads 3 and 4 (V_{34}) and the contact area A_c simply using the relation $\rho_c = V_{34} I^{-1} A_c$.

This technique is suitable for very low values of specific contact resistance (~10^{-7} Ωcm^2) and has found a wide diffusion in the case of ohmic contacts for Si. However, since the formation of a diffused or implanted layer may be a difficult process for SiC, and the typical values of ρ_c are higher than those obtained for contact on Si, the TLM is preferred to the Kelvin method for characterizing ohmic contacts to SiC.

4. Ohmic Contacts to n-type SiC

Although in their review papers Porter et al.[9] and Saxena et al.[12] reported on very early literature works on ohmic contacts to n-type 6H-SiC fabricated using chromium or tungsten, ohmic contacts to n-type SiC have been extensively investigated only in the last decade. In fact, during this time, the improvement of material quality, as long as that of doping techniques during epitaxial growth, resulted in the availability of good quality highly doped epitaxial layers. Hence, it is nowadays not uncommon to find works reporting on the fabrication of ohmic contacts with specific contact resistance in the range of ~10^{-6} Ωcm^2 for n-type heavily doped material (>10^{19}cm^{-3}).

Table 1 lists several ohmic contacts, reporting the values of the specific contact resistance ρ_c, for n-type 6H- and 4H-SiC material. The contacts have been deposited on SiC material with a wide range of doping concentration (obtained either by epitaxial doping or by ion-implantation techniques) and were subjected to post-deposition annealing processes under a large variety of conditions (temperature, time, annealing atmosphere). In most of the cases, the values of ρ_c were determined by means of the TLM technique.

Table 1: Ohmic contacts to n-type SiC. The deposition sequence of multilayer alloyed contacts is indicated by a slash (" / "), where the metal on the left side was deposited into direct contact with the SiC substrate.

Metal	*Annealing*	*N_D (at/cm^3)*	*ρ_c (Ω cm^2)*	*Polytype*	*Ref.*
Al	none	>10^{20} P-impl	0.54-1.2×10^{-6} (TLM)	4H	[21, 33]
Al/Ni	1000°C 2 min in Ar	2×10^{20} P-impl.	4.8×10^{-5} (TLM)	4H	[20]
Co/Si/Co	500°C 5 min + 800°C 2 min in vacuum	1.1×10^{19} epi	1.8×10^{-6} (TLM)	4H	[34]
Mo	none	>10^{20} P-impl	2×10^{-6} (TLM)	4H	[21]
Nb	1100°C 10 min	1.3×10^{19} epi	<1×10^{-6} (TLM)	4H	[35]
Nb	1100°C 10min	1.4×10^{18} epi	3×10^{-6} (TLM)	6H	[35]
Ni	none	>10^{20} P-impl	3×10^{-6} (TLM)	4H	[21]
Ni	1000°C 30 s in forming gas	1-3×10^{19} epi	<1.7×10^{-5} (Cox and Strack)[36]	3C/6H	[37]
Ni	1000°C 1 min	4.2×10^{15} epi	2.8×10^{-3} (c-TLM)	4H	[38]
Ni	1000°C 2 min in Ar	>10^{20} P-impl	1.2×10^{-6} (TLM)	4H	[33]
Ni	950°C 10 min in Ar	1×10^{19} epi	2.8×10^{-6} (TLM)	4H	[17]
Ni	1000°C 2 min in Ar	2×10^{20} P-impl.	6×10^{-6} (TLM)	4H	[39]
Ni	950°C 2 min in vacuum	7-9×10^{18} epi	<5×10^{-6} (TLM)	6H	[14]
Ni	none	5×10^{19} epi	1×10^{-4} (Cox and Strack[36])	6H (000-1)C face	[40]
Ni	1000°C 5 min in Ar	4.5×10^{20} epi	1×10^{-6}	6H (000-1)C face	[40]
Ni	950°C 10 min in Ar	1.8×10^{18} epi	~1×10^{-5} (TLM)	6H	[17]
Ni	950°C 60 s in N_2	7.4×10^{18} epi	3.9×10^{-5} (TLM)	6H	[18]
Ni/Si	950°C 10 min in Ar	1×10^{19} epi	2.7×10^{-5} (TLM)	4H	[17]
Ni/Si	950°C 10 min in Ar	1.8×10^{18} epi	1-3×10^{-4} (TLM)	6H	[17]
NiCr	1100°C 3 min in vacuum	1.3×10^{19} epi	1.2×10^{-5} (TLM)	4H	[41]
NiCr	1100°C 3 min in vacuum	3.2×10^{17} epi	2.5×10^{-6} (TLM)	6H	[41]

Si/Ni	900°C 10 min Ar-H_2	2×10^{19} epi	1.9×10^{-6} (TLM)	4H	[42]
Si/Ni	300°C 9h in N_2	1.5×10^{19} epi	6.9×10^{-4} (TLM)	6H	[43]
TaC/Au	1000°C 15 min in vacuum	2.3×10^{19} epi	1.4×10^{-6} (TLM)	6H	[44]
TaC/W/WC	1000°C 15min in vacuum	7.8-8.1×10^{18} epi	3×10^{-5} (TLM)	6H	[45]
Ti	none	>10^{20} P-impl	2.7×10^{-7} (TLM)	4H	[21]
Ti	900°C in vacuum	7.4×10^{18} epi	1×10^{-4} (TLM)	6H	[46]
Ti	1000°C in vacuum	7.4×10^{18} epi	6.7×10^{-5} (TLM)	6H	[46]
Ti/Al	1000°C 5 min in Ar	4.5×10^{20} epi	<1×10^{-3}	6H (000-1)C face	[40]
Ti/$TaSi_2$/Pt	600°C 30 min in N_2	2×10^{19} epi	4.7×10^{-4} (TLM)	4H	[19]
Ti/$TaSi_2$/Pt	600°C 30 min in N_2	7×10^{18} epi	1.68×10^{-4} (TLM)	6H	[19]
TiC	none	1.3×10^{19} epi	9.28×10^{-6} (TLM)	4H	[47]
TiC	950°C 2 min in 10% H_2/Ar	1.3×10^{19} epi	4.01×10^{-5} (TLM)	4H	[47]
TiC	1300°C 15 min in H_2	4×10^{19} epi	1.3×10^{-5} (TLM)	6H	[48]
$TiSi_x$	1150°C 2 min in Ar	5×10^{18} N-impl.	7×10^{-6} (TLM)	6H	[49,50]

It is interesting to notice that much work on ohmic contacts started on the polytype 6H, which was the less expensive material at the beginning of the 90s. However, the present works are mainly focused on 4H-SiC, which is the material of choice for power devices because of its higher mobility.

As pointed out in section 2, the Schottky barrier height $q\Phi_B$ is the most important parameter for the current transport both for ohmic and rectifying contacts. Therefore the knowledge of the its values permits the appropriate selection of the metals for fabricating ohmic contacts. According to Eqs. (5-7), metals with low Schottky barriers are expected to give low values of ρ_c.

Waldrop et al.[51,52] first studied the Schottky contacts to n-type 6H-SiC for several metals, finding low values for unannealed Ti (0.73 eV) and Al (0.26 eV). As a matter of fact, for Ti [21] and Al [21,33] very low values of ρ_c in the range of 10^{-7} Ωcm^2 were recently reported on heavily doped 6H- and 4H-SiC. In particular, Al and Mo contacts on heavily doped ion-implanted n-type 4H-SiC ($N_D>10^{20}cm^{-3}$) exhibited attractive values of ρ_c (0.5-2×10^{-6} Ωcm^2) without post-deposition annealing treatments[21] and, for that reason, they were proposed for the fabrication of the source/drain and gate contacts in MESFETs on 4H-SiC. However, the Al and Mo Schottky contacts gates were significantly side attacked during the wet etch processing of the interconnects and therefore Ni and Ti were preferred to these metals for the device fabrication.

A low specific contact resistance in the same range was also found for Nb contacts on 4H-SiC (average value of 6.9×10^{-7} Ωcm^2).[35] These contacts were rectifying after deposition (with a barrier height of 1.3-1.4 eV) and become ohmic by annealing at 1100°C. Although no extensive intermixing at the Nb/SiC interface could be observed by means of RBS analysis,[35] the formation of Nb_5Si_3C at the interface was suggested as responsible for the ohmic behavior. A common problem remains the oxygen incorporation in the Nb film during sputter deposition whose content, however, decreases after the annealing processes for ohmic contact formation.

4.1. *Ti- and Ta-based ohmic contacts*

Titanium-based contacts cover a wide portion of the literature data dealing with ohmic contacts to n-type SiC. Extensive works concerning the microstructural characterization of reacting Ti/SiC interfaces have been reported by Goesmann et al.,[53] Gorsse et al.,[54] and Porter et al..[55]

Ti is known to react with SiC forming different phases such as silicides, carbides and ternary phases.[53,54] In particular, at elevated temperatures, a reaction dominated by the formation of a two-layer structure (Ti_5Si_3 + TiC_{1-y}) takes place.[55] The detailed sequence of phase formation during reaction of a Ti/SiC junction depends on the annealing temperature, while the electrical properties depend on the phase which is into direct contact with the SiC substrate.

Ti already forms an ohmic contact to n-type SiC after deposition, but with a high resisitivity.[53] As an example, it has been reported that Ti/SiC contacts may become ohmic without post-deposition annealing using appropriate surface preparation methods, which result in the formation of band gap defects states near the interface.[56] On the other hand, annealing processes below 900°C lead to Schottky contacts, while above this temperature ohmic contacts are formed.[53] Plausibly, the coexistence of the silicides and carbides phases results into rectifying contacts, whilst the formation of a ternary phase (Ti_3SiC_2) lead to ohmic behavior of the contacts at higher temperatures.

La Via et al.[46,57] studied the structural and electrical properties of Ti contacts on n-type 6H-SiC samples with a doping concentration N_D=7.4×10^{18}cm^{-3}, observing that the formation of different phases after annealing in vacuum in the range 900-1000°C significantly changes the electrical properties of the contact. In fact, while a double layer of TiC and Ti_5Si_3 was formed at 900 and 950°C, the ternary phase Ti_3SiC_2 was observed at 1000°C. In particular, Fig. 5 reports the XRD spectra of Ti/6H-SiC samples after annealing at 950°C and 1000°C. At 950°C the ternary phase Ti_3SiC_2 starts to appear and coexists with TiC and Ti_5Si_3 (Fig. 5a). After annealing at 1000°C, the ternary phase is dominant, while a small amount of silicide is still detectable (Fig. 5b). At these high temperature processes the specific contact resistance, determined by TLM, decreased from 1×10^{-4} Ωcm^2 at 950°C to 6.7×10^{-5} Ωcm^2 at 1000°C.[46]

Because of the high thermal budget required to form ohmic contacts and the ease of oxidation of Ti, several studies have been carried out to investigate more stable Ti-based compounds, such silicides ($TiSi_x$) and carbides (TiC) as ohmic contacts to n-type SiC.

Titanium silicide ($TiSi_x$) contacts were reported by Schmid, Getto et al.[49,50] on 6H-SiC material, doped by nitrogen implantation followed by a thermal process at 1770°C (active donor concentration N_D=5×10^{18}cm^{-3}). The contacts were obtained by direct sputtering of

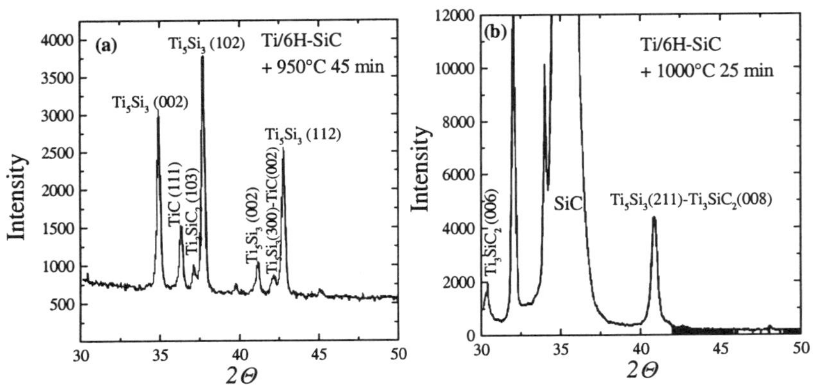

Fig. 5: XRD spectra of Ti/6H-SiC contacts after annealing in vacuum (a) at 950°C for 45 min and (b) at 1000°C for 25 min, showing the formation of the ternary phase Ti_3SiC_2 at high temperatures.

a $TiSi_2$ compound target. Also in this case, the formation of Ti_3SiC_2 at the interface, after annealing of the contacts at 1150°C, was indicated as the reason for the reduction of the specific resistance up to a value of $\rho_c=7\times10^{-6}\ \Omega cm^2$. Indeed, during contact formation at temperatures higher than 1180°C, structural characterization performed by secondary neutral mass spectroscopy confirmed the growth of the ternary phase Ti_3SiC_2 between a $TiSi_2$ overlayer and the SiC substrate.[50] The values of ρ_c for the implanted material were comparable to thoese obtained by the same authors on epitaxial SiC layers.

Titanium carbide (TiC) contacts have shown promising advantages for ohmic contacts on n-type SiC. In fact, because of the small lattice spacing mismatch (<0.9% between the basal plane lattice parameter of 6H-SiC and the [111] lattice spacing of TiC) TiC contacts can be epitaxially grown by CVD or co-evaporation methods.[47,48] Moreover, the deposition temperatures of about 500°C act simultaneously as a pre-annealing process for the contacts, thus lowering the thermal budget needed to reduce the value of ρ_c. On 6H-SiC, TiC contacts annealed at 1300°C exhibited values of $1.3\times10^{-5}\ \Omega cm^2$, along with a good chemical and thermal stability up to 1400°C. Values of $\rho_c<1\times10^{-5}\ \Omega cm^2$ were obtained on 4H-SiC ($N_D=1.3\times10^{19}cm^{-3}$) using unannelaed TiC contacts.[48]

Alloyed Ti-Al ohmic contacts were successfully demonstrated by Uemoto [40] onto the $(000\bar{1})$C face of n-type heavily doped 6H-SiC ($4.5\times10^{20}cm^{-3}$) layers, grown by liquid phase epitaxy. These contacts displayed ohmic characteristics after heat treatments in Ar at 1000°C, thus resulting into values of $\rho_c<1\times10^{-3}\ \Omega cm^2$. However, as it will be seen in the section 5, Ti-Al alloys are more suitable for p-type material, where lower values of ρ_c are obtained.

Tantalum carbide (TaC) was also investigated as an ohmic contact to n-type SiC. In fact, because of its thermodynamic compatibility with SiC (the ternary phase diagram Ta-Si-C predicts TaC to be in equilibrium with SiC at 1000°C [58]), TaC can form a stable interface with SiC. Additionally, the high melting temperature improves the adhesion to SiC when contact deposition occurs at temperatures above 100°C. All these properties, along with the small work function of TaC (< 4 eV), make it a desirable candidate for low specific resistance ohmic contacts to SiC. A detailed structural and electrical characterization of TaC contacts with and without Au, Pt and W/WC overlayers on n-type 6H-SiC was performed by Jang et al..[44,45] These contacts were non-ohmic after deposition and became ohmic after annealing in vacuum at temperatures between 800 and 1075°C. For a doping concentration of 1.3×10^{19}cm^{-3}, the best results were obtained after annealing in vacuum for 15 min at 1000°C, with values of ρ_c measured at room temperature for TaC contacts and for TaC with Pt and Au overlayers of 2.1×10^{-5} Ωcm^2, 7.4×10^{-6} Ωcm^2 and 1.4×10^{-6} Ωcm^2, respectively.[44]

4.2. *Ni-based Ohmic Contacts*

As can be seen from Table 1, annealed Ni has been the most widely used metal for ohmic contacts to n-type SiC. The Schottky barrier height for Ni/SiC contacts strongly depends on the surface preparation prior to metal deposition.[59] Values of $q\Phi_B$ in the range of 1.25-1.29 eV on 6H-SiC [59,60] and of 1.40-1.59 eV on 4H-SiC [61,62,63] are reported for unannealed Ni. These high values of $q\Phi_B$ lead to the formation of Ni/SiC rectifying contacts after deposition. Hence, post-deposition annealing processes are required to form ohmic contacts, generally at temperatures in the range 900-1000°C.

The I-V characteristics of Ni contacts on n-type 6H-SiC, annealed at different temperatures, measured between two adjacent pads of an array of TLM structures, are reported in Fig. 6. The contacts were syntered by rapid thermal annealing in N_2 between 600 and 950°C for 60 s. It can be observed that after annealing at 600 °C the I-V characteristics show a rectifying behavior, while only after annealing at 950°C an ohmic contact is obtained. The contact resistance after annealing at 950°C, determined by the TLM method, was 3.9×10^{-5} $\Omega\cdot$cm^2 for a doping level of $N_D=7.4\times10^{18}$ cm^{-3}.[18]

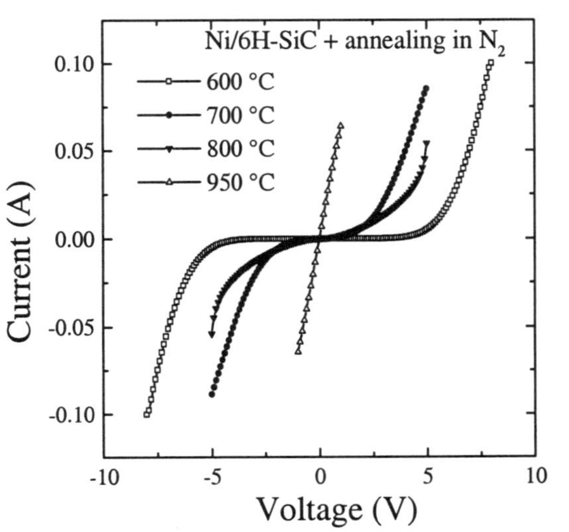

Fig. 6: Current-Voltage (I-V) characteristics of Ni/6H-SiC contacts, annealed at different temperatures between 600 and 950°C. Ohmic behaviour is observed after rapid thermal annealing at 950°C in N_2.

As in the case of Ni contacts on silicon, where silicides form upon annealing already at low temperatures (250-

400°C),[64] the interaction between the metal and SiC determines the electrical properties of the contact. Therefore, the understanding of the structural changes of the Ni/SiC system upon annealing is central to the development of reliable contacts.

The first papers reporting on the Ni/SiC system focused on the structural characterization of Ni/SiC interfaces. Pai et al.[65] reported the first structural investigation of annealed Ni films on SiC, carried out by combining RBS and XRD measurements. After annealing up to 400°C no detectable reaction was observed, while complete reaction of the metal film occurred at 500°C, with the formation of silicides phases.

Accordingly, we have recently shown by chemical analysis that nickel becomes mobile above 400°C and then can react with SiC by forming silicides.[66] However, ohmic behavior is observed only at higher temperatures. Therefore, a structural characterization of the reacted layer was carried out by acquiring X-ray diffraction spectra of the contacts annealed at 600°C (Schottky contact) and 950°C (ohmic contact). The XRD spectra, reported in Fig. 7, show the presence of the two silicide phases ($Ni_{31}Si_{12}$ and Ni_2Si) at 600°C. On the other hand, after annealing at 950°C Ni_2Si is the only phase which could be detected.

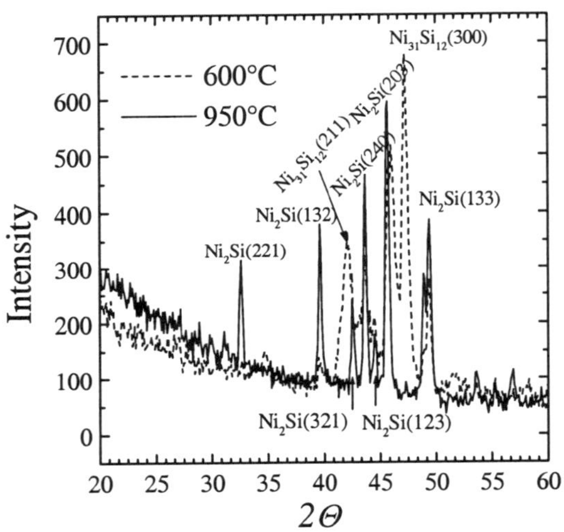

Fig. 7: XRD spectra of Ni/6H-SiC contacts after annealing at 600°C and at 950°C. At 600°C the coexhistence of two silicides phases ($Ni_{31}Si_{12}$ and Ni_2Si) occurs, at 950°C only the Ni_2Si phase is present.

Ohdomari et al.[67] investigated the Ni/SiC interfacial reaction, with particular attention to the role of carbon during silicide formation. Auger spectroscopy indicated that elementary carbon, i.e. graphitic carbon, was present in the film and no carbides were formed. Moreover, from XRD analysis, the formation of polycrystalline Ni_2Si after annealing at 600°C was confirmed, even though the nickel rich phase ($Ni_{31}Si_{12}$) was not detected.

Ni/SiC interfaces were also studied by Höchst et al.,[68] who observed the initial formation of NiSi after deposition at room temperature of a thin Ni overlayer on SiC(100). The system maintained the stability up to 600°C, and the formation of Ni_2Si was observed above this temperature. By thermodynamic considerations, i.e. from an estimation of the free energy changes at the reaction temperature, the authors concluded that Ni-carbide (Ni_3C) is unlikely to form because it requires an input of heat, since the reaction enthalpy is positive (25 kcal/mol). Indeed, in the interaction Ni/SiC, the silicide phase $Ni_{31}Si_{12}$ is the first to be formed because the reaction is characterized by the largest negative enthalpy change,[69] while thereafter the formation of Ni_2Si becomes thermodynamically favourable.

The reaction of thin Ni films was observed at lower temperatures, already at around 300°C, by Slijkerman et al.,[70] who also observed that all the carbon released during reaction accumulated on the sample surface as a continuos graphite layer. Other works on the structural characterization of the interfacial reaction of Ni with amorphous SiC also reported the first signs of reaction at around 300°C. [71,72]

As can be deduced from the aforementioned literature data, there is no good agreement on the onset temperature for reaction between Ni and SiC, on the sequence of phase formation and on the redistribution of carbon in the reacted layer.

The first quantitative electrical data on the specific contact resistance of nickel silicide ohmic contacts to n-type SiC were reported by Crofton et al.[14] who measured values of $\rho_c<5\times10^{-6}$ Ωcm^2, in Ni_2Si contacts formed by rapid annealing in vacuum at 950°C. However, in this work the only structural characterization was RBS analysis, thus being unsuitable to understand the role played by carbon during silicide formation.

Uemoto[40] reported a lower value of ρ_c ($\sim1\times10^{-6}$ Ωcm^2) after annealing in Ar atmosphere for 5 min at 1000°C Ni films deposited onto heavily doped 6H-SiC substrates (4.5×10^{20} cm^{-3}), grown by LPE method. The doping concentration values obtained by LPE are about 10 times higher than those normally achieved by CVD techniques. However, no structural characterization was reported. In this work the specific contact resistance was determined by measuring the resistance at various contact areas on the top surface with a large contact on the back of the substrate (Cox and Strack method). [36] However, this technique is less accurate than the TLM, especially for measuring small values of ρ_c.

Values of the specific contact resistance as a function of the carrier concentration for nickel silicides ohmic contacts to n-type SiC, obtained by different authors by annealing Ni films above 900°C, are reported in Fig. 8. As can be seen, a significant decrease of ρ_c occurs for concentrations close to $1\times10^{19}cm^{-3}$. The solid line is the fit reported by Crofton et al.[11] obtained using the field emission transport model and assuming a barrier height of 0.35 eV for the nickel silicide. This value, however, is significantly lower than the typical experimental values of the Schottky barrier of Ni_2Si on 6H-SiC

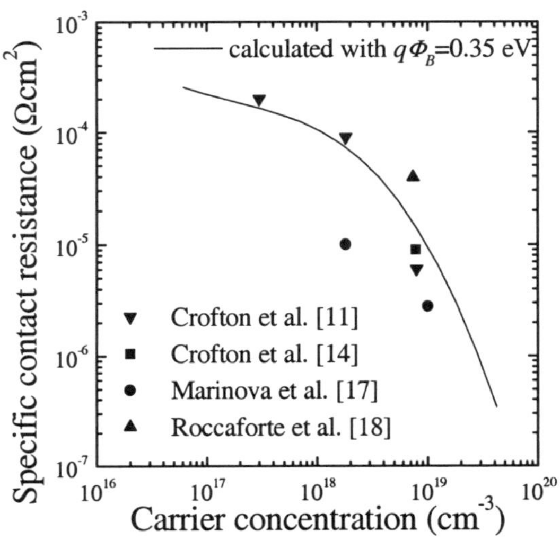

Fig. 8: Specific contact resistance as a function of the carrier concentration for nickel silicides contacts to n-type 6H-SiC. The solid line is the fit obtained by Crofton et al.[11] assuming a barrier height of 0.35 eV.

reported in the literature (1.3-1.4 eV).

One of the most controversial arguments concerning the nickel silicide formation is the redistribution of the carbon which is released after consumption of SiC. This process, in fact, may play a crucial role in ohmic contact formation. We have combined TEM and Energy Filtering TEM (EFTEM) analysis in order get more insights into the carbon redistribution in the silicide layer. The EFTEM technique[73] allows to map the elemental chemical composition of the sample, with the spatial resolution of the TEM analysis, by subjecting the transmitted electron beam to an energy filtering process after the interaction with the specimens present in the matrix.

Fig. 9 shows the cross section TEM pictures (left) and the corresponding carbon maps determined by EFTEM analysis (right), for the nickel silicides/SiC samples formed at 600°C and 950°C. In the EFTEM maps, a bright contrast indicates the presence of the element.

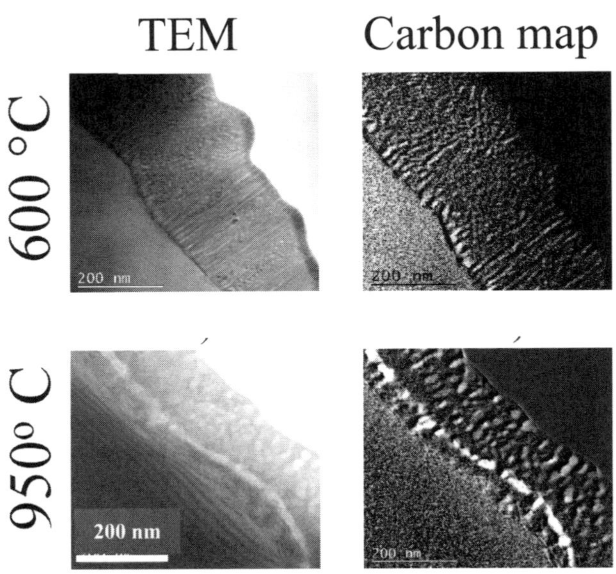

Fig. 9: Cross section TEM (left) and carbon maps determined by EFTEM analysis (right) for nickel silicides/SiC contacts formed at 600°C and at 950°C.

In the sample annealed at 600°C the silicide/SiC interface is quite flat, while carbon form nanometer-size clusters, almost uniformly distributed in the silicide. On the other hand, after annealing at 950°C, i.e. where the contact became ohmic, while the silicide structure remains nearly unchanged, a strong change in the carbon distribution occurs. In fact, as can be deduced by EFTEM, the carbon clusters increased in size and agglomerated to a depth of about 20 nm from the silicide/SiC interface. From X-ray diffraction analysis, the average size of the clusters was estimated using the Debye–Scherrer equation,[74] finding an increase of the cluster diameter from 13 to 30 nm during the Schottky (600°C) to ohmic (950°C) transition. Instead, no significant variation of the carbon clusters size (~5 nm) formed at the surface region of the silicide was detected by Raman spectroscopy.[59]

The mechanism of ohmic contact formation in the nickel silicide/SiC system represents one of the most interesting topics concerning the physics of ohmic contacts to n-type SiC and was widely discussed.

A possible reason of the formation of ohmic contacts could be a reduction of the silicide's Schottky barrier height during annealing. Indeed, Han and Lee[75] found that the effective Schottky barrier height of nickel silicides contacts on lightly doped 4H-SiC increased from 1.55 to 1.81 eV after annealing at 600°C, but again decreased to 1.25 eV after annealing at 800°C. They pointed out that the atomic composition of Si in nickel silicides at the interface increased with increasing annealing temperature and then that the silicidation process occurs through the outdiffusion of Si, with an increase of the Schottky barrier height. However, the temperature of 900°C for ohmic contact formation is higher than the onset temperature for silicide formation (about 600°C). Hence, an excess of C atoms outdiffuse towards the surface of contact layer even after silicidation is complete, thus leaving C vacancies (V_C) below the contact. These V_C act as donors for electrons, determining an increase in net electron concentration and leading to the formation of the ohmic contact through the reduction of depletion layer width.[75]

Conversely, Calcagno et al.[76] reported the distributions of the Schottky barrier heights for a set of several Ni/6H-SiC diodes, after deposition and after annealing between 600°C and 950°C. From these data an increase of $q\Phi_B$ from 1.25 eV to 1.40 eV was observed, that cannot explain the Schottky to ohmic transition in the nickel silicide/SiC system. Accordingly, Roccaforte et al.[77] also reported a value of barrier height of 1.42 eV for Ni_2Si/6H-SiC Schottky diodes formed under similar annealing conditions at 950°C. Moreover, from capacitance voltage (C-V) measurements no change in the uncompensated donor concentration of the substrate was observed.[76]

Furthermore, the presence of carbon vacancies in the SiC substrate suggested by Han and Lee,[75] which should introduce a level at 0.5 eV below the conduction band edge, was not experimentally revealed by deep level transient spectroscopy.[76]

More recently, La Via et al.[78] observed that a deviation from the ideal behavior in the forward I-V characteristics of Ni_2Si/SiC contacts occurs in the diodes annealed at 950°C. This behavior, which is also accompanied by a strong increase of the leakage current under reverse bias, is characteristic of the presence of interface state defects, i.e. of the formation of an inhomogenous Schottky barrier.[79] The formation an inhomogeneous Schottky barrier, characterized by the presence of low barrier regions embedded in a uniform Ni_2Si background, may explain the ohmic behavior. In fact, in the presence of inhomogeneities through the nickel silicide/SiC interface, the current will preferentially flow throughout the low barrier regions,[80] thus determining the ohmic properties of the contact. The presence of inhomogeneites of the barrier can be attributed to the complicated silicidation process and/or to the presence of fabrication-induced defects or residual oxide in the contact area.

4.3. *Interface Morphology of Silicides Contacts*

Another interesting issue related to the Ni_2Si/SiC ohmic contacts is the possibility to improve the silicide morphology and/or reduce the annealing temperature for ohmic contact formation, which are both fundamental requirements for optimizing the device fabrication and performances.

Although the values of the resistivity in Ni_2Si/SiC contacts seem to be not strongly dependent on the interface morphology, the dissociation of SiC during reaction may result into a poor interface quality, with the accumulation of carbon and the presence of a large amount of voids,[17] all being potential sources of contact degradation, particularly under high temperature operation.

An improvement of the Ni_2Si/SiC interface morphology was achieved by Roccaforte et al.,[81] by means of ion irradiation of the near interface region after deposition of the Ni films. In this way, a better adhesion of the Ni film and an enhancement of the Ni atoms mobility in the amorphized material was obtained, which resulted in a better morphology of the silicide/SiC interface after annealing at 800°C.

To avoid the common structural problems related to reacted Ni_2Si contacts (graphite precipitates and voids in the interfacial region) novel $NiSi_2$ ohmic contacts were obtained by Nakamura et al. [42,82] by annealing Ni/Si/SiC multilayer structures. These contacts had an abrupt interface and a value of ρ_c as low as 1.9×10^{-6} Ωcm^2 for a substrate doping of $N_D=2.5\times10^{19}$ cm^{-3}.[42,82]

The high temperatures (~950°C) required to obtain nickel silicide ohmic contacts often represent an inordinate thermal budget during devices fabrication. A significant simplification in the device fabrication can be obtained by using thermodynamically stable Ni silicides as ohmic contacts to n-type SiC formed with a modest thermal budget. This latter was achieved by Deeb and Heuer,[43] who obtained nickel monosilicide (NiSi) ohmic contacts by annealing at 300°C an amorphous Si film with a Ni overlayer. For a doping concentration of 1.5×10^{19} cm^{-3} these contact showed a $\rho_c=6.9\times10^{-4}$ Ωcm^2, which decreased in the ~10^{-5} Ωcm^2 range upon annealing in air at 600°C, probably due to the reconstruction of the NiSi/SiC interface, and remained unchanged even after 300 h annealing.

Cho et al.[34] used Co/Si/Co multilayer structures to obtain low resistance ohmic contacts to 4H-SiC. During film deposition, a ratio Si:Co=2:1 was maintained, in order to achieve the stoichiometry of the silicon rich cobalt silicide phase ($CoSi_2$). In this way, the deposition of the Si layer was expected to prevent the precipitation of carbon and limit the formation of voids at the metal/SiC interface, which can strongly degrade the contact performances in binary silicides layers. Furthermore, the reduction of carbon accumulation can improve the bonding capability. The contacts did not show ohmic behavior after deposition but became ohmic after annealing treatments above 800°C, performed in vacuum or in Ar with H_2(10%), which resulted in the formation of $CoSi_2$. A single-step annealing at 800°C, however, was cause of oxidation problems. Therefore, the authors proposed a two step annealing, to improve the electrical characteristics of the contact. The first step annealing (500°C for 3 min) was used as low temperature process for intermixing of Si and Co, whereas the second one (800°C for 2 min), at higher temperature, led to the final silicide phase formation. Using this two-step annealing procedure for the formation of $CoSi_2$ significantly improved the electrical properties of

the contacts reducing the value of ρ_c by more than one order of magnitude (from 4.1×10^{-5} Ωcm^2 after the single-step at 800°C to 1.8×10^{-6} Ωcm^2 after the two-step process).
In spite of the variety of alternatives proposed to avoid the structural problems related to the silicide formation, rapid thermal annealing above 900°C of Ni films in N_2 or Ar atmosphere remains the standard procedure used for the formation of ohmic contacts to n-type SiC for device fabrication.

4.4. *Problems Related to Bonding*

For applications to practical devices, good bonding connections between the ohmic contact and the die package are required.
Nickel silicide (Ni_2Si) contacts are generally formed by annealing Ni films at 950°C. As a consequence of the silicidation, free carbon remains inside the silicide layer and may move towards the surface,[70] thus leading to practical problems to device bonding applications such as the poor adhesion of cap layers (Au, Pt) to silicide, Moreover, an intermediate diffusion barrier layer can be required in order to prevent the interdiffusion between the cap layer and the contact.
The use of Nichrome, NiCr (Ni 80 % wt.), was proposed by Luckowski et al.[41] to limit the accumulation of carbon on the surface of the sample and favour the device bonding. Annealing processes at 1100°C in vacuum resulted in the formation of ohmic contacts with performance comparable to those of the contacts formed using pure Ni. During annealing, an increase of the atomic concentration of Cr relative to Ni occurred at the surface, thus limiting the interdiffusion between the ohmic contact layer and the Au cap layer. Thin (50 nm) reacted NiCr layers gave success rate of bonding to Au cap layer of almost 100 %, contrary to thick layers (200 nm) which showed a poor adhesion due to an excess of carbon at the surface.
Moreover, while in the case of pure Ni interdiffusion between the Au and the reacted layer occurs, the Au and NiCr contact layers show essentially no mixing after long term annealing at 300°C.

5. Ohmic Contacts to p-type SiC

Due to the wide band gap of SiC, the formation of ohmic contacts to p-type SiC is a crucial issue because of the difficulty to obtain low Schottky barrier heights values. In fact, the large forbidden energy gap of SiC (2.8-3.2 eV) together with the electron affinity of the material (around 4 eV) leads to a position of the valence band more than 6 eV away from the vacuum level. Since the most metals have workfunction in the range 4-5.5

eV, a significantly high energy difference arises between the conducting carriers of the metal and p-type SiC.

Less works are reported on ohmic contacts to the p-type with respect to the n-type material. Table 2 reports a list of ohmic contacts to p-type 6H- and 4H-SiC, using different metals under several annealing conditions.

Table 2: Ohmic contacts to p-type SiC. The deposition sequence of multilayer alloyed contacts is indicated by a slash (" / "), where the metal on the left side was deposited into direct contact with the SiC substrate.

Metal	*Annealing*	N_A *(at/cm³)*	ρ_c *(Ω cm²)*	*Polytype*	*Ref.*
Al	1000°C 2 min in vacuum	4.8×10^{18} epi	4.2×10^{-4} (TLM)	4H	[16,83]
Al/Mo	1200°C 40 sec in N_2-H_2	1×10^{18} epi	4.5×10^{-5} (TLM)	6H	[95]
Al/Ni	1000°C 2 min in Ar	7×10^{20} Al-impl	5.2×10^{-4} (TLM)	4H	[20]
Al/Ta	400°C 3 min + 1000°C 20 sec in Ar	1×10^{18} epi	$1\text{-}4\times10^{-4}$ (TLM)	6H	[95]
Al/Ti	950°C 2 min in N_2	$1\text{-}3\times10^{19}$ epi	$2\text{-}3\times10^{-5}$ (TLM)	3C/6H	[37]
Al/Ti	900°C 3 min in vacuum	1×10^{19} epi	6.4×10^{-4} (TLM)	4H	[94]
Al/Ti	900°C 4 min in N_2	1.6×10^{19} epi	4×10^{-4} (TLM)	6H	[84]
Al/Ti/Al	1000°C 2 min in vacuum	4.8×10^{18} epi	3.3×10^{-4} (TLM)	4H	[16,83]
Al/Ti/Pt/Ni	1000°C 2 min in vacuum	$6\text{-}8\times10^{18}$ epi	9×10^{-5} (TLM)	4H	[85]
$AlSi_{(2\%)}Ti_{(0.15\%)}$	950°C 5 min in Ar	$3\text{-}5\times10^{19}$ epi	9.6×10^{-5} (TLM)	4H	[86]
Al-Ti	1000°C 2 min	1.3×10^{19} epi	$5\times10^{-6}\text{-}3\times10^{-5}$ (TLM)	4H 6H	[15, 87]
Al-Ti	1000°C 5min in Ar	2×10^{19} epi	1.5×10^{-5} (c-TLM)	6H	[88]
Al-Ti	1000°C 2 min in Ar	4×10^{19} Al-impl.	3×10^{-5} (TLM)	6H	[89]
Co/Si	500°C 5 h + 900°C 2h in vacuum	2×10^{19} epi	$< 4\times10^{-6}$ (TLM)	6H	[90]
CoAl	900°C 5 min in vacuum	9×10^{18} epi	4×10^{-4} (TLM)	4H	[91]
Ge/Ti/Al	600°C in vacuum	4.5×10^{18} epi	1×10^{-4} (c-TLM)	4H	[92]
Mo	1000°C 20 sec in N_2-H_2	1×10^{18} epi	4.08×10^{-3} (TLM)	6H	[95]
Ni	1000°C 2 min in Ar	2×10^{20} Al-impl.	7×10^{-3} (TLM)	4H	[39]
Ni/Al	1000°C 5-30 min in vacuum	$3\text{-}9\times10^{18}$ epi	9.5×10^{-5} (c-TLM)	4H	[8]
Ni/Ti/Al	800°C 5-30 min in vacuum	$3\text{-}9\times10^{18}$ epi	6.6×10^{-5} (c-TLM)	4H	[8]
Pd	700°C 5-40 min in N_2	5×10^{19} epi	5.5×10^{-5} (TLM)	4H	[93]
Si/Al	700°C 5-40 min in N_2	5×10^{19} epi	3.8×10^{-5} (TLM)	4H	[93]
Si/Pt	1100°C 3 min in vacuum	1×10^{19} epi	5.8×10^{-4} (TLM)	4H	[94]
Ta	1100°C 10 min in N_2-H_2	1×10^{18} epi	2.13×10^{-3} (TLM)	6H	[95]
Ti	800°C 1 min in vacuum	1.3×10^{19} epi	$2\text{-}4\times10^{-5}$ (TLM)	4H 6H	[15]

Ti	None	1.3×10^{19} epi	3.44×10^{-4} (TLM)	4H	[47]
Ti	950°C 2 min in 10% H_2/Ar	1.3×10^{19} epi	7.70×10^{-4} (TLM)	4H	[47]
Ti/Si/Co	500°C 5 min + 800°C 1min in vacuum	3.9×10^{18} epi	4×10^{-4} (TLM)	4H	[96]
Ti-Al	1000°C 2 min in vacuum	4.8×10^{18} epi	2.5×10^{-4} (TLM)	4H	[16,83]
TiC	none	$>10^{20}$ epi	1.08×10^{-4} (TLM)	4H	[47]
TiC	950°C 2 min in 10% H_2/Ar	$>10^{20}$ epi	5.62×10^{-5} (TLM)	4H	[47]
TiC	500°C 3 min in Ar-H_2	2×10^{19} Al-impl	2×10^{-5} (TLM)	4H	[97]
TiN	grown at 600°C	1×10^{19} epi	4.4×10^{-5} (TLM)	6H	[98]

A survey of metals contacts on 6H-SiC (Pd, Ni, Au, Ag, Mg, Al, Ti) was investigated by Waldrop,[99] as preliminary work for an appropriate selection of the metal for low resivity ohmic contacts to p-type SiC. Obviously, high values of the barrier $q\Phi_B$ were measured, in the range 1.17-2.56 eV after metal deposition. Among the annealed contacts, the lowest value of $q\Phi_B$ was found for Al contacts annealed at 600°C (1.04 eV on the C face 6H-SiC).[99]

Because of the low Schottky barrier and of the possibility of p-type doping of the SiC substrate by Al-indiffusion during annealing, Al-based alloyed contacts received much attention as ohmic contacts to p-type SiC. Indeed, many literature works reported on Al-based ohmic contacts, and in particular Al/Ti, to p-type SiC.

5.1. *Al/Ti Contacts*

One of the first practical application of Al/Ti contacts for p-type SiC was already reported in 1991 by Suzuki et al.,[100] who used Al/Ti alloyed materials as an ohmic contact to the p^+-type region of 6H-SiC light emitting diodes. However, in this work neither contact fabrication details were reported nor structural or electrical investigations were performed to study the physical mechanism of the Al/Ti ohmic behavior.

Al/Ti contacts are generally non-ohmic after deposition, while annealing temperatures above 900°C are required to obtain an ohmic behaviour of the contact. As an example, Fig. 10 shows the I-V characteristics of Al(150nm)/Ti(30nm) layers onto highly doped p-type 6H-SiC substrates, reported by Pécz et al..[84] These contacts were rectifying after deposition and became ohmic after annealing at 900°C in vacuum for 4 min.

The first quantitative investigation on the electrical properties of Al/Ti contacts on p-type SiC was reported by Crofton et al.,[88] who measured the values of ρ_c by means of circular TLM in a wide range of doping concentration of 6H-SiC epitaxial layers. The contacts were annealed at 1000°C, which resulted into an ohmic behavior with a low value of $\rho_c=1.5\times10^{-5}$ Ωcm^2 at a high doping level ($N_A=2\times10^{19}$ cm^{-3}). Fig. 11 reports the specific contact resistance for Al/Ti contacts on p-type 6H-SiC as a function of the carrier

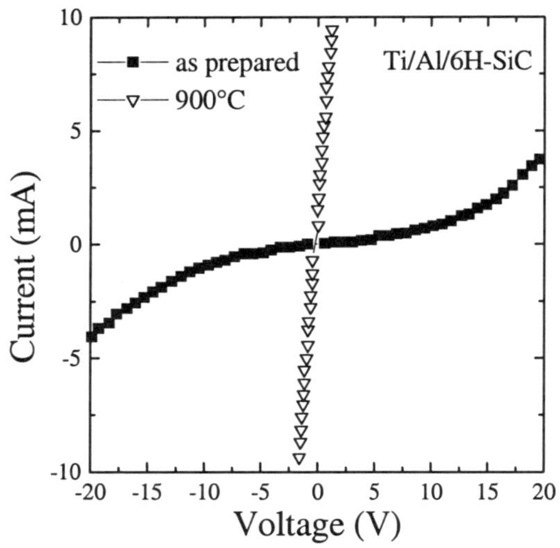

Fig. 10: Current-Voltage (I-V) characteristics of Al/Ti contacts on p-type 6H-SiC after deposition (as prepared) and after annealing at 900°C, showing the transition from the rectifying to the ohmic behavior. The SiC doping concentration was 1.6×10^{19} cm^{-3} (Data from Ref. [84]).

Several studies report on the correlation between the structural and electrical properties of Al/Ti contacts on p-type SiC, giving the possible mechanisms for ohmic contact formation.

A typical Al/Ti alloy consisting of 90% Al and 10% Ti, annealed at 1000°C, was characterized by Crofton et al..[15] By etching away the annealed metal layer the presence of many pits on the SiC surface was revealed. It was suggested that these pits, whose lateral dimensions varied between 1 and 10 μm and whose depth was around 30 nm, are intrusions that may enhance the field emission at metal/SiC interface, thus lowering the contact resistivity. According to this interpretation, p-concentration of the substrate, showing a strong decrease of ρ_c in the doping range of 4×10^{18}-2×10^{19} cm^{-3}. The dependence of ρ_c on N_A enabled the authors to model the Al-Ti/6H-SiC contact by assuming a barrier height of 0.37 eV.[11]

A similar value (0.53 eV) of the barrier for Al(70%)/Ti alloys on p-type ion-implanted 6H-SiC was recently determined by Moscatelli, Scorzoni et al.[89,101] from the measurement of the specific contact resistance as a function of the temperature, basing on the thermoionic field emission model.[24,25] In the same work, they also found a value of 0.95 eV on p-type ion implanted 4H-SiC.

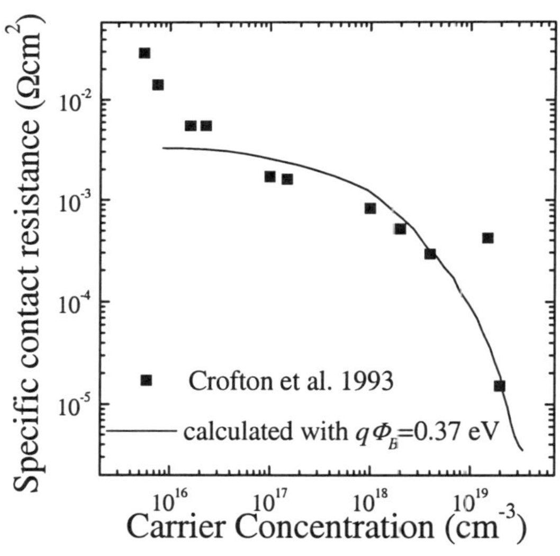

Fig. 11: Specific contact resistance as a function of the carrier concentration for Al/Ti contacts, annealed at 1000°C, on p-type 6H-SiC, reported by Crofton et al..[88] The solid line is the fit obtained assuming a barrier height of 0.37 eV.

type doping by Al indiffusion into the SiC substrate in Al/Ti-based contacts does not necessarily occur but rather intrusions-induced field emission is the factor responsible for the ohmic behavior.

The same authors[87] extensively studied different Al/Ti alloys to find the optimal composition for ohmic contacts. After observing that the binary Al-Ti phase diagram shows many different phases prior to any reaction of the metal with SiC, the ohmic behavior was correlated with the presence of a liquid phase at the annealing temperature of 1000°C. The best results were obtained for a composition of Al(70%)/Ti(30%), which gave a value of ρ_c=4.9×10^{-5} Ωcm^2 at a doping level of N_A=1×10^{19} cm^{-3}. At the optimal alloy composition, however, the intrusions produced in the SiC substrate may lead to contact degradation. Hence, limiting the degree of spiking remains the challenge for Al/Ti ohmic contacts in order to obtain shallow reproducible contacts to p-type SiC. Optimization of the contacts can be achieved by controlling the thickness of the p-type epilayer and of the metal film, the Al/Ti composition and the annealing conditions.[102]

Parisini et al.[103] performed a detailed structural characterization of this optimized Al(72%)/Ti contacts, alloyed at 1000°C, on p-type 6H-SiC. Also in this case, protrusions of the reacted metal layer into the SiC substrate were identified, by means of TEM analysis, and were indicated as the main reason for the ohmic contact formation rather than the formation of the ternary compound Ti_3SiC_2.

From all the aforementioned results, it is not clearly established whether the spiking due to the intrusions in SiC leads to an increase of the doping by Al indiffusion or rather the intrusions themselves act as field emitters increasing the leakage current and leading to the formation of an ohmic contact.

On the other hand, the heavy doping of SiC by Al indiffusion and electric field enhancement by intrusions were recently ruled out in an experimental work by Johnson and Capano.[16] In this work, it was demonstrated that after removing the annealed Al/Ti/Al contacts (31wt % Ti) and subsequent gold deposition, the specific contact resistance increased by one order of magnitude. However, if heavy doping of the SiC substrate by Al diffusion had occurred, the contact resistance should not have become worse after metal etch.

Instead, the formation of Al_4C_3 in annealed pure Al contacts and of Ti_3SiC_2 in Ti/Al/Ti alloyed contacts was demonstrated by XRD analysis, as can be observed in Figs. 12a and 12b, respectively.[16] These contacts were formed by annealing in vacuum at 1000°C for 2 min. The formation of these phases was indicated as the cause of ohmic contact formation. Clearly defined layers of $TiAl_3$ and of the ternary phase Ti_3SiC_2, after high temperature annealing of Al/Ti layers on SiC, were also observed by TEM analysis by Pécz et al.,[84] who found a perfectly epitaxially oriented Ti_3SiC_2 layer at the interface with 6H-SiC. In particular, it was argued that the Ti_3SiC_2 alloy behaves like a narrow-gap semiconductor with an electron affinity close to that of SiC, thus lowering the Schottky barrier height and allowing the formation of an ohmic contact.[16]

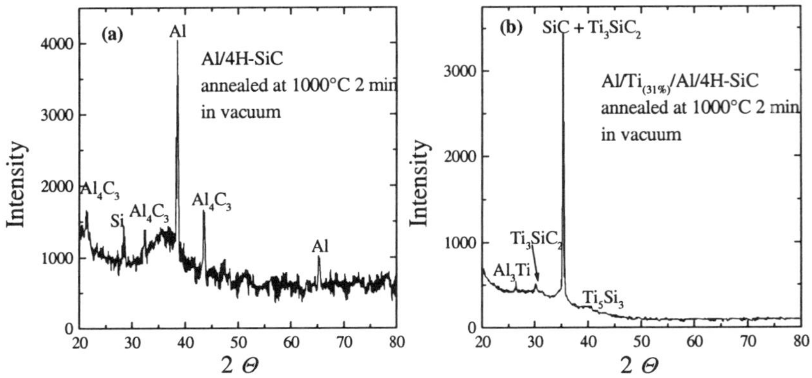

Fig. 12: XRD spectra of (a) annealed Al contacts, showing the formation of Al_4C_3; (b) annealed $Al/Ti_{(31\%)}/Al$ contacts, showing the formation of Ti_3SiC_2. The contacts were formed on 4H-SiC, with an annealing at 1000°C in vacuum for 2 min (Data from Ref. [16]).

5.2. *Alternatives to Al/Ti Contacts*

Since the formation of low resistance ohmic Al/Ti contacts requires annealing temperatures as high as 1000°C, with the consequent high driving force of Al for oxidation, alternative approaches were investigated, with the aim to reduce the annealing temperature and have a more suitable industrial process for device fabrication. Several authors observed oxidation of the top layer of the Al contact after annealing at temperatures higher than 800°C,[89,104,105] even after rapid thermal annealing under gas flow. Therefore, the use of cap layers may improve the reproducibility of the contact. Pd is attractive because of the high resistance to oxidation with the possibility of a low specific contact resistance. Stack sequences as Al/Ti/Pd/Ni or Al/Ti/Pt/Ni were proposed by Vassilevski et al.[85,105] to form Al/Ti based ohmic contacts to p-type SiC resistant to oxidation. Also Luo et al.[104] reported Al contacts on p-type 4H-SiC covered by a Pd cap layer which showed a ρ_c=5.7×10^{-5} Ωcm^2 with a reduced surface insulating problem during annealing at 1050°C.

Moreover, the high annealing temperatures required to obtain low resistance ohmic contacts with Al/Ti alloys to p-type SiC could reduce the capacitance of the gate oxide layers grown on the SiC substrates, by inducing a reaction between the metallic contact and the oxide, deteriorating the reliability of MOSFETs devices. Also the contact roughness can give serious problems during photolithography alignments. Therefore, in order to prevent these side effects a reduction of the annealing temperature for ohmic contact formation, i.e. below 800°C, becomes mandatory.

Zhao et al.[106] demonstrated the formation of good ohmic contacts to p-type 6H-SiC below 1000°C using pure Al. In this work, an innovative method, consisting of C and Al co-implantation, was used. P-type doping of the samples over a depth of 2 μm was achieved by Al ion implantation at 600°C or with C implantation followed by Al implantation at room temperature. In both cases, the same Al concentration of $4\times10^{20}cm^{-3}$ was implanted. After implantation, the dopant activation was performed by annealing at 1550°C in Ar. On the p-type doped material, Al contacts were deposited and annealed at 950°C in Ar ambient for 5 min. Whereas in the Al-implanted samples the specific contact resistance was in the 10^{-2} Ωcm^2 range, the contacts obtained using C-Al co-implantation gave values of ρ_c in the middle 10^{-5} Ωcm^2 range, i.e. with an improvement of about three orders of magnitude. The excess of Si vacancies introduced in the SiC matrix by C pre-implantation was indicated as the key factor for the improved contact performances. In fact, since Al acceptors tend to occupy the Si sites, more Al dopant can be activated on the vacancies sites, thus leading to a reduction of the specific contact resistance.[106]

Sakai et al.[92] demonstrated a significant reduction of the annealing temperature for ohmic contact formation by using GeTiAl contacts on p-type epitaxial 4H-SiC layers with a doping concentration of $4.5\times10^{18}cm^{-3}$. The average composition of the deposited contact was $Ge_{10\%}Ti_{15\%}Al_{75\%}$. The Ge/Ti/Al contacts showed ohmic behavior (with an average specific contact resistance of 1×10^{-4} Ωcm^2) already after annealing at 600°C in ultrahigh vacuum, i.e. with a reduction of about 400°C with respect to the typical temperatures required to obtain ohmic contact in Al/Ti alloys. The authors indicated in the presence of Ge the factor which facilitates the reaction between Ti/Al and SiC. After cooling at room temperature, different phases were observed at the interface, such as Al_4C_3 and the ternary phase Ti_3SiC_2, which determine the reduction of the contact resistance, consistently with the explanation given in Ref. [16]. In spite of these promising results, Ge/Ti/Al contacts are still far to be currently used in SiC technology because of their high surface roughness, which is undesired for the lithography steps.

In order to reduce the annealing temperature, also transition metals such as Ni and Co were used as metallic contacts since they can react with SiC at relatively low temperatures forming silicides. Moreover, refractory metal silicides have a low resistivity, a high thermal stability as long as an overall manufacturability, thus making them promising for the high temperature operation

Lundberg et al.[90] demonstrated low specific ohmic contacts using cobalt silicide ($CoSi_2$). However, $CoSi_2$ contacts formed by annealing Co films on SiC only led to a specific contact resistance in the range of 1×10^{-4} Ωcm^2. The poor electrical performance as long as the precipitation of carbon during reaction suggested that the Co/SiC structure is not indicated for reliable ohmic contacts. Hence, the authors fabricated a novel structure by sequential evaporation of a Si/Co structure, syntered by a two step annealing process in vacuum (at 500°C and 900°C respectively). In this way a significant improvement in the specific contact resistance was achieved ($\rho_c<4\times10^{-6}$ Ωcm^2). However, a degradation of the electrical properties of the syntered Co/Si/SiC contacts was observed after a high temperature reliability test at 1100°C, most probably as a consequence of morphological changes of the $CoSi_2$ layer.

Because of the relatively low temperatures required for reaction with SiC (~600°C), Co was also used by Nakatsuka et al.[91] to improve the surface morphology of Al-based ohmic contacts to p-type 4H-SiC. In fact, typical Al/Ti alloyed contacts often exhibit a rough surface after annealing at high temperature, because of an excess of $TiAl_3$. The use of a CoAl metallization annealed at 900°C resulted into good ohmic contact with $\rho_c=4\times10^{-4}$ Ωcm^2 and led to smooth surfaces, when compared with the conventional Al/Ti contacts and with cobalt silicide contacts without Al.

Another method for reducing the temperature of ohmic contact formation was investigated by Konishi et al.,[8] who studied the electrical performance of Ni/Al and Ni/Ti/Al contacts to p-type 4H-SiC with $N_A=3\text{-}9\times10^{18}$ cm^{-3}. The aim of their work was using Ni as material for ohmic contact formation, which reacts with SiC by forming Ni_2Si, selecting Al as an acceptor for SiC. In contrast to the conventional Ti/Al contacts, the Ni/Al contacts already became ohmic after annealing at 800°C, with a contact resistance of 5×10^{-3} Ωcm^2. This value could be reduced by more than one order of magnitude (9.5×10^{-5} Ωcm^2) after annealing at 1000°C, as can be observed in Fig. 13. Microstructural analysis performed by XRD showed the formation of several Ni-based compounds, with a predominance of the Ni_2Si phase after annealing at 1000°C. The formation of Ni_2Si leads to an increase of the Al concentration close to the metal/SiC interface, which reduce the specific contact resistance. On the other hand, the addition of a Ti interlayer to the contacts (i.e. Ni/Ti/Al), avoided the accumulation of unreacted carbon at the interface and further improved the ohmic properties of

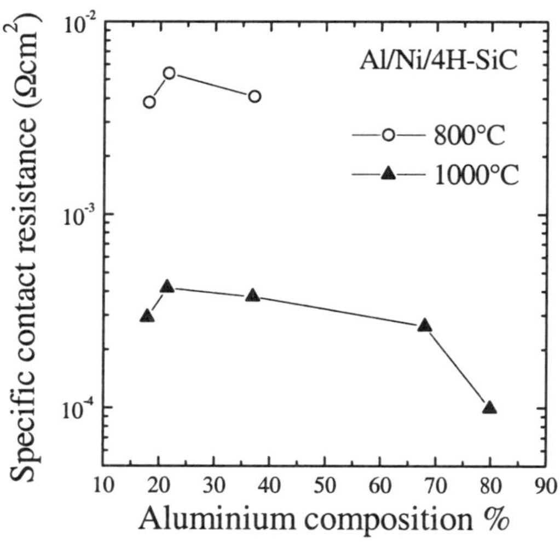

Fig. 13: Specific contact resistance of Ni/Al contacts on p-type 4H-SiC vs the Al composition after annealing at 800 or 1000°C. Ohmic behavior is already achieved at 800°C (Data from Ref. [8]).

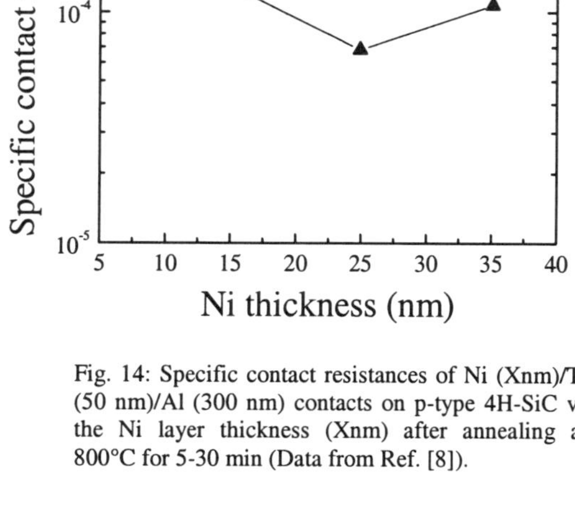

Fig. 14: Specific contact resistances of Ni (Xnm)/Ti (50 nm)/Al (300 nm) contacts on p-type 4H-SiC vs the Ni layer thickness (Xnm) after annealing at 800°C for 5-30 min (Data from Ref. [8]).

the contact. In particular, $Al_{(300nm)}/Ti_{(50nm)}/Ni_{(Xnm)}$ contacts annealed at 800°C gave ρ_c=6.6×10^{-5}Ωcm^2 for a Ni thickness of 25 nm, as can be observed in Fig. 14. These contacts also exhibited an excellent thermal stability at 400°C. In this case, microstructural analysis of the samples annealed at 800°C, beyond the formation of Ni_2Si and $NiAl_3$, indicated the presence of a highly textured Ti_3SiC_2 layer at the interface with SiC, which prevented the accumulation of unreacted carbon. According to the interpretation given by the authors, during reaction of metal with SiC, an increase of the p-type doping concentration (Al) at the interface occurred by a "snow-plow" mechanism.[8] Although the values of specific contact resistance of conventional Al/Ti contacts were lower, Ni/Al and Ni/Ti/Al showed an improved surface morphology and thermal stability. Moreover, the theoretical calculation showed that Ni/Ti/Al contacts can potentially give values of ρ_c below 1×10^{-5} Ωcm^2 for doping concentrations higher than 1×10^{19}cm^{-3}.

As in the case of the n-type material, also for the p-type material Ti-based contacts are expected to produce low resistance and thermal stability but their application is strongly limited by the easy oxidation of Ti in air even at room temperature. In order to overcome this problem, covering layers of noble or nearly noble metals (Pt, Co) can be deposited onto the Ti layer.[96] Moreover a diffusion barrier can be also deposited between the two metals to avoid intermixing which may cause an increase of the contact resistivity. Co/Ti contacts onto p-type Al-doped (N_A=3.9×10^{18}cm^{-3}) 4H-SiC exhibited an encouraging value of the specific contact resistance (4×10^{-4} Ωcm^2) after annealing at 850°C.

Lee et al.[47,97] reported low resistance ohmic contacts on p-type epitaxial and implanted SiC layers, using TiC epitaxially grown by coevaporation with an e-beam for Ti and a Knudsen cell for C_{60}. The as-deposited Ti/4H-SiC contacts exhibited a good ohmic behavior with ρ_c=1.08×10^{-4} Ωcm^2, which was reduced by a factor of 2 by annealing at 950°C.

Finally, Si/Pt bilayer contacts were reported by Papanicolau et al.[94] as an alternative to the Al-based contacts, which are often affected by a high driving force for oxidation and limited by the low melting temperature of Al. Annealing processes at 1100°C in vacuum resulted in the formation of a ohmic contact with a specific contact resistance in the low 10^{-4} Ωcm^2 range. A metallization layer as mixture of Pt, Si and C, probably in the form of Pt- silicides and carbides, as long as the depletion of C at the Si/SiC interface were responsible for the ohmic behavior after high temperature annealing.

6. Long-Term Thermal Stability of Ohmic Contacts to SiC

Although Ni_2Si or Al/Ti have superior properties for low resistance ohmic contacts to n- and p-type SiC, they exhibit the tendency to oxidize when exposed to air at high temperature. Also the uppermost device metallization layer, namely the bond pad (Al, Au), can be unstable (in terms of oxidation and/or diffusion) when exposed to high temperatures in non-inert environments. All these factors can lead to contact degradation, resulting into an increase of the value of ρ_c and, ultimately, to an increase of the device on-resistance.

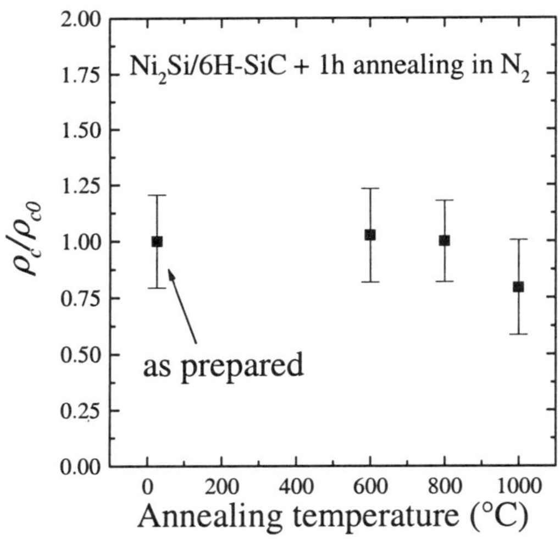

Fig. 15: Variation of the specific contact resistance ρ_c with respect to its initial value ρ_{c0} (=3.9×10^{-5} Ωcm^2) as a function of the annealing temperature in N_2 for 1 h for Ni_2Si contacts on n-type 6H-SiC ($N_D=7.4\times10^{18} cm^{-3}$).

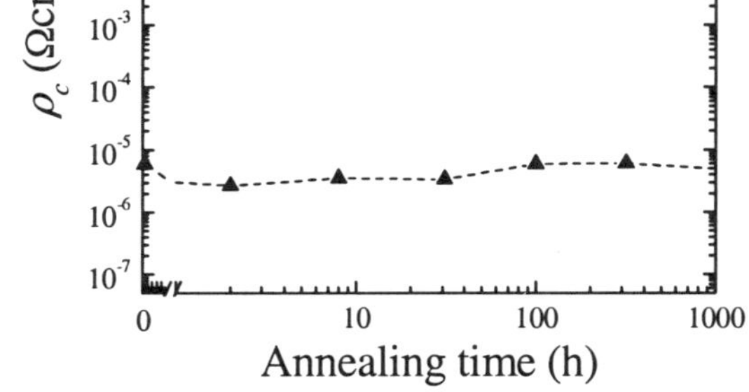

Fig. 16: Specific contact resistance ρ_c as a function of the annealing time for annealing cycles at 500°C in N_2 for Ni_2Si ohmic contacts on highly doped (~2×10^{20} cm^{-3}) n-type and p-type 4H-SiC (Data from Ref. [7]).

Moreover, the complete device fabrication flowchart may require the contact to be subjected to additional thermal budget for oxide densification, passivation, electrical activation of edge terminations, packaging etc. Then, fabricating thermally stable ohmic contacts for devices and sensors is a crucial process and, consequently, testing the long term reliability of the contacts at high temperatures (not only in vacuum but also in aggressive environment like air) is an important physical and technological issue.

It is has been reported that Ni_2Si contacts with low specific resistance on n-type SiC are stable over long period of time (hundreds of hours) in vacuum at 650°C.[14] More recently, we have demonstrated that the values of ρ_c of Ni_2Si contacts remain almost unchanged even after annealing for one hour up to 1000°C, in N_2 environment.[18] This latter is shown in Fig. 15, which reports the variation of the ρ_c with respect to its initial value ρ_{c0} (=3.9×10^{-5} Ωcm^2) as a function of the annealing temperature, for Ni_2Si contacts on n-type 6H-SiC.

Tanimoto et al.[7] demonstrated the long term thermal stability at 500°C for 1000h in N_2 of Ni_2Si contacts used both for n- and p-type highly doped SiC material in MOSFETs devices. The values of ρ_c for these contacts are shown in Fig. 16 as a function of the annealing time. The slight decrease of the contact resistance in the n-type material occurring in the initial stages of the annealing cycle was not well understood.

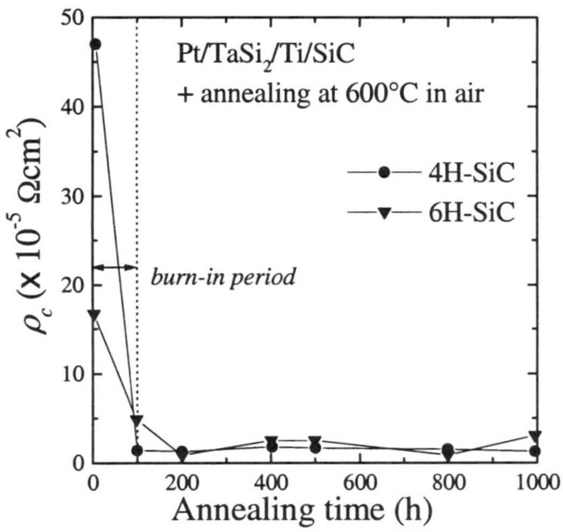

Fig. 17: Specific contact resistance ρ_c as a function of the annealing time in air at 600°C for Ti(100 nm)/$TaSi_2$(400 nm)/Pt(200 nm) contacts on the *n*-type 6H- and 4H-SiC. The first 100 hours represent the "burn-in" period needed to complete the dominant reactions (Data from Ref. [19]).

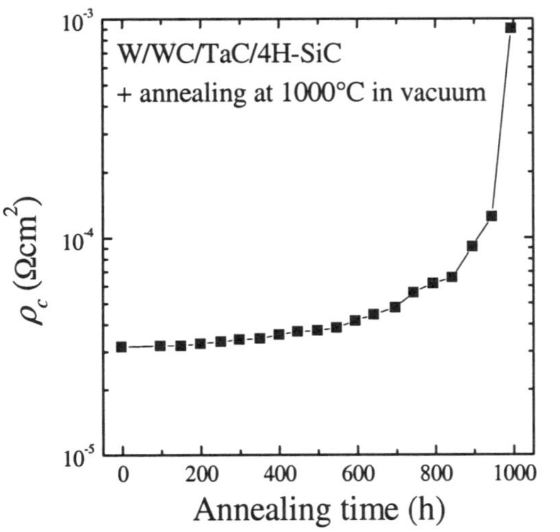

Fig. 18: Specific contact resistance ρ_c as a function of the annealing time in vacuum at 1000 °C for TaC(100nm)/WC(10nm)/W(50nm) contacts on 4H-SiC. The doping concentration in the *n*-type SiC epilayer was 8.1×10^{18} cm^{-3} (Data from Ref. [45]).

Beyond the oxygen intake during annealing in air, other common causes of contact degradation are the diffusion processes and interaction between the metals of the layer at higher temperatures. Therefore, for advanced metallizations, cap layers of noble metals like Au or Pt, as well as diffusion barriers for the metal are required.

Baeri et al. [107] studied the TiW diffusion barrier upon annealing cycles either in vacuum and in O_2 environment, in a metallization scheme based on the multilayer Au/TiW/Ni_2Si/6H-SiC. For short annealing times (i.e. three hours), this metallization showed electrical stability in O_2 up to 575°C, while contact degradation took place above this temperature. In vacuum, the specific contact resistance remained substantially unchanged upon a thermal cycles at 600°C. On the other hand, long term stability tests for 100 hours, performed in O_2 environment, showed that ρ_c remained constant up to the temperature of 450°C.[108]

Basing on chemical analysis of the multilayer, performed by XPS, the Ti diffusion process is a combination between grain boundary diffusion and bulk diffusion, the latter starting at temperatures higher than 400°C. In particular, since no Ti was detected inside the Au layer upon 315°C, its out-diffusion occurs

throughout the Au grain boundaries, until it reaches the sample surface where TiO_x is detected. Conversely above 400°C Ti diffusion became a bulk diffusion through Au layer. In fact, besides the oxidized titanium at the surface, a considerable amount of Ti (up to 30% after annealing at 500°C) was also found inside the Au layer.[107,108]
Other diffusion barriers (like $TaSi_2$ or TaC) were investigated to achieve contact reliability up to high temperatures (~600°C) and for longer times (several hundreds of hours).
Thermally stable Ti-based ohmic contacts were fabricated by Okojie et al.[19] on both n-type 6H- and 4H-SiC, using a stacking sequence Pt/$TaSi_2$/Ti/SiC, and annealed in N_2 at 600°C after deposition. This thermal process resulted in the formation of Ti_5Si_3 at the interface with SiC and in the decomposition of $TaSi_2$ with the formation of platinum silicide in the uppermost layer. For doping levels in the range $0.7\text{-}2\times10^{19}$ cm^{-3}, the contact resistance was in the range $1\text{-}5\times10^{-4}$ Ωcm^2 after 30 min in N_2 and improved to $1\text{-}6\times10^{-5}$ Ωcm^2 after annealing at 600°C for 100 h in air, as can be observed in Fig. 17. The initial high values of ρ_c within the first 100 h annealing were ascribed to a "burn-in" period needed to complete the dominant reactions. These contacts could successfully withstand 1000 h at 600°C in air, with platinum silicide forming a protective overlay against oxygen diffusion.
TaC ohmic contacts reported by Jang et al.,[45] with a stack layer sequence W/WC/TaC/SiC, had a specific contact resistance of 3×10^{-5} Ωcm^2 and showed an excellent thermal stability at 1000°C for 600h in vacuum. This latter can be seen from Fig. 18, reporting the values of ρ_c as a function of the annealing time at 1000°C. Annealing at 1000°C only resulted in small changes after 600 h, primarily due to oxygen incorporation in the contacts, while after annealings longer than 800h metallurgical reactions (W_2C, TaO_x, TaC_xO_y) and contact oxidation were responsible of the contact degradation.

7. New Trends in Ohmic Contacts for Practical Devices

The possibility of using Ni for ohmic contacts on both n- and p-type SiC can significantly improve the present standard procedures for the fabrication of three terminal high-current devices. Indeed, the metallization technology will be extremely simplified by simultaneously fabricating ohmic contacts on n- and p-type material, i.e. with a single metal deposition and a unique post-annealing step.
However, obtaining low specific resistance ohmic contacts on both n- and p-type SiC with a single metal was commonly ruled out because it is known that a metal with a low barrier on n-type material results into a high value of $q\Phi_B$ on p-type, and vice versa.
The first demonstration of Ni ohmic contacts on both n^+- and p^+-type SiC was reported by Fursin et al..[109] Al and C co-implantation followed by annealing at 1550°C was used to create the p^+ doping region, while nitrogen was implanted for the n^+ doping. Nickel films,

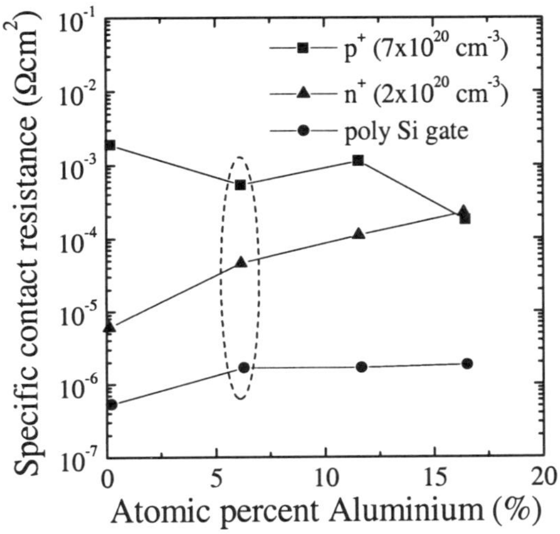

Fig. 19: Specific contact resistance as a function of the Al atomic percent in single-metal Al/Ni contacts for the n^+ source region, the contact of the p-well, and the poly Si gate in 4H-SiC MOSFETs (Data from Ref. [20]). The optimal values of ρ_c were obtained for an Al atomic percent of 6%.

200 nm thick, were annealed at 1050°C in forming gas for 10 min, in order to form ohmic contacts. Specific contact resistance values of 1.5×10^{-4} Ωcm^2 and 6×10^{-6} Ωcm^2 were measured for p- and n-type respectively, that were indicated as reasonable values for the fabrication of GTO SiC thyristors.

In the fabrication of SiC power MOSFETs, there is an urgent need to develop a single ohmic contact technology for n- and p-type SiC, that would directly contribute to the device shrinking and to the lowering of the specific on-resistance.

Kiritani et al.[20] demonstrated a 4H-SiC vertical MOSFET with a single material contacts to the n^+ source region, the p well contact and the polysilicon gate. The n^+ source and p^+ contact regions were fabricated by phosphorous and aluminium ion implantation, followed by an annealing at 1600°C for dopant activation. The contacts were simultaneously formed on the source, well and gate regions of the device, using an Al/Ni (Al 6%) layer, syntered by a rapid thermal annealing in Ar at 1000°C.

Fig. 19 reports the values of the specific contact resistance as a function of the aluminium

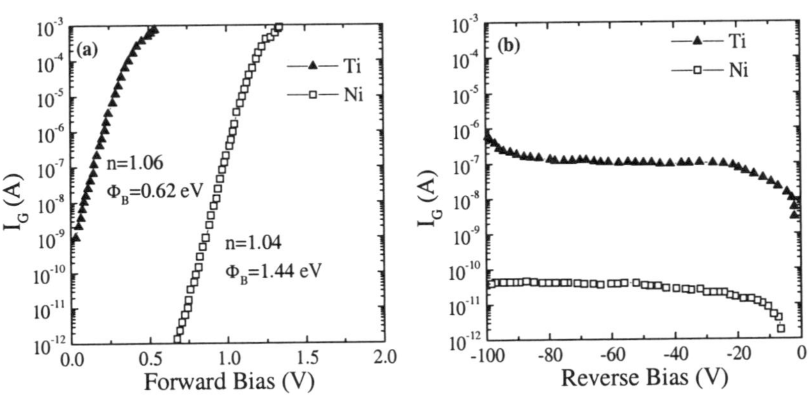

Fig. 20: Forward and reverse current-voltage (I_G-V_G) characteristics of the gate contacts of the MESFETs on 4H-SiC reported in Ref. [21], fabricated using Ti and Ni unannealed contacts. The values of the Schottky barrier height Φ_B were 0.62 eV for Ti and 1.41 eV for Ni, while the ideality factors were 1.06 and 1.04 respectively.

atomic percent in the Al/Ni contacts. A gradual increase of the specific contact resistance ρ_c for the n^+ source and n^+ polysilicon with increasing aluminium atomic percent in the Al/Ni layer is observed, while ρ_c decreased for the p^+ contact of the p-well. The optimal specific contact resistance values of 4.8×10^{-5} Ωcm^2 for the n^+ source region, 1.5×10^{-6} Ωcm^2 for the polysilicon gate and 5.2×10^{-4} $\Omega\ cm^2$ for the p-well contact region were obtained for an Al atomic percent of 6%.

Auger spectroscopy showed that, during annealing at 1000°C, Al diffuses to the surface through the Ni overlayer and traces of Al also diffuse into the SiC substrate. Then, while Ni reacts efficiently with the SiC to form nickel silicide, in the p^+ region, the addition of even a thin Al layer improved the ohmic properties due to the increased p-type carrier concentration near the interface. Further, in the n^+ region, Al/Ni contact materials also exhibited almost the same good ohmic properties of a pure nickel silicide.

For high-power microwave applications as SiC MESFETs, a gate Schottky contact with a sharp rectifying behavior and source/drain contacts with very low specific contact resistance are required. Usually, source/drain contacts are formed by depositing Ni in the heavily doped n^+ region followed by post-deposition annealing of the contacts at temperatures higher than 900°C. However, Schottky contacts may be unable to resist at these high temperatures and may be subjected to degradation. Therefore, in the conventional MESFET fabrication process, the gate contact formation must necessarily be preceded by source/drain contact formation. Recently, to overcome this limit, a single metal contact technology for MESFETs was developed by Tanimoto et al.,[21,33] who demonstrated the simultaneous fabrication of gate and source/drain contacts using a single contact material and without subjecting it to post-deposition annealings. Such a structure, could have a direct impact on furthering the miniaturization of MESFETs. Moreover, by applying this novel procedure, the device fabrication process can be shortened by eliminating the fabrication step for the source/drain contacts. MESFETs were fabricated using conductive 4H-SiC substrate with a 4.9 μm thick p^- insulating buffer layer ($N_A=5\times10^{15}$ cm^{-3}) and a 0.4 μm thick n channel ($N_D=1.7\times10^{17}$

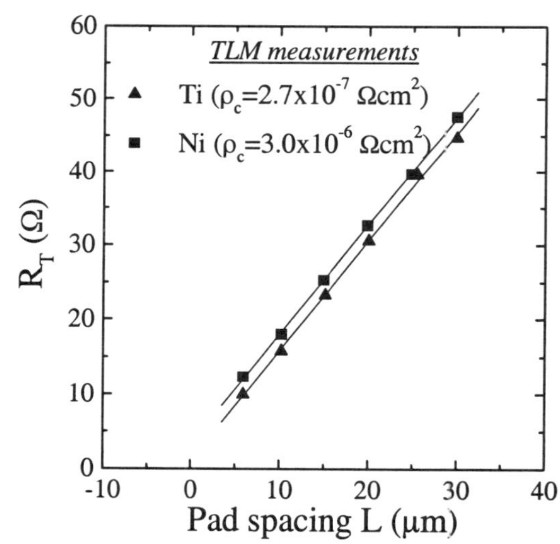

Fig. 21: Total resistance R_T as a function of the pad spacing L for the source/drain ohmic contacts of a MESFET fabricated on 4H-SiC with unannealed Ti and Ni contacts (Data from Ref. [21]). The n-type doping level of the source and drain region was $N_D>10^{20}$ cm^{-3}. The specific contact resistance for these contacts, determined by the TLM analysis, was ρ_c = 2.7×10^{-7} Ωcm^2 for Ti and 2.0×10^{-6} Ωcm^2 for Ni.

cm^{-3}). Heavily doped n^+ source and drain regions ($N_D > 10^{20}$ cm^{-3}) were created by phosphorous ion implantation followed by annealing at 1700°C. A sacrificial oxidation and its removal were performed in order to obtain a fresh surface for metallization. Either Ti or Ni were used as single contact material for source/drain and gate contacts of the device. An almost ideal (n=1.04-1.06) rectifying behavior was observed in the forward and reverse current voltage (I_G-V_G) characteristics of the gate contact, shown in Fig. 20 both for Ti and Ni, with barrier heights values of 0.62 eV and 1.44 eV, respectively. Concerning the source/drain ohmic contacts, Fig. 21 shows the total resistance R_T as a function of the pad distance L, determined in a set of TLM structures. From the analysis of these data, it was possible to determine specific contact resistance values of ρ_c = 2.7×10^{-7} Ωcm^2 for Ti and $\rho_c = 2.0\times10^{-6}$ Ωcm^2 for Ni, respectively.

8. Conclusion

The most significant results in the field of ohmic contacts to SiC reported in the last decade were reviewed. The formation of low resistance ohmic contacts to n-type SiC is achievable by Ni_2Si contacts, forming by annealing Ni films above 950°C. In this way, specific contact resistance values in the range of $\sim10^{-6}$ Ωcm^2 can be obtained. The carbon redistribution inside the silicide layer upon annealing as well as the inhomogeneous nature of the Schottky barrier are believed to be responsible of the mechanism of ohmic contact formation above 900°C. In the case of p-type material, Al/Ti alloyed contacts are commonly used and allow to obtain values of ρ_c in the $\sim10^{-5}$ Ωcm^2 range. The formation of low resistance ohmic contacts can be obtained after annealing processes at about 1000°C. In this case, Al intrusions or the formation of a ternary phase at the interface with SiC (Ti_2SiC_2) are the major factors responsible for the ohmic behavior. At these annealing temperature, however, there exists the risk of oxidation of the contact top layer. For that reason, alternative solutions have been proposed to reduce the thermal budget required to obtain ohmic behaviour of metal layers to p-type SiC (like Al/Ni/SiC).

However, for the device application point of view, p-type ohmic contacts without Al could be preferred because of the observed Al spiking, that represents a reliability concern. Hence other approaches were investigated as that of Ni contacts, where Al and C co-implantation was used to enhance the Al activation in the p^+-type region and reduce the specific contact resistance.

In order to simplify the device fabrication process, single material ohmic contacts for both n-type and p-type SiC were recently investigated. Adding a few percent of Al to a standard Ni contact, allowed to obtain values of 10^{-6} Ωcm^2 and 10^{-3} Ωcm^2 in the ion-implanted n^+ and p^+ regions, respectively. This technology was successfully applied to vertical power MOSFETs on 4H-SiC.

Moreover, a single metal technology was recently developed for high-power microwave application, demonstrating the simultaneous fabrication of Schottky (gate) and ohmic (source/drain) contacts using a single material (Ti or Ni) without post-deposition annealing processes.

References

[1] S. M. Sze, *Physics of Semiconductor Devices, 2nd Ed.*, John Wiley & Sons, New York, 1981.
[2] G.L. Harris, G. Kelner, M. Shur, Ohmic contacts to SiC, in *Properties of Silicon Carbide*, G.L. Harris ed. , EMIS Datareviews Series n° 13, INSPEC Publication, London, 1995.
[3] C.E. Weitzel, J.W. Palmour, C.H. Carter, K. Moore, K.J. Nordquist, S. Allen, C. Thero, Silicon carbide high-power devices, *IEEE Trans. Electron Devices* **43**, 1732-1741 (1996).
[4] K. Rottner, M. Frischholz, T. Myrtveit, D. Mou, K. Nordgren, A. Henry, C. Hallin, U. Gustafsson, A. Schöner, SiC power devices for high voltage applications, *Mat. Sci. Eng. B* **61-61**, 330-338 (1999).
[5] R.C. Clarke, J.W. Palmour, SiC microwave power technology, *Procedeeings of the IEEE* **90**, 987-992 (2002).
[6] J.A. Cooper, M.R. Melloch, J.M. Woodall, J. Spitz, K.J. Schoen, J.P. Henning, Recent advances in SiC power devices, *Mat. Sci. Forum* **264-268**, 895-900 (1998).
[7] S. Tanimoto, H. Okushi, K. Arai, Ohmic contacts for power devices on SiC, in *Silicon Carbide: Recent Major Advances*, W.J. Choyke, H. Matsunami, G. Pensl eds., Springer Verlag, Berlin-Heidelberg, 2004.
[8] R. Konishi, R. Yasukochi, O. Nakatsuka, Y. Koide, M. Moriyama, M. Murakami, Development of Ni/Al and Ni/Ti/Al ohmic contact materials for p-type 4H-SiC, *Mat. Sci. Eng. B* **98**, 286-293 (2003).
[9] L.M. Porter, R. F. Davis, A critical review of ohmic and rectifying contacts for silicon carbide, *Mat. Sci. Eng. B* **34**, 83-105 (1995).
[10] M.J. Bozack, Surface studies on SiC as related to contacts, *phys. Stat. Sol (b)* **202**, 549-580 (1997).
[11] J. Crofton, L.M. Porter, J.R. Williams, The physics of Ohmic Contacts to SiC, *phys. Stat. Sol. (b)* **202**, 581-603 (1997).
[12] V. Saxena, A.J. Steckl, Building blocks for SiC devices: ohmic contacts, Schottky contacts, and p-n junctions, in *SiC Materials and Devices*, Y.S. Park ed., Academic Press, San Diego, 1998.
[13] M.W. Cole, R.C. Joshi, Ohmic Contacts to SiC for High Power and High Temperature Device Applications, in *Silicon Carbide: Materials, Processing and Devices*, Z.C. Fen and J.H. Zhao eds., Taylor & Francis, 2003.
[14] J. Crofton, P.G. McMullin, J.R. Williams, M.J. Bozack, High-temperature ohmic contact to n-type 6H-SiC using nickel, *Appl. Phys. Lett.* **77**, 1317-1319 (1995).
[15] J. Crofton, L. Beyer, J.R. Williams, E.D. Luckowski, S.E. Mohney, J. M. Delucca, Titanium and Aluminum-Titanium ohmic contacts to p-type SiC, *Solid-State Electronics* **41**, 1725-1729 (1997).
[16] B.J. Johnson, M.A. Capano, Mechanism of ohmic behavior of Al/Ti contacts to p-type 4H-SiC after annealing, *J. Appl. Phys.* **95**, 5616-5620 (2004).
[17] Ts. Marinova, A. Kakanakova-Gerogieva, V. Krastev, R. Kakanakov, M. Neshev, L. Kassamakova, O. Noblanc, C. Arnodo, S. Cassette, C. Brylinski, B. Pécz, G. Radnoczi, Gy. Vincze, Nickel based ohmic contacts on SiC, *Mat. Sci. Eng. B* **46**, 223-226 (1997).
[18] F. Roccaforte, F. La Via, V. Raineri, L. Calcagno, P. Musumeci, Improvement of high temperature stability of nickel contacts on n-type 6H-SiC, *Appl. Surf. Sci.* **184**, 295-298 (2001).
[19] R.S. Okojie, D. Lucko, Y. L. Chen, D.J. Spray, Reliability assessment of Ti/$TaSi_2$/Pt ohmic contacts on SiC after 1000 h at 600°C, *J. Appl. Phys.* **91**, 6553-6559 (2002).

[20] N. Kiritani, M. Hoshi, S. Tanimoto, K. Adachi, S. Nishizawa, T. Yatsuo, H. Okushi, K. Arai, Single material ohmic contacts simultaneously formed on the source/p-well/gate of 4H-SiC vertical MOSFETs, *Mat. Sci. Forum* **433-436**, 669-672 (2003).
[21] S. Tanimoto, M. Inada, N. Kiritani, M. Hoshi, H. Okushi, K. Arai, Single contact-material MESFETs on 4H-SiC, *Mat. Sci. Forum* **457-460**, 1221-1224 (2004).
[22] E.H. Rhoderick, R.H. Williams, *Metal-Semiconductor contacts*, Oxford Science Publications, Oxford, 1988.
[23] D. K. Schroder, *Semiconductor material and device characterization*, John Wiley and Sons, New York, 1998
[24] F.A. Padovani, R. Stratton, Field and thermoionic-field emission in Schottky barriers, *Solid-State Electron.* **9**, 695-707 (1966).
[25] A.Y.C. Yu, Electron tunneling and contact resistance of metal-silicon contact barriers, *Solid-State Electron.* **13**, 239-257 (1970).
[26] H.H. Berger, Contact resistance and contact resistivity, *J. Electrochem. Soc.* **119**, 507-514 (1972).
[27] H.H. Berger, Model for contacts to planar devices, *Solid-State Electron.* **15**, 145-158 (1972).
[28] G.K. Reeves, H.B. Harrison, Obtaining the specific contact resistance from transmission line model measurements, *IEEE Electron Device Lett.* **3**, 111-113 (1982).
[29] M. Finetti M, A. Scorzoni, G. Soncini, Lateral current crowding effects on contact resistance: Measurements in four terminal resistor test patterns, *IEEE Electron Device Lett.* **5**, 524–526 (1984).
[30] G.K. Reeves, Specific contact resistance using a circular transmission line model, *Solid-State Electron.* **23**, 487-490 (1980).
[31] A.J. Willis, A.P. Botha, Investigation of ring structures for metal-semiconductor contact resistance determination, *Thin Solid Films* **146**, 15-20 (1987).
[32] S.J. Proctor, L.W. Linholm, A direct measurement of interfacial contact resistance, *IEEE Electron Device Lett.* **3**, 294-296 (1982).
[33] H. Na, H. Kim, K. Adachi, N. Kiritani, S. Tanimoto, H. Okushi, K. Arai, High-quality Schottky and ohmic contacts in planar 4H-SiC semiconductor field-effect transistors and device performance, *J. Electron. Mater.* **33**, 89-93 (2004).
[34] N.I. Cho, K.H. Jung, Y. Choi, Improved ohmic contact to the n-type 4H-SiC semiconductor using cobalt silicides, *Semicond. Sci. Technol.* **19**, 306-310 (2004).
[35] T.N. Oder, J.R. Williams, K.W. Bryant, M.J. Bozack, J. Crofton, Low resistance ohmic contacts to n-SiC using Niobium, *Mat. Sci. Forum* **338-342**, 997-1000 (2000).
[36] R.H. Cox, H. Strack, Ohmic contacts for GaAs devices, *Solid-State Electron.* **10**, 1213-1218 (1967).
[37] V.A. Dmitriev, K. Irvine, M. Spencer, G. Kelner, Low resistivity (~10^{-5} Ωcm^2) ohmic contacts to 6H silicon carbide fabricated using cubic silicon carbide contact layer, *Appl. Phys. Lett.* **64**, 318-320 (1994).
[38] S.Y. Han, K.H. Kim, J.K. Kim, H.W. Jang, K.H. Lee, N.K. Kim, E.D. Kim, J.L. Lee, Ohmic contact formation mechanism of Ni on n-type 4H-SiC, *Appl. Phys. Lett.* **79**, 1816-1818 (2001).
[39] S. Tanimoto, N. Kiritani, M. Hoshi, H. Okushi, Ohmic contact structure and fabrication process applicable to practical SiC devices, *Mat. Sci. Forum* **389-393**, 879-884 (2002).
[40] T. Uemoto, Reduction of Ohmic contact resistance on n-type 6H-SiC by heavy doping, *Jpn. J. Appl. Phys. Part 2* **34**, L7-L9 (1995).
[41] E.D. Luckowski, J.M. Delucca, J.R. Williams, S.E. Mohney, M.J. Bozack, T. Isaac-Smith, J. Crofton, Improved ohmic contact to n-type 4H and 6H-SiC using nichrome, *J. Electron. Mater.* **27**, 330-334 (1998).
[42] T. Nakamura, M. Satoh, $NiSi_2$ ohmic contact to n-type 4H-SiC, *Mat. Sci. Forum* **389-393**, 889-892 (2002).
[43] C. Deeb, A.H. Heuer, A low-temperature route to thermodynamically stable ohmic contacts to n-type 6H-SiC, *Appl. Phys. Lett.* **84**, 1117-1119 (2004).

[44] T. Jang, L.M. Porter, G.W.M. Rutsch, B. Odekirk, Tantalum carbide ohmic contacts to n-type silicon carbide, *Appl. Phys. Lett.* **75**, 3956-3958 (1999).
[45] T. Jang, B. Odekirk, L.D. Madsen, L.M. Porter, Thermal stability and contact degradation mechanisms of TaC ohmic contacts with W/WC overlayers on n-type 6H-SiC, *J. Appl. Phys.* **90**, 4555-4559 (2001).
[46] F. La Via, F. Roccaforte, A. Makhtari, V. Raineri, P. Musumeci, L. Calcagno, Structural and electrical characterisation of titanium and nickel silicide contacts on silicon carbide, *Microelectronics Engineering* **60**, 269-282 (2002).
[47] S.K. Lee, C.M. Zetterling, M. Östling, J.P. Palmquist, H. Högberg, U. Jansson, Low resistivity ohmic titanium carbide contacts to n- and p-type 4H-silicon carbide, *Solid-State Electron.* **44**, 1179-1186 (2000).
[48] A.K. Chadda, J.D. Parsons, G.B. Kruaval, Thermally stable, low specific resistance (1.30×10^{-5}) TiC Ohmic contacts to n-type 6Hα-SiC, *Appl. Phys. Lett.* **66**, 760-762 (1995).
[49] U. Schmid, R. Getto, S.T. Sheppard, W. Wondrak, Temperature behavior of specific contact resistance and resistivity on nitrogen implanted 6H-SiC with titanium silicide ohmic contacts, *J. Appl. Phys.* **85**, 2681-2686 (1999).
[50] R. Getto, J. Freytag, M. Kopnarski, H. Oechsner, Characterization of sputtered titanium silicide ohmic contacts on n-type 6H-silicon carbide, *Mat. Sci. Eng. B* **61-62**, 270-274 (1999).
[51] J.R. Waldrop, R.W. Grant, Y.C. Wang, R.F. Davis, Metal Schottky barrier contacts to alpha SiC, *J. Appl. Phys.* **72**, 4757-4760 (1992).
[52] J. Waldrop, Schottky barrier height and interface chemistry of annealed metal contacts to alpha 6H-SiC: Crystal phase dependence, *Appl. Phys. Lett.* **62**, 2685-2687 (1993).
[53] F. Goesmann, R. Schmid-Fetzer, Temperature-dependent interface reactions and electrical contact properties of titanium on 6H-SiC, *Semicond. Sci. Technol.* **10**, 1652-1658 (1995).
[54] S. Gorsse, Y. Le Petitcorps, A new approach in the understanding of the SiC/Ti reaction zone composition and morphology, *Composites Part A* **29A**, 1221-1227 (1998).
[55] L.M. Porter, R.F. Davis, J.S. Bow, M.J. Kim, R.W. Carpenter, R.C. Glass, Chemistry, microstructure, and electrical properties at interfaces between thin films of titanium and alpha (6H) silicon carbide (0001), *J. Mater. Res.* **10**, 668-679 (1995).
[56] T. Teraji, S. Hara, H. Okushi, K. Kajimura, Ideal Ohmic contact to n-type 6H-SiC by reduction of Schottky barrier height, *Appl. Phys. Lett.* **71**, 689-691 (1997).
[57] A. Makhtari, F. La Via, V. Raineri, L. Calcagno, F. Frisina, Structural characterization of titanium silicon carbide reaction, *Microelectronics Engineering* **55**, 375-381 (2001).
[58] J. S. Chen, E. Kolawa, M.A. Nicolet, R. P. Ruiz, L. Baud, Jaussaud, R. Madar, Reaction of Ta thin film with single crystalline (001) β-SiC", *J. Appl. Phys.* **76**, 2169-2175 (1994).
[59] F. Roccaforte, F. La Via, V. Raineri, L. Calcagno, P. Musumeci, G.G. Condorelli, Highly reproducible ideal SiC Schottky rectifiers: effects of surface preparation and thermal annealing on the Ni/6H-SiC barrier height, *Appl. Phys. A* **77**, 827-833 (2003).
[60] F. Roccaforte, F. La Via, V. Raineri, F. Mangano, L. Calcagno, Temperature dependence of the c-axis mobility in 6H-SiC Schottky diodes, *Appl. Phys. Lett.* **83**, 4181-4183 (2003).
[61] M. Treu, R. Rupp, H. Kapels, W. Bartsch, Temperature dependence of forward and reverse characteristics of Ti, W, Ta and Ni Schottky diodes on 4H-SiC, *Mat. Sci. Forum* **353-356**, 679-682 (2001).
[62] B.J. Skromme, E. Luckowski, K. Moore, M. Bhatnagar, C.E. Weitzel, T. Gehoski, D. Ganser, Electrical characteristics of Schottky barriers on 4H-SiC: the effects of barrier height nonuniformity, *J. Electron. Mat.* **29**, 376-383 (2000).
[63] V. Saxena, J.N. Su, A.J. Steckl, High-voltage Ni- and Pt-SiC Schottky diodes utilizing metal filed plate termination, *IEEE Trans. on Electron Devices* **46**, 456-464 (1999).
[64] S.P. Muraka, Self-aligned silicides or metals for very large scale integrated circuit applications, *J. Vac. Sci. Technol. B* **4**, 1325-1331 (1986).
[65] C.S. Pai, C.M. Hanson, S.S. Lau, X-ray diffraction and ion backscattering study of thermally annealed Pd/SiC and Ni/SiC, *J. Appl. Phys.* **75**, 618-619 (1985).

[66] F. Roccaforte, F. La Via, A. Baeri, V. Raineri, L. Calcagno, F. Mangano, Structural and electrical properties of Ni/Ti Schottky contacts on silicon carbide upon thermal annealing, *J. Appl. Phys.* **96**, 4313-4318 (2004).
[67] I. Ohdomari, S. Sha, H. Aochi, T. Chikyow, S. Suzuki, Investigation of thin Ni/single-crystal SiC interface reaction, *J. Appl. Phys.* **62**, 3747-3750 (1987).
[68] H. Höchst, D.W. Niles, G.W. Zajac, T.H. Fleisch, B.C. Johnson, J.M. Meese, Electronic structure and thermal stability of Ni/SiC(100) interfaces, *J. Vac. Sci. Technol. B* **6**, 1320-1325 (1988).
[69] M. Levit, I. Grimberg, B.Z. Weiss, Interaction of $Ni_{90}Ti_{10}$ alloy thin film with 6H-SiC single crystal, *J. Appl. Phys.* **80**, 167-173 (1996).
[70] W.F. Slijkerman, A.E.M.J. Fischer, J.F. van der Veen, I. Ohdomari, S. Yoshida, S. Misawa, Formation of the Ni-SiC(001) interface studied by high-resolution ion backscattering, *J. Appl. Phys.* **66**, 666-673 (1989).
[71] M. Nathan, J.S. Ahearn, On the nanometer-scale solid-state reactions at thin-film Ni/amorphous SiC and Co/amorphous SiC interfaces, *J. Appl. Phys.* **70**, 811-820 (1991).
[72] A.S. Edelstein, D.J. Gillespie, S.F. Cheng, J.H. Perepezko, K. Landry, Reactions at amorphous SiC/Ni interfaces, *J. Appl. Phys.* **85**, 2636-2641 (1999).
[73] R.F. Egerton, *Electron Energy-Loss Spectroscopy in the Electron Microscope*, second edition, Plenum Press, New York, 1996.
[74] F. La Via, F. Roccaforte, V. Raineri, M. Mauceri, A. Ruggiero, P. Musumeci, L. Calcagno, A. Castaldini, A. Cavallini, Schottky–ohmic transition in nickel silicide /SiC-4H system: is it really a solved problem?, *Microelectronic Engineering* **70**, 519–523 (2003).
[75] S.Y. Han, J.L. Lee, Effect of Interfacial Reactions on Electrical Properties of Ni Contacts on Lightly Doped n-Type 4H-SiC, *J. Electrochem. Soc.* **149**, G189-G193 (2002).
[76] L. Calcagno, E. Zanetti, F. La Via, F. Roccaforte, V. Raineri, S. Libertino, F.Giannazzo, M. Mauceri, P.Musumeci, Schottky-Ohmic transition in nickel silicide/SiC system: is it really a solved problem?, *Mat. Sci. Forum* **433-436**, 721-724 (2003).
[77] F. Roccaforte, F. La Via, V. Raineri, P. Musmeci, L. Calcagno, Electrical characterization of nickel silicide contacts on silicon carbide, *Mat. Sci. Forum* **389-393**, 893-896 (2001).
[78] F. La Via, F. Roccaforte, V. Raineri, M. Mauceri, A. Ruggiero, P. Musumeci, L. Calcagno, Schottky-Ohmic transition in nickel silicide/SiC-4H system: the effect of non uniform Schottky barrier, *Mat. Sci. Forum* **457-460**, 861-864 (2004).
[79] R.T. Tung, Recent advances in Schottky barrier concepts, *Mat. Sci. Eng. R* **35**, 1-138 (2001).
[80] F. Roccaforte, F. La Via, V. Raineri, R. Pierobon, E. Zanoni, Richardson's constant in inhomogeneous silicon carbide Schottky contacts, *J. Appl. Phys.* **93**, 9137-9144 (2003).
[81] F. Roccaforte, L. Calcagno, P. Musumeci, F. La Via, Ion-irradiation effect on the Ni/SiC interface reaction, *Mat. Sci. Forum* **353-356**, 255-258 (2001).
[82] T. Nakamura, M. Satoh, Schottky barrier height of a new ohmic contact $NiSi_2$ to n-type SiC, *Solid-State Electron.* **46**, 2063-2067 (2002).
[83] B.J. Johnson, M.A. Capano, The effect of titanium on Al-Ti contacts to p-type 4H-SiC, *Solid-State Electron.* **47**, 1437-1441 (2003).
[84] B. Pécz, L. Tóth, M.A. di Forte-Poisson, J. Vacas, Ti_3SiC_2 formed in Al/Ti contacts to p-type SiC, *Appl. Surf. Sci.* **206**, 8-11 (2003).
[85] K. Vassilevski, K. Zekentes, K. Tsagaraki, G. Constantinidis, I. Nikitina, Phase formation at rapid thermal annealing of Al:Ti:Ni ohmic contacts on 4H-SiC, *Mat. Sci. Eng. B* **80**, 370-373 (2001).
[86] R. Kakanakov, L. Kassamakova, I. Kassamakov, K. Zekentes, N. Kuznetsov, Improved Al/Si ohmic contacts to p-type 4H-SiC, *Mat. Sci. Eng. B* **80**, 374-377 (2001).
[87] J. Crofton, S.E. Mohney, J.R. Williams, T. Isaacs-Smith, Finding the optimal Al-Ti alloy composition for use as an ohmic contact to p-type SiC, *Solid-State Electron.* **46**, 109-113 (2002).
[88] J. Crofton, P.A. Barnes, J.R. Williams, J.A. Edmond, Contact resistance measurements on p-type 6H-SiC, *Appl. Phys. Lett.* **62**, 384-386 (1993).

[89] F. Moscatelli, A. Scorzoni, A. Poggi, G.C. Cardinali, R. Nipoti, Improved electrical characterization of Al-Ti ohmic contacts on p-type ion implanted 6H-SiC, *Semicond. Sci. Technol.* **18**, 554-559 (2003).
[90] N. Lundberg, M. Östling, Thermally stable low ohmic contacts to p-type 6H-SiC using cobalt silicides, *Solid-State Electron.* **39**, 1559-1565 (1996).
[91] O. Nakatsuka, Y. Koide, M. Murakami, CoAl ohmic contact material with improved surface morphology for p-type 4H-SiC, *Mat. Sci. Forum* **389-393**, 885-888 (2002).
[92] T. Sakai, K. Nitta, S. Tsukimoto, M. Moriyama, and Masanori Murakami, Ternary TiAlGe ohmic contacts for p-type 4H-SiC, *J. Appl. Phys.* **95**, 2187-2189 (2004).
[93] L. Kassamakova, R. Kakanakov, N. Nordell, S. Savage, Thermostable ohmic contacts to p-type SiC, *Mat. Sci. Forum* **264-268**, 787-790 (1998).
[94] N.A. Papanicolau, A. Edwards, M.V. Rao, W.T. Anderson, Si/Pt Ohmic contacts to p-type 4H-SiC, *Appl. Phys. Lett.* **73**, 2009-2011 (1998).
[95] J.O. Olowolafe, J. Liu, R.B. Gregory, Effects of Si and Al interface layers on the properties of Ta and Mo contacts to p-type SiC, *J. Electron. Mater.* **29**, 391-397 (2000).
[96] K.H. Jung, N.I. Cho, J.H. Lee, S.J. Yang, C.K. Kim, B.T. Lee, K.H. Rim, N.K. Kim, E.D. Kim, Titanium-based ohmic contact on p-type 4H-SiC, *Mat. Sci. Forum* **389-393**, 913-916 (2002).
[97] S.K. Lee, C.M. Zetterling, E. Danielsson, M. Östling, J.P. Palmquist, H. Högberg, U. Jansson, Electrical characterization of TiC contacts to aluminum implanted 4H-silicon carbide, *Appl. Phys. Lett.* **77**, 1478-1480 (2000).
[98] A.A. Iliadis, S.N. Andronescu, K. Edinger, J.H. Orloff, R.D. Vispute, V. Talyansky, R.P. Sharma, T. Venkatesan, M.C. Wood, K.A. Jones, Ohmic contacts to p-type 6H-SiC using focused ion-beam surface modification and pulsed laser epitaxial TiN deposition, *Appl. Phys. Lett.* **73**, 3545-3547 (1998).
[99] J.R. Waldrop, Schottky barrier height of metal contacts to p-type alpha 6H-SiC, *J. Appl. Phys.* **75**, 4548-4550 (1994).
[100] A. Suzuki, Y. Fujii, H. Saito, Y. Tajima, K. Furukawa, S. Nakajima, Effect of the junction interface properties on blue emission of SiC blue LEDs grown by step-controlled CVD, *J. Crystal Growth* **115**, 623-627 (1991).
[101] A. Scorzoni, F. Moscatelli, A. Poggi, G.C. Cardinali, R. Nipoti, Contact resistivity and barrier height of Al/Ti ohmic contacts on p-type ion implanted 4H- and 6H-SiC, *Mat. Sci. Forum* **457-460**, 881-884 (2004).
[102] S.E. Mohney, B.A. Bull, J.Y. Lin, J. Crofton, Morphological study of the Al-Ti ohmic contact to p-type SiC, *Solid-State Electron.* **46**, 689-693 (2002).
[103] A. Parisini, A. Poggi, R. Nipoti, Structural characterization of Al/Ti and Ti contacts on SiC, *Mat. Sci. Forum* **457-460**, 837-840 (2004).
[104] Y. Luo, F. Yan, K. Tone, J.H. Zao, J. Crofton, Searching for device processing compatible ohmic contacts to implanted p-type 4H-SiC, *Mat. Sci. Forum* **338-342**, 1013-1016 (2000).
[105] K.V. Vassilevski, G. Costantinidis, N. Papanicolau, N. Martin, K. Zekentes, Study of annealing conditions on the formation of ohmic contacts on p+ 4H-SiC layers grown by CVD and LPE, *Mat. Sci. Eng. B* **61-62**, 296-300 (1999).
[106] J.H. Zhao, K. Tone, S.R. Weiner, M.A. Caleca, H. Du, S.P. Withrow, Evaluation of ohmic contacts to p-type 6H-SiC created by C and Al coimplantation, *IEEE Electron Dev. Lett.* **18**, 375-377 (1997).
[107] A. Baeri, V. Raineri, F. La Via, V. Puglisi, G.G. Condorelli, Environment influence on Ti diffusion and layer degradation of a SiC/Ni_2Si/TiW/Au contact structure, *J. Vac. Sci. Technol. B* **22**, 966-970 (2004).
[108] A.Baeri, V.Raineri, F. Roccaforte, F.La Via, E. Zanetti, Study of TiW/Au thin films as metallization stack for high temperature and harsh environment devices on 6H Silicon Carbide, *Mat. Sci. Forum* **457-460**, 873-876 (2004).
[109] L.G. Fursin, J.H. Zhao, M. Weiner, Nickel ohmic contacts to p- and n-type 4H-SiC, *Electronics Lett.* **37**, 1092-1093 (2001).

International Journal of High Speed Electronics and Systems
Vol. 15, No. 4 (2005) 821–866

World Scientific
www.worldscientific.com

SILICON CARBIDE SCHOTTKY BARRIER DIODE

Jian H. Zhao and Kuang Sheng
SiCLAB, Rutgers University, 94 Brett Road, Piscataway, NJ 08854, USA

Ramon C. Lebron-Velilla
NASA Glenn Research Center, 21000 Brookpark Road, Cleaveland, OH 44135, USA

This chapter reviews the status of SiC Schottky barrier diode development. The fundamentals of Schottky barrier diodes are first provided, followed by the review of high-voltage SiC Schottky barrier diodes, junction-barrier Schottky diodes and merged-pin-Schottky diodes. The development history is reviewed and the key performance parameters are discussed. Applications of SiC SBDs in power electronics circuits as well as other areas such as gas sensors, microwave and UV detections are also presented, followed by discussion of remaining challenges.

1. Introduction

Silicon Carbide was one of the earliest semiconductor materials discovered. However, due to the difficulties of growing good quality crystals, the progress of SiC devices has lagged behind those of its counterpart materials. In recent years, SiC devices development have enjoyed remarkable progress and demonstrated significant advantages over Si devices in the area of high power electronics due to its superior physical and electrical properties.

With a significantly wider band gap (2.3eV for 3C-SiC, 2.9eV for 6H-SiC and 3.2eV for 4H-SiC) than silicon, the critical field of SiC is approximately eight to ten times higher than that of the latter. As a result, SiC high voltage devices can be realized on much thinner drift layers with higher doping, leading to orders-of-magnitude lower device on-state resistance.

Silicon Carbide (SiC) Schottky Barrier Diode (SBD) was the first SiC power device demonstrated. It was studied as early as 1974 [1]. Throughout the years, as single-crystalline SiC material became commercially available and its quality improved steadily, substantial amount of work and study has been done on SiC SBDs as well as Merged P-i-N Schottky diodes (MPS) and Junction Barrier Schottky (JBS) diodes. Schottky contacts on 3C, 6H and 4H SiC with a large variety of metals were made and their performance studied. Device designs on SBD, MPS and JBS diodes were also extensively investigated. Impressive progresses and developments have been made. Recently, SiC SBDs with breakdown voltage higher than 10kV has been reported [2]. SiC MPS diodes capable of 4.3kV were also reported [3]. SiC SBD has proved itself to be an important device as it became the first commercially available SiC power device. It is poised to replace existing

Si power fast recovery diode in most power electronics applications because of their greatly improved speed, substantially reduced system losses and improved system efficiency.

In this chapter, recent developments will be reviewed on the studies of Schottky contact physics, the design and development of power and high voltage SiC SBD, MPS and JBS diodes and their applications in power electronics as well as SiC SBDs for other applications.

2. SiC Schottky Contacts

2.1 Theory of SiC Schottky contact

When a metal is making intimate contact with SiC material, the Fermi levels in the two materials must be coincident at thermal equilibrium and the vacuum level must be continuous across the interface. In addition, the metal work function is different from that of the semiconductor. These two requirements determine a unique energy band diagram for an ideal metal-semiconductor contact where surface states are absent, as shown in Fig. 1 [4]. For this ideal case, the potential barrier, known as Schottky Barrier Height (SBH), at the metal-semiconductor interface is simply the difference between the metal work function $q\phi_M$ and the electron affinity of the semiconductor $q\chi_S$:

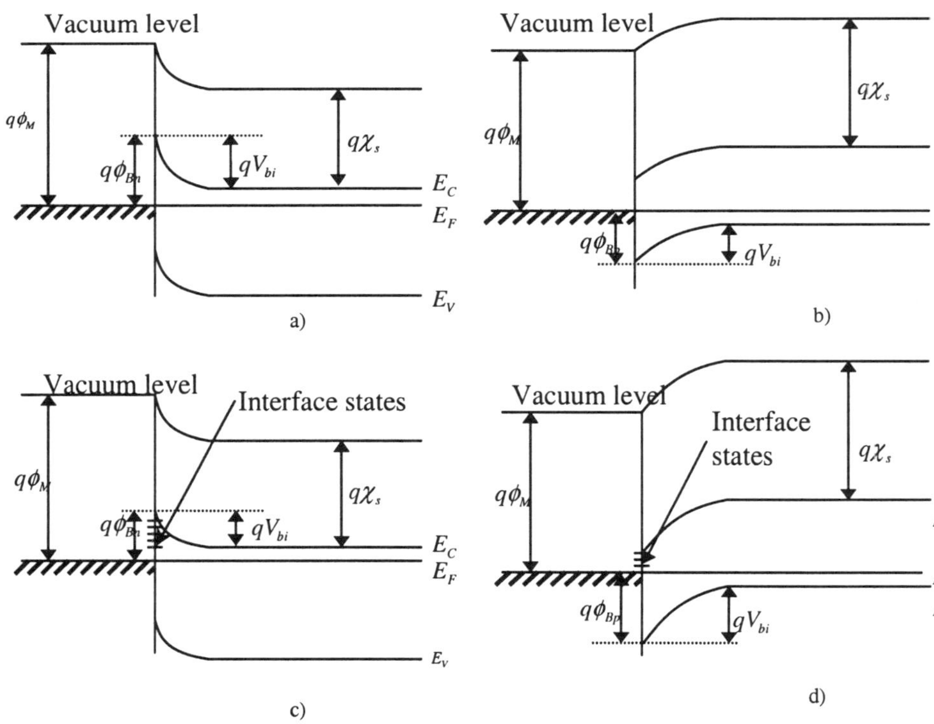

Fig. 1 Energy band diagram of metal-semiconductor contact. a) n-type SiC contact, b) p-type SiC contact without interface states, c) n-type SiC contact and d) p-type SiC contact with a large amount of interface states.

$$q\phi_{Bn} = q(\phi_M - \chi_s) \tag{1}$$

If the semiconductor material is p-type, the barrier height can be determined using a similar procedure:

$$q\phi_{Bp} = E_g - q(\phi_M - \chi_s) \tag{2}$$

In these equations, q is the electron charge and E_g is the band gap energy. The barrier potential determines the characteristics of a metal-semiconductor contact. Note that these equations apply to the ideal MS contact where the interface states have been neglected (parts a and b of Fig. 1).

The other extreme case would be that when the interface state density is large, the position of the Fermi level is pinned down to a certain level within the forbidden gap of the semiconductor, as shown in parts c) and d) of Fig. 1. In this case, the barrier height

$$q\phi_{Bn} = E_g - E_F \tag{3}$$

will be independent of the metal work function. This case is often referred to as Bardeen's limit. Most metal-semiconductor contacts in reality do have interface states but they may not be large enough to pin down the Schottky contact barrier height. Therefore, their barrier heights will be affected by both the metal work function as well as the interface state density. In practice, the barrier heights are found to be dependent on the SiC polytype, the metal work function, the surface condition before contact formation and the interfacial chemistry.

If the presence of a thin layer of insulator between the metal and the SiC material is taken into account, the barrier height can be given as follows [5].

$$\phi_{Bn} = [c_2(\phi_M - \chi) + (1-c_2)(\frac{E_g}{q} - \phi_0) - \Delta\phi_b]$$

$$+\left\{\frac{c_2^2 c_1}{2} - c_2^{3/2}\left[c_1(\phi_m - \chi) + (1-c_2)(\frac{E_g}{q} - \phi_0)\frac{c_1}{c_2} - \frac{c_1}{c_2}(V_n + \frac{kT}{q}) + \frac{c_1^2 c_2}{4}\right]^{1/2}\right\}$$

$$\text{where } c_1 = \frac{2q\varepsilon_s(N_d - N_a)\delta^2}{\varepsilon_i^2}, \text{ and } c_2 = \frac{\varepsilon_i}{\varepsilon_i + q^2\delta D_s} \tag{4}$$

Where ϕ_{Bn} denotes the barrier height at flat band condition, ϕ_0 is the energy level at the surface, N_d is the donor doping, N_a is the acceptor doping, δ is the thickness of the insulating layer, ε_i and ε_s are the permittivity of the insulator and the semiconductor, respectively, D_s is the density of the interface states, $\Delta\phi_b$ is the image force lowering of the potential barrier:

$$\Delta\phi_b = \sqrt{\frac{q\mathrm{E}_m}{4\pi\varepsilon_s}} \tag{5}$$

where E_m is the field intensity at the surface of the semiconductor.

There are four mechanisms that contribute to the current transport across a metal-semiconductor interface: (i) thermionic emission over the potential barrier, (ii) quantum-mechanical tunneling of the carriers through the potential barrier, (iii), carrier recombination in the depletion region and (iv) carrier recombination in the neutral region of the semiconductor [5]. For wide band gap semiconductor materials such as SiC, the dominant contribution to the current transport comes from the thermionic emission (TE) of electron over the barrier. If the SiC is not heavily doped and the operating temperature is not very low, the thermionic emission (TE) term will be dominant. The tunneling term will only be significant if the SiC is heavily doped and the device is operating under low temperatures. The thermionic emission current density can be expressed as:

$$J_n = [A^*T^2 \exp(-\frac{q\phi_{Bn}}{kT})][\exp(\frac{qV_F}{kT}) - 1]$$
$$= J_{ST}[\exp(\frac{qV_F}{kT}) - 1] \qquad (6)$$

where $J_{ST} \equiv A^*T^2 \exp(-\frac{q\phi_{Bn}}{kT})$ and $A^* = \frac{4\pi qm^*k^2}{h^3}$ is the effective Richardson's constant (for 4H-SiC $146 \text{Acm}^{-2}\text{K}^{-2}$ [43], h is the Plank's constant and m* is the effective mass).

2.2 Schottky barrier height of different metal contacts to SiC

Extensive work has been carried out in investigating the Schottky contact of different metals to the SiC [6-42]. As pointed out in the previous section, the behavior of the Schottky contact is dependent on the contacting metal, the polytype of SiC, surface condition as well as the interfacial chemistry. Therefore, although basic material properties are the same, Schottky barrier heights reported from different labs can be quite different.

Some basic material properties of the three main polytypes of SiC and silicon are listed in table 1 and the work functions of some metals are shown in table 2.

Table 1 Material property comparison of Silicon and SiC with different polytypes

	Silicon	3C-SiC	6H-SiC	4H-SiC
Bandgap (eV)	1.1	2.3	2.9	3.2
Relative dielectric Constant	11.9	9.7	9.7	9.7
Breakdown field at (N_D=5e15cm^{-3}) (MV/cm)	0.3	1.5	2.2	2.3
Electron mobility (cm^2/V/s)	1350	900	370	800
Hole mobility (cm^2/V/s)	480	40	80	120
Saturated electron velocity (10^7cm/s)	1.0	2.0	2.0	2.0
Electron affinity (eV)	4.05	3.8	3.3	3.1

Table 2 Work functions of various metals being investigated for Schottky contact to SiC.

Metal	Mg	Mn	In	Ag	Al	Ti	Mo	Cu	Au	Pd	Ni	Pt
ϕ_M(ev)	3.65	4.15	4.20	4.25	4.28	4.33	4.6	4.65	5.1	5.12	5.15	5.65

The electron affinities of the 3C, 6H and 4H SiC (0001) are 3.8eV, 3.3eV and 3.1eV, respectively. In the ideal contact case, as the metal work function increases, the barrier height for a certain SiC polytype will increase with the same amount. In practice, the dependence of the barrier height on the metal work function is weaker, as can be seen in Eq. 4.

The barrier height of a practical Schottky contact can be extracted from either current-voltage curves or the capacitance-voltage curves.

For moderately doped semiconductors, when recombination current can not be neglected, the current density (J) given in Eq. 6 will be modified to the following:

$$J = J_S\left[\exp\left(\frac{qV}{nkT}\right)-1\right],\quad J_S = A^*T^2\exp\left(-\frac{q}{kT}\phi_B\right) \tag{7}$$

where J_S is the saturation current density, n is the ideality factor and ϕ_B is the barrier height. When the forward voltage (V_F) is larger than $3\eta kT/q$, the above equation can be re-written as:

$$\ln\left(\frac{J_F}{T^2}\right) = \ln A^* - \frac{q}{k}\left(\phi_B - \frac{V_F}{n}\right)\frac{1}{T} \tag{8}$$

where $q(\phi_B\text{-}V_F/n)$ is referred to as the activation energy. This equation shows that, if $\ln(J/T^2)$ is plotted against 1/T for a given forward voltage drop, a linear dependence can be observed with a slope of $-q(\phi_B\text{-}V/n)/k$. By measuring the slope of such a plot, the barrier height of a Schottky contact can be extracted. One advantage of extracting Schottky barrier height from activation energy measurement is that it does not depend on the electrically active area of the contact which sometimes can be difficult to obtain. The ideality factor (n) in the equation can be obtained by measuring the slope of $\ln J_F$ versus forward voltage V_F.

$$n = \frac{q}{kT}\left[\frac{\partial V_F}{\partial(\ln J_F)}\right] \tag{9}$$

The ideality factor is a good indication of contribution from different current mechanisms. The thermionic emission current components has an ideality factor of 1 while that of the generation-recombination current component will be higher than 1. In practice, an ideality factor extracted to be very close to 1 is an indication that current is dominated by thermionic emission. Deviations from the ideal behavior can also be explained by the presence of lateral inhomogeneities of the Schottky barrier, as have been reported in [193, 194].

The barrier height can also be extracted from the C-V curves [5].

$$\phi_B = V_I + \frac{kT}{q}\ln\left(\frac{N_{DA}}{N_C}\right) + \frac{kT}{q} - \Delta\phi_b \tag{10}$$

$$N_{DA} = \frac{2}{q\varepsilon_S\left[-\frac{d(1/C^2)}{dV}\right]A^2} \tag{11}$$

where N_C is the conduction band density of states for SiC ($1.7\times10^{19}\text{cm}^{-3}$ for 4H-SiC) [44], V_I is the intercepting voltage of the $1/C^2$ versus voltage curve, as seen from Fig. 2, and A is the area of the diode. In addition to the two methods described above, the X-ray photoemission spectroscopy (XPS) can also be used to extract the barrier height.

A list of Schottky barrier height from n-type SiC Schottky contact studies reported in the literature for the recent years is shown in table 3. Only the most commonly used 3C, 6H and 4H SiC polytypes are included. In the table, the SiC polytype, the terminating surface layer of SiC, the contacting metal, its resultant barrier height extracted from the I-V and/or C-V method and the ideality factor are shown. For the same SiC polytype, the entries are listed with increasing metal work function. While the p-type SiC Schottky contacts are of less interest, there are also some reported works in the literature. Table 4 listed the SBH of these reported p-type Schottky contacts.

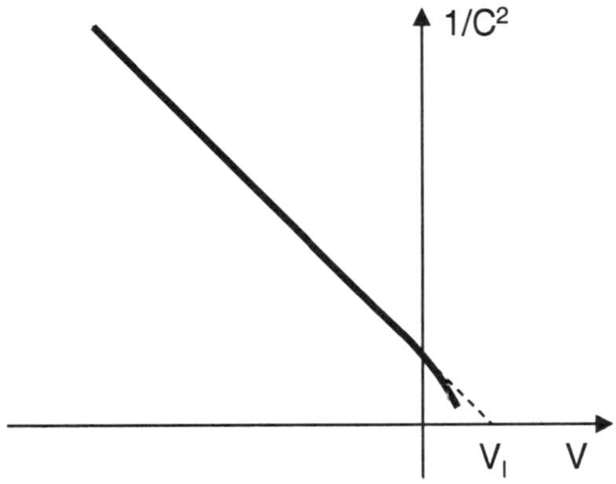

Fig. 2 Extraction of Schottky barrier height from the capacitance versus voltage curve, the intercepting voltage (V_I) is used in Eq. 11 for SBH calculation

As explained in the previous paragraphs, the barrier height does not simply increase with increasing metal work function, mainly due to the highly process dependent interface state density. Table 3 shows that, a fairly wide range of SBH (0.3eV to 2.0eV) can be achieved with different metals and processing conditions for the three SiC polytypes shown.

In the following paragraphs, SBH as a function of various influencing factors will be discussed.

Table 3 Barrier height of n-type Schottky contact to different SiC polytypes for various metals (300K unless otherwise stated)

Polytype	face	Metal	Barrier Height (eV)		Ideality Factor (n)	Refs
			I-V	C-V		
3C		Au		1.15		[45]
		Au		1.2	1.5	[46]
		Au	0.4-0.7		1.6-2.3	[28]
		Pt		0.95	1.5	[47]
6H	Si-	Mg	0.34^1		1.22	[14]
		Mg	0.69		1.3	[18]
	Si-	Mn	0.79	0.96	1.05	[14]
		In	0.84		1.3	[18]
	C-	Ag	1.10	1.21	1.08	[14]
	Si-	Ag	0.83	0.97	1.07	[14]
	C-	Al	0.89	0.96	1.06	[14]
	Si-	Al	0.26^2		1.72	[14]
		Al	0.92		1.3	[18]
	C-	Al	0.25		1.6	[20]
	Si-	Al	0.25		1.6	[20]
		Ti	0.85	0.85	<1.1	[15]

	C-	Ti	1.03	1.09	1.08	[48]
	Si-	Ti	0.73	0.75	1.05	[48]
		Ti		0.3	1.0	[22]
		Ti/Pt		0.8		[49]
		Mo		0.7	1.0	[22]
	C-	Au	1.19	1.3	1.08	[14]
	Si-	Au	1.37	1.48	1.07	[14]
		Au	1.12		1.15	[50]
	C-	Pd	1.62	1.58	1.05	[14]
	Si-	Pd	1.11	1.26	1.09	[14]
		Pd	1.01		1.3	[18]
	C-	Ni	1.68	1.76	1.04	[48]
	Si-	Ni	1.29	1.29	1.03	[48]
	Si-	Ni	1.41			[21]
	C-	Ni	1.30			[21]
		Ni		1.19	1.0	[22]
		Ni	1.39		1.02	[51]
		Pt	1.04	1.1		[52]
		Pt	1.05	1.05	<1.1	[15]
		$NiSi_2$	0.40			[53]
	Si-	Ti	0.95	1.17	1.02~0.2	[23]
4H	C-	Ti	1.16	1.3	1.02~0.2	[23]
	Si-	Ti	0.8		1.15	[54]
	Si-	Ti	0.85[3]		1.1	[54]
		Ti/Au/Pt/Ti	1.17		1.09	[55]
		Ti	1.17		1.06	[56]
	Si-	TiW	1.22	1.23	1.05	[57]
		TiW	1.18	1.19	1.1	[57]
		Ni_2Si	1.4		<1.1	[38]
		Cu	1.6		<1.1	[58]
		Cu	1.8		<1.1	[58]
	Si-	Au	1.73	1.85	1.02~0.2	[23]
	C-	Au	1.8	2.1	1.02~0.2	[23]
		Au		1.59		[59]
	Si-	Ni	1.62	1.75	1.02~0.2	[23]
	C-	Ni	1.6	1.9	1.02~0.2	[23]
		Ni		1.67		[60]
		Ni		1.64		[61]
	Si-	Ni	1.3		1.21	[54]
	Si-	Ni	1.59		1.05	[63]
		Ni	1.35		1.05	[63]
		Ni		1.7	1.07	[64]
		Ni	1.63		1.1	[65]
		Ni		1.63		[66]
	Si-	Pt	1.39		1.01	[63]

[1]Measured at 175K, [2]Measured at 150K, [3]Measured at 395K

Table 4 Barrier height of p-type Schottky contact to different SiC polytypes for various metals (300K unless otherwise stated)

	face	Metal	Barrier Height (eV)		Ideality Factor (n)	Comments	Refs
			I-V	C-V			
3C		Au	1.12	1.11		1~5×10^{16}cm^{-3}	[67]
6H		Ti/Al		1.95		5×10^{16}cm^{-3}	[25]
		Al	1.23	2.87	2.18	2~9×10^{15}cm^{-3}	[37]
		Mo	1.6				[19]
		Cu	1.22	1.39	1.01	2~9×10^{15}cm^{-3}	[37]
		Au	1.18	1.45	1.51	2~9×10^{15}cm^{-3}	[37]
4H		Ti/Al		1.5		1×10^{16}	[25]
		Ti	1.94	2.07	1.07	1.2×10^{16}	[35]
		Au	1.35	1.49	1.08	1.2×10^{16}	[35]
		Au	1.57				[19]
		Ni	1.31	1.56	1.29	1.2×10^{16}	[35]
		Ni	1.43		1.29	3×10^{15}	[68]

As shown in Fig. 1, the energy band diagram of the SiC is very important in determining the nature of Schottky contact. For SiC, the energy bandgap value increases with the degree of hexagonality in the polytype. Among the three polytype listed, 3C-SiC (pure cubic structure) has the narrowest and 4H-SiC (4H has 50% hexagonality and 6H has 33.3% haxagonality) has the widest bandgap [69].

While limited reports are available for 3C-SiC Schottky contact, it is still observable from the table that, for the same metal, the polytype with a wider bandgap and hence lower electron affinity (i.e., 4H-SiC) has a higher SBH than that for the polytype with narrower bandgap (3C-SiC). For Au, the barrier height is ~1.15eV, ~1.4eV and ~1.8eV for 3C, 6H and 4H SiC, respectively. Similar trend can be seen for Pt.

In addition to the polytype bandgap dependence, it has been established previously that the SBH is also dependent on the interface state density which in turn depends heavily on the polytype and the process conditions for surface treatments. In an attempt to quantify the degree of dependence of the interface state density on SiC polytype, Jacob and coworkers [70] performed a detailed analysis of the interface slope parameters for various metallization on 6H and 3C-SiC. The interface slope parameter is defined as $S=d\phi_B/d\phi_M$ or $S=d\phi_B/d\chi_M$, depending on whether the work function ϕ_M or electronegativity χ_M of the metal is used. A value of 1 for the interface slope parameter (S) means that an ideal contact following Equ. 1 (also named the Schottky-Mott theory) is achieved while S<1 indicates certain extend of Fermi-level pinning. Plots of the n-type contact SBH versus different metal work functions of 6H and 4H-SiC are shown in Fig. 3. While the SBH of different reports can vary in a wide range, there is a general trend of an increasing SBH with increasing metal work function. A few observations can be made from Fig. 3. Firstly, SBH of the Schottky contact is dependent on the SiC surface polarity, i.e., C-face of Si-face. It can be clearly seen that SBH of the same metal is significantly higher when making contact to C-face when compared to Si-face. This dependence of SiC SBH on the SiC surface polarity was extensively investigated in [14, 48, 71, 72]. Secondly, the SBH data extracted from measurement is also slightly dependent on the extraction methods. The SBH extracted with the C-V method is generally higher than that extracted with the I-V method by 0.05~0.15eV. This difference is likely caused by the

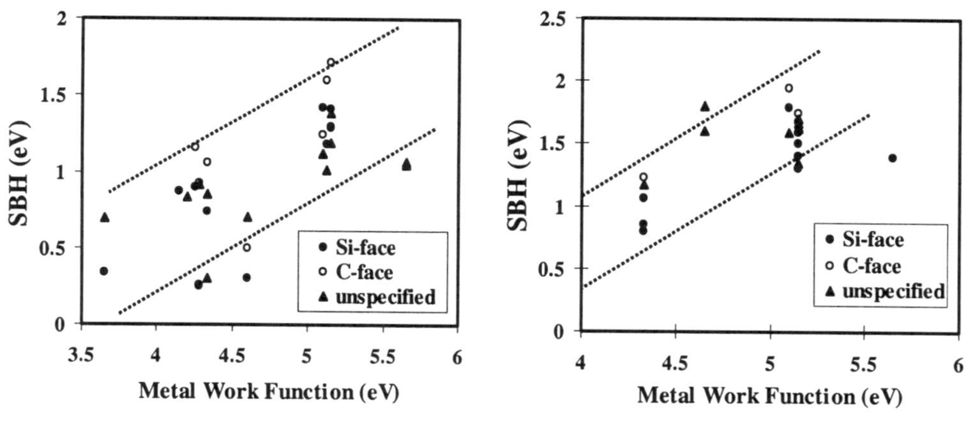

Fig. 3 The dependence of n-type Schottky barrier height on metal work functions, (a), 6H-SiC and (b), 4H-SiC. Solid circle: Si-face, open circle: C-face, solid triangle: unspecified. Both polytypes have similar interface slope parameters.

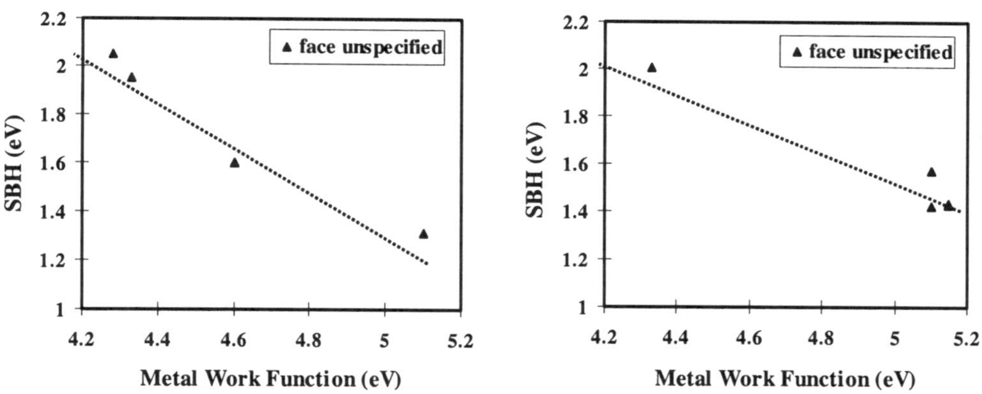

Fig. 4 The dependence of p-type Schottky barrier height on metal work functions, (a), 6H-SiC and (b), 4H-SiC. The limited data show that 4H-SiC has more severe interface Fermi-level pinning

presence of a thin intervening insulating layer between the SiC surface and the Schottky metal. This thin insulating layer decreases the junction capacitance, causes an increase in the intercepting voltage (V_I in Fig. 2) and the extracted SBH. On the other hand, the presence of the thin insulating layer has little effect on the I-V characteristics of the contact as current can easily tunnel through it. Thirdly, the SBH is generally higher for 4H-SiC than for 6H-SiC, as explained previously. Lastly, due to the interface state density, the overall interface slope parameter shown in Fig. 3 is found to be around 0.6 for 6H-SiC and 0.9 for 4H-SiC, indicating a stronger Fermi-level pinning for 6H-SiC. The interface state density was estimated to be in the range of 10^{13}~$10^{14}cm^{-2}eV^{-1}$ for 6H-SiC [73].

A similar plot for p-type SiC Schottky contact can be seen in Fig. 4. The interface slope parameter S can be extracted to be around 0.85 for 6H-SiC and 0.65 for 4H-SiC.

The Schottky contact interface property is strongly influenced by the surface condition prior to metal deposition as well as the post deposition annealing. Therefore, contaminations and interfacial oxide layers before metal deposition can significantly alter the SBH as well as junction electrical characteristics. This can be seen from the variation in SBH for the same metal for a specific SiC material.

There have been reports on surface treatment methods that will allow the Fermi-level to be un-pinned, i.e., virtually eliminates the interface state density. It has been reported in [22] that by a sequence of sacrificial oxidation, 5% HF etching and dipping into boiling water for 10min, the interface states can be reduced drastically. As a result, the contact approaches the ideal Schottky contact described in Fig. 1.a and the interface slope parameter (S or C_2 in Eq. 4) can approach unity. Plots of SBH versus metal work functions are shown in Fig. 5. Schottky interface property is also strongly influenced by the post deposition annealing process. With high temperature annealing, chemical reaction takes place on the interface of SiC and the Schottky metal. A silicide can develop on the interface and hence change the band diagram and the barrier height of the whole contact. As an example, a study [38] reported that by annealing the Ni-SiC Schottky contact at 600-800°C in N_2 ambient, a layer of Ni_2Si will be formed, which consumes a layer of SiC on the surface and hence eliminates the material damage caused by preceding process steps. This gives rise to a silicide-SiC interface that is relatively independent of surface treatment and achieves a reproducible, close-to-ideal Schottky contact.

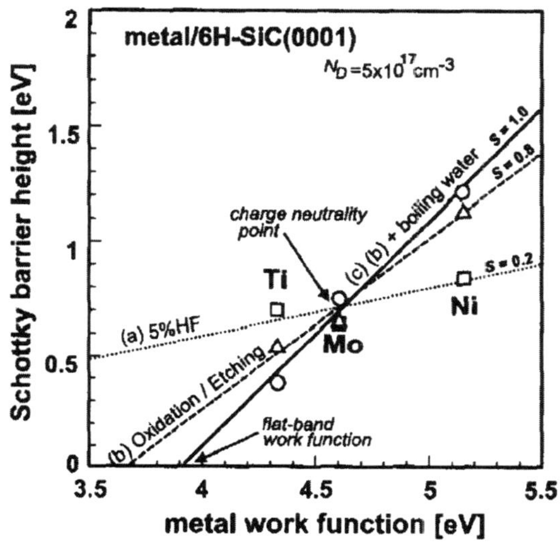

Fig. 5 Schottky barrier heights as a function of metal work functions. No treatment (short dashed line), oxidation followed by HF etching treatment (long dashed line), and dipping treatment in boiling water (straight line) are indicated. (after [22])

Moreover, it was also reported that the SBH of Pt Schottky contact on 6H-SiC will increase significantly after post deposition annealing at temperatures of 600°C and above [47, 15, 74]. As shown in Fig. 6, SBH of the Pt-SiC contact increased from 0.45eV to 1.35eV after annealing at 900°C. This was attributed to the formation of Platinum silicide at temperatures above 600°C, which forms a contact with higher Schottky barrier with 6H-SiC. Dependent on the metal and hence the silicide formed, post deposition annealing can also cause a barrier height to decrease [75]. Similarly, the SBH of a Ni contact to SiC with Si-face polarity is also found to decrease

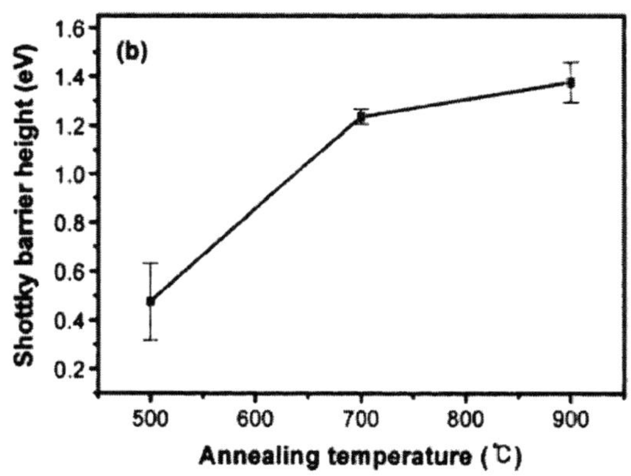

Fig. 6 Variation of Pt-SiC Schottky barrier height (SBH) as a function of annealing temperature. (after [74])

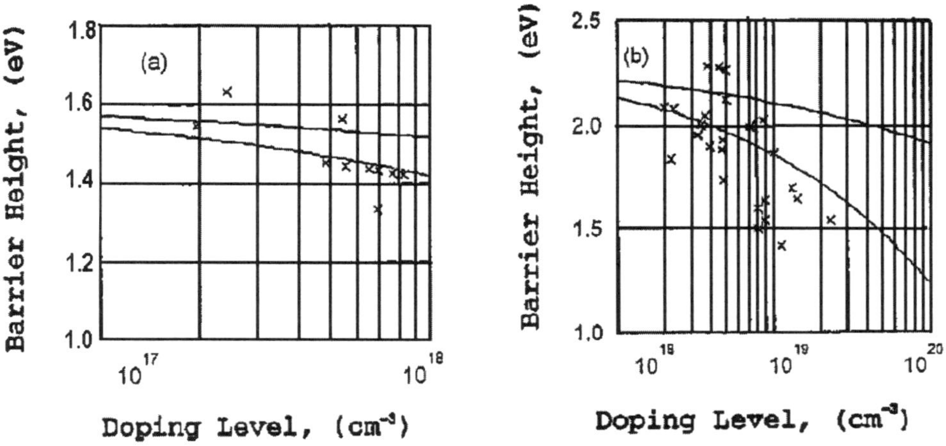

(a) (b)

Fig. 7 4H-SiC Schottky barrier heights against doping for (a), Au-n-4H-SiC contacts and (b), Au-p-4H-SiC contact. (symbol: experimental, top line: calculation according to image force barrier lowering and bottom line: calculation including total charge balance at the surface according to Eq. 4). (after [19])

with the substrate temperature during metal sputtering [21].

In [21], Ni barrier height is found to decrease with substrate temperature during metal sputtering for Si-face but increase for C-face 6H-SiC.

The SBH is also found to be dependent on the doping of the semiconductor for relatively highly doped materials. As shown in Fig. 7, the SBH tends to decrease as the donor concentration increases above the 2×10^{17}cm^{-3} level and the acceptor concentration increases above 1×10^{18}cm^{-3}. The barrier reduction is more significant than that caused by the image force lowering effect, as evident from Fig. 7. A calculation based on Eq. 4 taking into account the total surface charge balance gives a reasonable fit to the experimental data.

There have been some reports on the temperature dependence of the SBH [16, 76, 77, 78]. It was reported in [76] that, as the temperature increases, both the conduction band and valence band move downward relative to the semiconductor Fermi-level. As a result, the n-type SBH decreases and the p-type SBH increases with increasing temperature. Variation of the intercepting voltage of the $1/C^2$ versus voltage curve (V_I in Eq. 11, also known as the diffusion potential) against temperature for Ag contacts on n-type 6H-SiC is plotted in Fig. 8. A clear reduction in V_I, and hence the SBH can be seen.

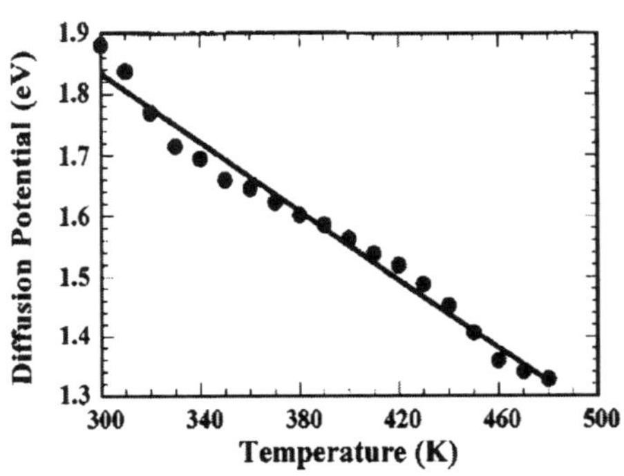

Fig. 8 Variation of the diffusion potential (V_I) against temperature for Ag contacts on n-type 6H-SiC. (after [76])

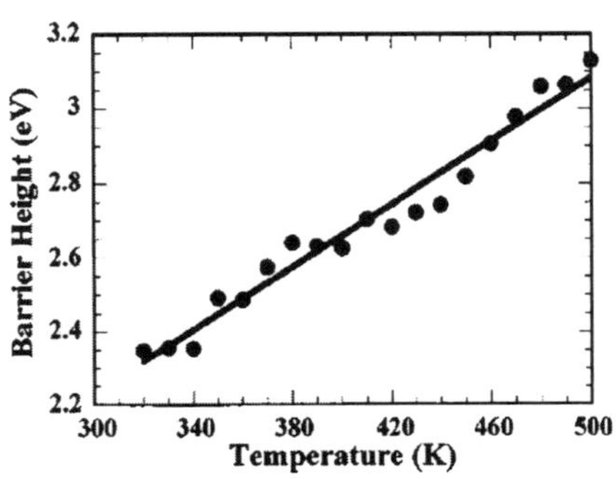

Fig. 9 Schottky barrier height of Al contacts on p-type 6H-SiC. A substantial increase with temperature is observed. (after [76])

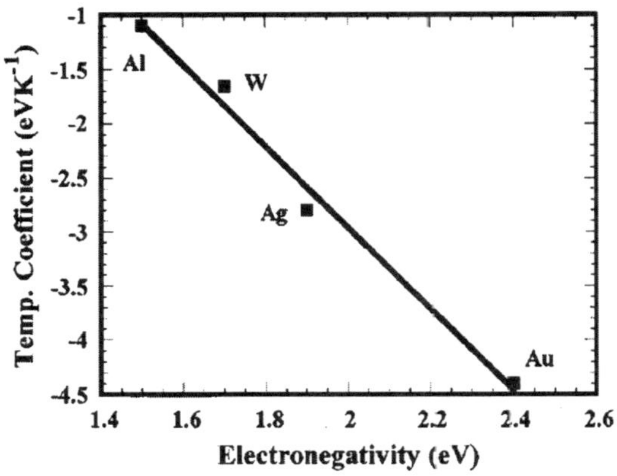

Fig. 10 Schottky barrier height temperature coefficient (in $meVK^{-1}$) versus electronegativity for different metals. (after [76])

On the other hand, the SBH of a p-type contact on 6H-SiC demonstrated positive temperature coefficient, indicating a shift of valence band away from the Fermi-level as temperature increases. The rate of change against temperature was found to be significantly higher for the p-type contacts when compared to the n-type contacts. While the bandgap is expected to narrow as temperature increases, which will lead to n-type SBH reduction, the magnitude of variation on the SBH reported is 10 to 20 times higher than that predicted by the bandgap narrowing. For Ag contacts on n-type 6H-SiC, temperature coefficient of the intercepting voltage (V_I) is 2.8meV/K. This V_I temperature coefficient was also found to be dependent on the electronegativity of type of metal being used. A metal with higher electronegativity will give a stronger SBH dependence on temperature.

Electronegativity describes the ability of an atom to attract an electron to itself. It is believed [76] that "as the temperature of a metal-semiconductor contact is increased, more carriers are freed in semiconductor. The Fermi level moves toward midgap. The barrier created at the interface initially separated carriers to achieve charge neutrality; as the temperature is increased, excess carriers will be drawn into the metal in order to maintain this charge neutrality at the interface" and hence lowering the barrier height. The better the metal's ability of attracting these carriers, the more barrier lowering occurs.

The temperature coefficients of the intercepting voltage (V_I) for n-type 6H-SiC contacts of four different metals are plotted in Fig. 10 against their electronegativity and a linear relation can be observed [76]. For metals with low electronegativity such as Al and Ti, the temperature coefficient of the intercepting voltage (V_I) and hence the Schottky barrier height is small and insignificant. A similar SBH reduction with increasing temperature was also reported in [77]. It is also recognized in [16, 78] that the temperature dependence of the SBH become significantly weaker at temperatures less than 300K. However, because SBH is very sensitive to the interface condition which can vary among different labs, lower SBH dependences on temperature have been reported elsewhere [195].

3. High Voltage SiC SBD, JBS and MPS diodes

For high power applications, it is well known that the specific on-resistance of unipolar semiconductor device, defined as product of device resistance and active device area, is a function of both the doping and thickness of the voltage-blocking layer. The specific on-resistance can be calculated by the following equation [79]:

$$R_{sp,on} = \frac{4V_B^2}{\mu \varepsilon_S E_C^3} \tag{12}$$

where E_C is the avalanche electric field and V_B is the corresponding avalanche breakdown voltage and μ is the electron mobility.

Due to its superior electrical material property, SiC as a semiconductor material has a unique advantage in high power and high temperature applications. According to Eq. 12, when compared to silicon, the roughly ten times higher critical electric field of 4H-SiC will give approximately three orders of magnitude improvement to the specific on-resistance for a given blocking voltage design. For this reason, a substantial amount of efforts has been invested in demonstrating and commercializing SiC unipolar device for power applications. SBD is the simplest unipolar device to fabricate and hence has been the first SiC power device commercialized. SiC SBD also has its unique advantage over SiC P-i-N diodes because of the high build-in voltage of the latter.

In this section, the development of SiC SBD will be discussed together with a few close variants such as SiC JBS diode, MPS diode, Dual Metal Trench (DMT) diode and Trench MOS Barrier Schottky (TMBS) diode. Termination techniques used to achieve a high blocking voltage; device structures used to minimize reverse leakage current; design approaches taken to obtain low on-state voltage/resistance and device performance under different temperatures will be discussed.

3.1 SiC SBD development

One of the earliest high voltage SBD was fabricated in 1975 [6] on 6H-SiC with an avalanche breakdown voltage of around 200V. More SiC devices were fabricated as high quality SiC become available in early 1990s and both the voltage and current capabilities of these SiC SBDs increase quickly in the years followed [54, 64, 66, 80-94]. Recently, a 4H-SiC SBD with a breakdown voltage of 10.8kV was reported. A forward voltage drop of 3.15V was achieved at $20A/cm^2$, corresponding to a specific on-resistance of $97.5m\Omega cm^2$ [2]. Considering the current spreading, a specific on-resistance of $187m\Omega cm^2$ can be estimated.

In table 5, some of the SBD reported in the literature is listed. They represent one of the highest voltage or current capability of SiC SBD at the time of the report. The progressive increase of voltage and current rating of SiC SBD can be clearly seen in the table. SiC SBDs with breakdown voltage of close to 1000V was first reported in [80] on $10\mu m$, $1\times10^{16}cm^{-3}$ Nitrogen doped 6H-SiC by using Pd in 1993. In mid-1990, as 4H-SiC material become available, it attracted most interests for SiC SBDs from 6H-SiC. This is because of its higher electron mobility and critical electric field resulted from its wider bandgap. 4H-SiC SBDs up to 1000 V were reported using a drift layer of $10\mu m$, $1\times10^{16}cm^{-3}$ Nitrogen doped 4H-SiC in 1995 [82]. The 1000-V 4H-SiC SBDs were later reported by using higher barrier metals of Ni and Pt, again based on a $10\mu m$ thick drift layer doped in the range of $6.1\times10^{15}cm^{-3}$ to $1.6\times10^{16}cm^{-3}$ in 1998 with improved current density [26, 54]. Breakdown voltage of 4H-SiC SBD reached 2.5kV in 1998 and 4kV in 1999. In 2002, SiC SBD with a breakdown voltage of 4.9kV was demonstrated on $7\times10^{17}cm^{-3}$ doped, $100\mu m$ thick 4H-SiC [94]. In the same paper, a 1.7kV, $0.64cm^2$, 130A SBD was also reported

The increase of SiC SBD breakdown voltage from the beginning of the 1990s is plotted in Fig. 11. SiC SBD with the highest breakdown voltage reported so far was reported with a $5.6\times10^{14}cm^{-3}$ doped, $115\mu m$ thick drift region on 4H-SiC [2]. While further improvements on the SiC SBD breakdown voltage are still possible, the rapidly increasing specific on-resistance at higher voltages makes unipolar devices like SBD less attractive.

Table 5 High voltage Schottky barrier diodes reported in the literature. Work listed here has one of the highest voltage or current at the time when the work was reported. (300K unless otherwise stated)

Polytype	Device	Metal	V_{br} (V)	$R_{sp,on}$ ($m\Omega cm^2$)	year	Comments	Ref
6H	SBD	Pt	205		1975	$1\times10^{16}cm^{-3}$,	[6]
6H	SBD	Pt	400		1992	$3.6\times10^{16}cm^{-3}$, 1.1V@100A/cm^2	[95]
6H	SBD	Ti	500		1992	$2\text{-}6\times10^{16}cm^{-3}$, 1.1V@100A/$cm^2$	[15]
6H	SBD	Pd	1000		1993	$1\times10^{16}cm^{-3}$, 10μm	[80]
6H	SBD	Au	1100	8	1993	$5.8\times10^{15}cm^{-3}$, 9.6μm	[81]
6H	SBD	Ti	1000		1994	$2\times10^{16}cm^{-3}$, 10μm, argon impl.	[96]
4H	SBD	Au	800	1.4	1995	$5\times10^{15}cm^{-3}$, 10μm	[83]
6H	SBD	AlTi	600		1995	$1.3\times10^{16}cm^{-3}$, 4.7μm, guard ring	[97]
4H	SBD	Ti	1000	2	1995	$1\times10^{16}cm^{-3}$, 10μm	[82]
4H	SBD		1750	5	1995	7e15cm^{-3}	[98]
4H	SBD	Ti	1100		1996	$0.7\sim2\times10^{16}cm^{-3}$, 10μm, B^+ impl.	[99]
4H	SBD	Ti	1300		1996	$8\times10^{15}cm^{-3}$, 10μm, argon impl.	[100]
4H	SBD		1400	1.5	1996	$1\times10^{16}cm^{-3}$, 10μm	[84]
4H	SBD		>1000		1998	Large area, 7mm^2, 30A/pulse	[101]
4H	SBD		2500		1998		[86]
4H	SBD	Ti	600	70	1998	1×10^{16} cm^{-3}, 2μm, P-type	[25]
4H	SBD	Ni	3000	34	1998	$0.7\sim2\times10^{15}cm^{-3}$, 42~47μm	[61]
4H	SBD		4000		1999		[93]
4H	SBD		3850		2000		[66]
4H	SBD		1700		2001	Large area, 25A	[102]
4H	SBD		1700		2002	$5\times10^{15}cm^{-3}$, 15μm, 0.64cm^2, 130A	[94]
4H	SBD		4900	43	2002	$7\times10^{14}cm^{-3}$, 50μm, B^+ impl.	[192]
4H	SBD		10800	97.5	2003	$5.6\times10^{14}cm^{-3}$, 115μm, MJTE	[2]

3.2 Termination techniques

With appropriately selected drift region and thickness, the most important factor in achieving a high blocking voltage for SiC SBD is the edge termination technique used to reduce electric field crowding at the periphery of the power device. Many works were reported on edge termination of SiC SBD to achieve high breakdown voltages [2, 15, 51, 63, 95-100, 103-119].

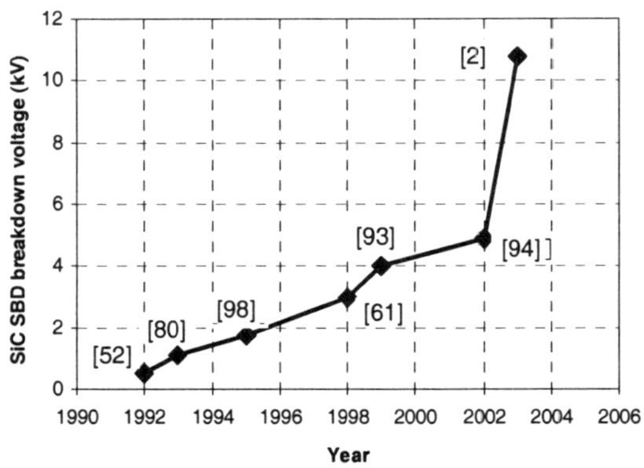

Fig. 11 The increase of SiC SBD breakdown voltage since 1992

Several techniques of edge terminations have been used to reduce the electric field crowding and to achieve a breakdown voltage close to the ideal parallel-plane avalanche breakdown voltage. These techniques include floating metal rings, metal field plating, argon/B^+ implantation termination, RESURF assisted field plate, modified field plate and multi-step junction termination extension (MJTE), etc.

The schematic structures of these termination techniques are presented in Fig. 12.

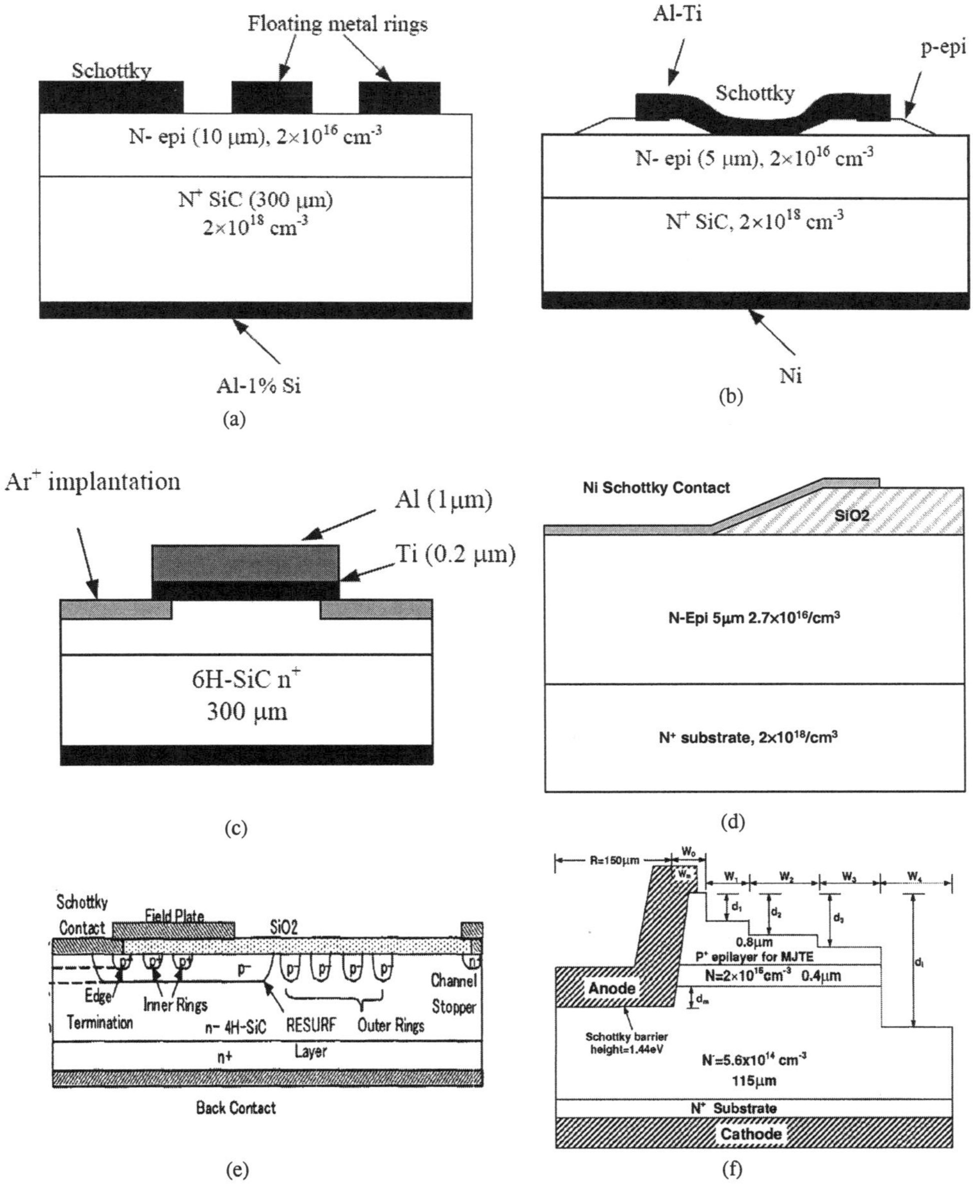

Fig. 12 Various termination techniques used for high voltage SiC SBDs. (a), floating metal rings, (b), p-epi guard rings, (c), argon implantation, (d), ramped oxide field plating, (e), guard ring assisted RESURF and (f), multi-step JTE. (after [2, 92, 117])

Floating metal ring method (Fig. 12.a) was used in [103] on 6H-SiC SBDs. A breakdown voltage of 600V was obtained with three rings. Similar approach was used in [114, 115] with multiple implanted P^+ and V^+ (Vanadium ion) regions as floating guard rings. Breakdown voltages of 1300V and 1630V were achieved on 4H-SiC wafers with drift regions of 10μm, 5e15cm^{-3} and 10μm, 1e16cm^{-3}, respectively. A p-epi guard ring termination approach was taken in [97, 104] on 6H-SiC and 600V was achieved. The shape of the p-epi (Fig. 12.b) was created by using a field oxidation process. It also gives rise to a smooth metal step over the p-epi, avoiding field crowding at the corner. In [108], a field plating technique was proposed and demonstrated in [51] on a ramped oxide (Fig. 12.d) to achieve a device breakdown voltage (800V) close to its ideal value. However, it was suggested in [119] that the method could give rise to electric field crowding in the passivation oxide at the edge of the field plate, raising concern of oxide reliability at high temperatures. The method of implanting a high dose of argon into the surface of the SiC area at the periphery of the device has been used and voltages close to ideal breakdown were also achieved [95, 15, 103, 96, 84]. By implanting a high dose of inert gas atoms into SiC, the exposed SiC will be damaged and turns into a high resistive region, which effectively reduces the field crowding at the device periphery (Fig. 12.c). The method was proven to be effective and SBDs with breakdown voltages of 1kV [96], 1.4kV [84] and 1.75kV [98] have been reported. However, Boron or Arsenic implanted edge termination may result in lower long term reliability of the SBD believed to be caused by the trapping effect of the defects in the implanted region. In [117], a method combining the floating guard ring and the RESURF effect was adopted (Fig. 12.e) and found to be able to achieve a good breakdown voltage (~1.2kV) for a wide range of implantation doses. On the other hand, the multi-step JTE (MJTE) approach was used in [116] and [2], achieving breakdown voltages of 1kV and 10.8kV, respectively. The MJTE structure shown in Fig. 12.f uses the multiple epitaxial p-extension zones to gradually reduce the electric field toward the edge of the SBD. Each step can be obtained through inductively coupled plasma (ICP) etching which can provide accurate etching control. It was demonstrated that close to ideal breakdown voltage can be achieved for both very high and medium voltage devices.

3.3 Reverse leakage current

Reverse leakage current is an important parameter for a diode used in power electronics systems. When blocking reverse voltage, a high leakage current can cause significant amount of heat dissipation. Because the reverse leakage current of a SBD is typically very sensitive to the junction temperature, if the leakage-current-caused heat dissipation reaches certain level, the resultant temperature increase can initiate a positive feedback in the thermal system and a thermal run-away will take place. The damage caused by a thermal run-away is normally permanent. Therefore, minimizing the reverse leakage current is always an important consideration for the device designers.

Two components out of the four listed in section 2.1 will contribute significantly to the reverse leakage current of a SiC SBD. The first is the thermionic emission current which is the dominating component for SBDs on silicon. The thermionic emission current can be calculated according to Eq. 6. It is worth noting that as the reverse bias voltage increase, the image force lowering effect described in Eq. 5 will continuously decrease the Schottky barrier (ϕ_B) and hence cause the reverse leakage current to increase. For a

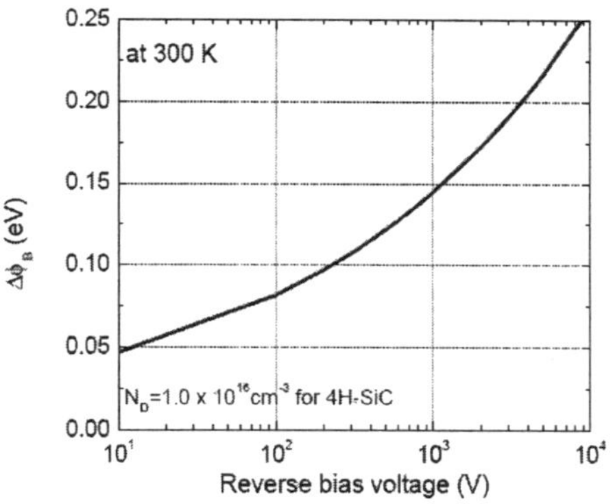

Fig. 13 Calculated Schottky barrier height reduction at room temperature due to the image force lowering versus reverse bias voltage. (after [92])

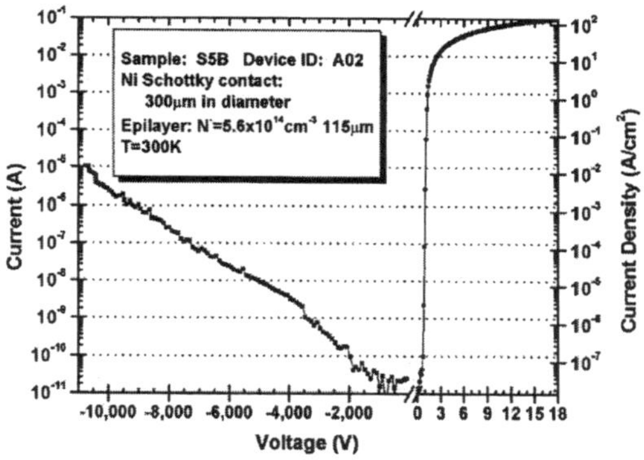

Fig. 14 Reverse and forward I-V and J-V curves for the 10.8kV Schottky barrier diode. (after [2])

one-dimensional Schottky junction, the surface electric field (E_m) in Eq. 5 can be written as:

$$E_m = \sqrt{\frac{2qN_D}{\varepsilon_S}(V_R + V_{bi})}$$

where V_R is the reverse bias voltage and V_{bi} is the built-in voltage for the Schottky junction. The amount of Schottky barrier lowering resulted from the increasing reverse bias voltage can be calculated as shown in Fig. 13. Because of the exponential dependence of the reverse leakage current on the Schottky barrier height, the barrier lowering can have significant effect on the increase of leakage current with increasing reverse bias voltage.

Because of the wide bandgap and hence high critical electric field of the SiC material, as the reverse bias voltage of the diode increases, the Schottky junction can experience electric field as high as >2MV/cm. This can contribute to the reverse leakage significantly.

Fig. 14 shows the reverse leakage of the 10.8kV 4H-SiC SBD in [2], which can be seen to increase over almost six orders of magnitude. Although the device has not reached avalanche, measurement of the device was stopped at 10.8kV because of the concerns on the device self-heating. It is common for SiC SBDs that the avalanche breakdown can not be safely measured because of the excessive reverse leakage current at high reverse blocking voltage, prompting serious self-heating problems.

Barrier height inhomogeneous has also been investigated as a possible reason of a much larger reverse leakage current than predicted by the thermionic emission current in the SiC SBD [120-123]. It was suggested in [120] that there are localized regions at the SiC/metal interface where Schottky barrier height is lowered due to the presence of epitaxial layer defects at the interface. Material inhomogeneous has also been suggested as possible causes.

3.3.2 Hybrid structures for leakage current reduction

In order to reduce the reverse leakage current of SiC SBDs, various methods have been proposed through the years [116, 124-141]. These include the Junction Barrier Schottky

(JBS) diode, Mergied P-i-N Schottky (MPS) diode, Dual Metal Trench Schottky (DMTS) diode, Dual Metal Planar (DMP) diode, Trench MOS Barrier Schottky (TMBS) diode.

In Table 6, various structures reported to reduce the reverse leakage current and increase on-off current ratio of the SBD are listed through recently years.

Table 6 High voltage JBS/MPS/DMT/TMBS diodes reported in the literature. (300K unless otherwise stated)

Polytype	Device	V_{br} (V)	V_F(V) /$R_{sp,on}$ (mΩcm^2)	year	Comments	Ref
4H	DMTS	450		1998	3×10^{15}cm^{-3}, 13μm, 2μm deep trench	[124]
4H	MPS	700	1.5V	1998	5×10^{15} cm^{-3}, 9μm, 100 A/ cm^2	[136]
4H	JBS	1000	3.1V/ 19mΩcm^2	1998	1×10^{15}cm^{-3}, 10μm, 100A/ cm^2	[137]
4H	JBS	2800	2.0V/7.5 mΩcm^2	1999	3×10^{15}cm^{-3}, 27μm, 100A/ cm^2	[138]
4H	JBS	3600	6.0V/43 mΩcm^2	2000	1.3-1.8$\times10^{15}$ cm^{-3}, 50μm, 100A/cm^2	[139]
4H	MPS	>1000	4.0V	2000	9.7×10^{15} cm^{-3}, 13μm, 638 A/cm^2	[140]
4H	JBS	700	2.5V	2000	10μm, 312A/cm^2, 50A	[141]
4H	MPS	800		2001	140A	[135]
4H	JBS	>1000	4.0V	2001	9.7×10^{15}cm^{-3}, 13μm, 630 A/ cm^2	[116]
4H	DMP	1250	1.0V	2002	3×10^{15} cm^{-3}, 10μm, 24A/ cm^2	[129]
4H	JBS	1500	3.0V	2002	6.4×10^{15}cm^{-3}, 10.5μm, 252A/cm^2	[130]
4H	JBS	3580V	2.8V	2002	1.5~3.0$\times10^{15}$cm^{-3}, 30μm, 100A/ cm^2	[131]
4H	MPS	1000V	3.2mΩcm^2	2003		[189]
4H	MPS	4308	4V/20.9mΩ cm^2	2004	2×10^{15} cm^{-3}, 30μm, 142 A/ cm^2	[3]
4H	MPS	1789	6.68mΩcm^2	2004		[190]

The structures of these various structures are plotted in Fig. 15.

The common idea behind the JBS, MPS, DMTS and TMBS structures is to protect the Schottky barrier from high electric field when the device is blocking reverse voltage. The image force lowering effect mentioned above are caused by an increasingly high electric field on the Schottky contact as reverse bias voltage increases. As can be seen in Fig. 15, JBS and MPS basically have the same structure except that the P^+ regions of these two devices are designed differently. In both structures, the P^+ regions are closely spaced so that when the device block reverse voltage, the depletion regions of two adjacent intruding P^+ regions overlap each other and shield the Schottky contact area between them from high electric field [79]. However, in the JBS, depth, width and spacing of the P^+ regions are designed such that the P^+-n junctions will not be forward biased in under normal operating conditions of the device. P^+ regions of the MPS, on the other hand, are designed differently so that when the device forward current density is high enough, the P^+-N junction can be forward biased, injecting minority carriers into the drift region and reduce the device forward conduction voltage drop. In order to forward

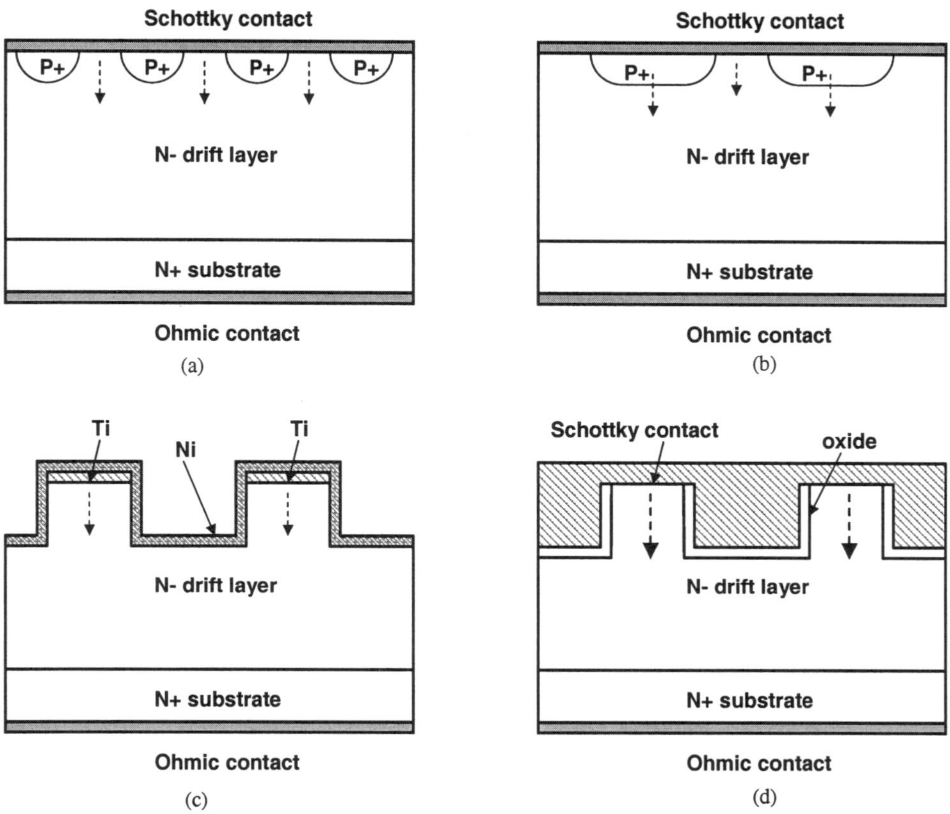

Fig. 15 Hybrid diode structures based on Schottky contact, (a), Junction Barrier Schottky (JBS) diode, (b), Merged P-i-N Schottky (MPS) diode, (c), Dual Metal Trench Schottky (DMTS) diode and (d) Trench MOS Barrier Schottky (TMBS) diode. All structure utilize the common principle of electric field shielding to reduce the field at the Schottky contacts.

bias the MPS P^+-n junctions, the voltage difference across the junction, which is caused by lateral current flow under the P+ regions, need to be close to the junction build-in potential. Because of the significantly higher build-in potential of SiC P^+-n junction, it will take a higher current density than its silicon counterpart to turn these junctions on.

The principle of electric field shielding also applies to the DMTS and TMBS diodes. In the DMTS diodes, closely located trenches are filled with metals with higher barrier heights while for TMBS diodes, similar trenches are oxidized before they are filled with metals. In both cases, the trenches play the role of protecting the Schottky contact in between from high electric field, and hence reducing diode reverse leakage current.

Similar principle was proposed for a lateral SiC Schottky based diode named Lateral Merged Double Schottky (LMDS) diode in [128].

On the other hand, the dual metal planar (DMP) structure works with a different principle [129]. It was suggested that reverse leakage current of a SiC SBD is mainly attributed by those at the edge of the device because of the higher electric field there caused by imperfect field termination. The DMP diode was therefore proposed whereby a second Schottky metal with higher SBH is deposited at the periphery of the lower barrier Schottky metal. Since the reverse leakage current of a Schottky contact depends exponentially on the SBH, the higher SBH of the second metal reduces the reverse leakage current at the device edge where electric field are the highest. The majority of active device area are still covered by the main Schottky contact metal with a lower SBH, ensuring a low forward voltage drop. In contrast to the other four structure discussed above, the DMP diode structure remains planar and no region protrusion occurs.

3.3.3 Temperature dependence of reverse leakage current

Both thermionic emission and generation-recombination components of the reverse currents are sensitive to temperature.

Fig. 16 shows the reverse leakage currents density of Ti and Ni Schottky barrier diode on $1.6\times10^{16}cm^{-3}$ doped 10μm thick n-type Si-face 4H-SiC at temperatures of 20°C, 122°C and 255°C. It is clear that the leakage current increases quickly with increasing temperature and decreasing Schottky barrier height.

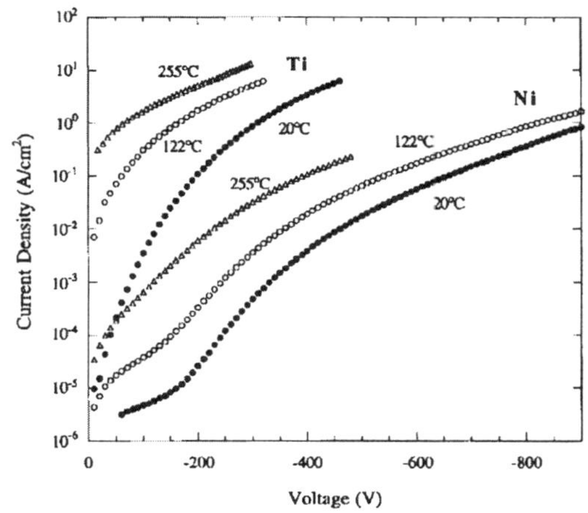

Fig. 16 Dependence of reverse leakage current density on temperature for Ti and Ni Schottky barrier diode on $1.6e16cm^{-3}$ doped 10μm thick n-type Si-face 4H-SiC (after [54])

3.4 Forward voltage drop

3.4.1 Forward voltage drop optimization

The forward voltage drop of a SiC diode with Schottky contact can be written based on Eq. 8 as:

$$V_F = \frac{nkT}{q}\ln\left(\frac{J_F}{A^*T^2}\right) + \eta\phi_B \tag{14}$$

Considering all parasitic resistances, the forward voltage drop of a Schottky barrier diode in reality takes the following form.

$$V_F = \frac{nkT}{q}\ln\left(\frac{J_F}{A^*T^2}\right) + \eta\phi_B + I_F R_{sp,on} / A \tag{15}$$

where I_F is the forward current, A is the device active area and $R_{sp,on}$ is the diode specific on-resistance which is an important parameter for a SBD. A low specific on-resistance is desirable since a smaller chip size will be needed to handle a given amount of current. The specific on-resistance is normally calculated as the differential resistance of the SBD forward J-V curve and its components comprises the resistance of the electrodes, the back-side ohmic contact, the substrate and the drift region. Besides a low specific on-resistance, the on-set voltage of a SBD should also be as low as possible. For high voltage n-type SiC SBD, electrode, ohmic contact and substrate resistances are typically negligible and the drift region resistance is the dominating factor.

Optimizing the drift region resistance is therefore critical in minimizing the forward voltage drop of a high voltage SiC SBD. Two critical parameters, namely the doping and thickness of the drift layer, determine both the voltage blocking capability and the specific on-resistance of the device. There is a clear trade-off between these two targeted performance parameters. While high voltage blocking capability requires a thick and lightly doped drift layer, a low specific on-resistance demands exactly the opposite. Detailed analysis can be found in [79] on this trade-off. Usually, the targeted blocking voltage rating and the effectiveness of the edge termination employed determines the minimum thickness and the maximum doping used for the design of drift layer. The specific on-resistance of the drift layer used for an n-type SiC SBD can be written as:

$$R_{drift} = \frac{W_D}{q\mu_n N_D} \tag{16}$$

where W_D is the drift layer thickness and μ_n is the electron mobility which is itself dependent on N_D. A non-punch-through (NPT) design of the drift layer corresponds to the case where drift layer thickness is equal or larger than the parallel plane avalanche breakdown depletion width. Similarly, a punch-through (PT) design corresponds to a drift layer thinner than the parallel plane avalanche breakdown depletion width. Design rules on the doping and thickness of the drift layer for NPT structures have been explained in [79]. However, it has been demonstrated that an optimized PT design can give a lower specific drift layer resistance for a given targeting breakdown voltage [94].

For a punch through structure, the drift region thickness can be derived as:

$$W_D = \frac{\varepsilon E_C}{qN_D} - \sqrt{\left[\frac{\varepsilon E_C}{qN_D}\right]^2 - \left[\frac{2\varepsilon V_B}{qN_D}\right]} \tag{17}$$

The critical electric field E_C in the above equation is dependent on drift layer doping (and is weakly dependent on drift layer thickness when the thickness is very low) and can be approximated by [191]:

$$E_C = \frac{2.49\times10^6}{1-\frac{1}{4}\log(N_D/10^{16})} \tag{18}$$

With the equations above, the optimization of drift layer doping and thickness can be performed to achieve the lowest possible specific on-resistance of a unipolar device for a targeting breakdown voltage.

In Fig. 17, the drift layer thicknesses and specific on-resistances are plotted against the drift layer doping for a targeted voltage blocking capability of 1200V. In this figure, it is assumed that the SBD edge termination can obtain 80% of the ideal parallel breakdown voltage of the drift layer. The result shows that the optimum design is a punch-through (PT) structure with approximately 90% of the NPT doping and 87% of the NPT drift layer thickness, giving 7% lower specific on-resistance than the NPT design.

As has been described in Eq. 12, the theoretical limit of the specific on-resistance of a SiC unipolar power device is around two to three orders of magnitude lower than that of silicon. In Fig. 18, experimentally demonstrated SiC diodes with one of the lowest specific on-resistances are plotted versus the device breakdown voltage.

The forward voltage drops of the JBS, MPS, DMTS, TMBS structures described in the previous section can all be written as the sum of the Schottky barrier voltage (V_{SB}) and the voltage drop on the channel/grid resistance (R_{grid}) between field-shielding regions,

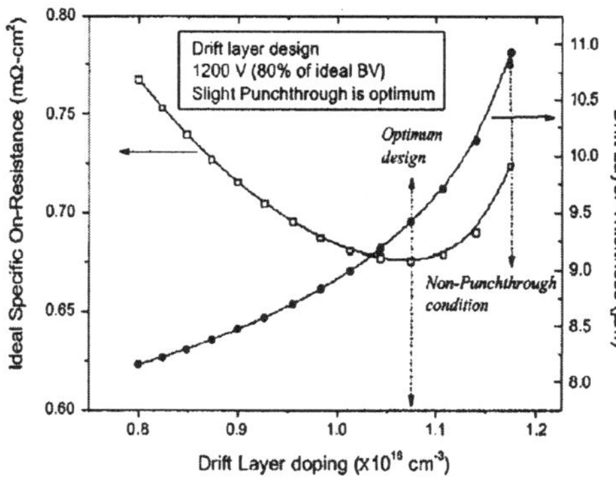

Fig. 17 Drift layer specific on-resistance and thickness versus doping for a breakdown of 1200V, assuming an 80% edge termination efficiency. The result shows that the optimum design can be achieved at approximately 90% of the NPT doping and 87% of the drift layer thickness. (after [66])

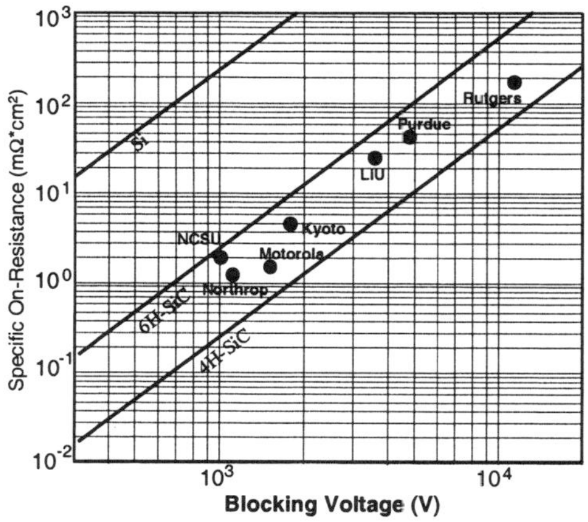

Fig. 18 Specific on-resistance versus breakdown voltage plot for Schottky barrier diodes

the drift region resistance (R_{dr}), the substrate resistance (R_{sub}) and the backside ohmic contact resistance (R_{SB}).

$$V_F = V_{SB} + I_F(R_{grid} + R_{dr} + R_{sub} + R_{contact}) \tag{19}$$

For all four structures, the forward conduction voltage will be somewhat increased because of the reduction in Schottky contact area through which current flows. (The only exception is if the MPS diode has its P^+-n junction forward biased, injecting minority carriers and modulating the drift layer conductivity.) The amount of increase in forward voltage drop is a function of relative area of the P^+ regions to the total active device area [116, 124]. The smaller the relative area of the P^+ regions, the less penalty on the forward voltage drop will incur. For devices with high blocking voltage, the drift region resistance component in Eq. 16 normally dominates. The increase of forward voltage, which arises from R_{grid}, will only be a small percentage of the total forward voltage drop [116, 124].

3.4.2 Temperature coefficient of the forward voltage drop

The variation of the forward drop of a SBD (or JBS, MPS) at different temperature can be determined by its three components given in Eq. 15. For a given current density within the range of practical interest, the first term in the equation is negative and decreases with increasing temperature. The second term in Eq. 15, i.e., the barrier height, also decreases with increasing temperature, as discussed in section 2.2. The third term in the forward drop equation accounts for the contributions of resistances of drift region, substrate region and the ohmic contact region. For SBD with a low substrate resistance and a good ohmic contact on the cathode (anode for p-type SBD), the resistance is dominated by that of the drift region resistance, which increases substantially because of mobility degradation as temperature increases.

As a result, the temperature coefficient of the total SBD forward voltage drop depends on the relative magnitude of the three terms in Eq. 15 against temperature. When the current density is low, the first two components in the equation dominate and the temperature coefficient of V_F is negative. For the range of current density of practical interest (e.g., 1~500A/cm^2), because of the exponential dependence of current on the voltage for the first two components, they remain relatively constant. On the other hand, the third resistive term increases linearly with current density and starts to dominate at higher current densities. The temperature coefficient of the total device forward voltage will therefore change from negative to positive as the current density increases above certain level. Fig. 19 shows the forward voltage drop of a SBD with relatively low voltage blocking capability (1.2~1.4$\times 10^{16}$cm^{-3}, 4μm) against temperature for different current densities. It can be seen that at around 300A/cm^2, the temperature coefficient of V_F becomes positive. For a SiC SBD with thicker and more lightly doped drift region, hence a higher blocking voltage, the contribution of the resistive term will be more significant and the V_F temperature coefficient will be positive at a lower current density.

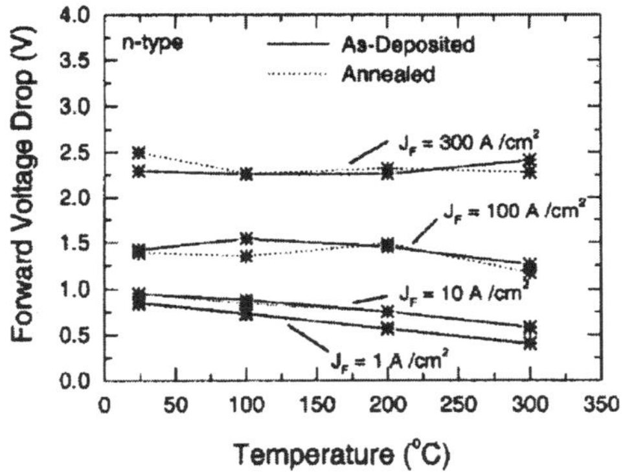

Fig. 19 Forward voltage drops of a Schottky barrier diode as a function of temperature for different current densities. (after [92])

4. Applications in Power Electronics Circuits

4.1 Importance of power diodes and Silicon limitation

A major application of SiC Schottky barrier diodes is in power electronics systems. Performance of a power electronics system is largely determined by those of the power semiconductor devices, namely, the power switches and the fast diodes. During recent years, there have been substantial improvements on Si switches. These include the commercialization and optimization of the Insulated Gate Bipolar Transistor (IGBT) in the last 10 to 15 years. IGBTs now dominate the power semiconductor device market in the range of 600V to 3.3kV for most power electronics applications. Fast IGBTs have a switching speed of less than 100ns while still having conductivity modulation. The commercialization of CoolMOS™ [142] device offers 600V to 800V rated power MOSFETs with specific on-resistance ($R_{sp,on}$) 5 times lower than the best traditional power MOSFETs, breaking the theoretical $R_{sp,on}$ limit of Si unipolar devices. Unipolar devices like CoolMOSs can be switched with a switching speed of less than 10ns.

The improvements on the Si fast power diodes have been limited. In the range of 600V to 1.5kV, which represents the most commercially lucrative market portion, Si Schottky diodes are known as unsuitable for practical use because of their excessive on-state resistances and high leakage currents. Si P-i-N diodes are therefore invariably used. When conducting forward current, a P-i-N power diode structure features strong minority carrier injection which modulates the conductivity of the device drift region and hence drastically reducing its resistance. The injected carriers have to be removed from the drift region of the P-i-N diode when the diode enters reverse blocking mode. This charge-removal procedure significantly slows down the transition of the diode from forward conduction mode to reverse blocking mode when compared to that of a Schottky diode which is a unipolar device. The fastest Si power diodes available in the range of 600V to

1200V have reverse recovery times of around 50ns to 100ns. Such reverse recovery speeds are lower than desirable and result in a substantial amount of switching losses in both the power switch and the diode itself. The speed of the diode is slower than most power switches used in the circuit and has become the limiting factor preventing further improvements in many power electronics systems [143].

As mentioned in previous sections, a two to three orders of magnitude advantage in specific on-resistance is expected for Schottky-based diodes in SiC over those in Si. It is therefore possible to develop SiC Schottky-based unipolar diodes for voltages up to 2-3kV with a reasonably low on-state resistance. Unipolar diodes in the range of 600V to 3kV, when available, are very attractive for power electronics applications.

While high voltage (~1000V) SiC Schottky-based diodes have been made as early as 1993, most diodes reported in early to mid-90's are of small device sizes and therefore are unable to conduct sufficient amount of current to be practically useful in a power electronics circuit. As the defect density of high quality SiC wafers decreases, SiC Schottky barrier diodes able to conduct significant amount of current (>a few amperes) were first reported in 1998 [101] and later in 2001 [102, 135]. It was followed quickly by the successful commercialization of SiC SBDs by power semiconductor device companies including Cree Inc and Infineon AG. Currently, SiC SBDs from Infineon are available for voltage ratings of 300V and 600V with current capability of 20A and 12A, respectively. Cree currently offers SiC SBDs of 300V, 600V and 1200V with current capability of up to 20A. The commercial availability of these SiC SBDs provides a new opportunity of development for power electronics applications including high frequency switch mode power supply, power factor correction as well as low to medium power motor drive applications.

There have also been reports in the literature studying the circuit performances of commercial SiC SBDs as well as those fabricated in research labs, which demonstrated largely consistent results [131, 144-154]. In this section, performances of the SiC Schottky barrier diodes (or large area hybrid Schottky diodes) will be compared to silicon power diodes. The advantages and disadvantagse of the SiC diodes will be discussed.

4.2 Semiconductor device losses in a power electronics circuit

In a power electronics circuit, power semiconductor devices (diodes and switches) largely dominate the whole circuit performance. A major part of the losses of the circuit arises from those of the semiconductor devices. They include the losses of power diodes and switches:

$$\begin{aligned} P &= P_{Diode} + P_{Switch} \\ &= P_{D\_CON} + E_{D\_rr} f_{sw} + P_{S\_CON} + (E_{S\_on} + E_{S\_off}) f_{sw} \end{aligned} \tag{20}$$

where P_{D_CON} is the diode conduction loss, E_{D_rr} is the diode reverse recovery energy loss, P_{S_CON} is the switch conduction loss, E_{S_on} is the switch turn-on energy loss, E_{S_off} is the switch turn-off energy loss and f_{sw} is the switching frequency. Losses caused by the leakage current of both diodes and switches can normally be neglected unless the devices are operating at the upper limit of their temperature capability (~175°C for Si and >300°C for SiC).

Switching frequency of a power electronics circuit is an important parameter which dominates the value as well as physical size and weight of energy-storage passive components, namely, power inductors and capacitors. The size and weight of these passive components, which dominate the size and weight of the system, are inversely proportional to the device switching frequency. While a higher switching frequency is therefore desirable, a trade-off exists between the switching frequency and the switching losses (second and fourth terms in Eq. 20) which need to be kept below a certain level for system efficiency and thermal management considerations. Switches and diodes with low switching losses are therefore the key to size and weight reduction of power electronics systems.

4.3 Static characteristics comparison of commercial SiC and Si diodes

In this section, the static characteristics of some commercial SiC power diodes are compared to those of silicon power diodes with similar voltage and current ratings. The commercial SiC diodes measured include SDT10S30 (300V, 10A) and SDT06S60 (600V, 6A) from Infineon and CSD20060 (600V, 10A) and CSD1020 (1200V, 5A) from Cree. Silicon devices used for comparison are some of the best fast recovery diodes available in the market: DSEP8-03 (300V, 10A) and DSEP30-12A (1200V, 30A) from IXYS [150].

The forward voltage drop of the diode accounts for the diode conduction loss, the first term in Eq. 20. The forward I-V characteristics of the Infineon SDT10S30 and IXYS DSP8-03 are plotted in Fig. 20 for different temperatures. Both devices, rated at 300V and 10A, are measured at 25°C, 100°C and 200°C up to 10A. As 300V silicon Schottky diodes are not commercially available for their excessive specific resistance, the IXYS device chosen is a P-i-N device.

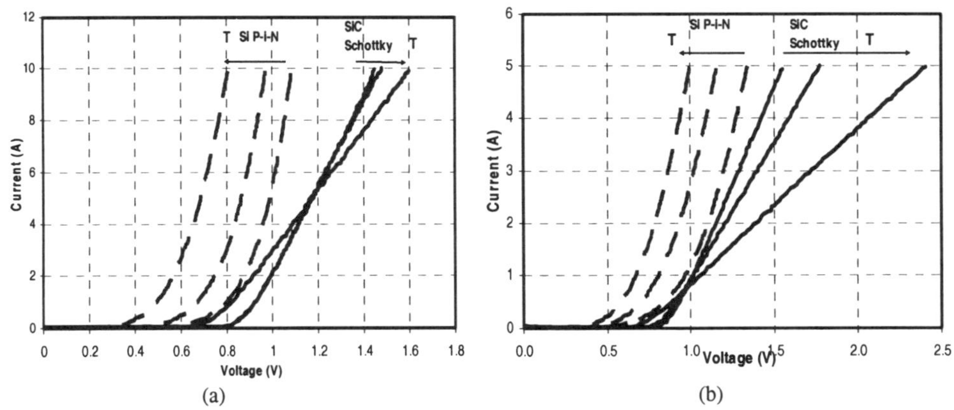

Fig 20 Forward I-V characteristic at 25°C, 100°C and 200°C for SiC Schottky diode and Si P-i-N diode, (a), Infineon SDT10S30 and IXYS DSEP8-03, both rated at 300V, (b), Cree CSD1020 and IXYS DSEP20-12, both rated 1200V. Solid lines: SiC SBD. Dashed lines: silicon diode. Direction of the arrows indicates the increase of temperature

It can be seen that at room temperature, the SiC devices has a forward drop of approximately 1.5V and the silicon device has 1.1V. A clear difference can be seen in Fig. 20.(a) between the two devices. As a bipolar power diode with high-level minority carrier injection in the epi base region, forward voltage drop of device DSEP 8-03 decreases significantly with increasing temperature. On the other hand, forward voltage drop of the SiC Schottky diode demonstrates a negative temperature coefficient at currents below 5A and a positive temperature coefficient at currents above 5A. The mechanism behind the variations of forward drop temperature coefficient has been discussed and agrees well with the analysis in section 3.4.2.

The silicon device has a lower on-state voltage drop for all operation current levels and temperatures. A lower voltage drop will give less power loss and therefore is advantageous in the circuit. However, the negative forward voltage temperature coefficient of the silicon power diode is undesirable for connecting multiple devices in parallel when handling large load current. It is understood that when devices are connected in parallel, such a negative coefficient will more likely result in uneven current sharing among different devices. When such unbalance reaches a certain level, the whole system can become thermally unstable. The SiC device has a positive forward voltage temperature coefficient for the current levels above 50% of the rating where the devices is most likely to be used. It will therefore be more stable in case of parallel connection.

A similar comparison can be found in Fig. 20.(b) where the forward I-V curves of CSD1020 (1200V, 5A SiC diode from Cree) and DSEP30-12A (1200V, 30A silicon diode from IXYS) are shown. Temperature coefficients of the forward voltage have shown similar trend for both devices with the silicon bipolar device demonstrating negative value and the SiC diode showing different polarity at different current levels.

It is worth noting that since the silicon device is rated at a higher current, its forward voltage drop at its rated current level (30A) is slightly higher than that of the SiC device at room temperature. It can also be seen that for the SiC device, temperature coefficient for the forward voltage becomes positive at a lower current when compared to the 300V SiC device. The reason is explained as follows. While the knee voltage (first term in Eq. 15) remains the same for the 300V and 1200V SiC devices, the drift region resistance becomes more dominant for the device with a higher voltage rating because of a thick and more lightly doped drift layer needed to block the required voltage. The temperature coefficient of the resistive component of the voltage drop will therefore also become more dominant. As explained in section 3.4.2, since the knee voltage decreases with increasing temperature and the drift layer resistance increases with temperature, the overall forward voltage temperature coefficient of the 1200V SiC diode is more positive than a 300V SiC diode.

While an increasing forward voltage is beneficial for device parallel connection, it does increase the conduction loss of the diode at high temperatures. The conduction losses of the SiC diodes are therefore generally higher than those of their silicon counter parts.

Another important static parameter of a diode is its reverse-blocking characteristics. In Figs 21, variations of the reverse-blocking I-V curves at different temperatures are plotted for the 300V devices and 1200V devices, respectively.

The reverse leakage currents of the SiC Schottky devices are in sub-μA level for most part of reverse blocking characteristics. It can be seen that, for both 300V and 1200V comparisons, the SiC Schottky diodes have lower leakage current except when the voltage increases to a level close to or exceeding the voltage rating. In practical applications, devices are not normally stressed with such voltages. Both figures also show that the leakage currents of the silicon P-i-N diodes are orders of magnitude higher than their SiC counterparts at 200°C. The excessive amount of leakage current for the silicon devices prevents them from being reliable at this high temperature.

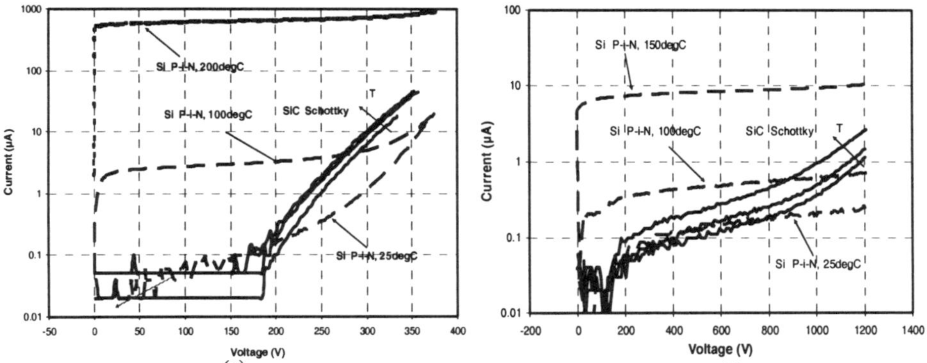

(a) (b)

Fig. 21 Reverse I-V characteristic of SiC Schottky diode and Si P-i-N diode at different temperatures, (a), Infineon SDT10S30 and IXYS DSEP8-03, both devices are rated at 300V, and (b) Cree CSD1020 and IXYS DSEP20-12, both rated at 1200V. Solid lines: SiC SBD. Dashed lines: silicon diode. Direction of the arrows indicates the increase of temperature for the SiC SBD.

Comparisons of forward and reverse I-V curves of SiC SBDs from different manufacturers with similar voltage and current ratings show that differences are insignificant. The small differences can be accounted for by different manufacturing process and die sizes.

4.4 Dynamic characteristics comparison of commercial SiC and Si diodes

Dynamic characteristics of the diodes are important for power electronics applications that require high frequency switching such as switch mode power supply, power factor correction, power inverter for induction heating and some motor drives. Low switching energy losses (E_{D_rr}, E_{S_on} and E_{S_off} in Eq. 20) are the key for high frequency switching which will enable a small system size and high efficiency.

Typical building block of a power electronics system is shown in Fig. 22. This circuit is also known as the 'clamped inductive load switching circuit'.

When the switch turns on, the load current diverts from the diode (normally named a freewheeling diode) to the switch. During the transient of switch turning on, the diode changes from forward conduction to the reverse blocking status. Reverse recovery time of the diode is the most important parameter for this transition since it determines the peak currents of diode and switch as well as the amount of losses incur on these two devices during this transition (E_{D_rr}, E_{S_on} in Eq. 20) [151].

Reverse recovery transients of the following three SiC diodes were evaluated at 25°C, 75°C and 150°C: UPSC200 (200V, 1A from Microsemi), SDT06S60 (600V, 6A from Infineon) and CSD10120 (1200V, 5A from Cree) [150].

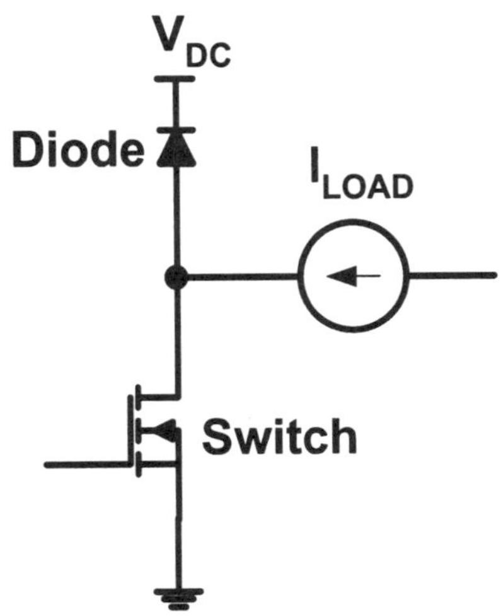

Fig. 22 Typical building block of a power electronics system comprises a switch, a free-wheeling diode.

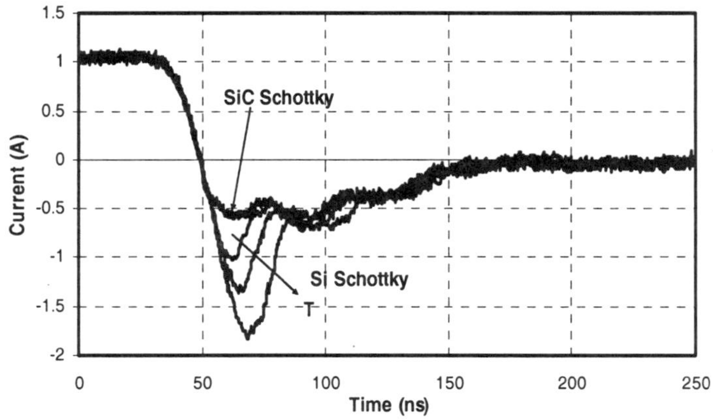

Fig. 23 Reverse recovery current waveforms at 25°C, 75°C and 150°C for the IR 10CTQ150 (150V/5A) Si Schottky diode and the Microsemi UPSC200 (200V/1A) SiC Schottky diode with V_{DC}=126V. Direction of the arrow indicates the increase of temperature.

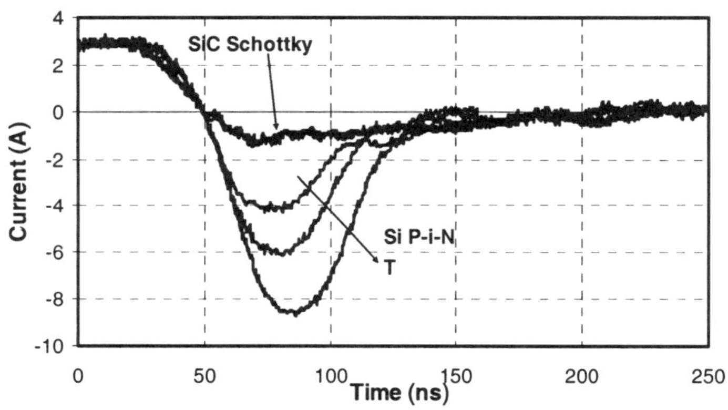

Fig. 24 Reverse recovery current waveforms at 25°C, 75°C and 150°C for Infineon SDT06S60 (600V/6A) SiC Schottky diode and the IXYS DSEI 8-06A (600V/8A) ultra fast Si diode with V_{DC}=400V. Direction of the arrow indicates the increase of temperature.

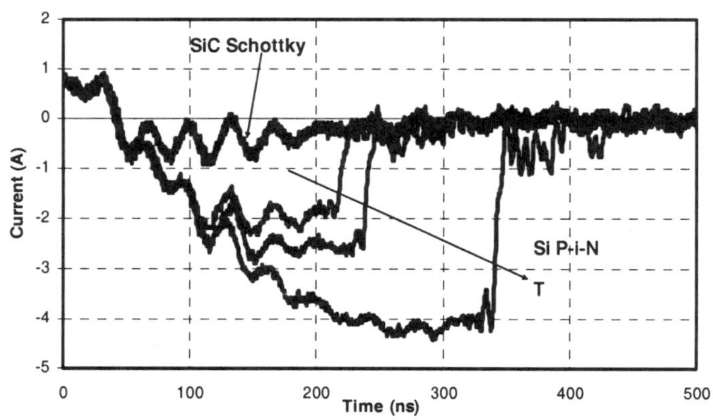

Fig. 25 Reverse recovery current waveforms at 25°C, 75°C and 150°C for Cree CSD10120 (1200V/5A) SiC Schottky diode and the IXYS DSEP30-12A (1200V/30A) ultra-fast silicon diode with V_{DC}=1000V. Direction of the arrow indicates the increase of temperature.

The reverse recovery current waveforms are shown in Figs. 23, 24 and 25. To leave some margin for safety, the DC voltage used in for these groups of measurements are 126V, 400V and 1000V, respectively.

In Fig. 23, the silicon device used is a 150V, 5A Schottky diode and both devices are tested at 1A forward current. It is clear that both reverse recovery time and peak current for the Si diode increase considerably with increasing temperature. The reverse recovery charge for the Si Schottky diode is seen to increase with temperature because thermal generation of minority carriers in Schottky diodes increases exponentially with temperature in the form of exp ($-E_g/2kT$). On the other hand, reverse recovery time and peak current of the SiC Schottky diode show no noticeable change as the temperature was increased from 25C to 150C, as seen by the reverse recovery curves lying on top of each other. Also, the reverse recovery current of the SiC Schottky diode is considerably lower than that of the Si Schottky diode.

The silicon diode used to compare with SiC diode SDT06S60 was DSEI8-06A (600V, 8A) from IXYS. Both devices were tested at 400V and 3A. It can be seen that the peak reverse recovery current of this silicon device increases from 4A at room temperature to over 8A at 150°C. Its reverse recovery time also increases substantially. The reverse recovery current waveforms of the SiC SBD remains almost unchanged for the different temperatures with a peak reverse current well below 2A. Similar comparison can be seen in Fig. 25 where the 1200V SiC diode was compared to a 1200V ultra-fast silicon diode (DSEP30-12A from IXYS).

It is seen that for SiC SBDs with these three blocking voltage, the reverse recovery current waveforms do not increase with increasing temperature. Similar findings have been reported in [145, 149, 152, 153]. This is due to the fact that as a unipolar device, the reverse recovery currents are mainly caused by internal device junction capacitance and, to a certain extend, some packaging capacitances. In contrast, there is a substantial increase of reverse recovery time and peak current at higher temperature in the silicon devices. This can be explained by the increased minority carrier injection and hence storage charge at higher temperatures. The storage charge of a silicon power diode also increases with device voltage rating because of a thicker charge storage layer. This can be used to explain that the SiC SBD offers more substantial advantage at higher voltage ratings.

The drastically reduced reverse recovery current and charge virtually eliminated the dynamic losses of the diode. In addition, the close-to-zero recovery time of the SiC SBDs can also reduce the turn-on energy (E_{S_on} in Eq. 20) of the corresponding switch substantially. Because the turn-on energy loss of the power switch is typically larger than the dynamic loss of the diode itself, more loss reduction can actually be obtained on the switch than on the diode itself. Switching loss reduction with either SiC MPSs or Si P-i-N diodes as freewheeling diodes has been demonstrated with the highest power in [170] for up to 2.5kV and 30A. Fig. 26 shows the voltage, current and power waveforms of an inductively loaded half-bridge inverter switching transient during switch turn-on with SiC MPS diodes or Si diodes as the freewheeling diodes. The turn-off energy loss of the diode was reduced from 8.0mJ to 4.6mJ while the turn-on energy loss of the IGBT was reduced from 54mJ to 34mJ.

Many high frequency power electronics circuit uses power MOSFETs as the switch. In these circuits, when silicon P-i-N diodes are used, switching speed is typically limited by that of the diodes. By replacing the bipolar silicon P-i-N diode with a unipolar SiC diode, the whole power electronics comprises of only unipolar semiconductor devices and

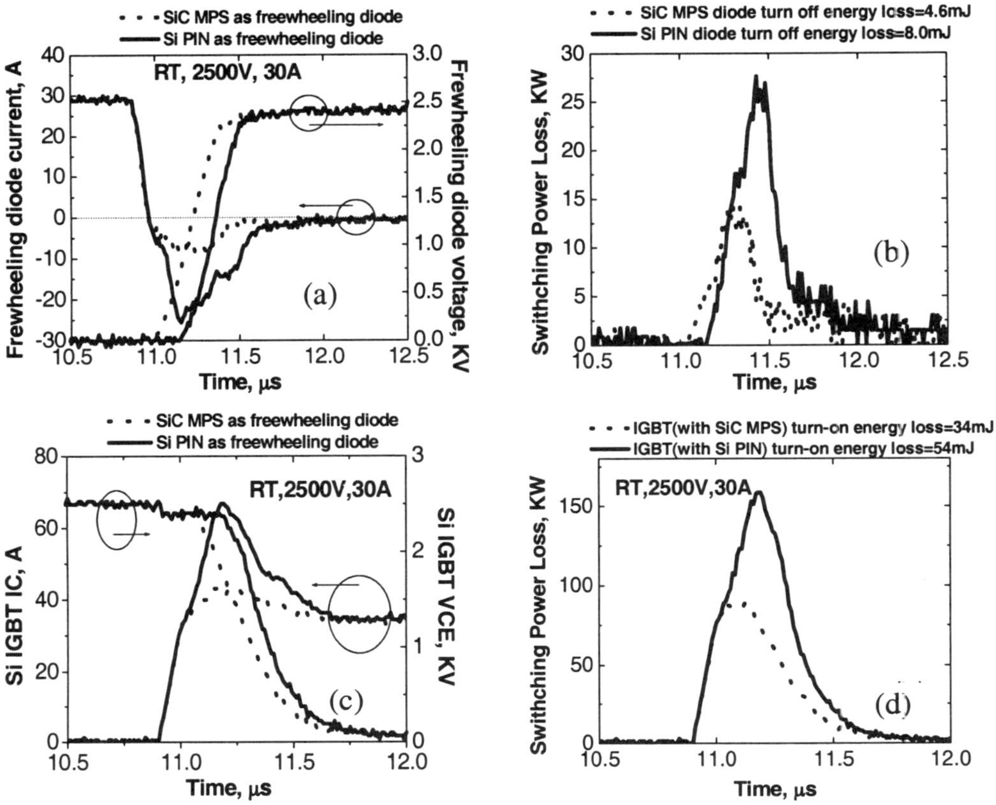

Fig. 26 Diode turn-off and Si IGBT turn-on waveforms of an inductively-loaded half-bridge inverter using either Si MPS diode or Si P-i-N diode as the freewheeling diode. (a), current and voltage of the diode, (b) power loss of the diode, (c), current and voltage waveforms of the IGBT and (d) power loss of the IGBT. (after [170])

the switching speed can be substantially increased. Power factor correction circuit using a combination of SiC Schottky diode and superior 600V CoolMOS has been reported to operate at close to 400kHz without significant amount of switching losses [154], as shown in Fig. 27. From the figure, it is reasonable to project that such combination can switch at frequencies above 1MHz. SiC SBDs as the zero-recovery diode will have great impacts on high frequency power electronics applications.

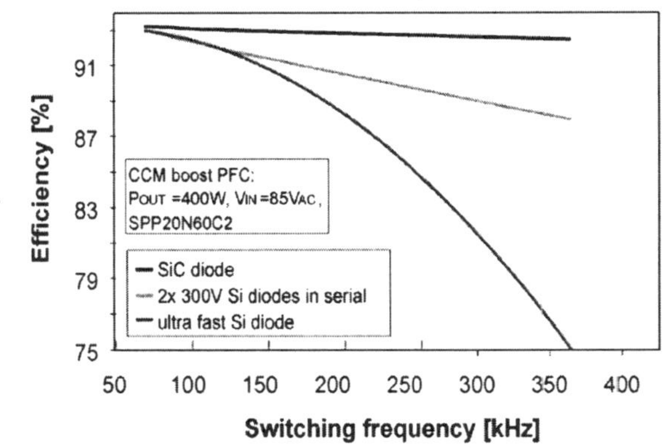

Fig. 27 Efficiency comparison of a power factor correction circuit with a 600V MOSFET and SiC diode or silicon diodes. With superior unipolar switch and SiC diode, the switching losses are kept low and efficiency remains almost unchanged for frequencies up to 400kHz. (after [154])

The unique high temperature capability of the SiC diode also offers potential application for power electronics in high temperature environments.

However, because of the lack of commercially available SiC three terminal switches, the high temperature capability of the commercial SiC diodes has so far not been fully taken advantage of.

5. Other Applications of SiC SBD

In addition to their applications in high frequency power electronics, SiC Schottky barrier diode structure can also be used in other applications such as gas sensors, in microwave circuits and as UV detectors. In this section, these applications of SiC SBDs will be briefly introduced.

5.1 SiC SBD as gas sensor

Increasing regulations on the release of gases or other chemicals into the environment have led to the increased attention on development of advanced sensors. There is a strong interest in SiC-based gas sensors for applications including fuel leak detection in automobiles and aircraft, fire detectors, exhaust diagnosis and emissions from industrial processes. Because of its wide bandgap, SiC is capable of operating and remain stable at much higher temperatures than many of the conventional semiconductors such as Si. Simple Schottky diode or field-effect transistor structures fabricated in SiC are sensitive to a number of gases, including hydrogen, hydrocarbons and oxygen [155-168]. Gas sensitive SiC SBDs with palladium gates were reported in 1992 [169]. Operation of SiC SBDs as a gas sensor has been reported at above 550°C [158].

Hydrogen atoms can be dissociated from hydrogen molecules and diffuse through the thin metal layer and form a polarized layer on the metal-semiconductor interface. This polarized layer changes the barrier height of the Schottky contact and shifts the I-V curve of the diode, typically toward lower voltage. A typical response in the I-V curve to the ambient gas is shown in Fig. 28. A significant shift in the I-V curve can be seen, which can be easily detected by an electronics circuit. Such SiC SBD gas sensors can also be used to sense carbon monoxide (CO). The response time of the sensor is in the order of milliseconds [158].

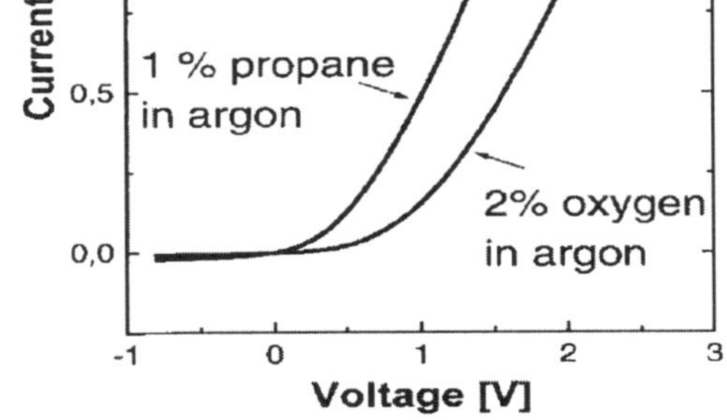

Fig. 28 Typical I–V curve for a Schottky diode at 550°C. The sensor response is indicated by an arrow. Composition of the device: Pt (150 nm); TaSix (15 nm); 6H SiC. (after [158])

One additional attractive attribute of SiC is the fact that gas sensors based on this material could be integrated with high temperature electronic devices on the same chip.

5.2 SiC SBD in microwave applications

The application of SiC SBD in microwave are mainly related to those require high voltage or high power, for example in limiters and high level mixers. There have been some limited works reported for such applications of SiC SBDs [171-173].

The cut-off frequency of the SiC SBD used as a varistor in a mixer circuit can be defined as:

$$f_C = \frac{1}{2\pi C_{j0} R_S} \tag{21}$$

where C_{j0} is the junction capacitance at zero bias and R_S is the series resistance of the device. The cut-off frequency is regarded as the figure-of-merit for such applications. For similar reasons as mentioned in Eq. 12 of section 3, when a relatively high voltage is needed, SiC SBDs can offer significant advantage over conventional materials such as Si and GaAs.

A theoretical calculation based on physical parameters of 4H-SiC gave a set of optimization curves for circle-shaped 4H-SiC SBDs, as shown in Fig. 29 [173].

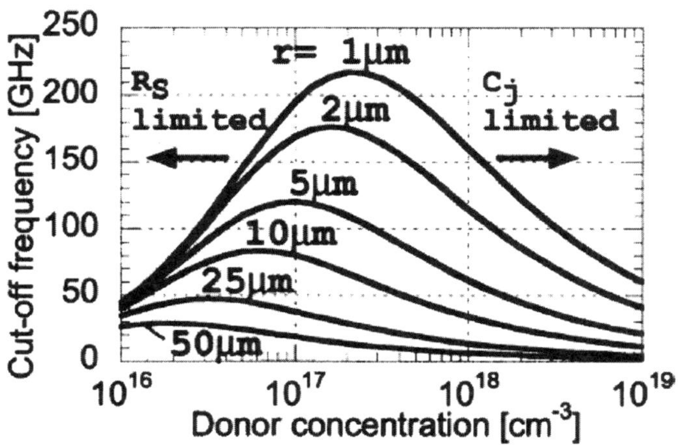

Fig. 29 Cut-off frequency versus drift layer doping of a circular 4H-SiC SBD diode for microwave mixer application. Results for different radius are plotted. (after [173])

The first SiC SBD mixer was published in [171] on a 0.4μm, 4×10^{17}cm^{-3} doped 4H-SiC. The device uses Au/Ti as Schottky contact metal and has an area of 40×40μm^2. The same group reported a similar SiC SBD based on a 0.38μm, 2.8×10^{17}cm^{-3} doped drift layer [173]. Limited by large ohmic contact resistance and designed for relatively low frequency application, these fabricated diodes has the lowest conversion loss of 5.2dB at a frequency of 800MHz.

5.3 SiC SBD as UV detector

SiC SBDs have also been used as detectors for radiations such as UV light [174-177], neutrons [178] and others [179-183].

The accurate measurement of UV radiation exposure has become increasingly important as a result of increasing environmental pollution. The main problem with Si

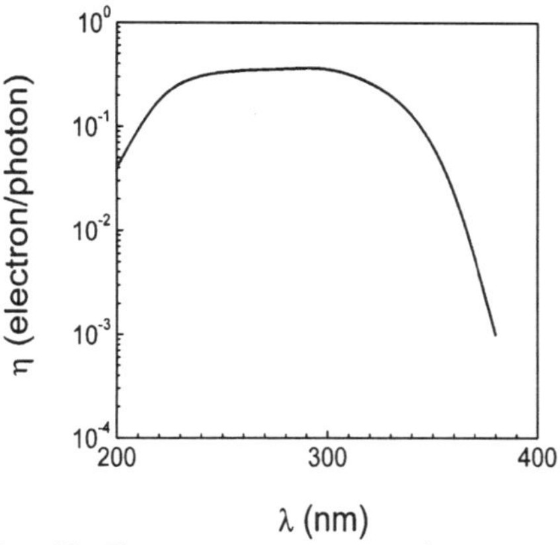

Fig. 30 Photo response spectra in quantum efficiency of Pt/4H-SiC Schottky photodiodes (after [177])

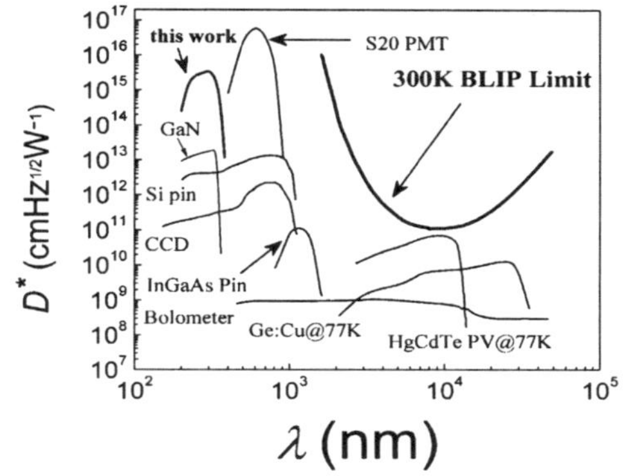

Fig. 31 Sensitivity comparison between SiC Schottky 5mm×5mm photodiodes and some common detectors. The 300K blackbody radiation limited D*, 300K BLIP limit, is also inset as a reference. (after [177])

UV detectors occurs as a result of the supplementary radiation filtering required in order to eliminate the visible and infra-red part of the light, which blinds the detector. The 3.2eV bandgap of 4H-SiC means a SiC detector responds to light only in the range of 380nm and below. The longer wavelength from visible and infra-red radiation can not bridge the wide bandgap, and hence the SiC detector is insensitive to these wavelengths and consequently does not necessitate supplementary radiation filtering. This is a great advantage in detecting UV by using SiC detectors in the presence of the substantial visible and infra-red background which generally is the case. Furthermore, as a wide bandgap material, SBDs made on SiC can have extremely low leakage current, substantially enhancing the sensitivity of the device.

SiC SBD-based UV detectors with detectivity orders of magnitude higher than Si photo-detectors and approaching that of photomultiplier tubes have been reported [177]. They were fabricated based on Schottky barrier structures by depositing a semitransparent (75Å) layer of Pt on top of a 3.7μm N- 4H-SiC epi layer (3×10^{14}~$3\times10^{15}cm^{-3}$) [177]. Schottky photodiodes with the device areas of 0.25mm×0.50mm, 2mm×2mm, 5mm×5mm, and 1cm×1cm were fabricated. The photodiodes demonstrated extremely low leakage current. The leakage current of a 5mm×5mm device is less than 1.2×10^{-14}A at a voltage bias of -1V. The quantum efficiency of the SiC Schottky photodiodes is shown in Fig. 30 for the spectra between 200nm and 400nm. Unlike GaN, 4H-SiC is indirect semiconductor, and does not have a sharp cutoff edge at the band edge. The absorption coefficient increases slowly from 385 nm as the wavelength decreases. As a result, the quantum efficiency of 4H-SiC Shcottky diodes increases gradually from less than 0.1% at 380 nm to 37% at 300 nm. The maximum quantum efficiency is around 37% and nearly constant from 240 to 300 nm.

The sensitivity of the SiC SBD UV detector was also evaluated. For a detector at zero bias, the noise of the detector is dominated by the Johnson noise. The specific detectivity D^* is one of the most frequently used figure-of-merits to evaluate the sensitivity of photodetectors and is defined as [177]

$$D^* = \frac{q\eta}{h\nu}\left[\frac{R_0 A}{4k_B T}\right]^{1/2} \tag{22}$$

when the Johnson noise dominates, where η is the quantum efficiency, h is the Planck constant, v is the radiation frequency, R_0 is the dynamic resistance at zero bias, and A is the detector area.

The specific detectivity D^* of 4H-SiC Schottky photodiodes is calculated based on the directly measured results of the 5mm×5mm photodiodes and compared with other common photo detectors in Fig. 31. The maximum sensitivity of 4H-SiC Schottky is 3.6×10^{15}cmHz$^{1/2}$/W at 300nm and the sensitivity is above 10^{15}cmHz$^{1/2}$/W from 210 to 350nm. This record-high detector sensitivity is two orders of magnitude higher than that of Si photodiodes and three orders of magnitude higher than the D^* of Si CCD.

Sensitivity of the SiC SBD is limited by many different types of defects on the surface of the material [184] that will cause inhomogeneous barrier height across the device. The effective SBH of the device fabricated is expected to be significantly lower than the theoretical value of 2.0eV for Pt/4H-SiC. Therefore, as the defect density decreases and material quality improves, the leakage current of the SBD can be further reduced and the sensitivity further improved.

More details on the applications of SiC SBDs as radiation-hardened detectors for neutrons [178] and others [179-183] can be found in the literature.

5. Summary and Future Challenges

5.1 Summary

In summary, basic physics of metal to SiC Schottky contact that determines the Schottky barrier height (SBH) and the carrier transport mechanisms have been explained. Barrier height of a SiC Schottky contact is dependent on factors such as metal used, SiC polytype, face polarity, surface treatment, post deposition annealing and SiC material doping. With increasing metal workfunction, the SBH of an n-type contact increases while that of a p-type contact decreases. The rate of change in SBH with metal workfunction (interface slope parameter, S) can be between 0 and 1, depending on the density of interface states. Different interface state density could result from different surface treatments and conditions. Unpinning of the SiC Fermi level by a drastic reduction in interface state density has been reported with special surface treatment techniques. The interface slope parameter is also found to be dependent on the SiC polytype with a lower value for 6H as compared to that for 4H for n-type contacts. Overall, SHB in the range of 0.4eV to 2.0eV has been obtained for n-type SiC. Although less investigation on p-type SiC Schottky contact has taken place because of its less practical usefulness resulted from the lower mobility of holes in SiC, studies have reported p-type SBH between 1.3eV to 2.0eV.

Because of its wider bandgap, contacts on the 4H polytype have shown a SBH 0.2~0.4eV higher than that of those on 6H polytype. For both 6H and 4H polytypes, Schottky contacts on carbon-terminating SiC materials have shown higher barrier heights than those on silicon-terminating SiC materials. Since metal-semiconductor interface chemistry is critical for a SiC Schottky contact, pre-deposition SiC surface treatment, metal deposition conditions and post-deposition annealing conditions have all found to influence the resultant SBH significantly. For instance, the Pt-SiC SBH can increase significantly with post-deposition annealing because of the formation of silicide. The substrate temperature during metal deposition has also been reported to affect the resultant SBH because of different chemical reactions taking place.

The SBH of a SiC Schottky contact has also been found, as expected, to decrease as the semiconductor doping increases above $3\times10^{17}cm^{-3}$ level. The reduction can be explained by the combined effects of image force lowering and charge distribution across the contact interface. At temperatures above room temperature, SBH of n-type SiC Schottky contact has been reported to decrease with increasing temperature while the p-type SBH shows an increase. The SBH sensitivity to temperature was linked to the electronegativity of the metal used. The sensitivity was found to be low for metals with a low electronegativity.

Extensive efforts in the development of SiC Schottky barrier diodes for high power application in the last ten years or so have resulted in substantial improvements in both the voltage and current capabilities. While most studies have been done on 6H-SiC before 1995, the efforts have been largely shifted to 4H-SiC after 1995 when high quality 4H-SiC material became widely available because 4H-SiC has higher electron mobility. 4H-SiC SBDs with a breakdown voltage of 10.8kV was reported with a specific on-resistance of $97.5m\Omega cm^2$ before considering current spreading effects [2] and single dies with an area of $1cm^2$ (expected to be able to handle >200A) have been fabricated with low leakage current [187]. In achieving a good voltage blocking capability, various edge-termination techniques have been used. Techniques such as floating metal rings, P^+ region rings, p-epi guard rings, inert gas implantation and metal field-plating have been proven to be effective. Muti-step junction termination extension (MJTE) technique was demonstrated at voltages above 10kV.

A higher SBH will give lower diode reverse leakage while sacrificing forward conduction voltage. The reverse leakage current of a SiC SBD has, however, been reported to be significantly higher than predicted by single-barrier thermionic emission theory. Material inhomogeneous has been suggested as possible causes. The leakage current also increases quickly with increasing reverse bias voltage due to the effects of electric field induced barrier lowering. To reduce the electric field at the Schottky contact area and hence the diode reverse leakage current, many structures utilizing the electric field shielding effect of high-barrier-junction depletion regions have been studied. These include JBS, MPS, DMTS, DMP, TMBS and LMDS. With different processes and approaches, these variants aimed at achieving a low leakage current with the same electric field shielding principle. Careful design trade-offs are needed between reverse leakage current and forward voltage drop for all these SBD variants except DMP diodes. MPS diodes with large current carrying capability were successfully used in PWM DC-AC inverter systems for motor control at a level of up to 7.4HP and with 20kHz switching frequency [188]. Half-bridge inverter switching based on 4H-SiC MPS diodes at 2.5kV bus voltage and 30A load current has also been reported [170].

The developments of SiC SBDs have enabled their successful commercialization for power electronics market. SiC SBDs are now commercially available with voltage up to 1200V and current up to 20A.

Comparisons on the performances of commercial SiC SBDs were made to existing silicon parts with comparable ratings. While the commercial SiC SBDs do show higher forward conduction voltages than their silicon counterparts at the 300V rating, their forward voltage become comparable to that of the silicon device for the 1200V class at room temperature. The SiC SBDs are found to offer significant advantages in all other aspects of the required performances. SiC SBDs have clear positive temperature coefficient on their forward voltage drop, enabling easy and reliable parallel connection of multiple chips for higher current application. They also demonstrated orders of magnitude lower leakage current, especially at temperatures higher than 150°C. This shows clearly their capability of operation at a temperature well above 150°C, which is widely expected due to the wide bandgap nature of the material.

The greatest advantage of SiC SBDs in a power electronics application is its close-to-ideal dynamic performance. During the crucial reverse recovery transient when the diode changes from a forward conduction mode to a reverse blocking mode, it demonstrated very low recovery time, which remains unchanged over the full range of operating temperatures. This is mainly due to the fact that SiC SBD is a unipolar device without minority carrier injection and hence free of storage charge. The only charge involved in the recovery transient is the junction depletion charge which has been found to be more than one order of magnitude lower than the recovery charge of a similar silicon device. The advantage is especially drastic for devices with higher blocking voltage and/or at higher temperatures. This represents a breakthrough in power electronics device development and solves a major bottleneck in high frequency power electronics application where the speed of the diode seriously limits the whole circuit performance. Applications of the commercial SiC SBDs in high frequency switch mode power supplies, power factor corrections and some motor drive applications have been reported. Combined with the best available power MOSFETs, hard-switching frequencies up to 400kHz have bee demonstrated and a frequency of above 1MHz can be expected. This is substantially higher than the typical 100kHz frequency used for devices at this high voltage level (>600V).

In addition to their application in power electronics systems, SiC SBDs can also be used in other applications such as gas sensing, microwave applications and ultra-violet light detection. Their wide bandgap and exceptional stability at high radiation environment and at temperatures as high as 500°C give SiC SBDs their unique advantages for these applications.

5.2 Future trends and challenges in SiC SBD development

The development of SiC SBDs has been exciting in the last few years with devices able to handle >10kV [2] or >100A being reported by research labs. Commercially available SiC SBDs have already demonstrated exceptional performance advantage over existing silicon fast-recovery diodes for high end power electronics applications. Commercial devices are currently limited to 1.2kV and 20A in their ratings.

It is expected that SiC SBDs with higher voltage and current handling capabilities will continue to be pursued by both research labs and commercial companies. Commercial devices capable of handling up to 3-5kV will be able to provide significant system advantage for relatively high power electronics applications.

In order to achieve SiC SBDs with 3-5kV breakdown voltage for commercial production, one of the challenges ahead is a reliable edge termination technique that gives high reproducibility, high yield, simple fabrication procedure and low leakage current. While various termination techniques have been used to obtain a high blocking voltage, as mentioned in previous sections, most of those involves an ion implantation step which generally results in increased leakage and reduced reliability. Issues associated with ion implantation caused crystal damages will need to be address or an implantation-free method will be needed for a termination technique that will satisfy criteria mentioned above. In addition, an appropriate passivation process is also critical in achieving a high-voltage-blocking device with a low leakage current and a long-term high reliability.

Another critical challenge for continuous development of SiC SBDs will be the SiC wafer substrate and epitaxial quality, which is also the basis for continuous and successful development of all SiC devices. While great improvements have been obtained in recent years on SiC substrate wafer and epitaxial layer quality, ultra-high quality SiC wafers with lower defect densities and larger sizes are needed. A breakthrough in ultra-high quality SiC substrate growth has been recently reported [185] by using a repeated a-face (RAF) growth process, demonstrating 2" SiC substrates with zero micro-pipe density and an etch pit density two to three orders of magnitude lower than the best existing materials available. This is an important development in single crystalline SiC material growth. Extending this method to wafers with larger diameter (3", 4" and beyond) at a competitive price will promise an exciting future in SiC device development and applications.

While SiC SBDs have been able to successfully compete with existing Si devices in high end power electronics applications such as server and telecom power supplies, the significantly higher product cost has been the major obstacle that has so far prevented their penetration into the mass consumer market (e.g., PC power supplies). The SiC device cost is mainly dominated by the orders-of-magnitude higher base material cost when compared to silicon. Per-device cost can be reduced by reducing the wafer cost as well as increasing wafer diameter. Currently, commercial SiC substrates are available with 3" diameter which gives significant cost advantage over the widely-used 2" wafers. This results from both the increased device yield and more alternatives in processing equipment when compared to 2" wafers. Many relatively modern and automated 4" production facilities used for silicon device can be modified for the 3" SiC wafers, giving high reproducibility, low fabrication defect density and better device reliability. Good device yield and performance uniformity have been reported [186]. With larger SiC wafers, devices with higher current capability can also be fabricated with better yield. Power modules capable of handling hundreds of amps with SiC SBDs can be expected in the reasonably near future.

A unique market for SiC SBDs is the high temperature electronics market because of their capability of reliable operation at high voltage and high temperature. While SiC SBDs are able to handle well beyond 200°C, the development in realizing the potential of these devices for high temperature (>200°C) application has been limit by available packaging approaches. Commercially available SiC SBDs are therefore currently only rated for up to 175°C. Identifying a viable low-cost packaging technology that can remain reliable at temperatures well above 200°C becomes the main challenge in applying SiC SBDs for high temperature and high power applications.

References

1. S.Y. Wu and R.B. Campbell, "Au-SiC Schottky Barrier Diodes", Solid-State Electronics, ***17***, 1974, pp. 683-687
2. J.H. Zhao, P. Alexandrov, X. Li, "Demonstration of the first 10-kV 4H-SiC Schottky barrier diodes" Electron Device Letters, IEEE, ***24***, 2003, pp. 402-404
3. J. Wu, L. Fursin, Y. Li, P. Alexandrov and J.H. Zhao, "4,308V, 20.9mΩcm^2 4H-SiC MPS diodes based on a 30μm drift layer", Mat. Sci. Forum, 2004, pp. 1109-1112
4. Z.C. Feng and J.H. Zhao, *Silicon Carbide, Materials, Processing and Devices*, pp. 205, 2004
5. S.M. Sze, "Physics of Semiconductor Devices", 2nd edition, 1999
6. G. H. Glover "Charge multiplication in Au-SiC (6H) Schottky junctions." Journal of Applied Physics, ***46***, 1975, pp.4842-4844.

7. Q. Zhang, V. Madangarli, M. Tarplee, T S. Sudarshan, "Comparison of current-voltage characteristics of n- and p- Type 6H-SiC Schottky diodes" J. Electr. Mat., 2001. pp. 196-201
8. H. J. Na, J. K. Jeong, M. Y. Um, B. S. Kim, C. S. Hwang, H. J. Kim. "Effect of annealing on electrical properties of Pt/beta-SiC contact" Solid-State Electronics, ***45***, 2001. pp. 1565-1570
9. H.S. Kong, JT. Glass, RF. Davis. "Chemical vapor deposition and characterization of 6H-SiC thin films on off-axis 6H-SiC substrates." Journal of Applied Physics, ***64***, 1988, pp.2672-2679
10. Z. Liu, S. Wang, F. Yu, Y. Zhang, H. Zhao, "Ti Schottky barrier diodes on n-type 6H-SiC" Proc. 6th Int. Conf. Solid-State and Integrated-Circuit Technology, ***2***, 2001, pp. 1183 – 1186
11. F. Yoshihisa, S. Mitsuhiro, F. Katsuki, S. Akira, N. Shigeo "Schottky diode characteristics of CVD grown beta-SiC single crystals on Si substrates" Sharp Tech. Journal. 1989 pp. 33-36
12. K, Das, H.S. Kong, J.B Pett, J.W. Bumgarner, Matus.L.G and Davis.R.F "Deep-Level Dominated Electrical Characteristics of Au Contacts on Beta-SiC" Material Research Society, ***162***, 1990, pp. 489-495.
13. Q. Wahab, M. Karlsteen, M. Willander, J –E. Sundgren. "Au Schottky barrier diodes on beta - SiC thin films deposited on silicon substrates by reactive magnetron sputtering technique" J. Electr. Mat., ***20***, 1991 pp. 899-901
14. J. R. Waldrop, R. W. Grant, Y. C. Wang, R. F. Davis.. " Metal Schottky barrier contacts to aplpha 6H-SiC", Journal of Applied Physics, ***72*** Issue 10, p4757-4760, Nov. 1992,
15. M. Bhatnagar, H. Nakanishi, P.K. Mclarty, B.J. Baliga, B. Patnaik, N. Parikh, "Comparison of Ti and Pt silicon carbide Schottky rectifiers", Int. Tech. Dig. Electr. Dev. Meet, 1992, 789
16. C. Fröjdh, and C. S. Petersson, "Comment on "Temperature dependence of the barrier height of metal-semiconductor contacts on 6H-SiC" J. Appl. Phys, ***80***, 1996, pp. 6570-6571
17. M. Mamor, JL. Perrossier, V. Aubry-Fortuna, , F. D. MeyerBouchier, S. Bodnar, J L. Regolini, "Fermi-level pinning in Schottky diodes on IV-IV semiconductors: effect of Ge and C incorporation" Thin Solid Films. ***294***, 1997. pp. 141-144
18. L. Magafas, N. Georgoulas, A. Thanailakis, "Influence of metal work function on electrical properties of metal/a-SiC:H Schottky diodes" Microelectronics. ***28***, 1997. pp. 107-114
19. A. L. Syrkin,. J. M. Bluet, G. Bastide, T. Bretagnon, A. A. Lebedev, M. G. Rastegaeva, N S. Savkina, v. Chelnokov, "Surface barrier height in metal-SiC structures of 6H, 4H and 3C polytypes" Mat. Sci. & Eng. B-Solid State Mat. Adv. Tech., ***46***, 1997. pp. 236-239
20. Anatoly. Strel'chuk, M. Rastegaeva, Marina G. "Characterization Schottky barriers occurring at the metal-6H-SiC contact based on results of studies of current-voltage characteristics" Mat. Sci. & Eng. B-Solid State Mat. Adv. Tech. 1997. pp. 379-382
21. M G. Rastegaeva, A N. Andreev, A A. Petrov, A I. Babanin, M A Yagovkina,. I P. Nikitina, "Influence of temperature treatment on the formation of Ni-based Schottky diodes and ohmic contacts to n-6H-SiC" Mat. Sci. & Eng. B-Solid State Mat. Adv. Tech. 1997, pp. 254-258
22. S. Hara, T. Teraji, H. Okushi, K. Kajimura, "Control of Schottky and ohmic interfaces by unpinning Fermi level" Applied Surface Science. ***117-118***, 1997. pp. 394-399
23. A. Itoh, H. Matsunami,. "Analysis of Schottky barrier heights of metal/SiC contacts and its possible application to high-voltage rectifying devices" Physica Status Solidi A-Applied Research, ***162***, 1997. pp. 389-408
24. T. Teraji, S. Hara, H. Okushi, K. Kajimura, "Ideal Ohmic contact to n-type 6H-SiC by reduction of Schottky barrier height" Appl. Phys. Lett., ***71***, 1997. pp. 689-691
25. R. Raghunathan, B J. Baliga, "High voltage Schottky barrier diodes on p-type 4H and 6H-SiC" Materials Science Forum. ***264-268***, 1998. pp. 933-936
26. V. Saxena, A. J. Steckl, "High voltage 4H SiC rectifiers using Pt and Ni metallization" Materials Science Forum., ***264-268***, 1998. pp. 937-940
27. L. A. Kosyachenko, V. M. Sklyarchuk, Ye. F. Sklyarchuk, "Electrical and photoelectric properties of Au-SiC Schottky barrier diodes." Solid-State Electronics, ***42***, 1998, pp. 145-151.
28. G. Constantinidis, J. Kuzmic, K. Michelakis, K. Tsagaraki, "Schottky contacts on CF4/H2 reactive ion etched beta –SIC" Solid-State Electronics. ***42***, 1998. pp. 253-256
29. N.A. Suvorova, , A.V. Shchukarev, , I.O. Usov, A.V. Suvorov, , "XPS study of dependence of Au/6H-SiC Schottky barrier height on carrier concentration" Semiconducting and Insulating Materials, 1998. Proceedings of the 10th Conference on, 1998 pp. 291 - 294

30. A. Kakanakova-Georgieva, Ts. Marinova, O. Noblanc, C. Arnodo, S. Cassette, C. Brylinski, "Characterization of ohmic and Schottky contacts on SiC" Thin Solid Films, ***343***, 1999, p. 637
31. G. Constantinidis, B. Pecz, K. Tsagaraki, M. Kayambaki, K. Michelakis, "Improvements in Pt-based Schottky contacts to 3C-SiC" Mat. Sci. & Eng. B-Solid State Mat. Adv. Tech., ***61***, 1999. 406
32. S. C. Kang, B.H. Kum, S.J. Do, J.H. Je, M.W. Shin, "Annealing effects of schottky contacts on the characteristics of 4H-SiC schottky barrier diodes" Mat. Res. Soc. Symp. Proc., ***572***, 1999. p. 141
33. S. Basu, , S. Roy, , R. Laha, , C. Jacob, , S. Nishino, , "Schottky behaviour of Pd//spl beta/-SiC junctions" Optoelectr. Microelectro. Mat. and Dev, 2000. Proceedings. 2000 pp. 328 – 331
34. D. J. Morrison, A. J. Pidduck, V. Moore, pp. J. Wilding, K P. Hilton, M J. C M. Uren, Johnson, N G. Wright, A G. O'Neill, "Surface preparation for Schottky metal - 4H-SiC contacts formed on plasma-etched SiC" Semiconductor Science & Technology, ***15***, 2000, pp. 1107-1114
35. S-K. Lee, C-M Zetterling, M. Ostling "Schottky barrier height dependence on the metal work function for p-type 4H-silicon carbide." TMS, IEEE. J. of Electr. Mat., ***30***, 2001, pp.242-6
36. D. Defives, O. Durand, F. Wyczisk, O. Noblanc, C. Brylinski, F. Meyer, "Electrical behaviour and microstructural analysis of metal Schottky contacts on 4H-SiC" Microelectr. Eng., ***55***, 2001. 369
37. K. Kamimura, S. Okada, H. Ito, M. Nakao, Y. Onuma, "Characterization of Schottky contact on p-type 6H-SiC", Mat. Sci. Forum. ***338***, 2000. pp.1227-1230
38. F. Roccaforte, La Via, F. Raineri, V. Musumeci, P. Calcagno, L. Condorelli, G G. "Highly reproducible ideal SiC Schottky rectifiers: Effects of surface preparation and thermal annealing on the Ni/6H-SiC barrier height" Appl. Phys. A: Mat. Sci. & Proc., ***77***, 2003. 827
39. S.Y. Han and J.L. Lee, "Characteristics of Schottky contacts on n-type 4H-SiC using IrO 2 and RuO2" Journal of Applied Physics., ***94***, 2003. pp. 6159-6166
40. F. La Via, F. Roccaforte, V. Raineri, M. Mauceri, A. Ruggiero, P. Musumeci, L. Calcagno, A. Castaldini, A. Cavallini, "Schottky-ohmic transition in nickel silicide/SiC-4H system: Is it really a solved problem?" Microelectr. Eng. ***70***, 2003. pp. 519-523
41. J. Kim, F. Ren, A.G. Baca, G.Y. Chung, S.J. Pearton, "Thermal stability of WSiX Schottky contacts on n-type 4H-SiC" Solid-State Electr. ***48***, 2004. pp. 175-178
42. A. Venter, M.E. Samiji, A.W.R. Leitch, "Formation of surface states during Schottky barrier fabrication on Al-doped p-type 6H-SiC" Diamond & Related Materials. ***13***, 2004
43. I.T. Kimoto and H. Matsunami, "Excellent Reverse Blocking Characteristics of High-Voltage 4H-SiC Schottky Rectifiers with Boron-Implanted Edge Termination," IEEE Electron Device Letters, ***17***, 1996, pp. 139-141
44. M. Bakowski, U. Gustafsson, U. Lindefelt. "Simulation of SiC high power devices". Phys Status Solidi 1997; A162:421–40.
45. S. Yoshida, K. Sasaki, E. Sakuma, S. Misawa, S. Gonda "Schottky barrier diodes on 3C-SiC." Applied Physics Letters, ***46***, 1985, pp.766-768. USA.
46. E. Ioannou. Dimitris, Papanicolaou.A.Nick, Nordquist.E.Paul.Jr "The Effect of Heat Treatment on Au Schottky Contacts on Beta-SiC" IEEE Trans. Electron Devices, ***34***, 1987, pp. 1694-1699.
47. N. A. Papanicolaou, A.Christou, M. L. Gipe, "Pt and PtSix Schottky contacts on n-type Beta-SiC" Journal of Applied Physics, ***65***, pp. 3526-3530 May. 1989
48. J.R. Waldrop and R.W. Grant, "Schottky barrier height and interface chemistry of annealed metal contacts to alpha 6H-SiC: Crystal face dependence", Appl. Phys. Lett. ***62***, 2685, 1993,
49. R.S. Okojie, A.A. Ned, A.D. Kurtz, W.N. Carr, "Electrical characterization of annealed Ti/TiN/Pt contacts on N-type 6H-SiC epilayer" IEEE Trans. Electr. Dev, ***46***, 1999 pp. 269 – 274
50. B.H. Li, L.H. Gao and J.H. Zhao, "Evaluation of damage induced by inductively coupled plasma etching of 6H-SiC using Au Schottky barrier diodes" Appl. Phys. Lett., ***73***, 1998, pp. 653-655.

51. G. Brezeanu, , M. Badila, , J. Millan, , Ph. Godignon, , M.L. Locatelli, , J.P. Chante, , A. Lebedev, , G. Dilimot, , I. Enache, , G. Bica, , V. Banu, , "A nearly ideal SiC Schottky barrier device edge termination" International Semiconductor Conference, 1999, pp. 183 - 186
52. M. Bhatnagar, P.K. McLarty, B.J. Baliga, "Silicon-carbide high-voltage (400 V) Schottky barrier diodes" Electron Device Letters, IEEE, ***13***, 1992, pp. 501 – 503
53. T. Nakamura and M. Satoh, "Schottky barrier height of a new ohmic contact NiSi2 to n-type 6H-SiC" Solid-State Electronics., ***46***, 2002. pp 2063-2067
54. K.P. Schoen, J.M. Woodall, J.A. Cooper and M.R. Melloch, "Design considerations and experimental analysis of high-voltage SiC Schottky barrier rectifiers" Electron Devices, IEEE Trans., ***45***, 1998, pp. 1595 – 1604
55. L. Kassamakova, A. Kakanakova-Georgieva, R. Kakanakov, T.S. Marinova, I. Kassamakov, Tz. Djambova, O. Noblanc, C. Arnodo, S. Cassette, C. Brylinski, "Thermostable Ti/Au/Pt/Ti Schottky contacts to n-type 4H-SiC" Semiconductor Science & Technology. ***13***, 1998. 1025
56. U. Zimmermann, A. Hallen and B. Breitholtz, "Current voltage characteristics of high-voltage 4H silicon carbide diodes" Materials Science Forum., ***338***, 2000. pp. 1323-1326
57. S.K. Lee, C.-M. Zetterling, M. Östling, "Schottky diode formation and characterization of titanium tungsten to n-and p-type 4H silicon carbide." J. Appl. Phys., ***87***, 2000, p. 8039
58. T. Hatayama, K. Kawahito, H. Kijima, Y. Uraoka and T. Fuyuki, "Electrical properties and interface reaction of annealed Cu/4H-SiC Schottky rectifiers" Materials Science Forum, ***389-393***, 2002. pp. 925-928
59. F. Nava, P. Vanni, C. Lanzieri, C. Canali, "Epitaxial silicon carbide charge particle detectors" Nuclear Instruments & Methods in Physics Research, ***437***, 1999. pp. 354-358
60. T. Kimoto, Q. Wahab, A. Ellison, U. Forsberg, M. Tuominen, R. Yakimova, A. Henry, E. Janzen, "High-voltage (greater than 2.5 kV) 4H-SiC Schottky rectifiers processed on hot-wall CVD and high-temperature CVD layers" Materials Science Forum, ***264-268***, 1998. pp. 921-924
61. Q. Wahab, T. Kimoto, A. Ellison, C. Hallin, M. Tuominen, R. Yakimova, A. Henry, J.P. Bergman, E. Janzen "A 3 kV Schottky barrier diode in 4H-SiC." Appl. Phys. Lett, ***72***, 1998, p.445
62. L. Kassamakova, A. Kakanakova-Georgieva, R. Kakanakov, Ts. Marinova, I. Kassamakov, Tz. Djambova, O. Noblanc, C. Arnodo, S. Cassette, C. Brylinski, "Thermostable Ti/Au/Pt/Ti Schottky contacts to n-type 4H-SiC" Semiconductor Science & Technology, ***13***, 1998. p 1025
63. V. Saxena, , Jian Nong Su, A.J Steckl, , "High-voltage Ni- and Pt-SiC Schottky diodes utilizing metal field plate termination", IEEE Trans. Electr. Dev., ***46***, 1999 pp. 456 – 464
64. R.K. Chilukuri, , B.J. Baliga, , "High voltage Ni/4H-SiC Schottky rectifiers", ISPSD, ***26-28***, 1999, p. 161
65. A. Kestle, , S.P. Wilks, P.R. Dunstan, , M. Pritchard, , P.A. Mawby, , "Improved Ni/SiC Schottky diode formation" Electronics Letters, ***36***, 2000 pp. 267 – 268
66. Q. Wahab, A. Ellison, J. Zhang, U. Forsberg, E. Duranova, A. Henry, L D. Madsen, E. Janzen, "Designing, physical simulation and fabrication of high-voltage (3.85 kV) 4H-SiC Schottky rectifiers processed on hot-wall and Chimney CVD films" Mater. Sci. Forum, ***338***, 2000. p. 1171
67. K. Kojima, M. Yoshikawa, T. Ohshima, H. Itoh and S. Okada, "Characterization of Au Schottky contacts on p-type 3C-SiC grown by low pressure chemical vapor deposition", Materials Science Forum, ***338***, 2000. pp.1239-1242
68. Q. Zhang, V. Madangarli, M. Tarplee, T S. Sudarshan, "Comparison of current-voltage characteristics of n- and p- Type 6H-SiC Schottky diodes" J. Electr. Mater., ***30***, 2001. p. 196
69. K. Jarrendahl and R.F. Davis, "SiC Materials and Devices", ed. Y.S. Park, 1998, p. 14
70. C. Jacob, P. Pirouz, H.I. Kuo and M. Mehregany, "High temperature ohmic contacts to 3C–silicon carbide films", Solid State Electronics, ***42***, 1998, pp. 2329-2334
71. A. Itoh, T. Kimoto, H. Matsunami. "Efficient power Schottky rectifiers of 4H-SiC", ISPSD'95, pp. 101-106
72. Lundberg. Nils, Ostling, Mikael. Tagtstrom, Per. Jansson, Ulf. "Chemical vapor deposition of tungsten Schottky diodes to 6H-SiC" J. Electrochemical Society., ***143***, 1996. pp. 1662-1667

73. M.W. Cole, P.C. Joshi, "Ohmic contact to SiC", Silicon Carbide Materials, Processing and Devices, ed. Z. Feng and J. Zhao, 2004, p. 212
74. H.J. Na, J.K. Jeong, M.Y. Um, B.S. Kim, C.S. Hwang and H.J. Kim, "Effect of annealing on electrical properties of Pt/beta-SiC contact" Solid-State Electronics, ***45***, 2001. pp. 1565-1570
75. K. Ip, S. Nigam, K.H. Baik, F. Ren, G.Y. Chung, B.P. Gila and S.J. Pearton, "Stability of SiC Schottky rectifiers to rapid thermal annealing" J. Electrochemical Society., ***150***, 2003. p G293
76. S.R. Smith, A.O. Evwaraye and W.C. Mitchel, "Temperature dependence of the barrier height of metal-semiconductor contacts on 6H-SiC", J. Appl. Phys., ***79***, 1996, pp. 301-304
77. S.K. Lee, C.M. Zetterling and M. Ostling, "Schottky diode formation and characterization of titanium tungsten to n- and p-type 4H silicon carbide", J. Appl. Phys., ***87***, 2000, pp. 8039-8044
78. S.R. Smith, A.O. Evwaraye, W.C. Mitchel. "Reply to "Comment on 'Temperature dependence of the barrier height of metal-semiconductor contacts on 6H-SiC" J. Appl. Phys., ***80***, 1996, p6572-6573
79. B.J. Baliga, "Power Semiconductor Devices", Boston, PWS, 1996
80. J.H. Zhao, K. Xie, W. Buchwald, and J. Flemish, "Near 1000V Schottky diodes and ECR plasma etching of 6H-SiC," ICSCRM, Paper II–61, 1993
81. T. Kimoto, T. Urushidani, S. Kobayashi, H. Matsunami, "High-voltage (>1 kV) SiC Schottky barrier diodes with low on-resistances" IEEE Electr. Dev. Lett., ***14***, 1993 pp. 548 – 550
82. R. Raghunathan, D. Alok, B.J. Baliga, "High voltage 4H-SiC Schottky barrier diodes" IEEE Electr. Dev. Lett., ***16***, 1995 pp. 226 – 227
83. A. Itoh, T. Kimoto, H. Matsunami, "High performance of high-voltage 4H-SiC Schottky barrier diodes" IEEE Electr. Dev. Lett., ***16***, 1995 pp. 280 – 282
84. C.E. Weitzel, , J.W. Palmour, , C.H. Carter, Jr, K. Moore, K.K. Nordquist, , S. Allen, C. Thero, M. Bhatnagar, "Silicon carbide high-power devices" IEEE Trans. Electr. Dev., ***43***, 1996 pp. 1732 – 1741
85. C.M. Zetterling, F. Dahlquist, N. Lundberg, M. Ostling, K. Rottner and L. Ramberg, "High voltage silicon carbide Junction Barrier Schottky rectifiers" Proc. IEEE Cornell Conf. Adv. Concepts High Speed Semiconductor Dev. and Circuits, ***80***, 1997. p. 256
86. T. Kimoto, Q. Wahab, A. Ellison, U. Forsberg, M. Tuominen, R. Yakimova, A. Henry, E. Janzen, "High-voltage (greater than 2.5 kV) 4H-SiC Schottky rectifiers processed on hot-wall CVD and high-temperature CVD layers" Materials Science Forum., ***264-268***, 1998. pp. 921-924
87. Q. Zhang, V. Madangarli, S. Soloviev, T.S. Sudarshan, "High voltage Schottky barrier diodes on p-type SiC using metal-overlap on a thick oxide layer as edge termination" MRS symposium, ***572***, 1999. pp. 75-80
88. R. Raghunathan, B.J. Baliga, "P-type 4H and 6H-SiC high-voltage Schottky barrier diodes" IEEE Electr. Dev. Lett., ***19***, 1998 pp. 71–73
89. R.K. Chilukuri and B.J. Baliga, "High voltage Ni/4H-SiC Schottky rectifiers" IEEE ISPSD, 1998, pp. 161-164
90. V. Khemka, R. Patel, T. P. Chow, R J. Gutmann, "Design considerations and experimental analysis for silicon carbide power rectifiers" Solid-State Electronics, ***43***, 1999. pp. 1945-1962
91. V. Dmitriev, S. Rendakova, N. Kuznetsov, N. Savkina, A. Andreev, M. Rastegaeva, M. Mynbaeva, A. Morozov, "Large area silicon carbide devices fabricated on SiC wafers with reduced micropipe density" Mater. Sci. & Eng. B-Solid State Mater. Adv. Tech., ***61***, 1999. p446
92. S.K. Lee, "Processing and Characterization of Silicon Carbide (6H- and 4H-SiC) Contacts for High Power and High Temperature Device Applications", PhD thesis, Roy. Inst. Tech, Sweden
93. H.M. McGlothlin, D.T. Morisette, J.A. Cooper and M.R. Melloch, "4 kV silicon carbide Schottky diodes for high-frequency switching applications" Dev. Res. Conf. Dig. 1999 p. 42
94. R. Singh, J.A. Cooper, M.R. Melloch, J.W. Palmour and T.P. Chow, "SiC power Schottky and PiN diodes" IEEE Trans. Electr. Dev., ***49***, 2002. pp. 665-672
95. M. Bhatnagar, P.K. McLarty, B.J. Baliga, "Silicon-carbide high-voltage (400 V) Schottky barrier diodes" IEEE Electr. Dev. Lett., ***13***, 1992, pp. 501 – 503

96. D. Alok, B.J. Baliga, P.K. McLarty, "A simple edge termination for silicon carbide devices with nearly ideal breakdown voltage" IEEE Electr. Dev. Lett., ***15***, 1994 pp. 394 – 395
97. K. Ueno, T. Urushidani, K. Hashimoto, Y. Seki, "The guard-ring termination for the high-voltage SiC Schottky barrier diodes" IEEE Electr. Dev. Lett., ***16***, 1995 pp. 331 – 332
98. H. Matsunami and A. Itoh, Extended Abst. 19th Conf. Solid State Dev. and Mater., 1995, p. 40
99. I.T. Kimoto. and H. Matsunami, "Excellent Reverse Blocking Characteristics of High-Voltage 4H-SiC Schottky Rectifiers with Boron-Implanted Edge Termination," IEEE Electr. Dev. Lett., ***17***, 1996, pp. 139-141
100. D. Alok, , R. Raghunathan, , B.J. Baliga, , "Planar edge termination for 4H-silicon carbide devices" Electron Devices, IEEE Trans., ***43***, 1996 pp. 1315 – 1317
101. G. M. Dolny, D. T. Morisette, P. M. Shenoy, M. Zafrani, J. Gladish, J. M. Woodall, J. A. Cooper, Jr, and M. R. Melloch, "Static and Dynamic Characterization of Large-Area High-Current-Density SiC Schottky Diodes," IEEE Device Research Conf, 1998.
102. D. Peters, K.O. Dohnke, C. Hecht and D. Stephani, "1700 v SiC Schottky diodes scaled to 25 A" Materials Science Forum, ***353-356***, 2001. pp. 675-678
103. M. Bhatnagar, H. Nakanishi, S. Bothra, P.K. McLarty, B.J. Baliga, "Edge terminations for SiC high voltage Schottky rectifiers", ISPSD '93. 1993 pp. 89 – 94
104. K. Ueno, T. Urushidani, K. Hashimoto, Y. Seki, "Al/Ti Schottky barrier diodes with the guard-ring termination for 6H-SiC" ISPSD, 1995 pp. 107 – 111
105. D. Alok, B.J. Baliga, "A planar, nearly ideal, SiC device edge termination" ISPSD'95, p. 96
106. Ueno. Katsunori, Urushidani. Tatsuo, Hashimoto. Kouichi, Seki. Yasukazu "Al/Ti Schottky barrier diodes with the guard-ring termination for 6H-SiC" ISPSD'95, pp. 107-111
107. R. Raghunathan, B.J. Baliga, "EBIC investigation of edge termination techniques for silicon carbide power devices" ISPSD, 1996, pp. 111 – 114
108. G. Brezeanu, J. Fernandez, J. Millan, J. Rebollo, M. Badila, G. Dilimot, P. Lungu, "MEDICI simulation of 6H-SiC oxide ramp profile (ORP) Schottky structure" International Semiconductor Conference, ***2***, 1996, pp. 531 – 534
109. R. Singh, J.W. Palmour, "Planar terminations in 4H-SiC Schottky diodes with low leakage and high yields", ISPSD'97, pp. 157 – 160
110. D. Alok, B.J. Baliga, , "SiC device edge termination using finite area argon implantation" Electron Devices, IEEE Trans., ***44***, 1997 pp. 1013 – 1017
111. A.P. Knights, M.A. Lourenco, K.P. Homewood, "Low temperature annealing of 4H-SiC Schottky diode edge terminations formed by 30 KeV Ar+ implantation." J. Appl. Phys., ***87***, 2000, p. 3973-7
112. D.J. Morrison, N.G. Wright and A.B. Horsfall, "Effect of post-implantation anneal on the electrical characteristics of Ni 4H-SiC Schottky barrier diodes terminated using self-aligned argon ion implantation." Solid-State Electronics, ***44***, 2000, pp. 1879-85.
113. V. Khemka, K. Chatty, T P. Chow, R J. Gutmann, "Breakdown voltage improvement of 4H-SiC Schottky diodes by a thin surface implant" Materials Science Forum, ***338***, 2000. pp. 1211-1214
114. D.C. Sheridan. G. Niu, J.N. Merrett, J.D. Cressler,. C. Tin, C.C. Ellis, R.R. Siergiej, "Simulation and fabrication of high-voltage 4H-SiC diodes with multiple floating guard ring termination" Materials Science Forum, ***338***, 2000. pp.1339-1342
115. T. Hatayama, T. Yoneda, T. Nakata, M. Watanabe, T. Kimoto and H. Matsunami, "Vanadium ion implanted guard rings for high-voltage 4H-SiC Schottky rectifiers" Jap. J. Applied Physics Part 2-Letters, ***39***, 2000. pp. L1216-L1218
116. K. Tone, J H. Zhao, M. Weiner, M. Pan, "4H-SIC junction-barrier Schottky diodes with high forward current densities" Semiconductor Science & Technology, ***16***, 2001. pp 594-597
117. K. Kinoshita, , T. Hatakeyama, , O. Takikawa, , A. Yahata, , T. Shinohe, , "Guard ring assisted RESURF: a new termination structure providing stable and high breakdown voltage for SiC power devices", ISPSD, 2002, pp. 253 - 256
118. K.V. Vassilevski, A. B. Horsfall, C. M. Johnson, N.G. Wright. "Edge Termination of SiC schottky diodes with Guard Rings Formed by High Energy Boron Implantation" Materials Science Forum, ***457-460***, 2004. pp 989-992.

119. S. Hu and K. Sheng, "A new edge termination technique for SiC power devices", International Journal on Solid State Electronics, ***48***, 2004, pp 1861-1866
120. M. Bhatnagar, B.J. Baliga, H.R. Kirk, G.A. Rozgonyi, "Effect of surface inhomogeneities on the electrical characteristics of SiC Schottky contacts" IEEE Trans. Electr. Dev., ***43***, 1996 p150
121. D. Defives, O. Noblanc, C. Dua, C. Brylinski, M. Barthula, F. Meyer, "Electrical characterization of inhomogeneous Ti/4H-SiC Schottky contacts" Materials Science & Engineering B-Solid State Materials for Advanced Technology, ***61***, 1999. pp. 395-401
122. D. Defives, , O. Noblanc, C. Dua, , C. Brylinski, , M. Barthula, , V. Aubry-Fortuna, , F. Meyer, , "Barrier inhomogeneities and electrical characteristics of Ti/4H-SiC Schottky rectifiers" IEEE Trans. Electron Devices, ***61***, 1999 pp. 449 – 455
123. R. Weiss, L. Frey and H. Ryssel. "Modeling of the Influence of Schottky Barrier Inhomogeneities on SiC Diode Characteristics" Materials Science Forum., ***457-460***, 2004. pp 973-976
124. K.J. Schoen, J.P. Henning, J.M. Woodall, J.A. Cooper, M.R. Melloch, "Dual-metal-trench Schottky pinch-rectifier in 4H SiC" Materials Science Forum, ***21***, 1998. pp.945-948
125. K.J. Schoen, J.P. Henning, J.M. Woodall, J.A. Cooper, M.R. Melloch, "Dual-metal-trench Schottky pinch-rectifier in 4H-SiC" IEEE Electr. Dev. Lett., ***19***, 1998. pp. 97-99
126. V. Khemka, V. Ananthan, T P. Chow, "Fully planarized 4H-SiC trench MOS barrier Schottky (TMBS) rectifier" IEEE Electr. Dev. Lett., ***21***, 2000. pp. 286-288
127. Q. Zhang, V. Madangarli, T S. Sudarshan, "SiC planar MOS-Schottky diode: A high voltage Schottky diode with low leakage current" Solid-State Electronic, ***45***, 2001. pp 1085-1089
128. Y. Singh, M.J. Kumar, "A new 4H-SiC lateral merged double Schottky (LMDS) rectifier with excellent forward and reverse characteristics" IEEE Trans. Electr. Dev., ***48***, 2001. pp. 2695-2700
129. K.V. Vassilevski, , A.B. Horsfall, , C.M. Johnson, , N.G. Wright, , A.G. O'Neill, , "4H-SiC rectifiers with dual metal planar Schottky contacts" IEEE Trans. Electr. Dev., ***49***, 2002 p947
130. J H. Zhao, P. Alexandrov, L. Fursin, Z C. Feng, M. Weiner, "High performance 1500 vol 4H-SiC junction barrier Schottky diodes" Electronics Letters., ***38***, 2002. pp 1389-1390
131. D. Peters, P. Friedrichs, R. Schorner and D. Stephani, "Comparison of 4H-SiC pn, pinch and Schottky diodes for the 3 kV range" Materials Science Forum, ***389-393***, 2002. pp. 1125-1128
132. M.J. Kumar and C.R. Linga, "A new, high-voltage 4H-SiC lateral dual sidewall Schottky (LDSS) rectifier: Theoretical investigation and analysis" IEEE Trans. Electr. Dev., ***50***, 2003. pp. 1690-1693
133. F. Roccaforte, F. La Via, A. La Magna, S. Di Franco and V. Raineri, "Silicon carbide pinch rectifiers using a dual-metal Ti-Ni2Si Schottky barrier" IEEE Trans. Electr. Dev., ***50***, 2003 p1741
134. F. Roccaforte, F. La Via, S. Di Franco, V. Raineri, "Dual metal SiC Schottky rectifiers with low power dissipation." Elsevier. Microelectronic Engineering, ***70***, 2003, pp.524-528
135. P. Alexandrov, W. Wright, M. Pan, M. Weiner, L. Jiao, J. Zhao, "Demonstration of a 140 A, 800 V, fast recover 4H-SiC P-i-N/Schottky barrier (MPS) diode", ISDRS, 2001, p13
136. R. Held, N. Kaminski and E. Niemann, "SiC Merged p-n/Schottky Rectifiers for High Voltage Applications" Materials Science Forum, ***246***, 1998, p 1057
137. F. Dahlquist C.M. Zetterling, M. Ostling and K. Rottner, "Junction Barrrier Schottky Diodes in 4H-SiC and 6H-SiC" Materials Science Forum, ***246***, 1998, p 1061
138. F. Dahlquist, J.O. Svedberg, C.M. Zetterling , M. Ostling, B. Breitholtz and H. Lendenmann, "A 2.8kV, Forward Drop JBS Diode with Low Leakage" Materials Science Forum, ***338-342***, 1999, p 1179
139. Y. Sugawara Y, K. Asano and R. Saito, "3.6kV 4H-SiC JBS Diodes with Low RonS" Materials Science Forum, ***338-342***, 1999, p 1183
140. K. Tone, J.H. Zhao, M. Weiner and M. Pan, "Fabrication and Testing of 1,000V-60A 4H-SiC MPS Diodes in an Inductive Half-Bridge Circuit" Materials Science Forum, ***338-342***, 1999, p 1187

141. R. Held R, M. Fullmann and E. Niemann, "SiC-Power Rectifiers" Materials Science Forum, ***338-342***, 1999 p 1407
142. CoolMOSTM Datasheets, Infineon, 2000
143. K. Sheng, "MOS-Controlled Diode (MCD) on Silicon-On-Insulator (SOI)," 29th Annual Conference of the IEEE Industrial Electronics Society, IECON, 2003, pp. 2602-2606.
144. H.R. Chang, R.N. Gupta, C. Winterhalter, E. Hanna, "Comparison of 1200V silicon carbide schottky diodes and silicon power diodes" Proceedings of the Intersociety Energy Conversion Engineering Conference, ***1***, 2000. IEEE, pp. 174-179
145. D.T. Morisette, J.A. Cooper, M.R. Melloch, G.M. Dolny, P.M. Shenoy, M..Zafrani, J. Gladish, "Static and dynamic characterization of large-area high-current-density SiC Schottky diodes" IEEE Trans. Electr. Dev., ***48***, 2001 pp. 349 – 352
146. D.T. Morisette, J.A. Cooper, "Performance of SiC bipolar (PiN) and unipolar (SBD) power rectifiers in current-voltage-frequency parameter space" Materials Science Forum, ***289-393***, 2002. p1181
147. B. Ozpineci, L.M. Tolbert, "Characterization of SiC Schottky diodes at different temperatures" IEEE Power Electr. Lett., ***1***, 2003. pp. 54-57
148. G.B. Spiazzi, S. Citron, M. Corradin, M. Pierobon, Roberto., "Performance Evaluation of a Schottky SiC Power Diode in a Boost PFC Application" IEEE Trans. Power Electr., ***18***, 2003. pp. 1249-1253
149. J. Lai, X. Huang, H. Yu, A.R. Hefner, D.W. Berning, R. Singh, "High current SiC JBS diode characterization for hard- and soft-switching applications", IAS, 2001, pp. 384 – 390
150. R.C. Lebron-Velilla*, G.E. Schwarze, B.G. Gardner and J. Adams, "Silicon Carbide Diode Characterization at High Temperature and Comparison with Silicon Diodes", unpublished
151. K. Sheng, S.J. Finney and B.W. Williams, "IGBT switching losses", Proceedings of the Second International Conference on Power Electronics and Motion Control, 1997, pp. 274-277
152. W. Wright, J. Carter, P. Alexandrov, M. Pan, M. Weiner, J.H. Zhao, "Comparison of Si and SiC diodes during operation in three-phase inverter driving ac induction motor", Electronics Letters, ***37***, 2001, Pages:787 – 788
153. A.R. Hefner, R. Singh, J. Lai, D.W. Berning, S. Bouche and C. Chapuy, "SiC power diodes provide breakthrough performance for a wide range of applications", IEEE Trans. Power Electronics, ***16***, 2001, pp. 273 – 280
154. L. Lorenz, G. Deboy and I. Zverev, "Matched Pair of CoolMOS Transistor With SiC-Schottky Diode—Advantages in Application", IEEE Trans. Ind. Appl., ***40***, 2004, pp.1265 – 1272
155. P. Tobias, S. Nakagomi, A. Baranzahi, R. Zhu, I. Lundstrom, P. Martensson, A Lloyd. Spetz, "Electrical characterization of chemical sensors based on catalytic metal gate-silicon carbide Schottky diodes" Materials Science Forum, ***264-268***, 1998. pp. 1097-1100
156. L.Y. Chen, G.W. Hunter, P.G. Neudeck and D. Knight, "Surface and interface properties of PdCr/SiC Schottky diode gas sensor annealed at 425 degree C" Solid-State Electronics, ***42***, 1998. pp. 2209-2214
157. C.K. Kim, J.H. Lee, Y.H. Lee, N.I. Cho, D.J. Kim and W.P. Kang, "Hydrogen sensing characteristics of Pd-SiC Schottky diode operating at high temperature" J. Electr. Mat., ***28***, 1999. pp. 202-205
158. A.L. Spetz, P. Tobias, A. Baranzahi, P. Martensson, I. Lundstrom, "Current status of silicon carbide based high-temperature gas sensors" IEEE Trans. Electr. Dev. ***46***, 1999. pp. 561-566
159. C.K. Kim, J.H. Lee, Y.H. Lee, N.I. Cho, D.J. Kim, "Study on a platinum-silicon carbide Schottky diode as a hydrogen gas sensor" Sensors & Actuators B, ***66***, 2000. pp. 116-118
160. C.K. Kim, J.H. Lee, S.M. Choi, I.H. Noh, H.R. Kim, N.I. Cho, C. Hong, G.E. Jang, "Pd- and Pt-SiC Schottky diodes for detection of H2 and CH4 at high temperature" Sensors & Actuators B, ***77***, 2001. pp 455-462
161. L. Uneus, S. Nakagomi, M. Linnarsson, M S. Janson, B G. Svensson, R. Yakimova, M. Syvajarvi, A. Henry, E. Janzen, L -G. Ekedahl, I. Lundstrom, A. Lloyd Spetz, "The effect of hydrogen diffusion in p- and n-type SiC Schottky diodes at high temperatures" Materials Science Forum, ***389-393***, 2002. pp. 1419-1422

162. S. Nakagomi, H. Shinobu, L. Uneus, I. Lundstrom, L.G. Ekedahl, R. Yakimova, M. Syvajarvi, A. Henry, E.L. Janzen and A. Spetz, "Influence of epitaxial layer on SiC Schottky diode gas sensors operated under high-temperature conditions" Materials Science Forum, ***389-393***, 2002. pp. 1423-1426
163. J.P. Xu, T. Lai, D.G. Zhong and C.L. Chan, "Improved hydrogen-sensitive properties of MISiC Schottky sensor with thin NO-grown oxynitride as gate insulator" IEEE Electr. Dev. Lett., ***24***, 2003. pp. 13-15
164. S. Roy, C. Jacob, C. Lang, S. Basu, "Studies on Ru/3C-SiC Schottky junctions for high temperature hydrogen sensors" J. Electrochemical Society, ***150***, 2003. pp. H135-H139
165. J. Kim, B.P. Gila, C.R. Abernathy, G.Y. Chung, F. Ren and S.J. Pearton, "Comparison of Pt/GaN and Pt/4H-SiC gas sensors" Solid-State Electronics, ***47***, 2003, pp. 1487-1490
166. S. Roy, C. Jacob, S. Basu, "Studies on Pd/3C-SiC Schottky junction hydrogen sensors at high temperature" Sensors & Actuators B, ***94***, 2003. pp. 298-303
167. S. Roy, C. Jacob and S. Basu, "Ohmic contacts to 3C-SiC for Schottky diode gas sensors" Solid-State Electronics, ***47***, 2003. pp. 2035-2041
168. L.A. Spetz, A. Baranzahi, P. Tobias and I. Lundstr€om, Phys Stat Sol A, ***162***, 1997, p. 493.
169. G. W. Hunter, P. G. Neudeck, G. D. Jefferson, G. C. Madzsar, C. C. Liu, and Q. H. Wu, "The development of hydrogen sensor technology at NASA Lewis Research Center," Ann. Space System Health Management Technology Conference, 1992
170. Y. Li, L. Fursin, J. Wu, P. Alexandrov and J.H. Zhao, "2.5kV-30A inductively loaded half-bridge inverter switching using 4H-SiC MPS free-wheeling diodes", Mat. Sci. Forum, 2004, pp. 1097-1100
171. J. Eriksson, F. Ferdos, H. Zirath, N. Rorsman, "Design and characterization of a SiC Schottky diode mixer" Materials Science Forum, ***338***, 2000. p1207
172. R.N. Simons and P.G. Neudeck, "Intermodulation-Distortion Performance of Silicon-Carbide Schottky-Barrier RF Mixer Diodes." IEEE Trans. Microw. Theory & Tech., ***51***, 2003, pp. 669-672.
173. J. Eriksson, , N. Rorsman, , H. Zirath, , "4H-silicon carbide Schottky barrier diodes for microwave applications" IEEE Trans. Microwave Theory and Techniques, ***51***, 2003 pp. 796 – 804
174. M.M. Anikin, A.N. Andreev, S.N. Pyatko, N.S. Savkina, A.M. Strelchuk, A.L. Syrkin and V. Chelnokov. "UV photodetectors in 6H-SiC". Sensors & Actuators A, ***33***, 1992. pp. 91-93
175. M. Badila, G. Brezeanu, J. Millan, P. Godignon, M. L. Locatelli, J P. Chante, A. Lebedev, P. Lungu, G. Dinca, V. Banu, G. Banoiu, "Lift-off technology for SiC UV detectors" Diamond & Related Materials, ***9***, 2000. pp. 994-997
176. G. Brezeanu, , M. Badila, , P. Godignon, , J. Millan, , F. Udrea, , A. Mihaila, , G. Amaratunga, , J. Rebollo, , I. Enache, , "UV detection properties of epitaxial 6H-SiC diodes with oxide ramp termination" CAS, 2001. pp. 345 - 348
177. F. Yan, X. Xin, S. Aslam, Y. Zhao, D. Franz, J.H. Zhao, M. Weiner, "4H-SiC UV Photo Detectors With Large Area and Very High Specific Detectivity", Quantum Electronics, IEEE Journal of, ***40***, 2004, pp.1315 - 1320
178. S. Seshadri, A.R. Dulloo, F. H. Ruddy, J. G. Seidel, L. B. Rowland, "Demonstration of an SiC neutron detector for high-radiation environments" IEEE Trans. Electr. Dev., ***46***, 1999. pp.567-571
179. A M. Ivanov, N B. Strokan, D V. Davydov, N S. Savkina, A A. Lebedev, Yu T. Mironov, G A. Riabov, E M. Ivanov, "Radiation hardness of SiC based ions detectors for influence of the relative protons" Applied Surface Science, ***184***, 2001. pp. 431-436
180. G. Verzellesi, P. Vanni, F. Nava, C. Canali, "Investigation on the charge collection properties of a 4H-SiC Schottky diode detector" Nuclear Instruments and Methods in Physics Research, Section A, ***476***, 2002. pp 717-721
181. M. Bruzzi, S. Lagomarsino, F. Nava, S. Sciortino, "Characterisation of epitaxial SiC Schottky barriers as particle detectors Diamond & Related Materials", ***12***, 2003. pp 1205-1208

182. F. Nava, G. Wagner, C. Lanzieri, P. Vanni, E. Vittone, "Investigation of Ni/4H-SiC diodes as radiation detectors with low doped n-type 4H-SiC epilayers" Nuclear Instruments & Methods in Physics Research Section A, ***510***, 2003. pp. 273-280
183. A. Le Donne, S. Binetti, M. Acciarri, S. Pizzini, "Electrical characterization of electron irradiated X-rays detectors based on 4H-SiC epitaxial layers" Diamond & Related Materials, ***13***, 2004. pp. 414-418
184. P. G. Neudeck, "Electrical impact of SiC structural crystal defects on high field devices," Material Science Forum., ***338***, 2000, pp. 1161–1166
185. D. Nakamura, I. Gunjishima, S. Yamaguchi, T. Ito, A. Okamoto, H. Kondo, S. Onda, K. Takatori1, "Ultrahigh-quality silicon carbide single crystals", Nature, ***430***, 2004, pp. 1009-1012
186. M. Treu, R. Rupp, H. Brunner, F. Dahlquist and C. Hecht, "Challenges and first results of SiC Schottky diode manufacturing using a 3 inch technology", Mat. Sci. Forum, ***457-460***, 2004, pp. 981-984
187. A. Syrkin, V. Dmitriev, V. Soukhoveev, M. Mynbaeva, R. Kakanakov, C. Hallin and E. Janzen, "4H-SiC power Schottky diodes on the way to solve size limiting issues", Mt. Sci. Forum, ***457-460***, 2004, pp. 985-988
188. J.H. Zhao, J. Zhang, Y. Luo, X. Hu, Y. Li, H. Yu, J. Lai, P. Alexandrov, L. Fursin, X. Li, J. Carter and M. Weiner, "The first 4H-SiC BJT-based 20kHz, 7HP PWM DC-to-AC inverter for induction motor control applications", Mat. Sci. Forum, ***457-460***, 2004, pp. 1137-1140
189. P. Alexandrov, W. Wright, M. Pan, N, Weiner, L. Jiao and J.H. Zhao, "Demonstration of high voltage (600-1300V), high current (10-140A), fast recovery 4H-SiC merged p-i-n/Schottky barrier (MPS) diode", Solid State Electronics, ***47***, 2003, pp.263-269
190. J.H. Zhao, L. Fursin, L. Jiao, X. Li and T. Burke, "Demonstration of 1789 V, 6.68 mΩ.cm2 4H-SiC merged-PiN-Schottky diodes", Electronics Letters, ***40***, 2004, pp390-391
191. A.O. Konstantinov, Q. Wahab, N. Nordell, and U. Lindefelt, "Ionization rates and critical fields in 4H silicon carbide", Appl. Phys. Lett., ***70***, 1997, 90-92
192. H.M. McGlothlin, D.T. Morisette, J.A. Cooper and M.R. Melloch, "4kV silicon carbide Schottky diode for high frequency switching applications", Proc. Dev. Res. Conf., 1999, p. 28
193. R.T. Tung, "Electron transport at metal-semiconductor interfaces: general theory", Physical Review B, ***45***, 1992, pp. 13509-13523
194. R.T. Tung, "Recent advances in Schottky barrier concepts", Mat. Sci. Eng. R, ***35***, 2001, pp. 1-138.
195. M.O. Aboelfotoh, C. Frojdh and C.S. Petersson, "Schottky-barrier behavior of metals on n-and p-type 6H-SiC", physical Review B, ***67***, 2003, 075312

International Journal of High Speed Electronics and Systems
Vol. 15, No. 4 (2005) 867–898

HIGH POWER SIC PIN RECTIFIERS

Ranbir Singh

GeneSiC Semiconductor Inc., ranbir@ieee.org

High voltage PiN rectifiers made using conventional semiconductor materials such as Silicon are restricted to less than 20 kHz and less 120°C operation, thereby severely limiting the availability of advanced electronic hardware used for power grid (also called electric utility), energy storage, pulsed power, intelligent machinery and ultra high voltage solid state power conditioning. Such applications require high power density, very high frequency, and high temperature rectifiers to realize reasonably sized systems. SiC PiN Rectifiers are expected to play an enabling role in a variety of such high voltage applications because they have been shown to offer 2 to 3 orders of magnitude faster switching, high junction temperature capability, high current density operation, and much higher power densities as compared to Silicon.

1. Introduction

Properties of SiC that enable dramatically improved capabilities for SiC PiN rectifiers include: (a) an order of magnitude higher breakdown electric field; (b) a ~3X wider bandgap; and (c) a ~3X higher thermal conductivity than Silicon. A high breakdown electric field allows the design of SiC power devices with thinner and higher doped blocking layers. The large bandgap of SiC results in a much higher operating temperature and higher radiation hardness. The high thermal conductivity for SiC (4.9°C/W) allows dissipated heat to be readily extracted from the device. Hence, a larger power can be applied to the device for a given junction temperature. However, SiC bipolar rectifiers suffer from a ~3X higher built-in junction voltage drop as compared to Si devices due to SiC's larger bandgap. However, since the on-state voltage drop decreases with increasing temperature, the junction voltage drop difference is reduced to less than 2X as compared to a Si PiN rectifier operating at room temperature for a SiC PiN rectifier operating at 500°C. An important property of SiC from a bipolar device perspective is its indirect bandgap (as opposed to GaN, which has a direct bandgap) which has the potential for high recombination lifetime, if the material is of excellent quality. The wide bandgap of SiC also allows negligible junction leakage currents up to 500 °C in PiN rectifiers. This allows high temperature operation without excessive leakage current or thermal runaway. Cooling requirements, and hence the size and weight, of power electronic systems could be reduced if the systems were designed to take advantage of the ability of SiC rectifiers to operate at elevated temperatures.

To increase the efficiency and decrease the cost, weight and volume of power conversion systems, it is important to reduce the heat generated during device operation. Most of the heating in a rectifier occurs either in the turn-on state or during switching transients. As in Si, SiC power rectifiers may be broadly classified into majority carrier rectifiers (Schottky rectifiers), in which on-state conduction is dependent on doping and bulk carrier mobilities; and minority carrier rectifiers (PiN rectifiers) which rely on conductivity modulation for on-state current transport. Schottky rectifiers offer extremely fast switching, and although their voltage drop at low current densities can be low, its magnitude increases dramatically at high current densities. This issue increases exponentially in its severity as these rectifiers are designed for higher voltage ratings. This is because the on-resistance of a voltage blocking layer increases exponentially with the designed breakdown voltage. On the other hand, bipolar PiN rectifiers offer low forward voltage drop at high current densities, but have higher switching losses as compared to Schottky rectifiers. Application engineers often ask which SiC rectifier may be suitable in the high voltage range – Schottky rectifier or PiN rectifier. If on-state voltage drop is the only consideration, the current density specification at which this choice can be made is for SiC Schottky and PiN rectifiers for ideal blocking voltage in the 2 to 30 kV range is shown in Figure 1 using generalized assumptions. Above this limit, PiN rectifiers may be more attractive than Schottky rectifiers because it offers lower on-state voltage (V_F) drop. Switching speed of rectifier also plays an important role in determining its appropriateness to a particular application. There often exists a design trade-off between the switching speed and the on-state voltage drop in a bipolar rectifier if lifetime is changed. There exists an upper limit to the current density for rectifiers, which is determined by commercial packaging technology. Conventional power device packaging technology can only dissipate 250-400 W/cm^2 continuously. Since the built-in voltage of 4H-SiC bipolar device is ~3 V, the maximum continuous current may be limited to only 100-150 A/cm^2 [1].

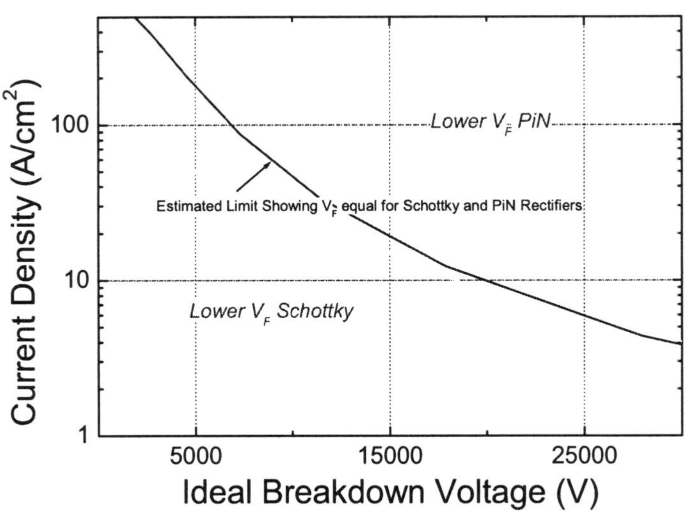

Figure 1: Approximate current density limit where SiC Schottky rectifiers and PiN rectifiers offer similar on-state voltage drop (V_F).

2. PiN Rectifier Design and Operation

SiC PiN rectifiers are fabricated using n^+ wafers with low doped, thick n^- epitaxial layers, which are typically grown over a n^+ epitaxial buffer layer. The p^+ Anode region may be formed by using a highly doped epitaxial layer or by using ion implantation of p-type dopants such as Aluminum or Boron. Typically, PiN rectifiers with epitaxial p^+ Anode as shown in Figure 2 may offer lower on-state voltage drop as compared to ion implanted PiN rectifiers because of high p-type dopant activation, which determines the injection efficiency of current carriers into a low doped, thick epitaxial layer. The blocking voltage of such a rectifier is determined by (a) the doping and thickness of voltage blocking epitaxial layer (i-region) and (b) the effectiveness of the edge termination technique used to fabricate the device. Generally speaking, the current carrying capability depends on the physical area of the device and the minority carrier lifetime in the i-layers. For the device to sustain a high electric field during the reverse bias operation, it is essential to reduce the substrate induced defects, such as micropipes and epitaxially induced defects.

2.1. *High voltage epitaxial layer design*

In order to realize high voltage rating on PiN rectifiers, the doping and thickness design of n^- epitaxial layer is crucial. During reverse bias application, the highest electric field occurs at the p^+/n^- junction, and the gradient of the electric field in the n^- layer is determined by its doping, and the depth to which the electric field extends is determined by its thickness. The doping and thickness of n- epitaxial layer determines the blocking voltage of the device. For a designed voltage blocking capability, an n^- epitaxial design is called a punchthrough design if a finite electric field exists at the n^- layer/n^+ buffer layer junction. On the other hand a non-punchthrough design corresponds to the case where i-layer thickness is equal to or larger than the parallel plane avalanche breakdown width [2]. In a typical ultra high voltage design, a punchthrough design is used in order to keep the thickness of the epitaxial layer reasonable. For such a case, the i-layer thickness

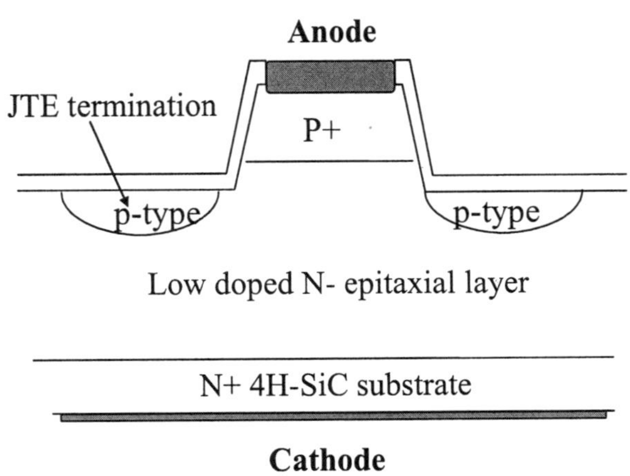

Figure 2: Cross-section of a high voltage 4H-SiC PiN rectifier using a highly doped epitaxially grown Anode layer and etched and implanted JTE structure.

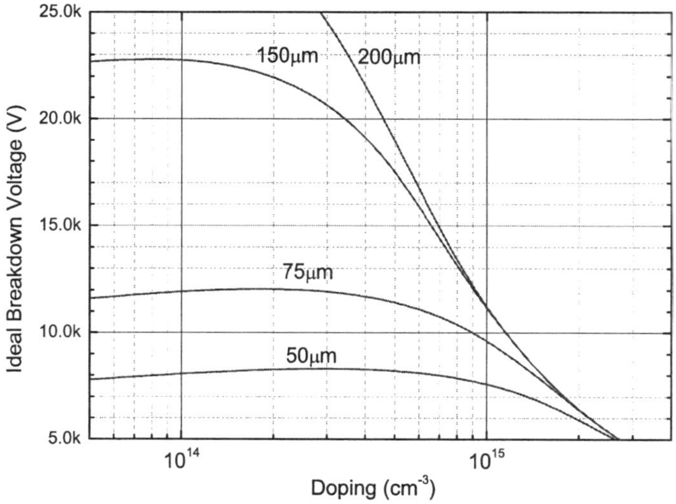

Figure 3 : Ideal blocking voltage as a function of i-layer doping and thickness.

can be derived from basic depletion equation as:

$$W = \frac{\varepsilon \cdot E_C(N_D)}{q \cdot N_D} - \sqrt{[\frac{\varepsilon \cdot E_C(N_D)}{q \cdot N_D}]^2 - [\frac{2\varepsilon \cdot V_B}{q \cdot N_D}]} \quad (1)$$

where V_B is the device blocking voltage, W is the epitaxial layer thickness, N_D is the doping of the epitaxial layer, q is the elementary charge, ε is the dielectric constant of SiC, and $E_C(N_D)$ is the critical electric field as a function of the i-layer doping. One of the commonly accepted empirically derived relationships showing the dependence of critical electric field with i-layer doping in 4H-SiC is given by [3]:

$$E_C = \frac{2.49 \times 10^6}{1 - \frac{1}{4}\log(\frac{N_D}{10^{16}})} \quad (2)$$

Using these formulae, the ideal blocking voltage is plotted as a function of i-layer doping and thickness in Figure 3. The decrease in blocking voltage with a decrease in doping is an artifact of the approximations used in the derivation of E_C, and is not accurate. Since the edge termination design as well as material defects influences the blocking voltage of ultra high voltage devices, very large blocking voltage margins are usually adopted in the design of ultra high voltage PiN rectifiers.

2.2. *Edge termination design*

From a device design standpoint, the biggest technological challenge in achieving high voltage PiN rectifiers is the design and implementation of an effective edge termination. Such a technique is expected to make the electric field distribution uniform

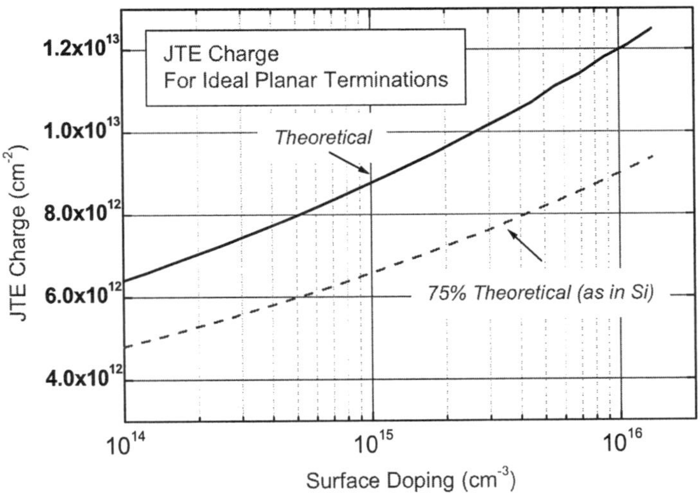

Figure 4: JTE charge vs. surface doping of the voltage blocking layer for a SiC device.

at the edge of the device, approaching the ideal breakdown voltage capability of the epitaxial layer used. Traditionally, many techniques like guard rings, floating field rings, and trench guard rings [4] have been used. Another promising edge termination design involves implanting the device edge with an optimum p-type charge in order to reduce the electric field gradually from the device edge to the outer periphery of the device structure. This technique is called junction termination extension (JTE) [2]. Using the approximations used to derive the critical electric field, the optimum JTE charge is expected to depend on the background doping concentration of the low doped n-type region. A plot showing the ideal total charge per unit area of the implanted JTE region is shown in Figure 4. The design of high voltage Silicon devices made using JTE typically utilize a JTE charge 25% below the ideal predicted charge by theoretical analysis. This is because JTE charge in excess of the ideal value results in a sharp reduction in the obtained breakdown voltage, while a charge smaller than the optimum does not drastically affect the breakdown voltage of the device. The JTE charge corresponding to 75 % is also plotted in this graph for SiC.

2.3. *Effect of lifetime on rectifier on-state voltage drop*

Achieving a high level of carrier lifetimes in the thick, SiC epitaxial layers is critical to obtaining acceptable on-state voltage drop in 4H-SiC PiN rectifiers. For a given blocking voltage design, which determines the thickness of i-layer, epitaxial growth experts need to deliver the required carrier lifetimes in these layers to obtain low on-state voltage drops in these rectifiers. Key lifetime limiting factors in thick epitaxial layers are: (a) compensating dopant species; (b) Unintentional metallic impurities; (c) morphology of the epitaxial layers during and after the growth; and (d) thickness and doping uniformity of thick epitaxial layers. For the extremely thick epitaxial layers used in these rectifiers, hot wall epitaxial reactors were used. Using these reactors, higher carrier lifetimes can be obtained.

The dominant components in the on-state voltage drop in a PiN rectifier are given by:

$$V_F = V_{P+Cont.} + V_M + V_{P+N-} + V_{N+N-} + V_{Subs} \tag{2a}$$

where V_F is the on-state voltage drop in a PiN rectifier, $V_{P+Cont.}$ is the p+ contact resistance V_M is the 'middle region' (i-region) voltage drop, V_{P+N} and V_{N+N-} are the junction drops at $p^+ n^-$ and $n^+ n^-$ junctions, and V_{Subs} is the resistive voltage drop in the substrate. The voltage drop in the i-layer (V_M) is determined by the extent of carrier modulation in the i-region of the PiN rectifier. Using a broad approximation, the i-region voltage drop is related to the carrier lifetime by the following relationship [2]:

$$V_M = \frac{3kT}{q}\left(\frac{W}{2L_a}\right)^2 \quad \text{for } W \leq 2\,L_a \tag{3}$$

$$V_M = \frac{3\pi\, kT}{8q} e^{\frac{W}{2La}} \quad \text{for W} \geq 2\,L_a$$

where k and T are Boltzmann's constant and operating temperature. L_a is the ambipolar diffusion length, which is given by $L_a = \sqrt{D_a \tau_{HL}}$. D_a is the ambipolar diffusion constant given by $D_a = \mu_a . \frac{kT}{q}$, and τ_{HL} is the high level carrier lifetime. The ambipolar carrier mobility, μ_a is given by $\mu_a = \frac{\mu_n \mu_p}{\mu_n + \mu_p}$. Here, μ_n and μ_p are the minority carrier electron and hole mobility in the voltage blocking layer.

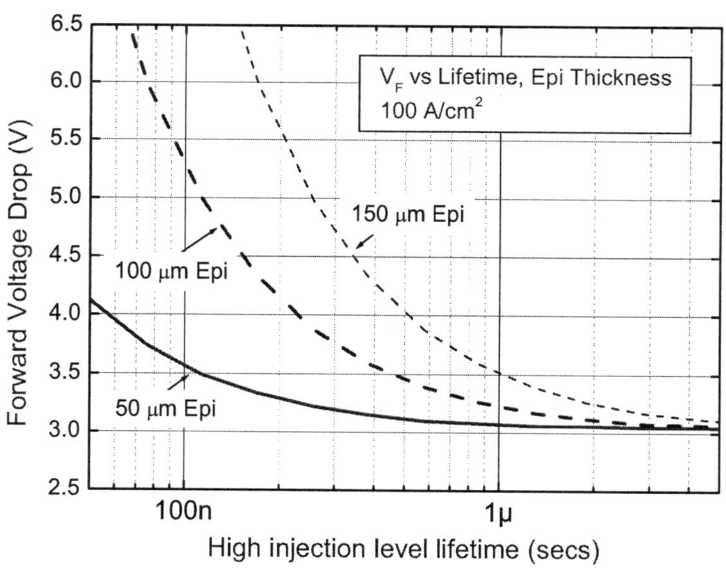

Figure 5: Forward voltage drop in PiN rectifiers as a function of lifetime and epitaxial thickness.

V_{P+N-} and V_{N+N-} are dependent on the minority carrier concentrations at the two end-regions of the n^- layer. A detailed analysis of these voltage drops requires an iterative solution [2], which is not very easily calculated. If a reasonably high injection level is assumed, the sum of these end-region voltage drops can be estimated to be equal to the turn-on voltage of the SiC p^+ n^+ junction. Resistive drops like V_{Subs} and $V_{P+Cont.}$ are not trivial in bipolar device like PiN rectifiers because of extremely high current densities that typically flow through these devices, as compared to majority carrier devices like Schottky rectifiers. The on-state voltage drop of a PiN as a function of carrier lifetime and epitaxial layer thickness is plotted in Figure 5. This figure shows that a higher carrier lifetime results in better conductivity modulation, with on-state voltage drop approaching the built-in voltage drop of a p^+ n^+ junction. A thicker epitaxial layer requires a higher carrier lifetime in order to achieve a lower on-state voltage drop PiN rectifier. However, it is encouraging to note that beyond a carrier lifetime of 3 μs, even a 150 μm epitaxial layer is well modulated, because the i-region voltage drop is negligible as compared to the total voltage drop of the diode. Since this carrier lifetime is more than two orders of magnitude smaller than Si devices with comparable blocking voltage ratings, SiC is expected to offer extremely high switching speeds.

2.4. *Carrier lifetime measurements in SiC PiN rectifiers*

Carrier lifetimes in the thick n^- voltage blocking layers of PiN rectifiers are frequently measured using the switching characteristics of these rectifiers. When the rectifier is switched from its current carrying on-state to its voltage-blocking off-state with the application of reverse bias, the current flowing through the rectifier gradually decreases, and then reverses before reducing to zero. This is referred to as the reverse recovery waveform of the rectifier. The magnitude and the rate of decay of the reverse current is determined by the thickness, doping and minority carrier lifetime of the n- voltage blocking layer as well as the on-state current flowing through the rectifier before it was switched. The technique to extract a value for carrier lifetime in the n- voltage blocking layer from the exponential decay tail of the current waveform is called the current recovery technique (CRT) or the Lax-Neustadter's technique [5].

However, the CRT fails frequently when applied to SiC rectifiers because it assumes a uniform carrier lifetime in the entire n-voltage blocking layer of a PiN rectifier [6]. It is often observed that the carrier lifetime in a thin layer close to the PN junction is significantly lower than that in the bulk of the n-voltage blocking layer. This observation may be explained with the assumption that the high-temperature growth of a heavily doped p^+-emitter leads to a dramatic reduction in the carrier lifetime in a thin layer near the p^+n^- junction, where the local lifetime $\tau*$ is much shorter than bulk carrier lifetime, τ_p. In [7], the minority carrier lifetime τ_p was measured by Open Circuit Voltage Decay (OCVD) technique. It is well known that CRT gives the local lifetime near the p^+n_o junction. Hence, $\tau*$, instead of τ_p, is measured by CRT in this case. Figure 6 shows a pulsed isothermal I-V characteristic of a 4H-SiC rectifier with an Anode area of 0.08 cm^2, voltage blocking n-layer thickness of 150 μm and a blocking capability of 10 kV at room temperature. At j_f = 180 A/cm^2, the rectifier differential resistance $r_d = dV/dj_f = 1.6\times10^{-2}$ Ohm cm^2. Meanwhile, the ohmic resistance of unmodulated base must be $r_n = W/q\mu_n n_o =$ 0.39 Ohm cm^2, i.e., 24 times

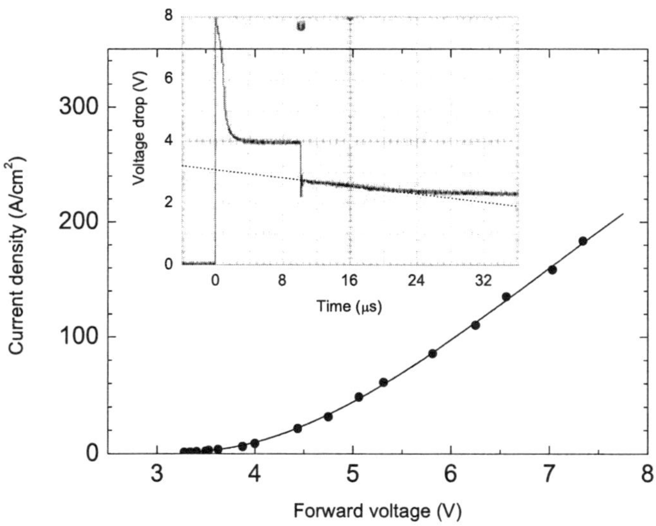

Figure 6: Forward current--voltage characteristic of a 10 kV 4H-SiC rectifier. Inset shows the time diagram of voltage drop across the rectifier during the OCVD measurements. Forward current $I_f = 0.6$ A ($j_f = 7.5$ A/cm^2).

the experimentally observed value at $\mu_n = 800$ cm^2/Vs and net donor concentration, $n_o = N_d - N_a = 3\times10^{14}$ cm^{-3}. Hence, the I-V characteristic demonstrates directly the high level of base modulation. Inset in Figure 6 shows voltage-time diagram of OCVD process. At $t = 0$, the pulse of forward current, with amplitude $I_f = 0.6$ A ($j_f = 7.5$ A/cm^2), is applied to the rectifier connected in series with a 50-Ohm load. It is easy to verify that this j_f value ensures a high injection level in the base. As seen in the inset, the steady state is attained in about 4 μs, which indicates that the τ_p value is 1.4-1.8 μs. After the termination of the current at $t = 10$ μs, all "classical" phases of the post-injection voltage decay are observed [8]: (i) abrupt change in voltage just after current termination, (ii) voltage decay linear in time, and (iii) exponential voltage decay at the final stage. At high injection level, simplified analysis of the OCVD curve [9] gives for τ_p the following expression:

$$\tau_p = -\frac{2kT}{q}\left(\frac{dV}{dt}\right)^{-1} \qquad (4)$$

where dV/dt is the slope of the linear part in the voltage decay curve. In this case, $\tau_p = 1.55$ μs at $T = 293$ K.

The ambipolar diffusion length L_a is defined by $L_a = \sqrt{D\tau_p}$, where $D = D_p 2b/(b+1)$. Here D_p is the hole diffusion coefficient, $b = \mu_n/\mu_p$ (μ_n and μ_p are the electron and hole mobilities, respectively). With $\mu_n = 800$ cm^2/Vs, $\mu_p = 100$ cm^2/Vs, and $D_p = 2.5$ cm^2/s, L_a is calculated to be about 26 μm at $T = 300$ K.

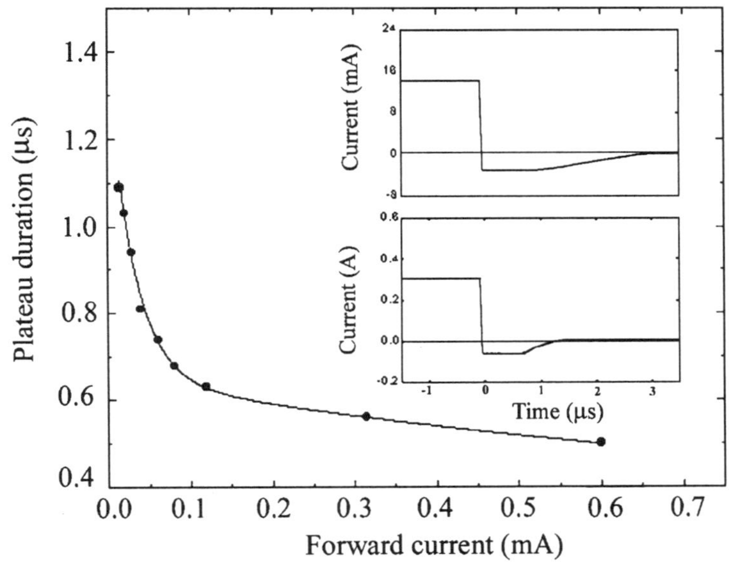

Figure 7: Dependence of the plateau duration in CRT measurements on the forward current I_f at $I_f/I_r = 5$. Inset shows the time dependences of current waveform at two forward current amplitudes: (a) $I_f = 12$ mA, (b) $I_f = 0.6$ A. $I_f/I_r = 5$ for both curves.

The inset in Figure 7 shows current-time diagrams during CRT measurements at $I_f =$ 12 mA and $I_f = 300$ mA. Pronounced reverse current "plateaus" with duration of 0.4 μs for $I_f = 300$ mA and 1.1 μs for $I_f = 12$ mA were observed. For both curves in the inset of Figure 7, the ratio $I_f/I_r = 5$. According to the "classical" theory of CRT [5], the plateau duration must be equal to the lifetime τ_p at this I_f/I_r ratio if the injection level is low. At high injection levels, the plateau duration must decrease monotonically with increasing I_f even at the same I_f/I_r ratio [10]. Figure 7 shows the dependence of the plateau duration on the forward current I_f at $I_f/I_r = 5$. As seen, the plateau duration decreases monotonically with increasing I_f, in agreement with the prediction of the theory. It is noteworthy that at the minimum current of 12 mA, the plateau duration (1.1 μs) is only slightly shorter than the τ_p value measured by OCVD.

It is well known that the following conditions should be satisfied for the classical CRT technique to be applicable: (i) the injection level is to be low, and (ii) the injection coefficient of the emitter should be equal, or very close to unity. To provide a low injection level, the condition $p(0) << n_o$ should be satisfied. For the injection coefficient to become close to unity, the recombination current in the Space Charge Region (SCR) of the p^+n^- junction is to be low enough, with the ratio $I_{rec}/I_{diff} << 1$. Assuming $p(0) = n_o$, we can obtain for rectifiers with $W/L >> 1$ the following expressions for j_{diff} and j_{rec} [11]:

$$j^*_{diff} = qN\frac{\sqrt{D_p}}{\sqrt{\tau_p}}, \qquad j^*_{rec} = \frac{N}{n_i} j_{ro}. \tag{5}$$

where n_i is the intrinsic carrier concentration, and j_{ro} is the pre-exponential factor of the SCR recombination current.

With $N = 3\times10^{14}$ cm^{-3}, $D_p = 2.5$ cm^2/s, and $\tau_p = 1.55$ μs, the condition (i) is satisfied at $j_{diff} << 6\times10^{-2}$ A/cm^2 ($I_{diff} << 5$ mA). It is noteworthy that the minimum current of 12 mA apparently corresponds to an intermediate injection level. The maximum current of 300 mA corresponds to a very high injection level. Assuming that:

$$\frac{j^*_{rec}}{j^*_{diff}} = \frac{j_{ro}}{qn_i}\sqrt{\frac{\tau_p}{D_p}} << 1 \tag{6}$$

Calculations based on these formulae show that j_{ro} (with $n_i = 10^{-8}$ cm^{-3}, $D_p = 2.5$ cm^2/s, and $\tau_p = 1.55$μs), must be much less than 2×10^{-24} A/cm^2. The reported j_{ro} values for 4H-SiC rectifier rectifiers fall within the range between 10^{-25} A/cm^2 and 10^{-22} A/cm^2 [12, 13]. So the condition (ii) can only be satisfied in rectifiers with j_{ro} close to the minimum observed values.

2.5. *PiN rectifier operation under superhigh current densities*

An important application area for SiC PiN rectifiers is under very high current density conditions in pulsed modes such as radars, directed energy weapons, space based lasers, space exploration, and electromechanical guns. In such applications, extremely high current density operation is required. At high current densities the forward voltage drop across the rectifier is determined by nonlinear processes, such as electron-hole scattering [14, 15] and Auger recombination [16], and the dependence of emitter efficiency on current density [17]. The contribution of various nonlinear processes to the current-voltage characteristic has been investigated in detail for Silicon thyristors and rectifiers (see, for example, Ref. [18]). However, for Silicon carbide devices, these processes were explained for the first time in Ref [19]. The non-ideality of the emitter p$^+$-n junction is considered to explain experimental results at superhigh current densities. In the framework of the theory [17], the properties of junctions are described by introducing saturation current densities, j_{ns} and j_{ps}:

$$j_{ns} = q\sqrt{\frac{D_n}{\tau_n}}\cdot\frac{n_i^2}{N_a} \tag{7}$$

where D_n, τ_n, N_a are the diffusion coefficient of electrons, electron lifetime, and acceptor concentration in the p$^+$-emitter, respectively; and n_i is the intrinsic concentration. The saturation leakage electron current density of the N-n$^+$ junction, j_{ps}, is defined similarly. The value $j_{ps} = j_{ns} = j_s$ is an essential fitting parameter of the theory, characterizing the quality of the emitter junctions.

In Silicon rectifiers, j_s lies commonly in the range 10^{-14}-10^{-12} A/cm^2 [17]; in GaAs rectifiers, j_s is within the range 10^{-18}-10^{-16} A/cm^2 [20] owing to the fact that the band gap in GaAs ($E_g = 1.4$ eV) is larger than that in Si ($E_g = 1.1$ eV). It can be calculated from the very wide SiC band gap, the leakage current j_s is as small as $\sim10^{-45}$A/cm^2 at 300 K. In the framework of the theory [17], the total voltage drop $V(j)$ at a given current density j is calculated as a sum of voltage drops across the contacts to a structure, V_c, the emitter junctions, V_e, and the rectifier base, V_b:

$$V = V_c + V_e + V_b. \tag{8}$$

The voltage drop across the contacts of the structure, V_c, is given by:

$$V_c = 2R_c j \tag{9}$$

where R_c is the contact resistivity. The voltage drop across the emitter junctions, V_e is given by:

$$V_e = \frac{kT}{q}\ln\frac{p_o p_W}{n_i^2} = \frac{2kT}{q}\ln\frac{j}{j_o} \tag{10}$$

where:

$$j_o = \frac{qDn_i}{2L}\cdot\frac{b+1}{\sqrt{b}}\left[\left(1+\sqrt{1+4jj_{ns}\frac{b}{b+1}\left(\frac{L}{qDn_i}\right)^2}\right)\cdot\left(1+\sqrt{1+4jj_{ns}\frac{1}{b+1}\left(\frac{L}{qDn_i}\right)^2}\right)\right]^{1/2}. \tag{11}$$

Here p_o (j) and p_W (j) are the boundary hole concentrations in the n-base at p^+-n and N-n^+ junctions, respectively; $D = [2b/(b+1)]D_p$ is the ambipolar diffusion coefficient, D_p is the hole diffusion coefficient, and $L = (D\tau_{ph})^{1/2}$. The voltage drop across the base, V_b, is given by:

$$V_b = \frac{j}{q(\mu_n+\mu_p)}\int_0^{W_n}\frac{dx}{p(x)+N_o} \tag{12}$$

$$p(x) = \frac{p_o\sinh\frac{W_n-x}{L} + p_W\sinh\frac{x}{L}}{\sinh\frac{W_n}{L}} \tag{13}$$

$$p_o = j\frac{b}{b+1}\cdot\frac{2L}{qD}\cdot\frac{1}{1+\sqrt{1+4jj_{ns}\frac{b}{b+1}\left(\frac{L}{qDn_i}\right)^2}} \tag{14}$$

$$p_W = j\frac{1}{b+1}\cdot\frac{2L}{qD}\cdot\frac{1}{1+\sqrt{1+4jj_{ns}\frac{1}{b+1}\left(\frac{L}{qDn_i}\right)^2}} \tag{15}$$

$$N_o = \frac{b}{b+1}\cdot n_o,\quad n_o = N_d.$$

where n_0 is free electron concentration in the base ($n_0 = N_d$); and W_n is the width of the i-region. The following values of 4H-SiC parameters are generally used at 300K: the electron to hole mobility ratio $b = 8$ ($\mu_n = 800$ cm^2/V·s, $\mu_p = 100$ cm^2/V·s), $D_p = 2.6$ cm^2/s, $n_i = 1.2\times10^{-8}$ cm^{-3}. As mentioned above, the leakage current j_s in 4H-SiC is as small as ~10^{-45} A/cm^2 at 300K. Moreover, j_s is severely dependent on temperature: in the temperature range 300-600 K, j_s varies between ~10^{-45} and ~10^{-17} A/cm^2. It was suggested to introduce instead of j_s a new fitting parameter a:

$$\alpha = j_s\left(\frac{L}{qDn_i}\right)^2. \tag{16}$$

Within the temperature interval 300-600K a varies (1.5-2)-fold. Physical limitations on a can be established by comparing (7) and (16) with the standard expression for the emitter injection efficiency γ [11, 21]:

$$\gamma = \frac{j_p}{j_p + j_n} = \frac{1}{1+\frac{N_d}{N_a}\left(\frac{D_n\tau_p}{D_p\tau_n}\right)^{1/2}}. \tag{17}$$

where τ_p is a hole lifetime which can be different at low and high injection levels. Equation (17) is valid at $W_n >> L_p$ and $W_p >> L_n$. In the simplest case γ_{max} is assumed to be close to unity (since $N_a >> N_d$), and $\gamma_{min} = 1/2$ (symmetrical p-n structure). Using (7), (16), and (17) we obtain:

$$\gamma = \frac{1}{1+\alpha q N_d\left(\frac{2b}{b+1}\right)^{1/2}\left(\frac{D_p}{\tau_p}\right)^{1/2}} \quad \text{for the condition:} \tag{18}$$

$$0 < \alpha < \left[qN_d\left(\frac{2b}{b+1}\right)^{1/2}\left(\frac{D_p}{\tau_p}\right)^{1/2}\right]^{-1}.$$

By using (16), we can simplify (11), (14), and (15):

$$j_o = \frac{qDn_i}{2L}\cdot\frac{b+1}{\sqrt{b}}\left[\left(1+\sqrt{1+4j\frac{\alpha b}{b+1}}\right)\cdot\left(1+\sqrt{1+4j\frac{\alpha}{b+1}}\right)\right]^{1/2} \tag{19}$$

$$p_o = j\frac{b}{b+1}\cdot\frac{2L}{qD}\cdot\frac{1}{1+\sqrt{1+4j\frac{\alpha b}{b+1}}} \tag{20}$$

$$p_W = j\frac{1}{b+1}\cdot\frac{2L}{qD}\cdot\frac{1}{1+\sqrt{1+4j\frac{\alpha}{b+1}}}. \tag{21}$$

This model has been shown to effectively model 5.5 kV 4H-SiC PiN rectifiers in [19] operation at superhigh current densities of up to 50,000 A/cm^2. Modeling up to such high current densities is an important subject for SiC power devices, because of an increasing military and commercial interest in pulsed power devices.

3. Experimental Results on PiN Rectifiers

3.1. *PiN rectifier structures measured*

The development of hot wall CVD reactors to obtain SiC epitaxial layers with high purity, low defect density, and with sufficiently high minority carrier lifetimes layers were crucial to the realization of PiN rectifiers with respectable blocking voltage while providing low on-state voltage drop. These reactors have enabled minority carrier lifetimes to a level that enables the realization of SiC rectifiers with <4.5 V on-state drop at 100 A/cm^2. Probably the first high voltage rectifiers with good performance were the 4.5 kV (width of the blocking n-base = 45 μm) [22] and the 3.4 kV (W=30 μm [23]) rectifiers.

The first successful attempt in the demonstration of a >5 kV 4H-SiC rectifier was done using a 4H-SiC n^- epitaxial layer with a thickness of 85 μm and a doping of 1 to 7 X 10^{14} cm^{-3} [24]. Other demonstrations [25, 26, 27] of >3 kV 4H-SiC PiN rectifiers show that extremely high switching speeds and an on-state voltage drop comparable to Si PiN rectifiers are achieved when operated at sufficiently high current densities. The highest voltage functional semiconductor device ever reported is the 19.3 kV SiC PiN rectifier [28]. After a long development process [30], the highest power single chip SiC device (a PiN rectifier) was demonstrated recently with a 7.4 kV, 330 A (pulsed) capability [31]. Similar devices have been put in active circuits to show the benefits of SiC PiN rectifiers for utility applications [32].

3.2. *PiN rectifier fabrication process*

Typically, the most critical part of the fabrication of PiN rectifiers is the implementation of the edge termination. PiN rectifiers Anodes usually have rounded edges to minimize concentration of electric field during their reverse bias operation. If this edge termination is implemented using ion implantation, the choice of implanted species is a critical decision. Whereas Aluminum offers higher and more precise activation than Boron because of its lower acceptor level, it's larger atomic size causes much more damage to the crystal lattice. If this damage is not annealed using suitable high temperature implant anneals, this damage may cause excessive leakage in large area SiC PiN rectifiers. For the mesa-JTE design (as shown in Figure 2), an optimum dose of Boron is implanted at >1000°C after reactive ion etching through the 2 μm p^+ Anode cap layer with more than 5×10^{18} cm^{-3} doping. The JTE dimension of 75 to 100 μm may be used for 5 kV devices, and 250 to 500 μm for 10 kV devices. The ion implant damage is usually annealed at >1600 °C under Si overpressure condition, followed by a 1-3 μm LTO SiO_2 deposition. Thereafter, backside ohmic contact may be formed using metals like Nickel and p-type ohmic contact metal like Titanium is deposited as Anode metal. These metals are annealed at temperatures in the 700 to 1000 °C range to form the PiN Anode and backside cathode contacts. These metals were followed by a thick 1-3 μm metal stack deposition (for example Ti/Pt/Au) to reduce the resistance and enable wire bonding. Devices are then diced, brazed and wire-bonded for further characterization.

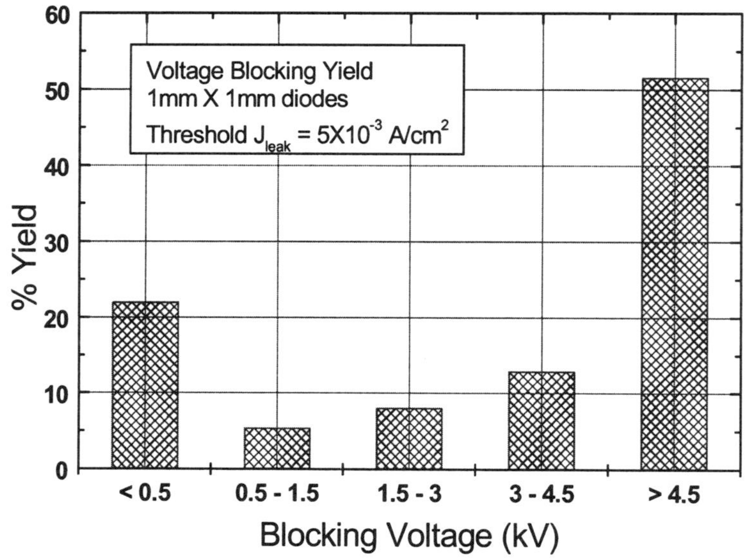

Figure 8: On-wafer yield of >50 % was obtained for 1 mm^2 rectifiers capable of blocking >4.5 kV for a leakage current density threshold of 5X10^{-3} A/cm^2.

3.3. *5 kV PiN Rectifiers*

At the time, the state-of-the-art characteristics of PiN rectifiers were given in ref [30]. In this report, a set of 1 mm^2 and 4 mm^2 rectifiers were designed for 5 kV operation using a 50 μm thick voltage blocking epitaxial layer with a doping of 9x10^{14} cm^{-3} on low micropipe density (<30 μP/cm^2) 4H-SiC n$^+$ substrates. The on-state voltage drop was usually uniform (i.e. ± 0.2 V) across the wafer and was not a yield-limiting factor. The mesa-JTE edge termination design outlined earlier and its associated processing allowed a yield of >50 % for 1 mm^2 rectifiers capable of blocking >4.5 kV when using a leakage current density threshold of 5X10^{-3} A/cm^2. The percentage distribution of all 1 mm^2 devices obtained from the wafer versus blocking voltage ranges is shown in Figure 8. As seen from this figure, 27 % of all devices block less than 1500 V, primarily due to materials defects such as micropipes. The reverse bias characteristics of a 2 mm X 2 mm rectifier are shown in Figure 9.

The measured leakage current density for these rectifiers was <10^{-3} A/cm^2 at 5 kV, and thereafter it increases dramatically at about 5.3 kV. The breakdown characteristics were not catastrophic; with the device surviving after the applied voltage was reduced. High temperature (up to 225 °C) measurements were performed only at 2 kV because of equipment limitations. The leakage current increases exponentially with temperature, but was still found to be less than 4x10^{-5} A/cm^2 at 225 °C at 2 kV for a 4 mm^2 rectifier, as shown in Figure 10. In standard commercial Si power devices, a leakage current density of <10^{-2} A/cm^2 is considered acceptable for defining blocking voltage. These results show that the edge termination design adopted for these devices is quite robust even at high temperatures.

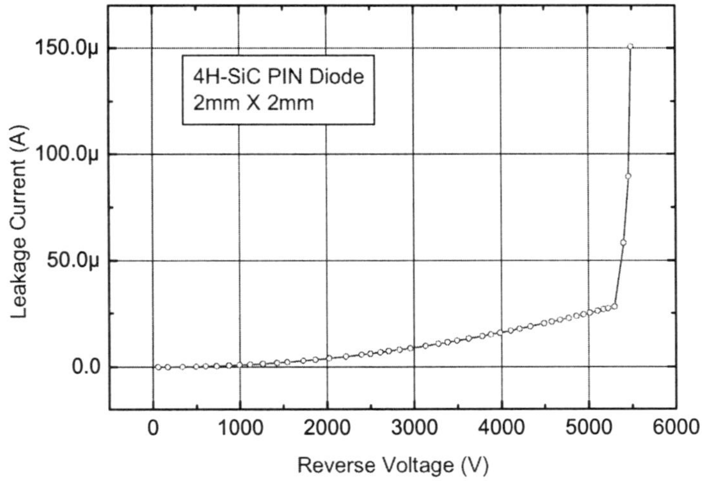

Figure 9: Blocking characteristics of a 5.3 kV rectifier capable of carrying >20 A.

The forward I-V characteristics show that at a high current density of 1250 A/cm^2 (50 Amperes on a 2 mm X 2 mm device), the on-state voltage drop was only 6.9 V, which was significantly affected by the wire bonding resistance. The measured temperature dependence of the on-state characteristics for a 5 kV SiC PiN rectifier device with a

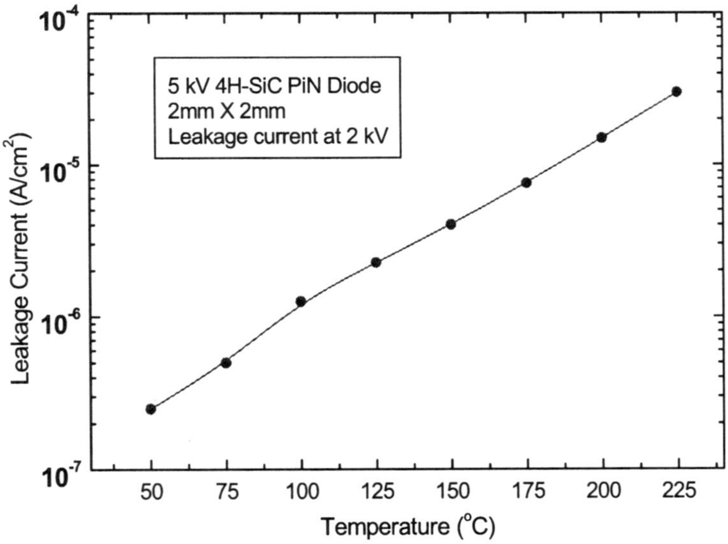

Figure 10: The leakage current density was less than 4x10^{-5} A/cm^2 at 225 °C at 2 kV for a 0.04 cm^2 rectifier, even after an exponential increase with temperature.

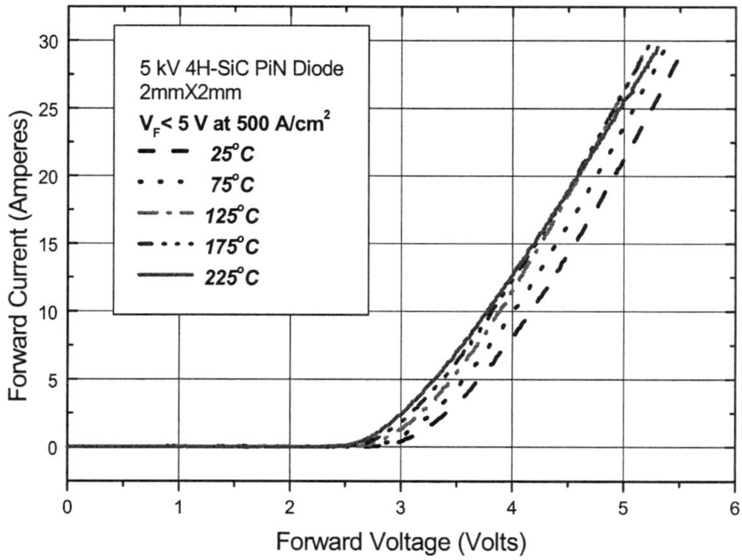

Figure 11: The measured temperature dependence of on-state characteristics.

typical current of 20 A (4 mm^2), is shown in Figure 11. The decrease in on-state voltage with temperature is indicative of the increase in lifetime with increasing temperature for a conductivity modulated device and a decrease in bandgap of the PN junction. However, at a high temperature of 225 °C, a reduction in carrier mobility starts to increase the differential on-resistance across the i-layer. This leads to a cross-over in the I-V characteristics at a high current density of 500 A/cm^2. In the entire 25 °C to 225 °C range, the change in on-state voltage drop remains in a somewhat insignificant 0.4 V range, as seen from Figure 11. This shows that SiC PiN rectifiers are stable with temperature. In case of Si rectifiers, over 40% reduction in on-state voltage drop was measured in the 25°C to 125°C range [8]. This results in poor current sharing when such devices are operated in parallel because of the generation of 'hot spots' that hog a large amount of current.

Detailed switching measurements were also conducted on some 5 kV blocking 4H-SiC PiN rectifiers. The most important dynamic characteristics for a rectifier are its reverse recovery characteristics, and their variation with operating temperature. The reverse recovery tests were performed for various values of di/dt. Figure 12 shows the current vs. time waveforms of the 20 A, 5 kV SiC PiN rectifiers for three different reverse di/dt values. At the conventional 40 A/μsec, the peak reverse current was only 65 % of the forward current. As the reverse di/dt was increased to 100 A/μsec and 1700 A/μsec, the peak reverse current to forward current ratio increased to 90 % and 150 %, respectively. As compared to high voltage Si PiN rectifiers, this is a relatively insignificant change, considering that extremely high reverse di/dt values were used. This effect could be explained by the fact that the amount of reverse recovery charge in SiC rectifiers is at least 3 orders of magnitude smaller than Si rectifiers.

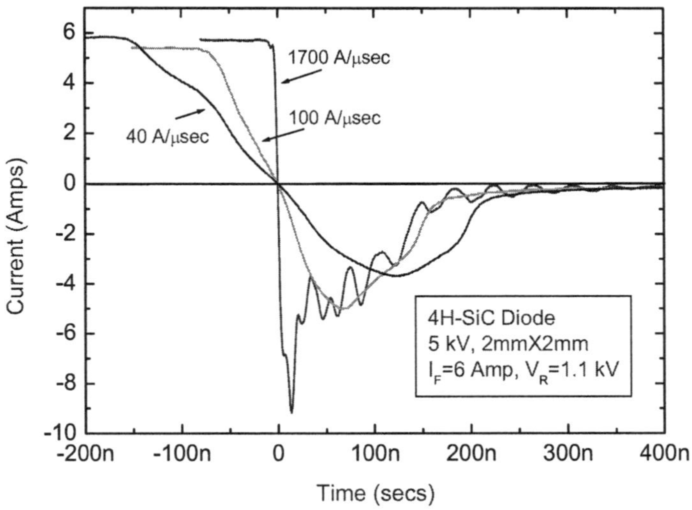

Figure 12: Reverse recovery characteristics show a 2X increase in the peak reverse recovery current when the reverse di/dt was increased from 40 A/μsec to an extremely high 1700 A/μsec.

The temperature dependence of the rectifier switching characteristics for a 2 mm X 2 mm 5 kV device is shown in Figure 13. These measurements are taken at a relatively high reverse di/dt of 175 A/μsec, when the rectifier is switching near zero voltage at 6.2 A. Usually, the reverse bias applied does not affect the reverse recovery characteristics. As seen from this figure, the peak reverse current increases by a modest 50% when the

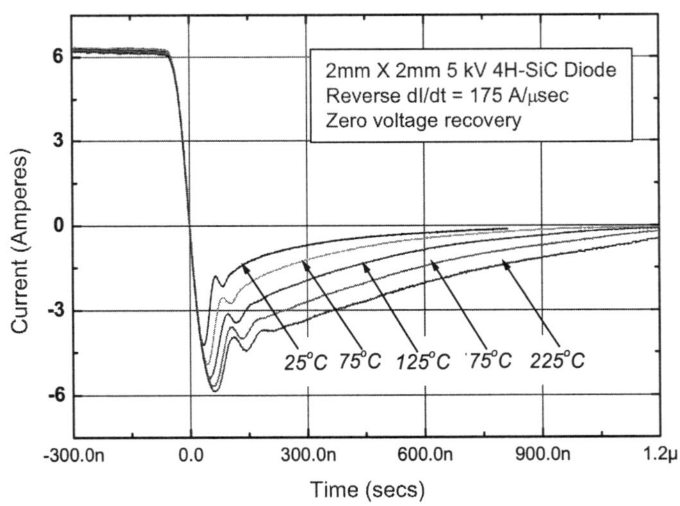

Figure 13: High temperature reverse recovery characteristics show a modest increase in the peak reverse current as the operating temperature is increased from 25 °C to 225 °C.

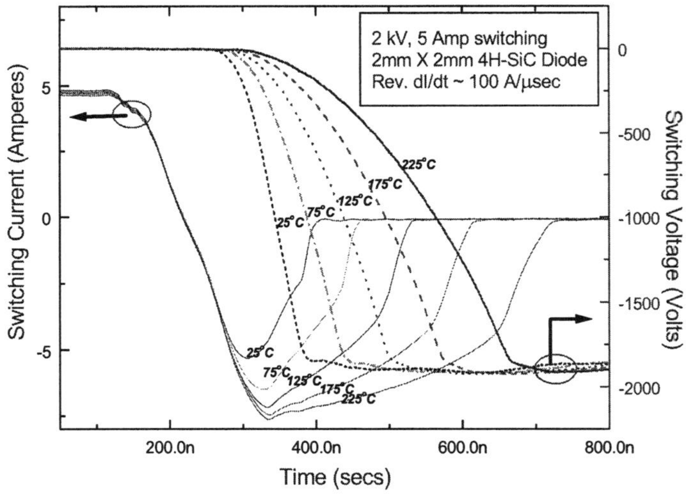

Figure 14: Current and voltage waveforms of a 5 kV rectifier switched at 25 °C, 75 °C, 125 °C, 175 °C and 225 °C.

temperature was increased from 25 °C to 225 °C. The total reverse recovery charge, which is the area under the I-t curve when the rectifier is undergoing reverse recovery, increases by approximately 100 %, as the operating temperature is increased from 25 °C to 225 °C. These rectifiers show repeatable switching characteristics as the operating temperature was increased from 25 °C to 225 °C. The turn-off time increases from 0.2 μsec to 0.65 μsec while switching 125 A/cm^2 (5 A) and 2 kV with a reverse di/dt of 100 A/μsec, as shown in Figure 14. A level of 2 kV was chosen as the highest voltage possible for the pulse generator used in this measurement. These rectifiers do not show a "snappy" recovery and have substantially smaller noise signatures when compared to 600 V Silicon PiN rectifiers.

Figure 15 shows that the on-state voltage drop changes from 3.9 V to 3.4 V at 150 A/cm^2 as the operating temperature is increased from 25 °C to 225 °C. Under the same conditions, the measured reverse recovery charge increases from 0.5X10^{-6} C to 1.3X10^{-6} C. This is a 10^{3}X reduction in Q_{rr} as compared to comparably rated Si (in terms of blocking voltage) rectifiers. It is worthwhile to note that reverse recovery charge shows a 4-6X increase with temperature from 25°C to 125°C, even for ultrafast Si rectifiers [33]. Two factors contribute to the dramatically smaller reverse recovery charge (Q_{rr}) in 4H-SiC rectifiers as compared to similarly rated Si rectifiers: (a) the 20-25X thinner voltage blocking layers with 20X higher doping dramatically reduce the total volume of excess charge in the i-layer and (b) the carrier lifetime required for these thinner voltage blocking layers can be >10X smaller than those required for Si devices for a similar i-region voltage drop. A much smaller carrier lifetime and thinner voltage blocking layer in 4H-SiC results in a very stable on-state voltage drop with temperature.

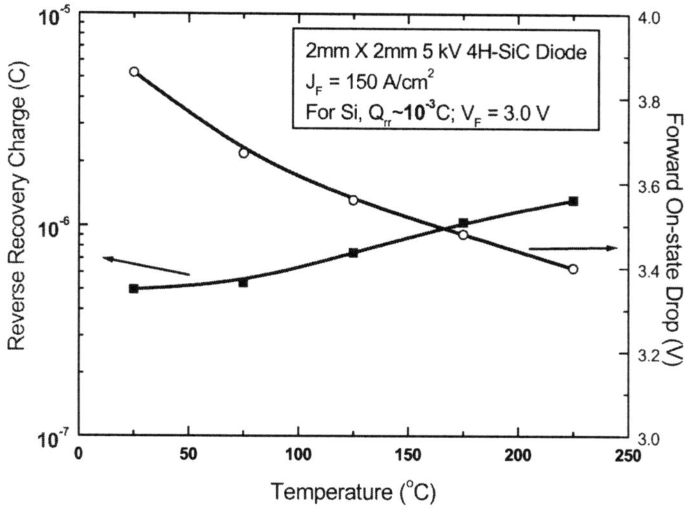

Figure 15: Reverse recovery charge and on-state voltage drop as a function of temperature for a 5 kV PiN rectifier.

3.4. *9mm^2, 10 kV 4H-SiC PiN Rectifiers*

Another set of PiN rectifiers were designed for 10 kV blocking with an active area of 9 mm^2 using a 150 μm thick voltage blocking epitaxial layer with a doping of 1 to 3×10^{14} cm^{-3}. Data obtained on these 4H-SiC PiN Rectifiers shows a very uniform on-state voltage drop across the wafers. At 2 kA/cm^2, the differential on-resistance was only 3 mΩ-cm^2, as shown in Figure 16. The reverse bias characteristics of a 3 mm X 3 mm rectifier are shown in Figure 17.

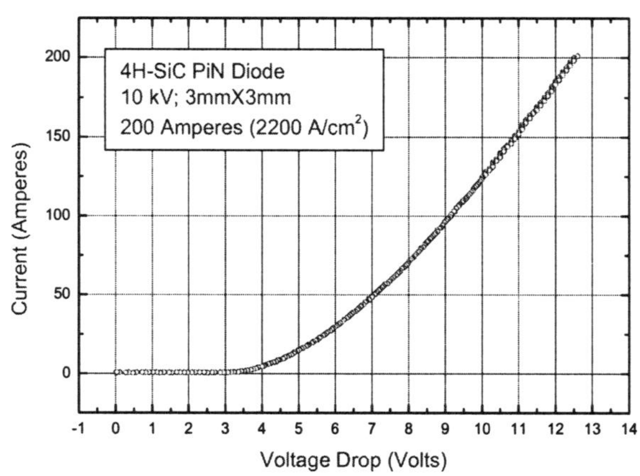

Figure 16: Pulsed (250 μsec) on-state characteristics of a 4H-SiC PiN rectifier capable of blocking 10 kV.

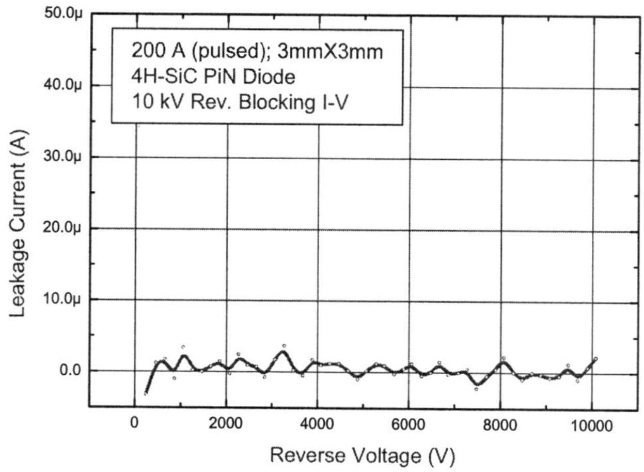

Figure 17: Blocking characteristics of a 10 kV rectifier capable of carrying >200 A (pulsed).

The measured leakage current density was $<10^{-4}$ A/cm^2 at 10 kV and increases dramatically thereafter. The device survived after the voltage was reduced, and then re-applied. High temperature (up to 200 °C) measurements on 8 kV capable packaged devices fabricated alongside the 10 kV rectifier shows that the leakage current increases with voltage beyond 150 °C, resulting in a blocking voltage of about 5.3 kV at the leakage current threshold of 20 μA, as shown in Figure 18. The choice of this leakage current is quite arbitrary, and was limited by the highest leakage current capability of the measurement equipment.

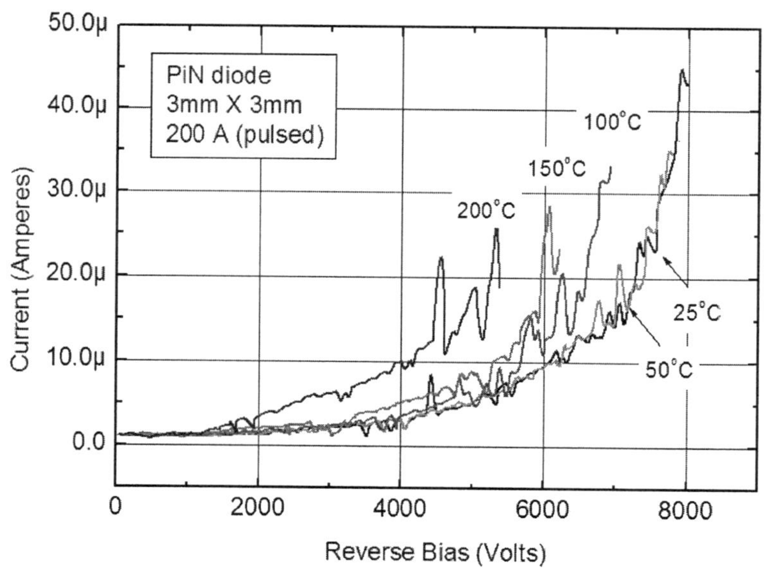

Figure 18: Reverse I-V characteristics of a packaged 8 kV PiN rectifier fabricated alongside the 10 kV blocking rectifier.

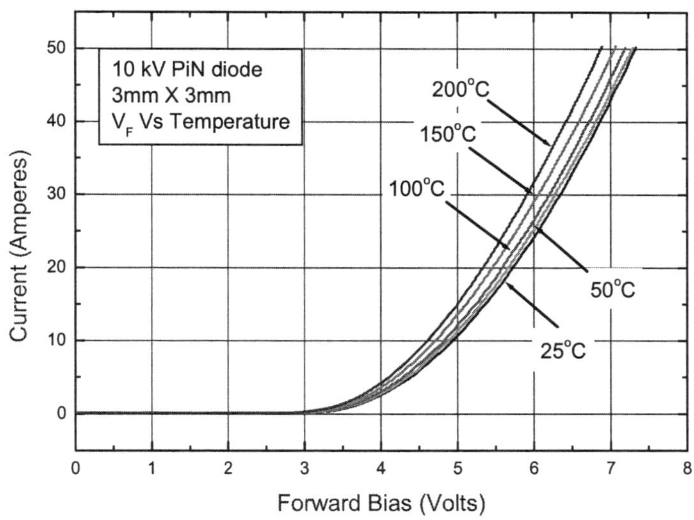

Figure 19: Forward I-V characteristics of a packaged 8 kV PiN rectifier fabricated alongside the 10 kV blocking rectifier.

The measured temperature dependence of the on-state characteristics for a 10 kV, 9 mm^2 SiC PiN rectifier is shown in Figure 19. The decrease in on-state voltage with temperature is indicative of the increase in lifetime with temperature for a conductivity modulated device, and a decrease in bandgap of the PN junction. However, at a high temperature of 200°C, a reduction in carrier mobility starts to increase the differential on-resistance across the i-layer (data not shown here). In the entire 25 °C to 200 °C range,

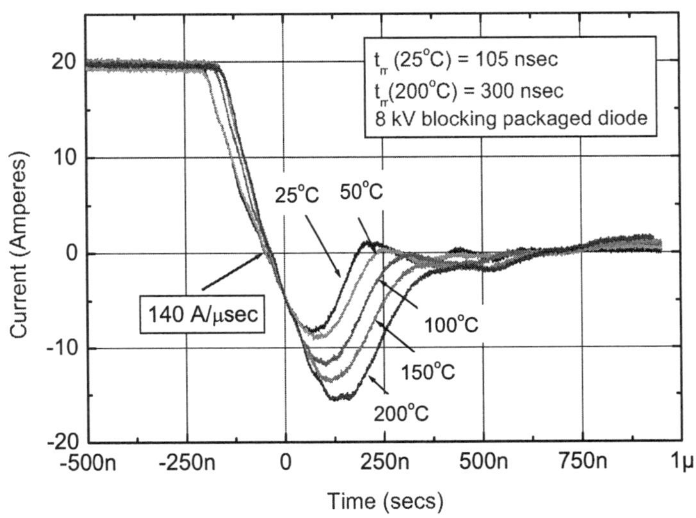

Figure 20: High temperature reverse recovery characteristics show a modest increase in the peak reverse current as the operating temperature is increased from 25 °C to 200 °C.

the change in the on-state voltage drop remains in a somewhat insignificant 0.4 V range.

Detailed switching measurements were conducted on some 8 kV blocking 4H-SiC PiN rectifiers fabricated alongside the 10 kV rectifiers. The temperature dependence of the rectifier switching characteristics for a 3 mm X 3 mm 10 kV device is shown in Figure 20. These measurements are taken at a relatively high reverse di/dt of 142 A/μsec, when the rectifier is switching from an on-current of 20 A to a blocking voltage of 600 V. As seen from this figure, the peak reverse current increases by a modest 110 % when the temperature was increased from 25 °C to 200 °C. The total reverse recovery charge, which is the area under the I-t curve when the rectifier is undergoing reverse recovery, increases by approximately 100 %, as the operating temperature is increased in this temperature range. These rectifiers show fairly stable switching characteristics as the operating temperature was increased from 25 °C to 200 °C. The turn-off time increases from 0.2 μsec to 0.7 μsec while switching 220 A/cm^2 (20 A). It is very encouraging to note that the total reverse recovery charge increases from an insignificant 1.17 μC to 3.8 μC as the temperature was increased from room temperature to 200°C. These rectifiers do not show a "snappy" recovery and have substantially smaller noise signatures when compared to Silicon PiN rectifiers.

4. Yield and Reliability of SiC Rectifiers

4.1. *Yield limiting factors in SiC rectifiers*

Usually, fairly uniform on-state characteristics are obtained across the wafers since the advent of hot-wall epitaxial reactors, and good anode ohmic metal processing. However, the yield due to blocking voltage on 4H-SiC devices is dependent on many material, processing and design related issues. Material related issues include: micropipes on the wafers, epitaxial growth related defects, and crystal defects like dislocations and stacking faults. It is difficult to quantize the effect of each of these factors on device yields, but great deal of data on these effects have been studied in reference [34]. Processing related issues are related to ion implant activation of JTE termination species, uniformity of the mesa etch, and quality of the dielectric used in passivation of edges. Design related effects on yield are junction edge diameter, and the choice of dose for JTE implanted species. As mentioned earlier, it is very difficult to separate the influence of all these parameters on device yields. Assuming a random distribution of defects, device yield is given by

$$Y = e^{-A.d} \tag{22}$$

where Y is the yield, A is the area (in cm^{-2}) of the device and 'd' is the defect density in cm^{-2}. Many researchers propose that the biggest yield-limiting factor in modern high voltage devices is micropipes. Figure 21 shows yield as a function of device current levels (assuming 100 A/cm^2) for different defect densities according to equation 22. As shown in an earlier section on the experimental results of PiN rectifiers, the yield on 1 mm^2 rectifiers was 52% at 4.5 kV. For 4 mm^2 devices, the obtained yield was approximately 20% at >4.5 kV; and for 9 mm^2 it was 22% at >7 kV. All these devices show quite abrupt breakdown characteristics at room temperature.

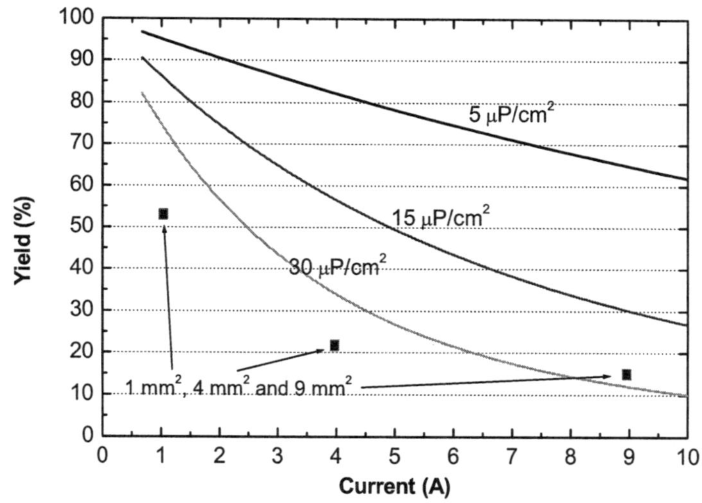

Figure 21: Theoretical and experimentally obtained yields on 1 mm^2, 4 mm^2 and 9 mm^2 devices at 4.5 kV, 4.5 kV and 7 kV.

All these devices were fabricated using similar processing techniques. However, the 9 mm^2 devices were fabricated roughly 18 months after the 1 mm^2 and 4 mm^2 devices, and might have benefited from an improvement in micropipe densities realized with time.

4.2. *Forward voltage degradation in SiC PN rectifiers*

While the reverse bias operation of SiC devices have been found to be relatively stable if good edge termination and passivation techniques are used, a curious

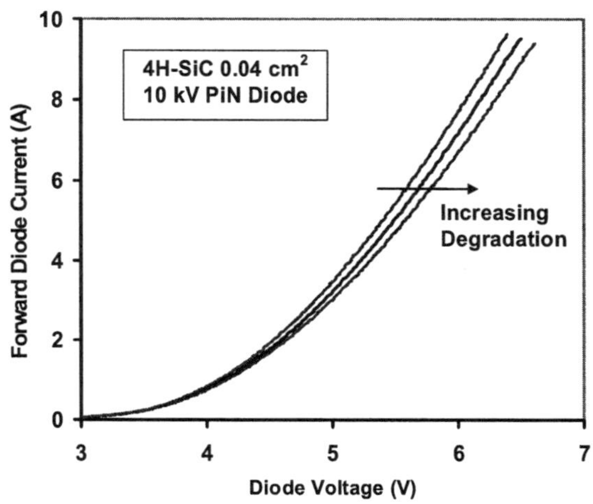

Figure 22: On-state characteristics of a high voltage SiC PiN rectifiers after various levels of forward bias stress.

phenomenon observed during the forward bias operation of SiC PiN rectifiers has caused a great deal of concern towards long term stability and reliability of these devices. It has been observed that as PN rectifiers are forward biased for an appreciable length of time, their on-state voltage drop increased with time, as shown in Figure 22. The duration over which these devices show this forward bias degradation varies from a few milliseconds to many hours [35]. A variation in on-state voltage drop (V_F) in PiN rectifiers has serious stability concerns because it can result in current filamentation and local current 'hogging'. If a portion of the rectifier has a lower on-state voltage drop than another region within the same rectifier, current will be diverted into the lower V_F region. This can cause excessive current densities in small portions of the rectifier, while leaving large portions of it with a low current density, leading to thermal instability of the entire rectifier. Such a situation will also prevent safe paralleling of devices to boost the total current required for typical high current applications for which these devices are targeted.

Optical observation of PN rectifiers undergoing VF degradation shows a concomitant formation of a certain material defect in these rectifiers, as shown in Figure 23. Probably the first report of this phenomenon, which were termed as 'Bright line defects', was made by Konstantinov et al [36], since they appear are mobile bright lines. This was compared to the previously studied formation of 'Dark line defects' in Gallium Arsenide light emitting devices [37], where dislocation growth due to non-equilibrium carrier injection and crystal strain results in a similar forward bias degradation phenomenon. Many researchers agree that mobile and propagating crystal stacking faults are the primary cause of forward bias degradation of PN rectifiers. This defect propagates through the entire n- base layer. It was initially proposed [38] that the increase in V_F is caused by reduced carrier lifetimes due to the formation of recombination centers from stacking faults. However, more recent results indicate that the increase in V_F causes the stacking faults to form a barrier to current flow and reduce the conduction area. A stacking fault defect is a two-dimensional error in the atomic stacking sequence of a polytype of SiC.

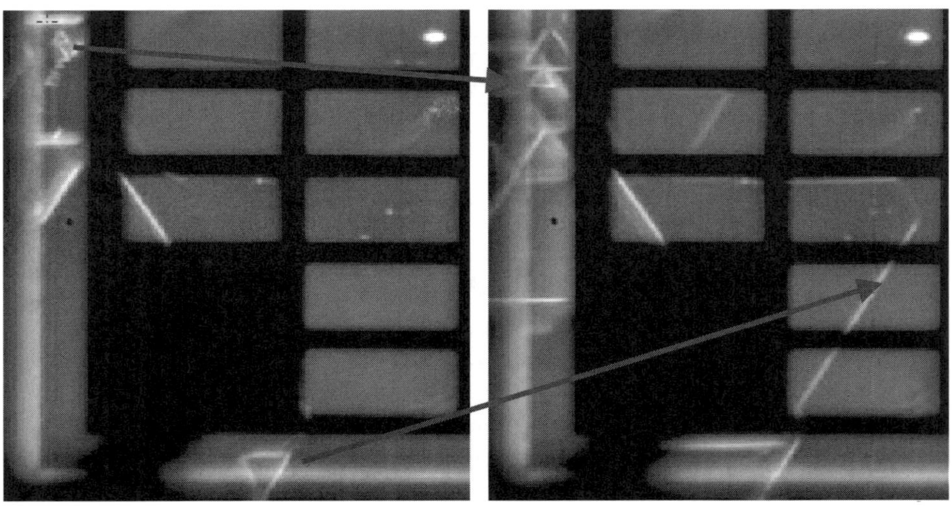

Figure 23: Light emission measurements of a PiN rectifier before (left) and after (right) forward bias stress indicating growth of stacking faults.

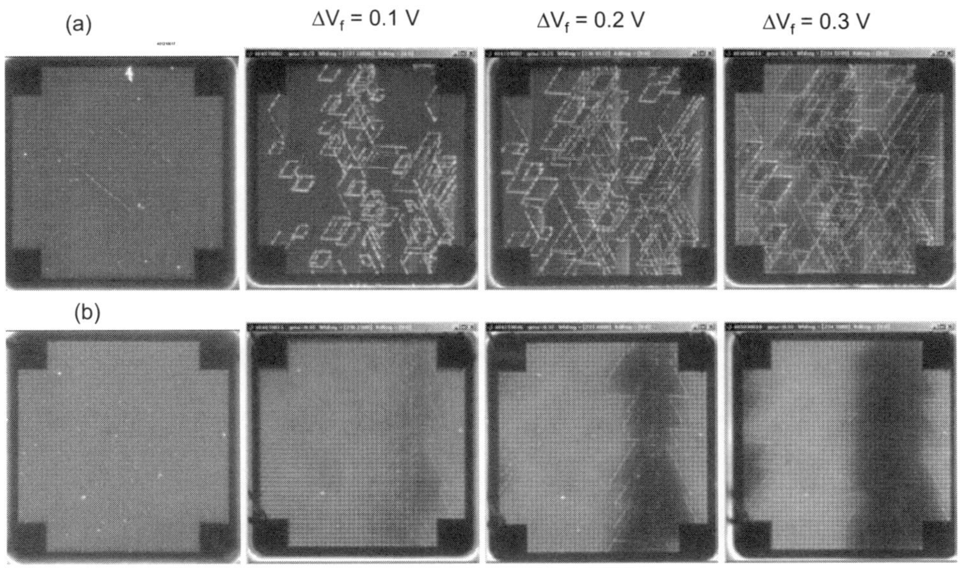

Figure 24: Light emission images at various intervals during degradation for two adjacent rectifiers (a) top and (b) bottom on the same chip.

Figure 24 shows a series of light emission images at various intervals during degradation for two adjacent 0.015 cm^2 rectifiers on the same chip. These optical phenomena were correlated with a lifetime measurement method that uses the turn-off reverse-recovery waveforms for conditions of high dI/dt and low dV/dt [39]. This result indicates that the effective i-region lifetime is not reduced and the V_F degradation is due to a reduction in conduction area. Figure 25 (a) and (b) show the forward bias voltage degradation and reverse recovery current for the same rectifiers as those shown in Figure 23 and Figure 24, where the arrows on Figure 25 (a) and (b) indicate the time for each frame in Figure 24 (a) and (b), respectively. The rhomboid shaped regions in Figure 24 (a) indicate stacking faults emanating from near the surface, presumably from the P-N junction. It is determined that these stacking faults are near the surface by the growth direction and shape as well as by changing of the focal plane of the CCD camera. The device degrades rapidly in Figure 25 (a) because the highest excess carrier concentration is near the P-N junction. The outlined triangular dark regions in Figure 24 (b) indicate stacking faults growing from near the n^+/n^- junction. The white outlines are from partial dislocations that bound the stacking faults and the dark regions indicate regions where current is reduced. The device degrades more slowly at first, as can be seen in Figure 25 (b), because the stacking faults are near the bottom of the i-layer where the excess carrier concentration is lower and the degradation rate increases as the stacking faults approach the P-N junction at the surface where the excess carrier concentration is larger.

The fundamental nature, origin and propagation of these dislocations have been extensively investigated by various researchers [38, 40, 41]. It is now established that these defects nucleate from existing substrate crystal defects in hexagonal (eg. 4H and 6H) SiC. The hexagonal crystal structures of these polytypes of SiC are formed when three distinct atomic patterns are stacked in a particular sequence. This is in contrast with cubic (3C) SiC, which has a fundamentally two dimensional structure with only two

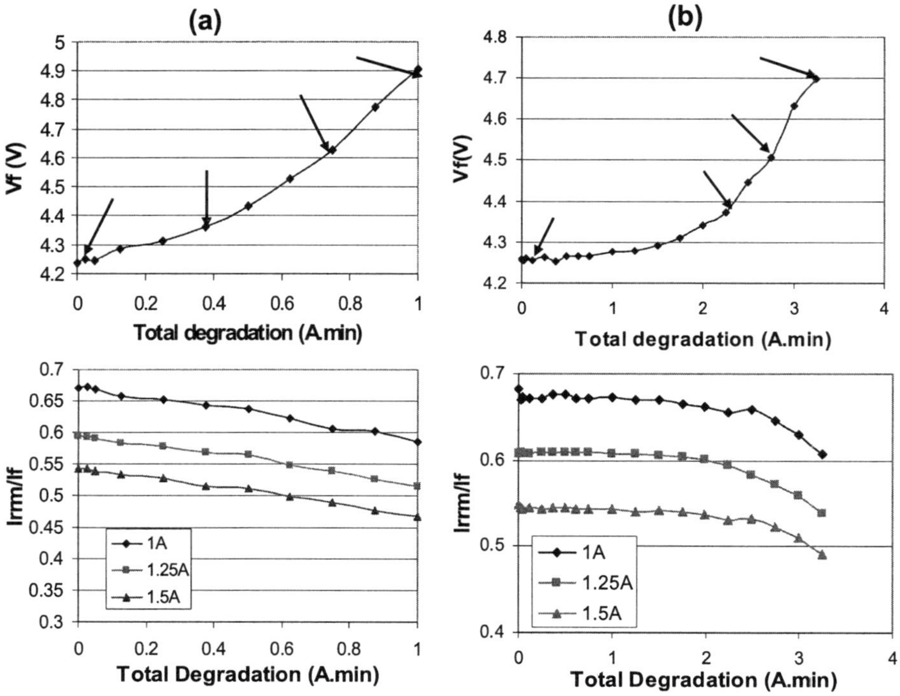

Figure 25: The on-state voltage VF (at 100 A/cm^2) degradation and reverse recovery current for the same rectifiers in Figure 24 (a) and (b), respectively.

atomic patterns stacked together. Hence, 4H-SiC and 6H-SiC crystals are metastable at room temperature and could convert to a locally faulted 3C-like structure if an appropriate nucleating site was present [42]. For this to occur, a proper type and density of nucleation sites and the activation energy in the form of electron-hole pair recombination energy are needed. The control of the origin and propagation of these defects may lie in the control of pre-existing defects serving as nucleation sites for these defects.

Spectral analysis of the electroluminescence shows that these bright line defects emit light in the red and near-infrared spectral regions, in contrast to near-bandgap luminescence of non-degraded PN rectifiers which is predominantly in the violet region [40]. These defects were determined to be an irreversible formation of a network of linear defects related to crystal dislocations that subsequently propagate through, and then beyond the rectifier area. The nucleating sites for these stacking faults appear to be primarily at low angle grain boundaries, among other substrate and surface defects. The resulting stacking faults have been observed to be bound by Schockley partial dislocations with a Burger's vector b=1/3<10-10> [41]. This implies that the faults propagate as triangular or rhombohedral structures with edges along the 11-20 directions. These looping structures multiply and propagate as the rectifier is kept in the forward bias state. Spectral measurements show that the stacking faults have a primary emission spectrum in the 450 +/-20nm peak range [40], and those of threading dislocations is in the 700 +/-20 nm range. The calculated activation energy for the gliding (propagation) of the partial dislocation that bounds the stacking fault is measured to be in the 0.27 +/- 0.02 eV range. In this experiment, it was estimated that the velocity of propagation of these defects was $7X10^{-5}$ m/s.

An activation energy of 0.27eV is small enough that most nucleating sites will result in the formation of these faults, and hence some degradation in the on-state voltage drop of PiN rectifiers. Hence, the solution for solving this problem lies in minimizing the defects in the active portion of the device. A novel approach to achieving this was recently demonstrated by growing of Lely crystals with no micropipes and only minimal defects on top of standard substrates [43]. In this experiment, an application of 200 A/cm^2 stress in the forward direction did not produce any degradation, which was observed with PN junction rectifiers using normal substrates. Recently, great strides have been made in 4H-SiC epitaxy to produce relatively drift-free PiN structures by reducing the material defect density [44]. From these preliminary results, it seems that reduction of material defects would be directly correlated to obtaining a high yield of drift-free PiN rectifiers in SiC.

Recently, various methods have been investigated for improving the V_F degradation [45]. The different methods produce different yields of degradation free devices with some processes having yields as high as 86% degradation free. In each degradation free process, some devices do not drift at all, while others drift a small amount and then stabilize. However, some of the processes also reduce the breakdown voltage yield. Most recently, a new Low basal plane dislocation (BPD) process demonstrates substantial increase in the overall yield with blocking and drift yields of 35% and 67%, respectively [46]. Figure 26 and Figure 27 show the degradation and monitoring results for a typical low degradation rate 50 A, 10 kV 4H-SiC PiN rectifier made with this process.

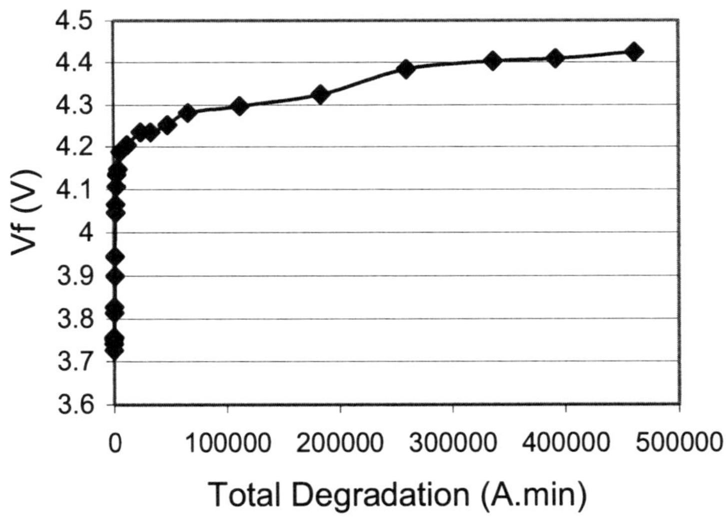

Figure 26: On-voltage voltage drop, V_F (at 50 A/cm^2) degradation as a function of total degradation for a 0.5 cm^2, 10 kV 4H-SiC PiN rectifier stressed at 50 A.

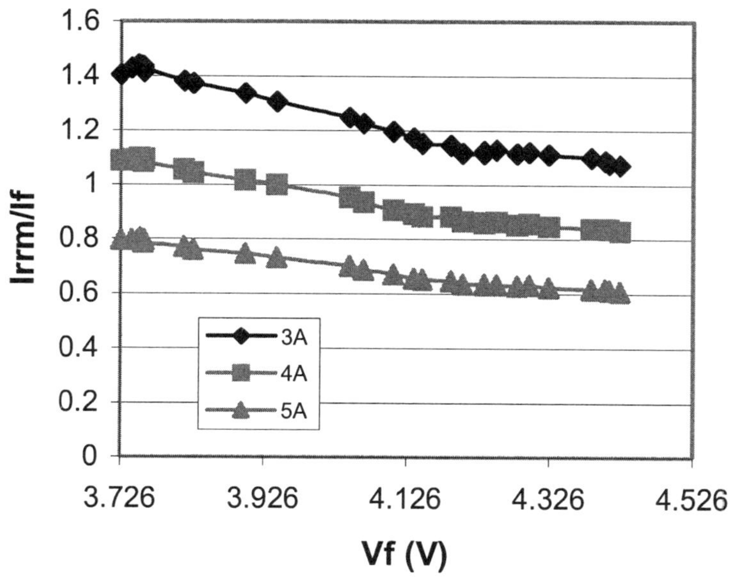

Figure 27: Reverse recovery peak as a function of total degradation for a 0.5 cm^2, 10 kV 4H-SiC PiN rectifier.

It is worthwhile to note that such a phenomenon is not observed in unipolar devices like Schottky rectifiers, even when the power density (on-current X V_F) approaches near levels at which the PN rectifiers show this phenomenon [47]. While detailed published reports are missing, anecdotal data suggests that the forward bias degradation associated with stacking faults occurs in bipolar SiC devices like Thyristors, GTOs, PiN rectifiers, and even bipolar junction transistors (BJTs), but is not observed in unipolar switches like MOSFETs and JFETs. The incidence of propagation of these dislocations was found to increase with higher current densities. Higher temperatures and thicker epitaxial thicknesses also resulted in a higher incidence of formation of dislocations. Probe scratches, chip scribe lines and even silicided contacts on PN rectifiers are considered to be nucleation sites for mobile and expanding dislocations.

5. Conclusions

Currently, there are significant efforts underway to accelerate the development and application insertion of new high voltage/high frequency SiC power devices including PiN rectifiers. These are needed for commercial and military power conversion and distribution applications. PiN rectifiers and switches are an enabling technology for alternative energy sources and storage systems. The emergence of such devices presents unique opportunities and challenges to the power electronics industry in specifying the device requirements and establishing pulse width modulation (PWM) converter topologies for high voltage applications. Rapid progress is continuing in the realization of ultra high voltage, high frequency PiN rectifiers with high yield and good reliability.

Acknowledgements

The author gratefully acknowledges the support of Dr. Allen R. Hefner and his group at the National Institute of Standards and Technology (Gaithersburg, MD, USA) for making many electrical measurements reproduced here. The author wishes thanks Dr. Robert Stahlbush and his group at Naval Research Labs (Washington, DC, USA) for supplying the optical photographs of PiN rectifiers undergoing V_F degradation.

References

1. R. Singh, "Silicon Carbide Bipolar Power Devices – Potentials and Limits," ***Mat. Res. Soc. Symp.*** Proc. Vol. 640, *Materials Research Society Fall Meeting, Boston, 2001.* pp. H4.2.1 – H.4.2.12.
2. B. J. Baliga, Power Semiconductor Devices, ***PWS Publishing Co***, 1996.
3. A. O. Konstantinov, Q. Wahab, N. Nordell, and U. Lindefelt, "Ionization rates and critical fields in 4H silicon carbide", ***Applied Physics Letters*, vol. 71, no. 1**, July 1997, pp. 90-92.
4. R. Singh, and J. W. Palmour, "Planar terminations in 4H-SiC Schottky rectifiers with low leakage and high yields," ***Proc. of International Symposium on Semiconductor Power Devices and ICs (ISPSD),*** *Weimar, Germany*, May 26-29, 1997, pp. 157-160.
5. B. Lax and T. Neustadter, "Transient response of a p-n junction," ***J. Appl. Phys.***, 1954, 25, pp. 1148-1154.
6. M. E. Levinshtein, T. T. Mnatsakanov, P. A. Ivanov, J. W. Palmour, S. L. Rumyantsev, R. Singh, and S. N. Yurkov, "'Paradoxes' of Carrier Lifetime Measurements in High-Voltage SiC Diodes," ***IEEE Transactions Electron Devices***, v. 48, no.8, pp. 1703-1710 (2001).
7. M. E. Levinshtein, T. T Mnatsakanov, P. Ivanov, R. Singh, K. G. Irvine, and J. W. Palmour, "Carrier lifetime measurements in 10 kV 4H-SiC diodes," ***IEE Electronics Letters***, vol. 39, no. 8, Apr. 2003 pp. 689 – 691.
8. B. R. Gossik, "About recovery characteristics of semiconductor rectifiers," ***J. Appl. Phys.***, 1956, 27, pp. 905-910.
9. H. Schlangenotto and W. Gerlach, "On the post-injection voltage decay of the p-s-n rectifiers at high injection levels," ***Solid State Electronics***, vol 15, 1972, pp. 393 – 402.
10. Yu. R. Nosov, "Physical foundations of the pulse semiconductor diode operation," 1968 (Moscow, "Nauka" Publ. House).
11. S. M. Sze, "Physics of Semiconductor Devices", ***John Wiley & Sons***, NY-Toronto-Singapore, 1981.
12. T. Kimoto, N. Miyamoto and H. Matsunami, "Performance limiting surface defects in SiC epitaxial p-n junction diodes," ***IEEE Trans. Electron Dev.***, vol 46, 1999, pp. 471-477.
13. A. M. Strel'chuk and N. S. Savkina, "Ideal 4H-SiC pn junction and its characteristic shunt," ***Material Science and Engineering***, B80, 2001, pp. 378 – 382.
14. R. A. Hall, "Power Rectifiers and Transistors," ***Proc. IRE***, vol. 40, pp. 512 - 1518, 1952.

15. V. A. Kuz'min, T.T. Mnatsakanov, and V.B. Shuman, "Effect of electron-holes scattering on the current-voltage characteristics at high current densities for silicon multilayer structures," ***Sov. Techn. Phys. Letters***, vol. 6, pp. 689-693, 1980.

16. N. G. Nilsson, "The influence of Auger recombination on the forward characteristics of semiconductor power rectifiers at high current densities," ***Solid State Electronics***, vol.16, pp. 681-688, 1973.

17. A. Herlet, "The forward characteristics of silicon power rectifiers at high current densities," ***Solid State Electronics***, vol. 11, pp. 717-742, 1968.

18. V. A. Kuz'min and T.T. Mnatsakanov, "Current-voltage characteristics of power semiconductor devices at high current densities," ***Elektrotekhnika***, vol. 20, pp. 39-44, 1984 (in Russian).

19. N. V. Dyakonova, P. A. Ivanov, V. A. Kozlov, M. E. Levinshtein, J. W. Palmour, S. L. Rumyantsev, and R. Singh, "Steady-State and Transient Forward Current- Voltage Characteristics of 4H-Silicon Carbide 5.5 kV Diodes at High and Superhigh Current Densities," ***IEEE Transactions on Electron Devices***, vol. 46, no.11, pp. 2188-2194, 1999.

20. L. A. Delimova, Yu.V. Zhilyaev, V.Yu. Kacharovskii, M.E. Levinshtein, and V.V. Rossin, "Forward current-voltage characteristics of gallium arsenide power diodes at high current densities," ***Solid State Electronics***, vol. 31, pp. 1101 - 1104, 1988.

21. M. Shur, "Physics of Semiconductor Devices," ***Prentice Hall Series in Solid State Physical Electronics***, Nick Holonyak, Jr. Editor, 1990.

22. O. Kordina, J.P. Bergman, A. Henry, E. Janzen, S. Savage, J.Andre, L.P. Ramberg, U. Lindefelt, W. Hermansson, and K. Bergman, "A 4.5 kV 6H silicon carbide rectifier," ***Appl. Phys. Lett.***, vol 67, pp.1561-1563, 1995.

23. K. Rottner, A. Schöner, M. Frischholz, J.-O. Svedberg, U. Gustafsson, A. Ellison, E. Janzen, and O. Kordina, "Design and implementation of a 3.4 kV ion implanted p-i-n rectifier in 4H SiC," ***Abstracts of Intern. Conf. on Silicon Carbide, III-nitrides and Related Materials***, pp. 136-137, Stockholm, Sweden, Aug. 31-Sept. 5, 1997.

24. R. Singh, K. G. Irvine, O. Kordina, J. W. Palmour, M. E. Levinshtein, and S. L. Rumyanetsev, "4H-SiC bipolar PiN rectifiers with 5.5 kV blocking voltage," ***Proc. Device Research Conference***, *Charlottesville, VA*, June 22-24, 1998, pp. 86-87.

25. H. Lendenmann, N. Johnasson, D. Mou, M. Frischholz, B. Astrand, P. Isberg, and C. Overn, "Operation of 2500 V, 150 A Si-IGBT/SiC rectifier module", ***Materials Science Forum*, 338-342,** 2000, pp. 1423-1426.

26. H. Mitlehner, P. Friedrichs, D. Peters, R. Schorner, U. Weinert, B. Weis and D. Stephani, "Switching behavior of fast high voltage SiC pn rectifiers," ***Proc. of International Symposium on Semiconductor Power Devices and ICs (ISPSD)***, *Kyoto*, June 3-6, 1998, pp. 127-130.

27. J. B. Fedison, N. Ramungul, T. P. Chow, M. Ghezzo, J. W. Kretchmer, "Electrical characteristics of 4.5 kV implanted anode 4H-SiC PiN junction rectifiers." ***IEEE Electron Device Letters***, Vol. 22 no. 3, March 2001. Page(s): 130 –132.

28. Y. Sugawara, D. Takayama, K. Asano, R. Singh, J. Palmour, T. Hayashi, "12-19 kV 4H-SiC pin diodes with low power loss," ***Proceedings of the 13th International Symposium on Power Semiconductor Devices and ICs***, 2001, pp. 27 –30.

29. Y. Sugawara, D. Takayama, K. Asano, R. Singh, H. Kodama, S. Ogata, T. Hayashi, "3kV 600A 4H-SiC high temperature diode module," ***Proceedings of the 14th International Symposium on Power Semiconductor Devices and ICs***, 2002. pp. 245 –248.

30. R. Singh, Kenneth G. Irvine, D. C. Capell, James T. Richmond, David Berning, A. R. Hefner, and J. W. Palmour, "Large Area, Ultra-high Voltage 4H-SiC p-i-n Rectifiers," ***IEEE Transactions on Electron Devices***, vol. 49, no. 12, (2002) pp. 2308 –2316.

31. R. Singh, D. C. Capell, K. G. Irvine, J. T. Richmond, and J. W. Palmour, "7.4 kV, 330 A (pulsed) single chip, high temperature 4H-SiC pin rectifiers," ***IEE Electronics Letters***, vol. 38, no. 25, Dec. 2002 pp. 1738 –1740.

32. Y. Sugawara, D. Takayama, K. Asano, R. Singh, H. Kodama, S. Ogata, T. Hayashi, "3kV 600A 4H-SiC high temperature diode module," ***Proceedings of the 14th International Symposium on Power Semiconductor Devices and ICs***, 2002. pp. 245 –248.

33. A. R. Hefner, R. Singh, J. S. Lai, D. W. Berning, S. Bouche, and C. Chapuy, "SiC Power Rectifiers Provide Breakthrough Performance for a Wide Range of Applications," ***in the extended abstracts of 1st International Workshop on Ultra-Low-Loss Power Device Technology***, pp. 140-147, June 2000, Nara Japan.

34. H. Lendenmann, F. Dahlquist, J. P. Bergman, H. Bleichner, and C. Hallin, "High power SiC rectifiers – characteristics, reliability and relation to material defects", ***Materials Science Forum*, 389-393,** 2002, pp. 1259-1264.

35. R. E. Stahlbush, J. B. Fedison, S. D. Arthur, L. B. Rowland, J. W. Kretchmer, and S. Wang, "Propagation of current induced stacking faults and forward voltage degradation in 4H-SiC PiN diodes," ***Materials Science Forum, vols. 389-393***, 2002. pp. 427-430.

36. A. O. Konstantinov, and H. Bleichner, "Bright-line defect formation in silicon carbide injection diodes," ***Journal of Applied Physics***, vol 71, no. 25, 1997. pp. 3700-3702.

37. K. Maeda, K. Suzuki, and M. Ichihara, ***Microsc. Microanal. Microstruct***, vol 4, 1993. pp. 211.

38. J. P. Bergman, H. Lendenmann, P. A. Nilsson, U. Lindefelt, and P. Skytt, "Crystal defects as source of anomalous forward voltage increase of 4H-SiC Diodes," ***Materials Science Forum, vols. 353-356***, 2001. pp. 299-302.

39. A. Hefner, IEEE Transactions on Industrial Applications, vol 30, 1994, p 394.

40. A. Galeckas, L. Linnros and P. Pirouz, "Recombination-enhanced extension of stacking faults in 4H-SiC p-i-n diodes under forward bias," ***Applied Physics Letters***, vol. 81, no. 5, 2002. pp. 883-885.

41. M. Skowrownski, J. Q. Liu, W. M. Vetter, M. Dudley, C. Hallin, and H. Lendenmann, "Recombination-enanced defect motion in forward-biased 4H-SiC p-n diodes," ***Journal of Applied Physics***, vol 92, no. 8, 2002. pp. 4699-4704.

42. S. Ha, H. J. Chung, M. Skowrownski, J. J. Sumakeris, "Solid state transformations of hexagonal SiC polytypes," ***Abstracts of International Symposium on Silicon Carbide and Related Materials 2003***. Lyons, France.

43. Y. Tanaka, S. Nishizawa, K. Fukuda, K. Arai, T. Ohno, N. Oyanagi, and T. Suzuki, "Fabrication of 1.4 kV mesa-type p+-n diodes with avalanche breakdown and without forward degradation on high-quality 6H-SiC substrates," ***Applied Physics Letters***, vol. 83, no. 2, 2003. pp. 377-379.

44. M. K. Das, J. J. Sumakeris, M. J. Paisely and A. R. Power, "High power 4H-SiC PiN diodes with minimal forward voltage drift," ***Proc. Intern. Conf. on Silicon Carbide and Related Materials (ICSCRM 2003)***, Lyon, France. Oct. 5-10. pp. 65-66.

45. J. J. Sumakeris, et al, "Approaches to Stabilizing the Forward Voltage of Bipolar Devices," ***International Conference on Silicon Carbide and Related Materials***, Lyon, France, October 2003.

46. M. K. Das, et. al., "Drift-Free, 50 A, 10 kV 4H-SiC PiN Diodes with Improved Device Yields," ***Proceedings of ECSCRM*** Bologna, Italy, 2004.

47. H. Lendenmann, "Forward Degradation in Bipolar SiC Devices: Phenomenon, Defect Origins, and Device Effects," ***ONR Workshop on Extended defects in Wide Band Gap Semiconductors: Electrical and optical Effects***, Belize, 27-31 January 2002.

International Journal of High Speed Electronics and Systems
Vol. 15, No. 4 (2005) 899–930

World Scientific
www.worldscientific.com

SILICON CARBIDE DIODES FOR MICROWAVE APPLICATIONS

Konstantin Vassilevski

School of Electrical, Electronic and Computer Engineering,

University of Newcastle, Merz Court, Newcastle upon Tyne, NE1 7RU,

United Kingdom

k.vassilevski@newcastle.ac.uk

Silicon carbide (SiC) offers significant advantages for microwave high power devises due to its unique electrical and thermal properties. This review presents a summary of the current position in the development of silicon carbide diodes operating at microwave frequencies: varactors, Schottky barrier mixer diodes, *p-i-n* and IMPATT diodes. Simplified theory of device operation is given for each kind of microwave diodes to derive the figures of merit, to estimate the potential silicon carbide performance for fabrication of these diodes and to compare SiC with conventional semiconductors. These analyses are followed by description of diode design, fabrication and measured characteristics.

Keywords: silicon carbide; microwave; Schottky barrier diode; mixer; varactor; detector; *p-i-n* diode; IMPATT diode

1. Introduction

Silicon carbide (SiC) is a very attractive material for high power, high voltage and high temperature applications due to its superior electrical and thermal properties.[1] Silicon carbide occurs in more than 170 different crystal structures (called polytypes) with each crystal structure having its own unique electrical and optical properties. The electrical properties of the commercially available 4H- and 6H-SiC polytypes are compared to the properties of silicon and GaAs in Table 1. The only disadvantage of silicon carbide evident from this data is the low electron mobility. However, very high breakdown field, wide band gap energy, high electron saturation velocity, and high thermal conductivity may lead to substantial performance gains in many device applications, as compared to the conventional semiconductors. Usually, different figures of merit are used to provide a basis for relative comparison of the device potential of different semiconductors. In particular, E. O. Johnson considered the high-frequency / high-power capability of materials and introduced Johnson's figure of merit, $(F_M U_S/4\pi)^2$, as a criteria of a maximum allowable power of bipolar transistor in the case of electrical limitation.[2] In 1973, R.W. Keyes introduced a figure of merit, $\lambda(U_S/4\pi\varepsilon)^{1/2}$, to estimate a maximum power limited by heat dissipation in the case when device size is defined by transit time.[3] He outlined for the first time that silicon carbide physical properties ensure its advantage for fabrication of high power and high frequency devices. In 1979, this general approach

Table 1. Selected semiconductor physical properties at room temperature [ref.[1,4,5]].

			Si	GaAs	4H-SiC	6H-SiC
Band gap energy	E_g	eV	1.12	1.424	3.23	3.09
Breakdown field	F_M	V/cm	$2\cdot10^5$	$3\cdot10^5$	$3\cdot10^6$ [ref.[6]]	$3\cdot10^6$
Electron saturation velocity	U_S	cm/s	$1.0\cdot10^7$	$0.9\cdot10^7$	$2.2\cdot10^7$ [ref.[7]]	$1.9\cdot10^7$ [ref.[7]]
Relative dielectric constant	ε		11.9	13.1	10.03	10.03
Thermal conductivity	λ	W/cm/K	1.5	0.46	4.9	4.9
Maximum operating temperature	T_M	K	~500	~630	~1000	~1000
Maximum temperature difference between the junction and the heat sink; T_M *-300K*	ΔT	K	200	~330	~700	~700
Intrinsic density	n_i	cm^{-3}	$9.7\cdot10^9$	$2.4\cdot10^6$	$1.6\cdot10^{-8}$	$1.8\cdot10^{-7}$
Electron mobility	μ_e	$cm^2/V/s$	1500	8500	800	360
Diffusion coefficient of electrons	D_e	cm^2/s	39	220	21	9.3
Electrons lifetime	τ_e	s	$3\cdot10^{-5}$	$2.5\cdot10^{-7}$	$8\cdot10^{-8}$ [ref.[8]]	$1\cdot10^{-9}$
Diffusion coefficient of holes	D_h	cm^2/s	12	10	3	2
Holes lifetime	τ_h	s	$1\cdot10^{-5}$	$3\cdot10^{-6}$	$2\cdot10^{-6}$ [ref.[9]]	$4.5\cdot10^{-7}$
Ambipolar diffusion coefficient, $D_hD_e/(D_h+D_e)$	D	cm^2/s	9.2	9.5	4.2 [ref.[10]]	1.7
Ambipolar lifetime, $\tau_h+\tau_e$	τ	s	$4\cdot10^{-5}$	$3.3\cdot10^{-6}$	$2.08\cdot10^{-6}$	$4.5\cdot10^{-7}$
Electron diffusion length, $\sqrt{D_e\tau_e}$	W_e	cm	3.4·10-2	$7.4\cdot10^{-3}$	$1.3\cdot10^{-3}$	$1\cdot10^{-4}$
Hole diffusion length, $\sqrt{D_h\tau_h}$	W_h	cm	$1.1\cdot10^{-2}$	$5.5\cdot10^{-3}$	$2.5\cdot10^{-3}$	$1\cdot10^{-3}$
Ambipolar diffusion length, $\sqrt{D\tau}$	W	cm	$1.9\cdot10^{-2}$	$5.6\cdot10^{-3}$	$3\cdot10^{-3}$	$8.6\cdot10^{-4}$

was worked out in detail by A.S. Tager to estimate limiting parameters of various microwave devices and their connection to the semiconductors properties.[11] Using an analytical approach, he derived the figures of merit of high-frequency / high-power bipolar and field-effect transistors, varactors, IMPATT and *p-i-n* diodes for both cases of electrical and thermal power limitation. He showed that silicon carbide is more convenient material than Si and GaAs for fabrication of these microwave devices, where a relatively high electric field or high specific power capability are usually needed.

This review presents a summary of the current position in the development of silicon carbide diodes operating at microwave frequencies. These frequencies, which cover by

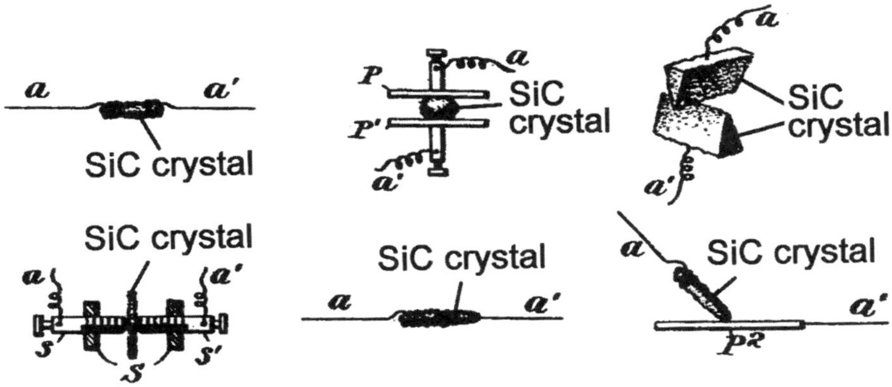

Fig. 1. Some kinds of carborundum "wave-responsive devices" patented by General H.H.C. Dunwoody.[13]

definition the range from about 100 MHz to 1000 GHz include millimeter-wave (30-300 GHz) and submillimeter-wave (>300 GHz) frequency bands.[12] The approach proposed by Tager will be used for each kind of microwave diodes to derive the figures of merit, to estimate the potential silicon carbide performance for fabrication of these diodes and to compare SiC with conventional semiconductors. These analyses will be followed by description of diode design, fabrication and measured characteristics.

Silicon carbide Schottky barrier mixer diodes, varactors, *p-i-n* diodes and IMPATT diodes will be considered, while numerical simulations and consideration of various SiC microwave diodes which were not demonstrated experimentally are beyond the scope of this review. The story of silicon carbide microwave diodes began almost a century ago when a carborundum "wave-responsive device" was invented and this diode is described first in this review.

2. Silicon Carbide Point-Contact Detectors

The point-contact detector diode, using a metal point pressed against a semiconductor surface, is one of the oldest semiconductor devices. In late 1906, General Henry Harrison Chase Dunwoody obtained a patent for wireless telegraph system.[13] A carborundum crystal, called by the inventor as a "wave-responsive device", was used to detect radiowaves in that system. Several ways of attaching the electrodes to the crystal covered by this patent are shown in Figure 1. Usually, one connection to this type of detector consisted of a small wire (a "catwhisker") that made a point contact to the crystal surface, while the other connection was a large area contact typically formed by a low-melting point alloy. The carborundum detectors needed a bias of a couple of volts and were less sensitive than made of galena, but they were more mechanically stable, and were sold adjusted and packaged in cartridges (see Figure 2). Devices made by this way might be considered as point-contact Schottky diodes. These kinds of diodes, fabricated from conventional semiconductors and operating either as mixers or as detectors, are still the most extensively used semiconductor devices in microwave receivers. That is why carborundum "wave-responsive device" has to be mention in this paper as a precursor of

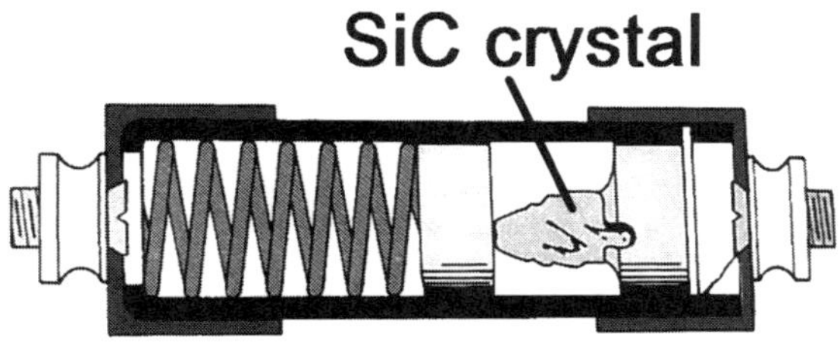

Fig. 2. The sketch of fixed carborundum detector manufactured by Carborundum Co in about 1925 year.[14]

SiC microwave diodes, although its (estimated) operating frequency was only about 100 kHz,[14] which is quite far from microwave frequency range.

3. Silicon Carbide Varactors

A varactor is a nonlinear reactive device used for harmonic generation, parametric amplification, mixing, detection and voltage-variable tuning. Present applications of varactors are mostly for harmonic generation at millimeter and submillimeter wave frequencies and as a tuning elements in various microwave applications. Commonly, Schottky or p^+-n junction diodes with moderately doped diode base which exhibit a voltage-dependent capacitance are used as varactors. Equation 1 gives the well-known variation of capacitance (C) with applied voltage (V) for a uniformly doped depletion region of Schottky barrier and abrupt asymmetrical p^+-n junction:[12]

$$C(V) = A\sqrt{\frac{q\varepsilon\varepsilon_0 N_D}{2(\phi_{bi} - V)}} \tag{1}$$

where ε_0 is the permittivity in vacuum; q is the elementary charge; A is the area of the diode, N_D is the doping level of the depleted region and ϕ_{bi} is the built-in barrier of the junction. The Quality factor of the varactor (Q-factor) can be defined as:[15]

$$Q = \frac{1}{2\pi\, f_1 R_S C(0)} = \frac{f_C}{f_1} \tag{2}$$

where f_1 is the operating frequency, and the static cut-off frequency f_C is defined as that frequency at which the capacitive reactance (X_C) is equal to the series resistance (R_S) of a zero biased diode:

$$f_C = \frac{1}{2\pi\, R_S C(0)} \tag{3}$$

The Q-factor has to be at least about ten for proper varactor operating at specific frequency. The value of f_C gives an indication of the loss and efficiency of the varactor independently of the operating frequency. The series resistance of the diode includes contributions from the diode substrate, ohmic contacts and the epitaxial layer. Properly designed varactors should have negligible resistances of the substrate and ohmic contacts. To minimize the epitaxial layer resistance, its thickness (d) has to be equal to the width of the space charge region (w_{SCR}) at maximum applied reverse voltage. Taking into account that the series resistance of epitaxial layer is $(d - w_{SCR}(0))/Aq\mu_e N_D$ and assuming that $d >> w_{SCR}(0)$, expression (3) for f_c can be rewritten as:

$$f_C = \frac{q\mu_e N_D w_{SCR}(0)}{2\pi\varepsilon\varepsilon_0 d} \approx \frac{q\mu_e N_D}{2\pi\varepsilon\varepsilon_0 \gamma} \tag{4}$$

where $\gamma = {C_{\max}}/{C_{\min}} \approx {d}/{w_{SCR}(0)}$ is another parameter which needs to be maximized for practical varactor design in almost all applications. Since silicon carbide has relatively low carrier mobility, Equation (4) clearly shows that SiC is not a preferable material for varactor fabrication when the value of N_D is not limited, e.g. for the use in parametric amplifiers.

In tuning applications and harmonic generators, however, varactors have to be designed for specific bias voltage (V_0). The advantages of high voltage and high power SiC varactors have been theoretically considered by Tager[11] and by Wright *et al.*[16] In this case the maximum applied voltage to the diode is limited by breakdown voltage ($V_M = F_M d/2$) and hence:

$$f_C \approx \frac{q\mu_e}{2\pi\varepsilon\varepsilon_0\gamma} N_D \le \frac{q\mu_e}{2\pi\varepsilon\varepsilon_0\gamma} \frac{F_M^2\varepsilon\varepsilon_0}{2qV_0} = \frac{\mu_e F_M^2}{4\pi\gamma V_0} \tag{5}$$

This expression shows that silicon carbide is a preferable material for fabrication of the varactors with given voltage rating V_0. The value of $\mu_e F_M^2$ is given in the Table 2 to compare SiC with conventional semiconductors. It is 14 times higher for 4H-SiC than for GaAs and hence, SiC varactor can operate at higher frequency as compared to the GaAs varactor designed for the same maximum voltage.

The maximum microwave power handled by varactor is electrically limited by the maximum amplitude of microwave voltage (V_1):[11]

$$P_M = \frac{V_1^2}{2X_C} \le \frac{1}{2X_C}\left(\frac{1}{2}V_M\right)^2 = \frac{1}{8X_C}\left(\frac{F_M^2\varepsilon\varepsilon_0}{2qN_D}\right)^2 \tag{6}$$

Taking into account Eq. (4), the electronic limit for maximum microwave power that can be applied to the varactor can be derived from this equation:

$$P_M X_C \le \frac{1}{32}\left(\frac{F_M^2\mu_e}{4\pi\gamma f_C}\right)^2 \tag{7}$$

In the case of thermal limitation, maximum thermal power (P_T) dissipated by varactor can be estimated as:[11]

$$P_T = {1}/{2X_C^2}\, V_1^2 R_S = {R_S}/{X_C}\, P_M \le \frac{\Delta T}{K_T} \tag{8}$$

where K_T is a thermal resistance and ΔT is the temperature difference between the junction and the heat sink. Assuming that

$$K_T \geq d/_{A\lambda} = \frac{X_C \varepsilon\varepsilon_0 2\pi f_1}{\lambda} \tag{9}$$

we can get a following expression for maximum thermally limited power of varactor:

$$P_M X_C \leq \frac{\Delta T \lambda}{\varepsilon\varepsilon_0 2\pi f_1} \tag{10}$$

The values of $(\mu_e F_M^2)^2$ and $\frac{\Delta T\lambda}{\varepsilon}$ are given in the Table 2 to compare SiC with conventional semiconductors. Equations (7) and (10) show that in both cases of thermal and electronic limitation, silicon carbide is a preferable material for fabrication of high-power varactors.

In practice however, SiC devices are dominated by the relatively high contact and substrate resistances which negate advantages in material properties at present. The only experimental attempt to prove the silicon carbide advantage for varactor fabrication was published by Vassilevski and Zorenko.[17] A p^+-n structure for varactors was grown by container-free liquid phase epitaxy[18] on Si face of n-type 6H-SiC Lely-crystal. The diode structures had moderately doped base with N_D-N_A~$4\cdot10^{17}$cm^{-3} and thickness of about 1 μm and 0.5 μm thick p^+-layer with uncompensated acceptor concentration N_A-N_D ~ $2\cdot10^{19}$ cm^{-3}. After the epitaxy, the substrate was ground down to a thickness of 100 μm. Mesa-structures 2 μm in height and 60 μm in diameter were formed by dry etching. Evaporated aluminum 1.5 μm in thickness was used to form the contacts to the p layer. Chromium contacts to the n-layers were formed on both sides of the wafer. Separated chips were maintained inside a ruby ring on a copper holder by soldering with the Au-Ge eutectic. A gold cross ribbon was welded to the aluminum contact by thermal compression.

Fabricated diodes revealed avalanche breakdowns at voltages of 140 V; capacitances at zero bias $C(0)$ in the range from 2.5 to 3.0 pF; and $\gamma \cong 8$. The diodes were used for tuning of the oscillating frequency of voltage controlled silicon IMPATT oscillator operating in cw mode at frequency of 140 GHz. The voltage-controlled frequency tuning of about 100 MHz was obtained but the diode caused an unacceptable power loss due to its relatively high resistance as it is shown in Figure 3. Since the time of that publication an essential improvement in silicon carbide growth and device processing took place. Namely, 4H-SiC epitaxial structures became commercially available,[19] and low resistivity ohmic contacts with R_C ~10^{-5} Ω·cm^2 formed on 4H-SiC with electron[20] and hole[21] type of conductivity were developed. Nevertheless, no more attempts to demonstrate high power silicon carbide varactors for microwave applications was

Table 2. Figures of merit for different parameters of high-power varactors (relative to the values for silicon).

Parameter	F.O.M.	Si	GaAs	4H-SiC	6H-SiC
Maximum frequency	$F_M^2\mu_e$	1	12.8	120	54
Electrical limit for maximum power	$(F_M^2\mu_e)^2$	1	163	14400	2920
Thermal limit for maximum power	$\frac{\Delta T\lambda}{\varepsilon}$	1	0.5	14	14

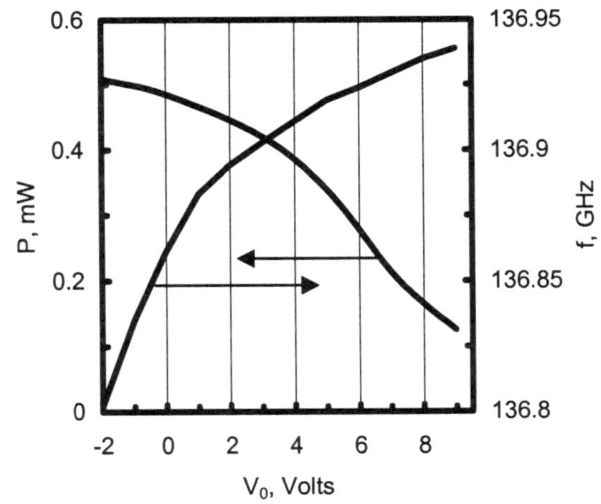

Fig. 3. Output power and frequency of Si IMPATT oscillator depending on tuning voltage applied to silicon carbide varactor.[17]

undertaken up to now.

4. Silicon carbide Schottky barrier mixer diodes

Microwave mixer diodes are used to mix a received microwave signal at frequency f_1 with a local oscillator (LO) signal in a non-linear resistance of mixer, so as to derive a difference signal at intermediate frequency $f_{IF} = f_1 - f_{LO}$, which can be easily amplified and detected. These diodes usually make use of the non-linear resistance of the metal-semiconductor contact (R_J), since a low minority carrier injection is required for their good performance and precludes the use of *p-n* junctions. The performance of a mixer is judged by its conversion loss (L_C) defined as:

$$L_C[in\ dB] = 10 \cdot \lg\left(P_1 / P_{IF} \right) \tag{11}$$

This conversion loss includes parasitic loss (L_1) and intrinsic down-conversion loss due to the mixing action (L_J):

$$\begin{aligned} L_C[in\ dB] = L_1 + L_J &= 10 \cdot \lg\left(1 + R_S / R_J + \omega_1^2 C_J^2 R_S R_J \right) + L_J \\ &\approx 10 \cdot \lg\left(1 + R_S / R_J + \left(f_1 / f_C \right)^2 R_J / R_S \right) + L_J \end{aligned} \tag{12}$$

where C_J is a Schottky barrier capacitance shunting R_J; R_S is a series resistance; and f_C was defined by Eq. (3). This expression shows, that the low junction capacitance C_J and low series resistance R_S are desired to reduce the loss in input signal power delivered to R_J. Unlike varactors, mixer Schottky diodes operate at zero or forward voltage biases and need to block only the relatively low reverse voltage of about several V_{LO} values. For this reason, SiC advantage over conventional semiconductors due to its high avalanche breakdown electric field is not significant. On the other hand, Schottky mixer diodes have a thin epitaxial base (usually 100-200 nm to ensure thermionic emission over the barrier). The base doping level is low enough so that tunneling does not occur and punch-through conditions are met at zero or low reverse bias. This reduces or even excludes the resistance of the un-modulated diode base from R_S and hence minimizes the negative effect of low electron mobility in silicon carbide. Furthermore, a planar or beam-lead design of Schottky mixer diodes can be employed to reduce the contribution of a substrate resistance in the R_S value, and to minimize the effect of the relatively low electron mobility in silicon carbide comparing with Si and GaAs.

The mixer diode is used for down-conversion of small power microwave signals but its intrinsic conversion loss L_J decreases with increasing of LO power level. Therefore, mixer diodes may take advantage of silicon carbide capability to operate at high electrical power in applications where improvement of the mixer characteristics at the cost of higher LO power is acceptable. Indeed, the current over the Schottky barrier at forward bias is described by thermionic emission theory as:[22]

$$I_F = AJ_S\left(\exp\left(\frac{qV}{nk_BT}\right)-1\right) \tag{13}$$

where

$$J_S = A^*T^2\exp\left(\frac{-q(\phi_{B0})}{k_BT}\right) \tag{14}$$

Here, A^* is the effective Richardson constant, k_B is the Boltzmann's constant, ϕ_{B0} and n are the barrier height and the ideality factor, respectively. For a voltage drop $V > 3k_BT/q$, Eq. (11) can be rewritten as:

$$I_F \approx AJ_S\exp(\Lambda V) \tag{15}$$

where $\Lambda = \dfrac{q}{nk_BT}$.

Over a small range of applied voltages about a dc bias voltage V_0, this expression can be represented by a Taylor expansion:

$$\begin{aligned} I_F(V_0+\delta V) \approx I_F(V_0) + \Lambda I_F(V_0)\delta V + \\ (\Lambda)^2\frac{I_F(V_0)}{2}(\delta V)^2 + (\Lambda)^3\frac{I_F(V_0)}{6}(\delta V)^3 + \ldots \end{aligned} \tag{16}$$

Substituting $\delta V = V_1 sin(\omega_1 t) + V_{LO} sin(\omega_{LO} t)$ in this equation gives the current amplitudes at input and LO frequencies:

$$\begin{aligned} I_1 &\approx \Lambda I_F(V_0)V_1 \\ I_{IF} &\approx (\Lambda)^2\frac{I_F(V_0)}{2}V_1V_{LO} \end{aligned} \tag{17}$$

and down-conversion loss can be estimated as:

$$L_J[in\ dB] = 10 \cdot \lg\left(P_1 / P_{IF} \right) \approx 20 \cdot \lg\left(\frac{2}{\Lambda V_{LO}} \right) \tag{18}$$

The above small signal analysis assumes an ideal Schottky diode operation - that of a pure non-linear resistance (varistor). More accurate large-signal theory of mixer diode[23] taking into account voltage dependence of both R_J and C_J gives the following expression for minimum conversion loss:

$$L_J[in\ dB] \cong 10 \cdot \lg\left(1 + \frac{2}{\sqrt{\Lambda V_{LO}}} \right) \tag{19}$$

The use of high-level LO signal allows to reduce the conversion loss L_J in accordance with Eqs. (18), (19) and to increase the mixer linearity, since the conversion loss increases when the power of input signal became comparable with the LO power.[24] Another advantage of a high-level diode mixer is a reduction of intermodulation distortion (IMD) in it at the cost of high LO power.[25] The IMD products takes place when a second signal at the frequency of f_2 arriving at the mixer RF port interacts with the desired incoming signal (f_1). The second-order IMD products arise at frequencies $f_{LO} \pm (f_1 - f_2)$, which are located far away on either side of the desired IF signal and are serious only in the case of wide-band systems. The signals resulting from the interaction at frequencies $f_{LO} - (2f_1 - f_2)$ and $f_{LO} - (2f_2 - f_1)$ are the third-order IMD products and are located very close to the desired IF signal. The IMD products are usually created when the signal level reaches the LO power level, and are commonly characterized by an input intercept point (IIP_2 or IIP_3 for second- and third-order IMD products, respectively) – the input power at which the IMD power at IF port reaches the power of a useful signal.[24] The higher the IIP is, the lower the IMD products are and, consequently, the spurious-free dynamic range is wider.

The value of V_{LO} is restrained electrically and thermally. Neglecting by resistances of the substrate and ohmic contacts, the V_{LO} value is electrically limited by maximum forward voltage bias of Schottky diode $V_{MAX} = V_0 + V_{LO}$ at which $R_J = R_S$:

$$\frac{1}{A \Lambda A^* T^2 \exp\left(-\frac{q\phi_{B0}}{k_B T} \right) \exp\left(\Lambda V_{MAX} \right)} = R_S \tag{20}$$

and hence:

$$V_{LO} \le V_{MAX} = n\phi_{B0} + \frac{1}{\Lambda} \ln\left(\frac{\pi\, f_C \sqrt{2\varepsilon\varepsilon_0 q N_D}}{\Lambda A^* T^2 \sqrt{\phi_{B0}}} \right) \approx \phi_{B0} - V_A \tag{21}$$

where V_A is about 0.1V at room temperature for all reasonable values of the other parameters. Silicon carbide has evident advantage over conventional semiconductors: it has lower electron affinity and hence higher Schottky barrier. Thus, Ti, Ni and Au form the Schottky barriers to n-type 4H-SiC with typical heights of 0.95, 1.62 and 1.73 eV, respectively.[26] These values significantly exceed typical barriers heights of Si and GaAs

Schottky diodes and make silicon carbide a preferable material for fabrication of high-level diode mixers in point of electrical limitation of the LO power.

Thermal limitation of the V_{LO} value can be derived from the expression for the total power absorbed by the diode (P_T). Assuming that the RF power P_I is negligible, the LO power absorbed by a Schottky diode may be related to V_A by following equation:[27]

$$P_T = \frac{\Lambda I(V_0)V_{LO}^2}{2}\left(1+ {R_S}/{R_J} + (\omega_{LO}C_J)^2 R_S R_J\right) = \frac{\Lambda I(V_0)V_{LO}^2}{2}\left(1+ {R_S}/{R_J} + \left({f_{LO}}/{f_C}\right)^2 {R_J}/{R_S}\right) \tag{22}$$

Taking into account that $f_{LO} << f_C$ and $R_S \approx R_J$ at high power level, and P_T is limited by thermal dissipation as $P_T \le \frac{\Delta T}{K_T}$, the thermal limitation for V_{LO} can be written as:

$$V_{LO}^2 \le \frac{2\Delta T}{K_T \Lambda I(V_0)\left(1+ {R_S}/{R_J} + \left({f_{LO}}/{f_C}\right)^2 {R_J}/{R_S}\right)} \approx \frac{2\Delta T}{K_T \Lambda I(V_0)(1+R_S \Lambda I(V_0))} \tag{23}$$

By considering that the thickness of epitaxial layer is small, we can neglect its thermal resistance and substitute the value of $K_T \approx {1}/{r\pi\lambda} = \frac{\sqrt{X_C \varepsilon\varepsilon_0 2 f_{LO}}}{\sqrt{\pi}\, d\lambda}$ into Eq. (23):

$$V_{LO} \le \left(\frac{\Delta T\lambda\sqrt{2\pi}\, d}{\Lambda I(V_0)(R_S \Lambda I(V_0)+1)\sqrt{X_C \varepsilon\varepsilon_0 f_{LO}}}\right)^{1/2} \tag{24}$$

The value of $(\Delta T\lambda)^{1/2}$ is given in Table 3 to compare silicon carbide with conventional semiconductors from the point of thermal limitations on high-level Schottky mixer diode performance.

In 1997, Fazi and Neudek proposed the first use of a silicon carbide Schottky diode for fabrication of high-level mixers.[28] They measured IF output spectra of a 500 MHz input single-balanced mixer with two commercial Si diodes and of the same mixer with SiC Schottky diodes and 20 dB higher LO level. These spectra are shown in Fig. 4. While

Table 3. Figures of merit for different parameters of high-level Schottky mixer diodes (relative to the values for silicon).

Parameter	F.O.M.	Si	GaAs	4H-SiC	6H-SiC
Maximum frequency	$\frac{\mu_e}{\varepsilon}$	1	5	0.63	0.28
Electrical limit for maximum amplitude of the LO signal	ϕ_{BO} (eV)	0.8 Au [ref.[12]]	0.9 Au [ref.[12]]	1.73 Au [ref.[26]]	1.56 Ni [ref.[26]]
Thermal limit for maximum amplitude of the LO signal	$(\Delta T\lambda)^{1/2}$	1	0.7	3.4	3.4

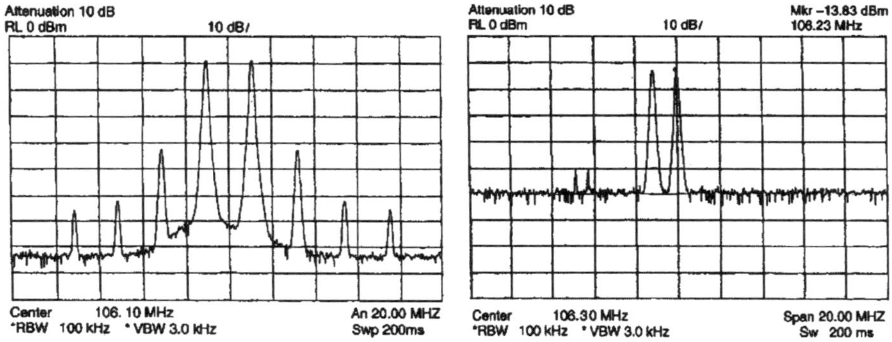

a b

Fig. 4. IF spectra of a single-balanced mixer (a) with two commercial Si diodes and (b) with two SiC diodes and 20 dB amplifier of LO signal.[29] (Reprinted with permission, © 2005 TTP)

conversion loss with the SiC mixer was about 2 dB worse than with Si diodes (12 dB vs. 10 dB), the SiC mixer has no observable IMD products. The high noise level of the LO generator did not allow the exact determination of IMD product levels for the SiC mixer, which was at least 20 dB lower than those observed with of Si mixers. No details concerning SiC Schottky diode design were given in that papers.

Fabrication and characterization of SiC mixer diodes with titanium Schottky barrier have been published by Eriksson *et al.* in a series of papers.[30-34] The structures of the diodes used in this work were grown on the Si-face of a highly conductive 340 μm-thick 4H-SiC substrate. The donor doping of the substrate was 1.1×10^{19}cm^{-3}. On the top of the 4H-SiC substrate, a highly doped spacer layer was grown, followed by the 0.38 μm-thick drift epitaxial layer with donor doping of 3.1×10^{17}cm^{-3}. Mesa structures were etched by ion beam etching (IBE) with Ar. Schottky barriers were formed by evaporation of Ti-Au using a lift-off technique. A backside contact was formed by Ni-evaporation and annealing at 950°C for 5 min in Ar/He (10:1) ambient. This diodes revealed Schottky barrier heights ϕ_{B0} = 1.08 eV (measured by *C-V*) and ϕ_{B0} = 0.88 eV (measured by *I-V*), with ideality factors below 1.10.

The diodes with contact area of 1.6×10^{-5}cm^2 (40×40 μm^2) have been installed and characterized in the single-balanced mixer shown in Fig. 5. The mixer had the conversion loss below 10 dB in 0.7-1.4 GHz frequency range for P_{LO}=15 dBm. A minimum conversion loss of 5.2 dB was measured at input signal frequency of 850 MHz and P_{LO}=25 dBm with diode bias current (I_0) of 15 mA. This value of conversion loss is typical for diode mixers.[35] The maximum applied LO power to this circuit was above 1 W. A two-tone measurement was used to determine the third-order intermodulation intercept point. The equal-power RF input signals were separated by 10 MHz (f_1 = 850 MHz, f_2 = 860 MHz). The IM and IF powers against the input RF power were measured as it is shown in Fig. 6. The highest IIP$_3$ for this mixer was found to be of 31 dBm at f_{LO}=800 MHz, P_{LO}=30 dBm, and I_0=15 mA.[34]

SiC Schottky mixer diodes with highest operating frequency were published by R. Simons and P. Neudeck.[36] The diodes were fabricated on a commercial (Si-face, 0.036 Ω·cm) n-type 4H–SiC substrate with homoepitaxial layers of doping and thickness

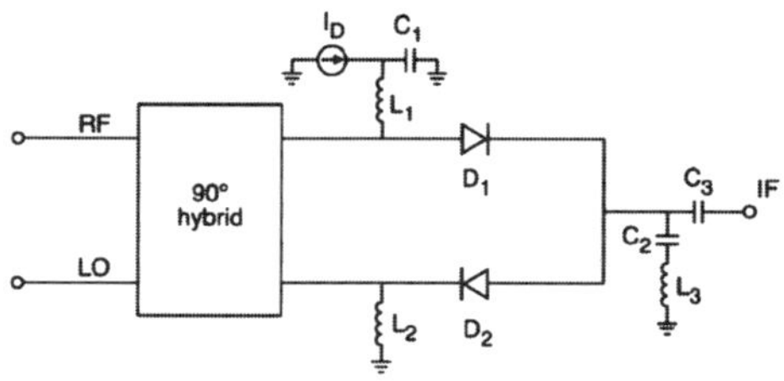

a

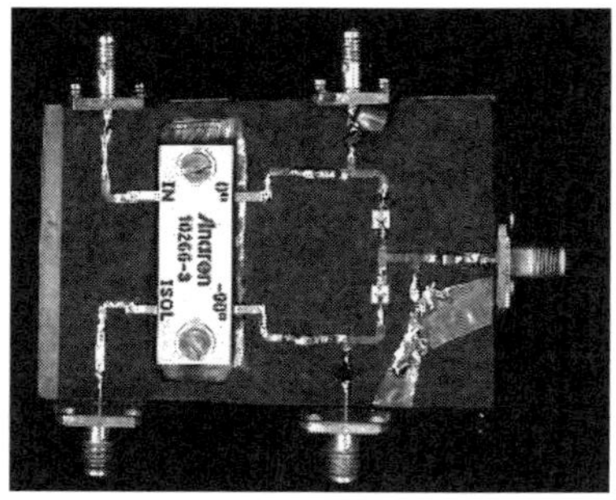

b

Fig. 5. The topology (a) and photograph (b) of the single-balanced mixer used for characterization of SiC Schottky barrier mixer diodes.[34] (Reprinted with permission, © 2005 IEEE)

depicted in Fig. 7. The as-received wafer was thinned from the backside to a thickness around 0.2 mm. In the first step, the opposite side of the wafer was metallized with nickel and furnace annealed at high temperature to form the cathode contact. The second step was the Schottky-barrier metal deposition (0.1 μm thick gold) and patterning using a lift-off process to form the anode contacts with diameters ranging from 5 to 50 μm. Third, an insulating layer to support a large anode contact pad was applied to a thickness of 1.0 μm by a multiple application of spin-on-glass (SOG). Fourth, using photo-resist and dry etching, the SOG over the anode was removed creating a circular via opening. Fifth, the large anode contact pad (100 μm × 100 μm) was fabricated using gold by a lift-off process. In this step, the circular opening was also metallized to ensure continuity between the anode and the contact pad.

The diodes with contact diameter of 25 μm were characterized for linearity by inserting them into single-balanced mixer circuits. The conversion loss was observed to remain constant at approximately 7.5 dB over the input power range from -40 dBm to +13 dBm. The intercept diagram of single-balanced mixer with SiC Schottky diodes is shown in Fig. 8. The spurious-free dynamic range was approximately 60 dB and the value of IIP_3 about 22 dBm can be estimated from this figure. The conversion-loss value

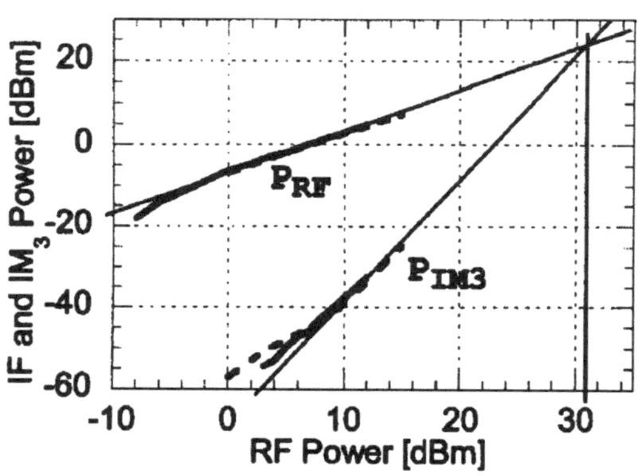

Fig. 6. IF power and IM power of single-balanced mixer with SiC Schottky diodes versus RF input power. P_{LO} = 30 dBm, I_0 = 15 mA, f_{LO} = 800 MHz. The supporting lines show the 31 dBm IIP3. Solid lines are measurements and dashed lines are simulations.[34] (Reprinted with permission, © 2005 IEEE)

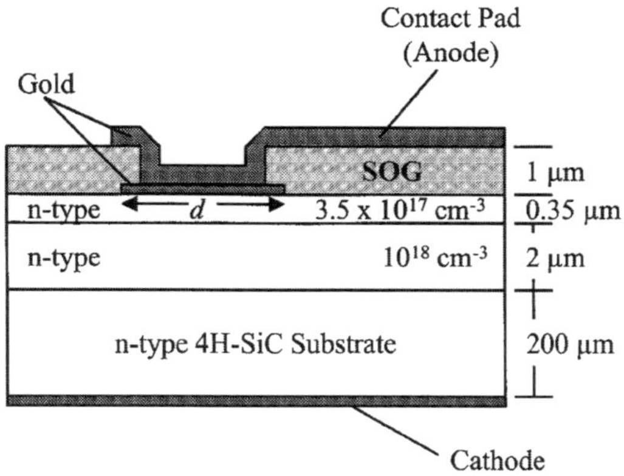

Fig. 7. Schematic showing the cross section of the SiC Schottky-barrier mixer diode.[36] (Reprinted with permission, © 2005 IEEE)

in the above measurements is about the same as that of a commercially available single-balanced mixer with silicon Schottky-barrier diodes. However, to achieve a comparable input performance with silicon Schottky-barrier diodes, a more complex design involving a double-balanced mixer with two diodes in each arm of a quad is required.

5. Silicon carbide *p-i-n* diode

P-i-n diodes are the most extensively used semiconductor components in microwave circuits for amplitude modulation, attenuation and leveling functions. They are basic components of RF power switches, phase shifters, pulse modulators, active protectors, electrically tuned attenuators. *P-i-n* diode has a very low doped or intrinsic region located between heavily doped *p*- and *n*- layers. At reverse bias the *i*-region results in high breakdown and low punch-through voltages of the diode. The *p-i-n* diode capacitance (C_J) is reduced by the increased separation of *n*- and *p*- layers and remains invariable in a wide range of reverse biases.

A main feature of the forward biased *p-i-n* diode is that the conductivity of *i*-layer is

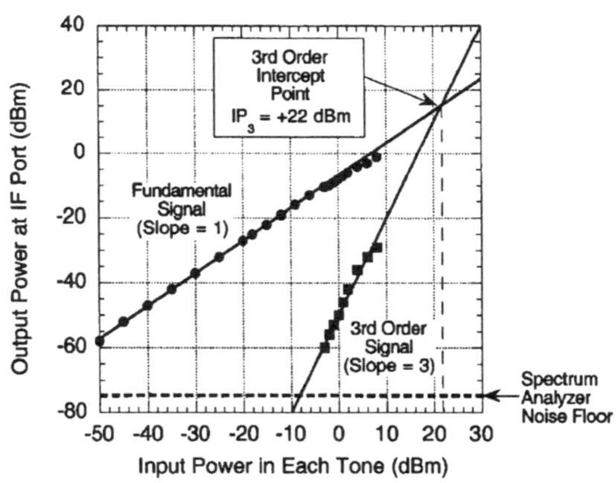

Fig. 8. Power output at the IF port as function of the input power in each tone: f_1 = 1.57542 GHz, f_2 = 1.57842 GHz, f_{LO} = 1.55542 GHz, P_{LO}= 19.5 dBm.[36] (Reprinted with permission, © 2005 IEEE)

controlled by the injection of charge carriers from both ends of i-region.[37] The thickness of i-layer has to be small compared with diffusion lengths for holes and electrons. This implies that the hole and electron concentrations may be assumed uniform throughout the i-layer. Assuming that i-layer doping is sufficiently light that the carrier concentrations without injection are negligible in comparison with the injected carrier concentrations and that all carrier recombination takes place within the i-layer, the dc bias current will be exactly the current needed to replace the carriers lost by recombination. The current crossing the boundary between p- and i-layers will be entirely hole current and can be written as:

$$I_F = A\frac{qpd}{\tau_h} \tag{25}$$

where p is a hole concentration in the i-layer. Similarly:

$$I_F = A\frac{qnd}{\tau_e} \tag{26}$$

where n is an electron concentration in the i-layer. The resistance of i-layer (R_i) may therefore be written as:

$$R_i = \frac{d}{Aq(p\mu_h + n\mu_e)} = \frac{kT}{qI_F}\frac{d^2}{(D_h\tau_h + D_e\tau_e)} = \frac{kT}{qI_F}\frac{d^2}{(W_h^2 + W_e^2)} \tag{27}$$

Eq. (27) shows that the resistance of forward biased p-i-n diode is inversely proportional to the current and does not depend on the area or the doping level of i-layer. Notice that, instead of being proportional to the i-layer thickness, the R_i value increases as the square of d.

The first consideration of SiC as a promising material for fabrication of high power and fast switching p-i-n diodes for handling of high power microwave signals was performed in 1979 by A.S. Tager.[11] Similar to the varactor, the reverse biased p-i-n diode is characterized by the cut-off frequency f_C:

$$f_C = \frac{1}{2\pi R_S C} = \frac{d}{2\pi R_S \varepsilon\varepsilon_0 A} \le \frac{\sqrt{D\tau}}{2\pi R_S \varepsilon\varepsilon_0 A} \sim \frac{\sqrt{D\tau}}{\varepsilon} \tag{28}$$

since the thickness of i-layer cannot appreciably exceed a diffusion length $\sqrt{D\tau}$. The series resistance R_S of the diode only includes resistances of ohmic contacts, diode substrate and heavily doped n- and p-regions, since the i-layer is completely depleted. Eq. (28) shows that SiC p-i-n diodes shall operate at lower frequencies in comparison to Si diodes, since silicon carbide has lower diffusion length. Nevertheless, the main application of p-i-n diodes is their use in high power switches, and silicon carbide has obvious advantages over silicon for fabrication of high-power p-i-n diodes. Indeed, the switched microwave power is limited electrically by the maximum amplitude of microwave voltage (V_1):

$$P_M = \frac{V_1^2}{2X_C} \le \frac{1}{2X_C}\left(\frac{1}{2}V_M\right)^2 = \frac{1}{2X_C}\left(\frac{dF_M}{2}\right)^2 \le \frac{1}{2X_C}\left(\frac{\sqrt{D\tau}F_M}{2}\right)^2 \tag{29}$$

and the electronic limit for maximum microwave power that can be applied handled by *p-i-n* diode can be derived from this equation as:

$$P_M X_C \leq \frac{D\tau F_M^2}{8} \tag{30}$$

Thermal limitations of the maximum microwave power dissipated by a *p-i-n* diode is given by Eq. (10) since the thermal resistivity of *p-i-n* diode is defined mainly by resistivity of the thick *i*-layer. Equations (30) and (10) tell us that in both thermal and electronic limitation cases, silicon carbide is a preferred material for the fabrication of high-power *p-i-n* diodes. Furthermore, the switching time (τ_r) of the *p-i-n* diode is defined by the lifetime of electron-hole plasma in *i*-region and can be estimated as $\tau_r \leq d^2/D$. If the *p-i-n* diode is rated for specific maximum bias voltage V_0, the *i*-layer has to be thicker than V_0/F_M and the switching time may be estimated as:

$$\tau_r \leq \frac{V_0^2}{DF_M^2} \tag{31}$$

The relative values of $\left(DF_M^2\right)^{-1}$; $D\tau F_M^2$; and $(\Delta T\lambda)/\varepsilon$ are given in the Table 4 to compare SiC with conventional semiconductors.

The first experimental investigations of silicon carbide microwave switches for operation at frequencies above 1 GHz were published recently.[38-40] 4H-SiC p^+-n-n^+ diodes used in the switcher were fabricated on epitaxial wafers with low resistivity (0.015-0.028 Ω·cm) substrate commercially available from Cree, Inc. The *n* base layer was 2 μm thick and the donor concentration was $\sim 1.0\times10^{17}$cm^{-3}. The acceptor concentration in the 1 μm thick p^+ layer and the donor concentration in the substrate were equal to 6×10^{18}cm^{-3} and 5×10^{17}cm^{-3} respectively. The diode fabrication procedure was described by Vassilevski *et al.*[41] Nickel, 200 nm thick, annealed at 1000°C for 120 sec was used as a back side ohmic contact. The geometry of the contacts to p^+ epitaxial layer (circular dots of 40 μm diameter) was defined by lift-off procedure. The multi-layer metal composition, Al(50 nm)/ Ti(100 nm)/ Pt(25 nm)/ Ni(50 nm), annealed at 1000°C for 120 sec was used to form ohmic contact to p^+ epitaxial layer. Gold over-layers of 200 nm thickness were deposited on both contacts. Mesa structures with diameters from 60 to 100 μm and heights of 4 μm were formed by reactive ion etching (RIE) to define the diode area. The diodes had a zero bias capacitance *C(0)* from 1.4 to 3.8 pF and

Table 4. Figures of merit for different parameters of high-power microwave *p-i-n* diodes (relative to the values for silicon).

Parameter	F.O.M.	Si	GaAs	4H-SiC	6H-SiC
Maximum frequency	$\varepsilon^{-1}\sqrt{D\tau}$	1	0.25	0.2	0.05
Switching time	$\left(DF_M^2\right)^{-1}$	1	0.45	0.01	0.025
Electrical limit for maximum power	$D\tau F_M^2$	1	0.2	5.4	0.5
Thermal limit for maximum power	$(\Delta T\lambda)/\varepsilon$	1	0.5	14	14

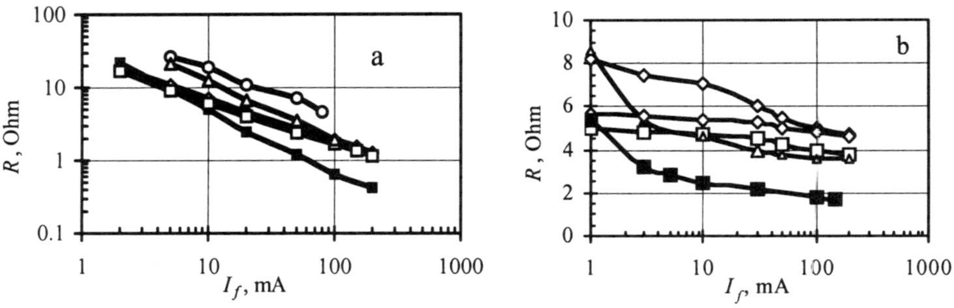

Fig. 9. 4H-SiC p^+-n-n^+ diodes resistances measured under forward bias at (a) low frequency and (b) at frequency of 10 GHz; open signs - 4H-SiC p^+-n-n^+ diodes; solid signs - typical Si *p-i-n* diode for 10 GHz.[39]

specific series resistance of $8.4 \cdot 10^{-5} \Omega \cdot cm^2$. Figure 9 shows diode resistance measured under forward bias at low frequency and at frequency of 10 GHz. The dependence of resistance with bias current in accordance with Eq. (27) was clearly observed at low frequency. This proves that the minority carrier lifetime in 4H-SiC was sufficient to get a high level of modulation of 2 μm thick base resistance.

The diodes have doping levels and base thicknesses which are too high to serve as a perfect *p-i-n* diodes. Nevertheless, the diodes with mesa-structure diameters of 80 μm have been used in a microstrip switch and characterized at frequencies from 1 to 10 GHz. Diode capacitance changed from 3.0 to 0.6 pF as reverse voltage varied between 0 and 40 V. The lead inductance measured at 10 GHz was about 0.6 nH. The time of diode resistance recovery at switching from forward current of 10 mA to reverse voltage of 15 V was about 3 ns. The switch was represented as a section of microstrip transmission line with a shunt-mounted diode. In this circuit configuration the maximum isolation and minimum transmission are obtained when the diode most closely approximates a short circuit, and maximum transmission is obtained at reverse bias of the diode. The modulator was connected to the measuring section via a microstrip-coaxial junction. The measured isolation and transmission losses depending on forward and reverse biases are shown in Fig. 10. The maximal modulator isolation value of 18 dB was obtained at a frequency of 10 GHz, while the minimal losses of 0.8 dB were obtained at a frequency of 1 GHz. It should be noted that the silicon *p-i-n* diode typically provides essentially better results for the same frequency range and circuit configuration (isolation higher than 28 dB and transmission losses less than 0.1 dB). Nevertheless, the fabricated switcher with 4H-SiC p^+-n-n^+ diode demonstrated a satisfactory operation at frequencies of 1 and 3 GHz which was in very accurate agreement with diode and switcher models.[38] Further improvement of 4H-SiC *p-i-n* diode characteristics can be easily achieved by use of epitaxial structure with low doping level which are commercially available.[42] This will allow the reduction of punch-through voltage for the diode and increase of breakdown voltage whilst maintaining *i*-layer thickness. It should also be noted that the top contact area in investigated diodes was four times smaller than the area of mesa structure, and hence the series resistance of 4H-SiC *p-i-n* diodes can be reduced by a factor of four just by simple change of diode geometry reaching the level expected for Si *p-i-n* diodes.

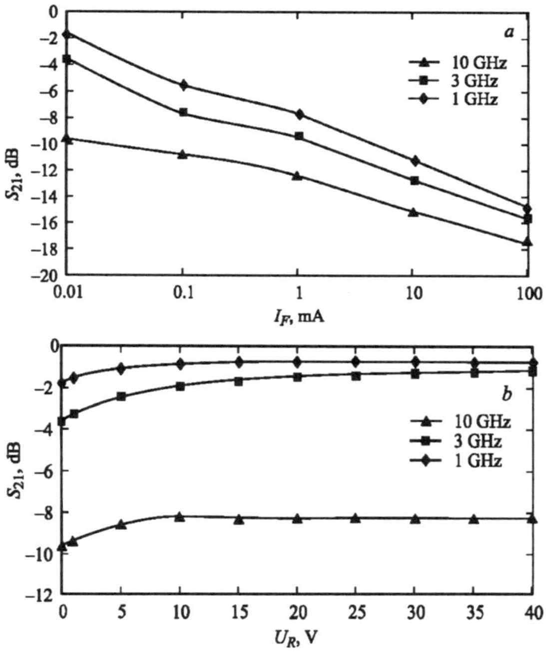

Fig. 10. Microwave characteristics of microstrip switcher with 4H-SiC p^+-n-n^+ diode measured at frequencies 1, 3, and 10 GHz.[38]
(a) - isolation losses depending on forward current bias;
(b) - transmission losses depending on reverse voltage bias.

6. Silicon carbide IMPATT diode

IMPact-ionisation Avalanche Transit-Time (IMPATT) diodes were proposed by Read[43] for microwave power generation in 1958. In 1959, the first IMPATT diodes fabricated from germanium p^+-n-n^+ structures formed by diffusion in the epitaxial layer were demonstrated by A.S. Tager *et al.*[44] An expediency of IMPATT fabrication with use of various wide-band semiconductors including silicon carbide was analyzed by A.S. Tager in 1974 for the first time.[45] Using the same approach as proposed by Johnson[2] and Keyes,[3] he estimated the influence of semiconductor material parameters on IMPATT performance and demonstrated the advantage of silicon carbide in comparison to conventional semiconductors. Two material parameters, saturated drift velocity of charge carriers (U_S for electrons) and avalanche breakdown electric field (F_M), define almost all IMPATT diode characteristics. Indeed, assuming the same value of efficiency (η) for all semiconductor materials, the maximum electrical power of IMPATT diode is limited electrically by the input dc current and the breakdown voltage:

$$P_M = \eta I_0 V_M \le A U_S qn\left(\frac{1}{2}F_M d\right) = \frac{dU_S}{4\pi X_C f_1} F_M^2 = \frac{U_S}{2f_{tt}} \cdot \frac{U_S}{4\pi X_C f_1} F_M^2 = \frac{\theta U_S^2}{8\pi^2 X_C f_1^2} F_M^2 \tag{32}$$

where transit-time frequency (f_{tt}) and transit time angle (θ) are defined as:

$$f_{tt} = \frac{U_S}{2d} \text{ and } \theta = \pi \frac{f_1}{f_{tt}} \tag{33}$$

For $\theta \cong \pi$, electrical limit for maximum IMPATT power can be derived from Eq. (32):

$$P_M X_C \le \frac{U_S^2 F_M^2}{8\pi f_1^2} \tag{34}$$

Thermal limitation of maximum microwave power dissipated by an IMPATT diode is given by Eq. (10) if one assumes that the thermal resistivity of IMPATT diode cannot be less than the K_T value of diode base. Equations (34) and (10) tell us that in both thermal and electronic limitation cases, silicon carbide is a preferable material for fabrication of high-power IMPATT diodes.

Oscillation frequency of IMPATT diode depends inversely on the thickness of drift region $f_1 = U_S\theta(2\pi d)^{-1}$ which is limited by the tunneling breakdown. Tunneling breakdown voltage usually is estimated[46] to be equal to $V_T \cong 6E_g/q$ and an electric field at fixed tunneling current to be directly proportional to $E_g^{3/2}$. Hence, the maximum oscillation frequency of IMPATT diode limited by tunneling breakdown can be written as:

$$f_{\max} \cong \frac{U_S}{2d_{\min}} \sim \frac{U_S q E_g^{3/2}}{2 \cdot 12 E_g} \sim U_S E_g^{1/2} \tag{35}$$

Another limitation of maximum operating frequency of IMPATT diode is caused by diffusion spreading of injected carriers in the drift region. The transit time, d/U_S, has to be less than the time of carrier diffusion through the drift region, $\tau_r \cong \frac{d^2}{D}$:

$$\frac{1}{U_S} < \frac{d}{D_e} \cong \frac{U_S}{2f_{\max} D_e} \tag{36}$$

and hence

$$f_{\max} < \frac{U_S^2}{2D_e} \tag{37}$$

Sufficiently accurate analytical estimations of SiC performance for IMPATT diode fabrication became available after the saturated drift velocity of the electrons (U_S) in 6*H*-SiC were measured.[47] The relative values of power and frequency limits for SiC, Si, and GaAs are compared in the Table 5. As a result, it may be concluded that SiC IMPATT diodes may have 14 times greater output power and 3 times greater theoretical oscillation frequency maximum in comparison to Si and GaAs diodes. This analysis does not take into account a difference in the IMPATT diode efficiency, which is strongly dependent on parasitic series resistance (R_S). Although the value of R_S can be reduced by a proper chip design, 4*H*-SiC is a preferable choice between two SiC polytypes for IMPATT diode fabrication due to the higher and almost isotropic electron mobility compared to 6H.

Also, an additional factor affecting the choice of polytype is the negative temperature coefficient of V_M in *p-n* junctions fabricated on a basal plain of 6H-SiC.[48]

The first numerical simulations of SiC IMPATT diodes were performed by Mehdi *et al.* in 1988.[49] Double-drift p^+-*p-n*-n^+ 6H-SiC structures with a *p-n* junction plain parallel to the *c* axis were analyzed at frequencies from 10 to 94 GHz. The experimental data[5] were extrapolated to higher electric fields to approximate the electron drift velocity and carrier ionization rates measured for this *p-n* junction plain orientation[50] have been used in the simulations. It was calculated that the output power of SiC IMPATT diodes exceeded the output power of Si and GaAs diodes in all considered frequency ranges in both the pulsed and cw modes of operation. It should be noted that a double-drift structure is known to be more efficient than a single-drift IMPATT only if optimally designed. However, the hole transport parameters which are required for optimally designing of double-drift structures are not very well known for SiC. In order to calculate more realistic results, a single drift p^+-*n*-n^+ 6H-SiC IMPATT structure with a *p-n* junction plane perpendicular to the *c* axis was modeled.[51] The insignificant contribution of holes to the negative resistance of this structure allowed a painless use of crude estimates of holes drift velocity in the simulation. The experimental data[5] were extrapolated to higher electric fields to approximate the electron drift velocity. Carriers ionization rates measured for this *p-n* junction plain orientation in 6H-SiC[48] were used to calculate pulsed diode characteristics and to optimize the diode structure for oscillation frequency of 140 GHz. It was found that both a maximum efficiency (about 7%) and dc current density at which the diode had the maximum efficiency (about 150 kA/cm^2) were very close to that values of Si singe-drift IMPATT diodes, providing a gain in output power in accordance to the difference of dc voltage drop across the structure. The optimized p^+-*n*-n^+ 6H-SiC IMPATT structure with the base region length of 0.7 μm and doping level of $4 \cdot 10^{17}$cm^{-3} had an operating dc voltage of 140 V. This value, and hence the output power was about 15 times higher than typical output power of Si IMPATT diode at this frequency. Results of this numerical modelling[49,51] quantitatively proved the analytical estimation described above, but were insufficiently accurate to serve as a guide for SiC IMPATT diode design and fabrication since the most important material

Table 5. Figures of merit for different parameters of IMPATT diodes (relative to the values for silicon).

Parameter	F.O.M.	Si	GaAs	4H-SiC	6H-SiC
Maximum frequency limited by tunneling	$U_S E_g^{1/2}$	1	1.0	3.7 (1.3)	3.2
Maximum frequency limited by diffusion	U_S^2 / D_e	1	0.14	9.1 (1.1)	15
Electrical limit for maximum power	$U_S^2 F_M^2$	1	1.8	1090 (140)	810
Thermal limit for maximum power	$\frac{\Delta T \lambda}{\varepsilon}$	1	0.5	14	14

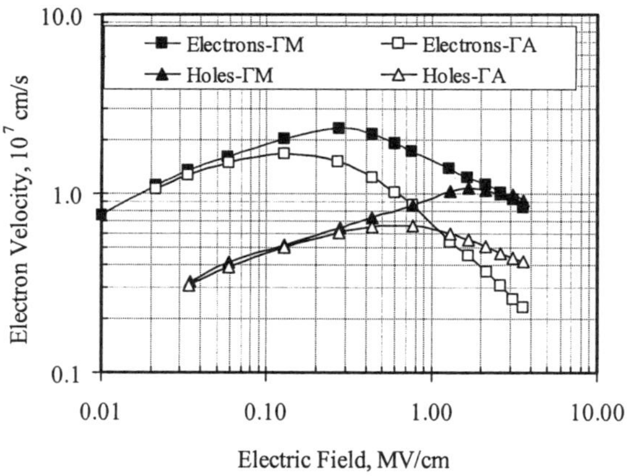

Fig. 11. MC simulated drift velocity as a function of electric field in the *ΓM* (perpendicular to the *c*-axis) and *ΓA* (along the *c*-axis) directions.[60]

parameter used in these numerical calculations remained undetermined. Namely, the U_S value was measured at electric field not exceeding $2{\cdot}10^5$ V/cm and in the direction perpendicular to the *c*-axis,[47] while the avalanche breakdown field in SiC is much higher.

More accurate simulation of SiC IMPATT diodes became available after intensive Monte-Carlo (MC) study of electron transport in SiC had been undertaken.[52-54] The carrier transport at high electric field was calculated using the model with three electron and two hole bands. It was discovered that the field dependence of the electron drift velocity had a maximum at about 400 kV/cm and negative differential drift velocity at higher fields. To investigate an effect of this negative differential drift velocity on SiC IMPATT diode performance, the operation of 4H-SiC IMPATT diode with hi-lo p^+-n-n^--n^+ structure at frequency of 200 GHz was simulated by MC method.[55,56] The impact ionization was accounted for by using the Keldysh formula[57] with fitting parameters obtained from comparing the simulation results with measured avalanche ionization rates in 4H-SiC *p-n* junctions lying in the basal plane.[6,58,59] It was discovered that this negative differential drift velocity was responsible for a steep increase of dc bias current (the IMPATT diode was simulated in voltage driving mode) and for additional amplitude modulation of output voltage oscillations. The simulations shown that a low-voltage (74 V dc) 4H-SiC IMPATT diode with total n-n^- base length of 0.25 μm and contact resistivity of $1{\cdot}10^{-6}\Omega{\cdot}cm^2$ can deliver a pulsed power up to 11 W at around 200 GHz with efficiency of 10%. It should be pointed out that these values of diode power and efficiency were significantly overestimated since the model did not include carriers tunneling as well as a temperature dependence of impact ionization coefficients.

The most comprehensive Monte Carlo study of high-field charge carriers transport and impact ionization in 4H-SiC was recently published by M. Hjelm *et al.*[60] Figure 11 shows simulated drift velocity as a function of electric field in the *ΓM* (perpendicular to the *c*-axis) and *ΓA* (along the *c*-axis) directions. At electric fields below 400 kV/cm, these results are in good agreement with the experimental dependencies measured by Khan and Cooper[7],[61] and shown in Fig. 12. The experimental[6,58,59] and simulated[60] impact-ionization coefficients as a function of inverse electric field in the *ΓA* (along the *c*-axis)

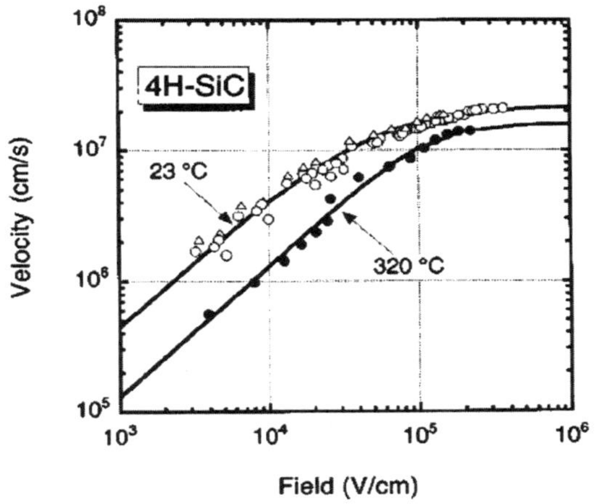

Fig. 12. Measured drift velocity of electrons parallel to the basal plane in (0001) 4H–SiC.[7] (Reprinted with permission, © 2005 IEEE).

direction are shown in Figure 13. This significant progress in simulation and measurements of carrier transport and ionization parameters in 4H-SiC as well as commercial availability of high quality 4*H*-SiC epitaxial layers with satisfactory tolerance in doping level and layer thickness[42] laid the firm foundation for development of SiC IMPATT diodes.

The first microwave oscillation generated by silicon carbide diodes have been observed by Yuan *et al.*[62-64] The hi-lo p^+-n-n^--n^+ 4H-SiC IMPATT diode shown in Fig. 14 was designed using commercial device simulation program Medici.[65] An electron saturation drift velocity of $2{\cdot}10^7$ cm/s was assumed, based on the extrapolation to higher electric fields of experimental values measured perpendicular to the *c*-axis.[7] The diode structure was optimized to operate at frequency of 10 GHz. The epitaxial structures were ordered from Cree Inc.[42] The process steps included mesa isolation by reactive ion etching, top- and back-side ohmic contacts fabrication and overlaying with gold layers.

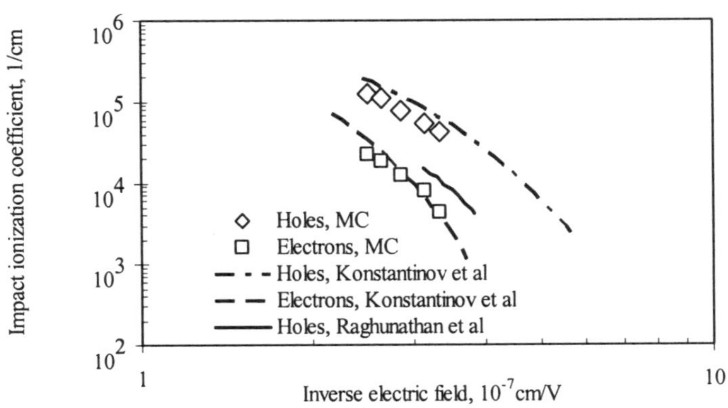

Fig. 13. Experimental and simulated impact ionization coefficients as a function of inverse electric field in the *ΓA* (along the *c*-axis) direction.[60]

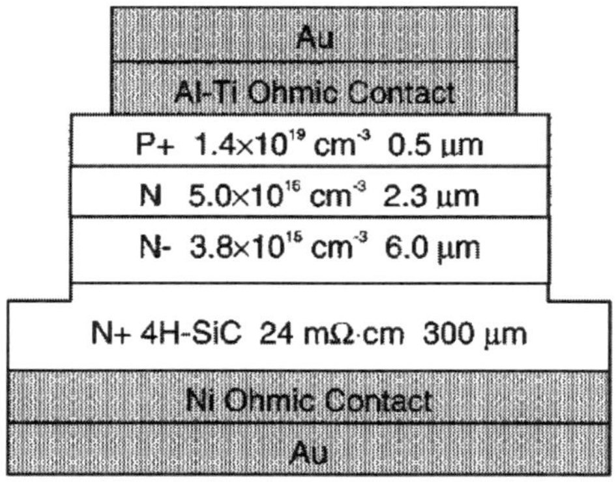

Fig. 14. Cross section of the SiC X-band IMPATT diode.[63] (Reprinted with permission, © 2005 IEEE).

The breakdown voltages of tested diodes ranged from 750 to 950 V due to a nonuniform doping concentration across the wafer. Microwave testing of packaged devices with mesa-structure diameter of 125 μm was performed in a reduced-height X-band waveguide cavity. The diode was biased in pulsed mode with a typical pulse length of 500 - 1000 ns and duty cycle below 0.5%. The microwave oscillations were observed at a frequencies from 7 GHz to 9 GHz depending on cavity tuning.[62] The maximum output power level of 1 mW was occurred at frequency of 7.75 GHz from the diode biased at current amplitude of 200 mA and bias voltage of 1000 V. Figure 15 shows the spectral output of tested SiC IMPATT diode.

Figure 16 shows a waveform of bias current pulse and the spectral output of a SiC diode with 75 μm mesa structure diameter.[64] The diode was operating at the pulsed current with the amplitude of 400 mA corresponding to the input current density of 9 kA/cm^2. The threshold current density of microwave oscillations appearance was found to be of 0.8 kA/cm^2. A different, mostly empirical approach was used by

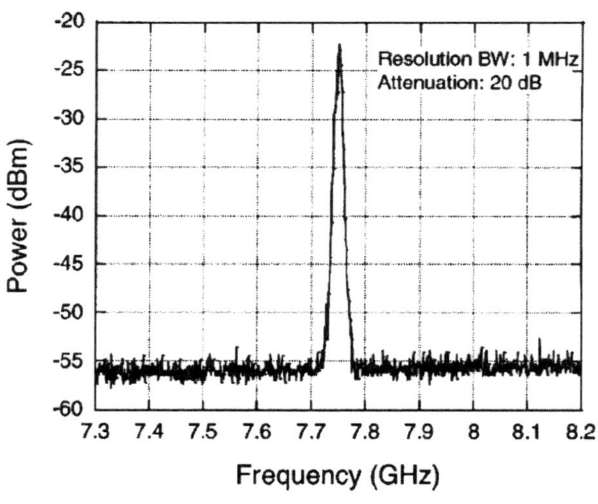

Fig. 15. The spectral output of a SiC IMPATT diode operating in pulsed mode, with 0.5 ms current pulse length and 0.5% duty cycle.[63] (Reprinted with permission, © 2005 IEEE).
The bias current amplitude is 200 mA, the area of the device is 1.227×10^{-4}cm^2.

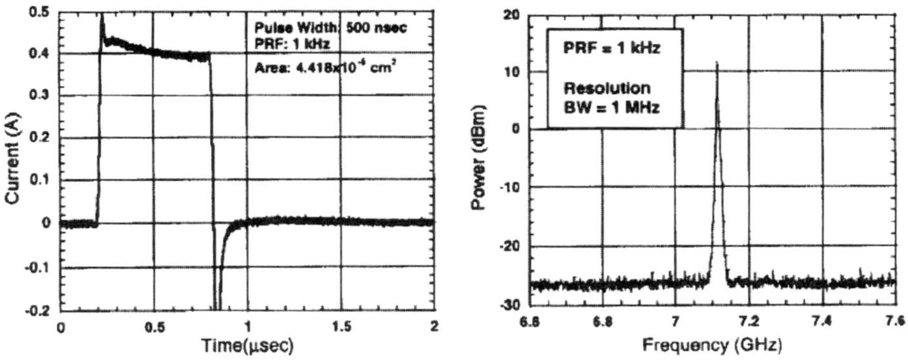

Fig. 16. The waveform of bias current pulse and the spectral output of a 4H-SiC IMPATT diode with 75 μm mesa structure diameter.[64] (Reprinted with permission, © 2005 TTP).

Vassilevski *et al.*[66-68] to demonstrate SiC IMPATT oscillators. A simple p^+-n-n^+ 4H-SiC structure was chosen for IMPATT's fabrication, since the diodes with this structure are less sensitive to the deviations of doping profile. A numerical design was not performed since the goal was to demonstrate the microwave oscillations, but not to reach a high efficiency operation. The only necessary condition which had to be ensured at fabrication of this structure was that the thickness of space charge region (SCR) at avalanche breakdown should be approximately equal to the thickness of n-layer. This was easily fulfilled by the calculation of electric field distribution in the structure with use of avalanche ionization rates measured by Konstantinov *et al.*[6] It was assumed that in this case, the diode shall have negative resistance in a sufficiently wide frequency range. The upper boundary of this range corresponds to the transit-time frequency of the structure and is defined by the value of U_S. The lower boundary of this range corresponds to the avalanche frequency (f_A) of the diode and may be estimated by measurements of a diode noise power spectral density (NPSD) which must exhibit an abrupt decreasing at this frequency.

The 4*H*-SiC p^+-n-n^+ structure with n-base length of 2 μm and doping level of about $1.1 \cdot 10^{17}$ cm^{-3} was chosen for further evaluation. The diodes with this doping profile should ensure the recoverable avalanche breakdown at voltage of about 300 V which may be realized in the diode with open sidewalls of mesa-structure. Commercial 4*H*-SiC p^+-n-n^+ epitaxial wafers from Cree Inc.[42] with minimum available resistivity of the substrate (~0.02 Ω·cm) and acceptor concentration of $8 \cdot 10^{18}$cm^{-3} in the 1 μm thick p^+ layer were used to fabricate the diodes. A rapid thermal annealing was used to form ohmic contacts to p- and n-type 4*H*-SiC. Nickel, 200 nm in thick, annealed at 950°C for 120 sec in vacuum was used as the back ohmic contact. The contacts containing nickel silicides and TiC were used to form ohmic contacts to p- type 4*H*-SiC.[69] They had low resistivity ($\cong 3 \cdot 10^{-5}$Ω·cm^2) and were suitable for further overlay deposition and bonding. Gold overlays, 200 nm thick, were deposited on both sides of the sample after contact formation. The mesa structure (80 μm in diameter) formed by reactive ion etching with no side wall surface passivation was used for edge termination. Finally, the wafer was separated into square chips of 0.6×0.6 mm.

Fig. 17. Microstrip oscillator with 4H-SiC IMPATT diode chip mounted in the break of 50 Ω stripline.[73]

The diodes had the avalanche breakdown voltages of about 300 V, low dynamic impedance of 170 Ω, and series resistivities of $6{\cdot}10^{-5}\Omega{\cdot}cm^2$. They were capable of operating at dc avalanche currents up to 60 mA. To define the transit-time frequency for these diodes, the electron saturated (or average, if negative differential drift velocity take place at high electric field) drift velocity was estimated from the slope of *I-V* characteristics in avalanche region measured in pulse mode of operation.[70,71] The value of U_S =$7.7{\cdot}10^6$cm/s corresponding to f_{tt}=19 GHz was extracted. To define the bottom boundary of the frequency range where the negative dynamic resistance may occur, the NPSD of the fabricated diodes was measured[72] and the value of $f_A \approx 6$ GHz was estimated.

Based on these estimations, the operating frequency of the 4*H*-SiC IMPATT diodes was expected to be from about 6 to 19 GHz. To find the microwave oscillation in this relatively broad band, the diodes were mounted in microstrip oscillators fabricated on duroid RT5880 substrate. The diode chip was mounted directly in the break of 50 Ω stripline as it is shown in Fig. 17. The diodes were biased with dc avalanche current of 100 nA to charge the capacitance of the *p-n* junction. Then, the current pulses with duty factor of 1:700 were applied to the diode. The shape of pulsed current was observed by the oscilloscope and is shown in Fig. 18.

The microwave oscillations appeared at a threshold current density of about 6 kA/cm^2. The frequency of oscillations was in X-band (8.2-12.4 GHz). A pulsed microwave power of about 300 mW was measured at the current of 0.35 A and pulse width of 40 ns. The corresponding power conversion efficiency was about 0.3%. The typical waveform of output power video pulse is shown in Figure 18.

Recently, the most powerful SiC IMPATT diodes were reported by another research group.[74,75] A schematic cross-section of the diode is shown in Figure 19. The single-drift p^+-n-n^+ 4H-SiC diode had a doping profile very similar to that one used by Vassilevski *et al.*[66-68] but the diode fabrication technology was essentially differ. Namely,

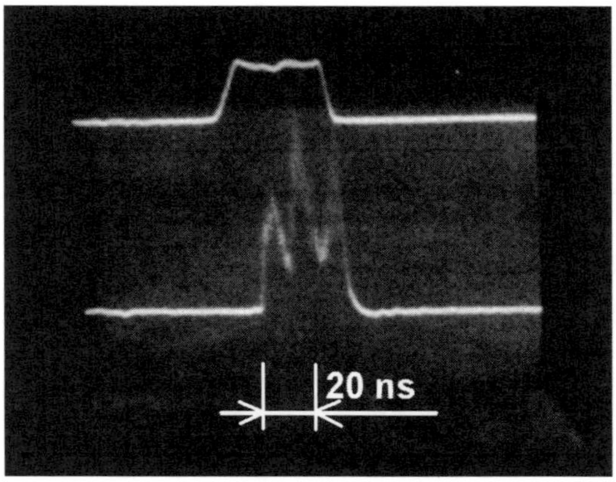

Fig. 18. A typical input current pulse of 0.35 A, 40 ns (top trace) and the corresponding output power video pulse at frequency of 9.9 GHz (bottom trace) of the 4H-SiC IMPATT oscillator.[67]

the p^+ layer was formed by aluminum ion implantation into 2.5 μm-thick epitaxial layer grown on top of a 400 μm-thick n-type 4H-SiC substrate with resistivity of 0.011 Ω·cm. The sample was then annealed at 1700°C in argon for 30 min to activate the impurities. The diodes edge termination was made with a high-resistivity amorphous 0.6 μm-thick layer formed by vanadium ion implantation. Nickel is then deposited on both the cathode and the anode, and the sample is annealed at 1000°C in argon for 2 min to form ohmic contacts. Bonding pads were formed by depositing Ti/Pt/Au on both sides of the top and bottom electrodes. The fabricated IMPATT diodes were diced and mounted on a copper heat-sink in a pill package. The diameters of the diode and Ni contact were 220 and 190 μm respectively.

The diode sample was placed in a waveguide resonant cavity 15.8 mm wide and 7.9 mm high. An impedance transformer and a stub tuner were used to reduce the impedance to the diode. A pulsed voltage with a 500 ns pulse width and 2 ms repetition period was supplied to the cathode through the bias post of the resonant cavity. When the voltage applied to the cathode exceeded 300 V, microwave oscillations were observed at

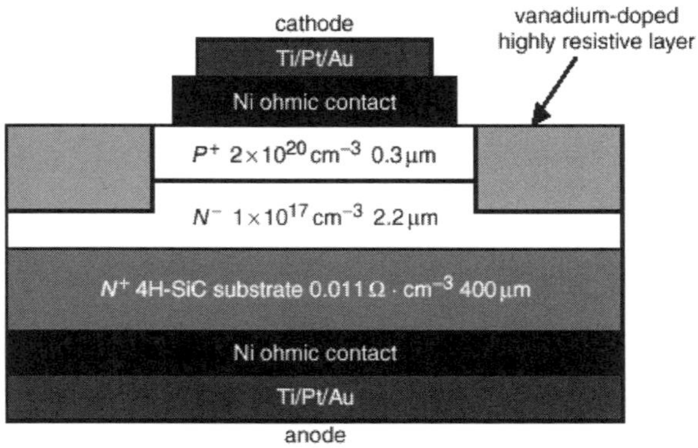

Fig. 19. Schematic cross-section of SiC single-drift IMPATT diode with vanadium doped guard ring.[74] (Reprinted with permission, © 2005 TTP)

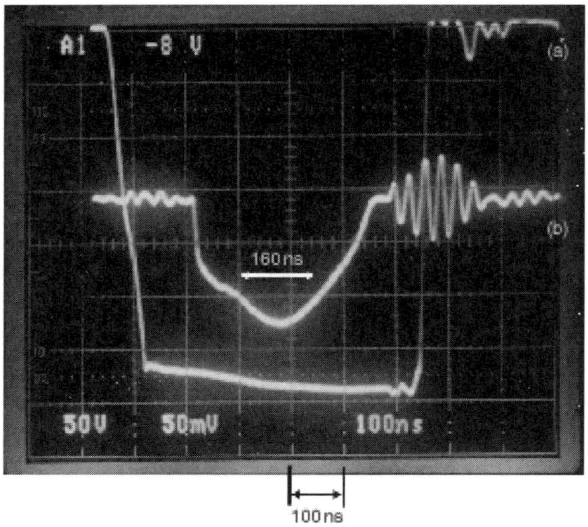

Fig. 20. Waveforms of input voltage pulse and corresponding output power at microwave detector taken from SiC single-drift IMPATT diode with vanadium doped guard ring; (a) input voltage pulse (50 V/div.); (b) video out from detector (a.u.).[74] (Reprinted with permission, © 2005 TTP)

11.93 GHz. The diodes were tested at a lower frequency in an S-band waveguide resonant cavity also to ensure that the observed oscillation was a fundamental oscillation. Figure 20 shows the waveforms of an input voltage pulse and the corresponding video output from a crystal microwave detector. An average output power of 150 mW was measured with a power meter, and the maximum peak output power obtained by dividing the averaged output power by the pulse duty (determined by dividing the 160 ns output power width by the 2 ms repetition) was 1.8 W. The pulsed current through the diode was about 3 A. This value corresponds to the current density of 8 kA/cm^2. The power conversion efficiency was about 0.2%.

All reported diodes had single-drift structures operating in a small signal mode. They were biased with relatively low dc current densities, so as the electron concentrations were much lower than the concentrations of doping impurities in any part of drift regions. In this case, the electric field in the diode structure can be approximated by its static distribution calculated at breakdown voltage without taking into account the charge of mobile carriers. Figure 21 shows the doping profiles and calculated electric field distribution at avalanche breakdown for 4H-SiC Read p^+-n-n^--n^+ structure[63] (Fig. 21.a) and for the p^+-n-n^+ structure with flat doping profile[67] (Fig. 21.b). The line marked by open circles represents the electron drift velocity calculated for the local electric field with use of $U_S(F)$ dependence shown in Fig. 11. It is clearly seen that the U_S value significantly varies in the drift region and hence, the values of f_{tt} and θ can not be defined according to Eq. (33) but need to be calculated by numerical integration of electron transit time through the diode drift region. Calculated values of f_{tt} and θ as well as other diode parameters and characteristics are summarized in the Table 6 for all three structures which generated microwave oscillations.

To compare the diodes with different mesa-structure areas and drift region lengths, we need to compare their efficiencies, specific characteristics (power and current

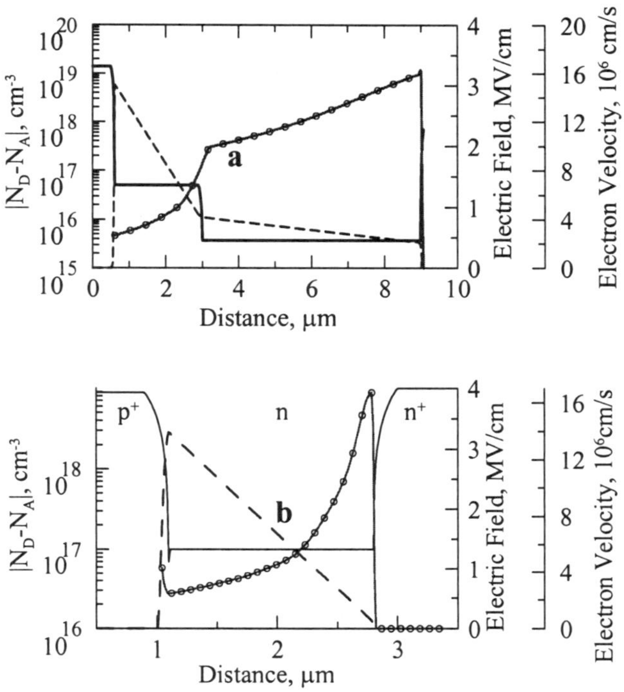

Fig. 21. Doping profile (solid line) and calculated electric field distribution at avalanche breakdown (dashed line). The line marked by open circles represents the local electron drift velocity.
(a) 4H-SiC p^+-n-n^--n^+ diode structure;[63]
(b) 4H-SiC p^+-n-n^+ diode structure.[67]

densities) and transit time angles. It is clearly seen that the diodes with flat doping profiles[67,75] have very similar values of current density, efficiency and transit time angle, despite significant differences in a diode structure fabrication method; device geometry, design, processing and packaging; as well as in the design of microwave oscillator. Furthermore, the measured frequencies of oscillations for these diodes are well consistent with the results of MC simulation since the transit time angles for both structures are very close to π value. On the other hand, the observation of microwave oscillations from the Read structure[63] at frequency of 7.75 GHz corresponds to the transit time angle about 2π. This is not in line either with the experimental results for the single-drift diodes with flat doping profiles[67,75] or with the results of MC simulation of carriers transport at high electric field.[60]

In any case, all 4H-SiC diodes fabricated by several independent research teams employing different device design, fabrication technologies and characterization methods revealed microwave oscillations at frequencies significantly lower than it was expected assuming that the electron saturated drift velocity is about $2\cdot10^7$cm/s. These results confirm the MC calculation[60] and experimental estimation[71] of relatively low drift velocity of the electrons at high electric field in 4H-SiC. This means that the F.O.M. values given in the Table 5 for 4H-SiC are overvalued. More realistic F.O.M. values calculated for 4H-SiC with U_S =$7.7\cdot10^6$cm/s are given in brackets in the Table 5. These

Table 6. Parameters and characteristics of 4H-SiC IMPATT diodes.

Diode doping profile	d, μm	f_{tt}, GHz	f_l, GHz	θ	V_M, V	I_0, A	J_0, kA/cm^2	P, mW	η, %
Read p^+-n-n^--n^+ structure[63]	8.3	4.1	7.75	1.94π	750-950	0.25	2	1	0.0004
flat p^+-n-n^+ structure[67]	2.0	12.3	9.9	0.80 π	290	0.35	7	300	0.3
flat p^+-n-n^+ structure[75]	2.2	12.1	11.93	0.98 π	300	3	8.3	1800	0.2

values show that the fabrication of 4H-SiC IMPATT diodes with operating frequencies higher than for Si diodes is unlikely, but silicon carbide remains to be very attractive material for fabrication of high power IMPATT diodes.

7. Conclusions

Rapid progress in SiC bulk and epitaxial growth technology as well as commercial availability of high-quality epitaxial layers encouraged the fabrication of silicon carbide microwave diodes. The first SiC varactors, Schottky barrier mixers, *p-i-n* and IMPATT diodes have been demonstrated, although their characteristics are still far below the predicted performance. Further improvement of SiC microwave diodes is not restrained by the material quality since the device area of microwave diodes typically does exceed 10^{-4} cm^2 to provide proper matching with microwave load. Nowadays, physical properties of silicon carbide are sufficiently well studied, accurately measured or calculated and this allows a very thorough numerical design of SiC microwave diodes to be performed. It may be concluded that the greatest obstacle in further development of SiC microwave diodes is a retarded state of its post-growth technology. All the diodes described in this paper have been fabricated by simple techniques, and their characteristics can be improved almost in all cases by simple thinning of the substrate, by fabrication of double side ohmic contact to the substrate, or by use of double side heat-sink. Development of low-resistivity ohmic contacts, design of new SiC processing flows which will incorporate a high-temperature treatment, design of substrate thinning methods, membrane and integrated heat-sink technologies are necessary for further progress in the development of SiC microwave diodes.

8. Acknowledgments

This work was supported by the UK Engineering and Physical Sciences Research Council (grant numbers GR/S20420/01 and GR/R96385/01).

Appendix A. List of symbols

A	area of the diode
A^*	effective Richardson constant
C	capacitance
C_J	Schottky barrier or *p-n* junction capacitance
D	ambipolar diffusion coefficient
d	epitaxial layer thickness
D_e	diffusion coefficient of electrons
D_h	diffusion coefficient of holes
E_g	band gap energy
f_1	operating or input frequency
f_2	frequency of second signal at the mixer RF port
f_A	avalanche frequency
f_C	static cut-off frequency
f_{IF}	intermediate frequency
f_{LO}	frequency of local oscillator
F_M	breakdown electric field
f_{tt}	transit-time frequency
I_0	diode dc bias current
I_1	current amplitude at input frequency
I_F	current over the Schottky barrier at forward bias
I_{LO}	current amplitude at LO frequency
J_0	dc current density
J_S	saturation current density
k_B	Boltzmann's constant
K_T	thermal resistance
L_1	mixers parasitic loss
L_C	mixers conversion loss
L_J	mixers intrinsic down-conversion loss
n	ideality factor of Schottky barrier
n	electron concentration
N_A	acceptor concentration
N_D	donor concentration
n_i	intrinsic density
p	hole concentration
P_1	mixers input power
P_{IF}	mixers output power
P_M	maximum microwave power
P_T	thermally dissipated power
q	elementary charge
Q	quality factor of the varactor
R_C	contact resistivity
R_i	resistance of *i*-layer
R_J	non-linear resistance of the metal-semiconductor contact
R_S	diode series resistance
T	temperature
T_M	maximum operating temperature
U_S	electron saturation velocity
V	voltage
V_0	diode dc bias voltage
V_1	amplitude of microwave voltage
V_{LO}	voltage amplitude of local oscillator signal
V_M	breakdown voltage
V_T	tunneling breakdown voltage
W	ambipolar diffusion length
W_e	electron diffusion length
W_h	hole diffusion length
w_{SCR}	width of the space charge region
X_C	capacitive reactance
ΔT	temperature difference between the junction and the heat sink
Λ	$q(nk_BT)^{-1}$
ε	relative dielectric constant
ε_0	permittivity in vacuum
ϕ_{B0}	Schottky barrier height
ϕ_{bi}	the built-in barrier of the junction
γ	C_{max}/C_{min}
λ	thermal conductivity
μ_e	electron mobility
μ_h	hole mobility
η	efficiency
θ	transit time angle
τ	ambipolar lifetime, $\tau_h + \tau_e$
τ_e	electron lifetime
τ_h	hole lifetime
τ_r	time of carrier diffusion through the drift region

9. References

1. M.E. Levinshtein, S.L. Rumyantsev, and M.S. Shur, *Properties of Advanced Semiconductor Materials: GaN, AlN, InN, BN, SiC, SiGe* (John Wiley & Sons, Inc., 2001).
2. E.O. Johnson, Physical limitations on frequency and power parameters of transistors, *RCA Rev.* **26**, 163-177 (1965).
3. R.W. Keyes, Silicon carbide from the perspective of physical limits on semiconductor devices, *Silicon Carbide - 1973, R.C. Marshall, J.W. Faust and C.E. Ryan, ed. (University of South Carolina Press, Columbia)*, 534-541 (1974).
4. M.E. Levinshtein, S.L. Rumyantsev, *Handbook Series on Semiconductor Parameters, vol.1, M. Levinshtein, S. Rumyantsev and M. Shur, ed.* (World Scientific, London), 77-103 (1996).
5. W. von Munch, Silicon Carbide, in *Physik der Elemente der IV Gruppe und III-V Verbindungen, K.-H. Hellwege, Ed.* (Berlin, Heidelberg, Springer Verlag, 1982).
6. A.O. Konstantinov, Q. Wahab, N. Nordell, U. Linderfelt, Ionisation rates and critical fields in 4H-SiC junction devices, *Mater. Sci. Forum* **264-268**, 513-516 (1998).
7. I.A. Khan and J.A. Cooper, Measurement of high-field electron transport in silicon carbide, *IEEE Trans. Electron Devices* **47**(2), 269-273 (2000).
8. M.E. Levinshtein, J.W. Palmour, S.L. Rumyantsev and R. Singh, Forward current–voltage characteristics of silicon carbide thyristors and diodes at high current densities, *Semicond. Sci. Technol.* **13**, 1006–1010 (1998).
9. P. Grivickas, A. Galeckas, J. Linnros, M. Syvajarvi, R. Yakimova, V. Grivickas, J.A. Tellefsen, Carrier lifetime investigation in 4H–SiC grown by CVD and sublimation epitaxy, *Mater. Sci. in Semiconductor Processing* **4**, 191–194 (2001).
10. P. Grivickas, J. Linnros and V. Grivickas, Free carrier diffusion measurements in epitaxial 4H-SiC with a fourier transient grating technique: injection dependence, *Mater. Sci. Forum* **338–342**, 671-674 (2000).
11. A.S. Tager, Utmost parameters of semiconductor microwave devices and their relations to properties of semiconductor materials, *Izvestiya VUZov SSSR* **22**(10), 56-16 (1979) [in Russian].
12. S.M. Sze, *Physics of Semiconductor Devices* (Wiley, New York, 1981) p. 513.
13. H.H.C. Dunwoody, *Wireless telegraph system, US Patent US837616*, (1906).
14. D.Thackeray, When tubes beat crystals: early radio detectors, *IEEE Spectrum* (March), 64-69 (1983).
15. H.V. Shurmer, *Microwave Semiconductor Devices* (Pitman, London, 1971).
16. N.G. Wright, A.P. Knights, A.G. O'Neill, C.M. Johnson, Evaluation of 4H-SiC varactor diodes for microwave applications, *Proceedings of EDMO-99, IEEE Symposium on High Performance Electron Devices for Microwave and Optoelectronic Applications*, 301-306 (1999).
17. K.V. Vassilevski, A.V. Zorenko, Silicon carbide varactor operating at frequency about 140 GHz, *Int. Seminar Semiconductor Silicon Carbide and SiC-Based Devices, Book of Abstracts, V.E. Udaltsov, ed. Novgorod State University*, 66-67 (1995).
18. V.A. Dmitriev, Silicon carbide and SiC-AlN solid-solution p-n structures grown by liquid-phase epitaxy, *Physica B* **185**, 440-452 (1993).
19. C.H. Carter, V.F. Tsvetkov, R.C. Glass, D. Henshall, M. Brady, S.G. Muller, O. Kordina, K. Irvine, J.A. Edmond, H.-S. Kong, R. Singh, S.T. Allen, J.W. Palmour, Progress in SiC: from material growth to commercial device development, *Mater. Sci. Eng., B* **61**, 1-8 (1999).
20 C. Arnodo, S. Tyc, F.Wyczisk and C. Brylinski, Nickel and molybdenum ohmic contacts to silicon carbide, *Inst. Phys. Conf.* **142**, 577 (1996).
21. K. Vassilevski, K. Zekentes, G. Constantinidis, I. Nikitina, Phase formation at rapid thermal annealing of Al/Ti/Ni ohmic contacts on 4H-SiC, *Mater. Sci. Eng., B* **80**(1-3), 370-373 (2001).
22. E.H. Rhoderick, R.H. Williams, *Metal-Semiconductor Contacts, 2nd ed.* (Clarendon Press, Oxford, 1988).
23. A.A.M. Saleh, Theory of resistive mixers, PhD. thesis, Massachusetts Institute of Technology (1970).
24. R.G. Harrison, Detectors and mixers, in: *Microwave Solid State Circuit Design, I. Bahl, P. Bhartia, ed.* (John Wiley & Sons, Inc., 2003).
25. S.A. Maas, Two-tone intermodulation in diode mixers, *IEEE Trans. Microwave Theory Tech.* **35**(3), 307-314 (1987).

26. A. Itoh, H. Matsunami, Analysis of Schottky barrier heights of metal/SiC contacts and its possible application to high-voltage rectifying devices, *Phys. Status Solidi A* **162,** 389-408 (1997).
27. K. Chang, *Microwave Solid-State Circuits and Applications* (John Wiley & Sons, Inc., 1994) p. 157.
28. C. Fazi and P. Neudeck, Wide dynamic range RF mixers using wide-bandgap semiconductors, *IEEE MTT-S International Microwave Symposium Digest* **1**, 49-51 (1997).
29. C. Fazi and P. Neudeck, Wide dynamic range RF mixers using wide-bandgap semiconductors, *Mater. Sci. Forum* **264-268,** 913-916 (1998).
30. J. Eriksson, F. Ferdos, H. Zirath and N. Rorsman, Design and characterization of a SiC Schottky diode mixer, *Mater. Sci. Forum* **338-342,** 1207–1210 (2000).
31. J. Eriksson, N. Rorsman and H. Zirath, Microwave silicon carbide Schottky diodes, *Electron. Lett.* **37**(4), 250–252 (2001).
32. H. Zirath, Design and fabrication of SiC microwave diodes and field-effect transistors, with applications, *ESCIES Wide Bandgap Technology Meeting, ESTEC, 17- 18th of October, 2001,* (https://escies.org/public/widebandgap/meeting2001/)
33. J. Eriksson, N. Rorsman, F. Ferdos, and H. Zirath, Design and characterization of a singly balanced silicon carbide Schottky diode high-level mixer, *Electron. Lett.* **37**(1), 54–55 (2001).
34. J. Eriksson, N. Rorsman, H. Zirath, 4H-silicon carbide Schottky barrier diodes for microwave applications, *IEEE Trans. Microwave Theory Tech.* **51**(3), 796-804 (2003).
35. I. Kneppo, J. Fabian, P. Bezousek, P. Hrnicko, M. Pavel, *Microwave Integrated Circuits* (Chapman & Hall, London, 1994) p. 246.
36. R.N. Simons, P.G. Neudeck, Intermodulation-distortion performance of silicon carbide Schottky barrier RF mixer diodes, *IEEE Trans. Microwave Theory Tech.* **51**(2), 669-672 (2003).
37. H.M. Olson, p-i-n diodes, *Microwave Semiconductor Devices and Their Circuit Applications, H.A. Watson, ed.* (McGraw-Hill Book Company, 1969).
38. A.V. Bludov, N.S. Boltovets, K.V. Vasilevskii, A.V. Zorenko, K. Zekentes, V.A. Krivutsa, T.V. Kritskaya, A.A. Lebedev, Microwave switches based on 4H-SiC p-i-n diodes, *Tech. Phys. Lett.* **30**(2), 123-125 (2004).
39. K.V. Vassilevski, P.B. Gamuletskaya, A.V. Kirillov, A.A. Lebedev, L.P. Romanov, V.A. Smirnov, Experimental study of SiC p-i-n diodes in the 3-cm range, *Semiconductors* **38**, 237- 238 (2004).
40. A.V. Bludov, M.S. Boltovets, K.V. Vassilevski, A.V. Zorenko, K. Zekentes, A.A. Lebedev, V.A. Krivutsa, Simulation and prototype fabrication of microwave modulators with 4H-SiC p-i-n diodes, *Mater. Sci. Forum* **457-460**, 1089-1092 (2004).
41. K. Vassilevski, K. Zekentes, G. Constantinidis, A. Strel'chuk, Fabrication and electrical characterization of 4H-SiC p^+-n-n^+ diodes with low differential resistance, *Solid-State Electron.* **44**(7), 1173-1177 (2000).
42. SiC substrates and epitaxy, *Product Specification, Cree, Inc.* (http://www.cree.com)
43. W.T. Read, A proposed high-frequency negative resistance diode, *Bell Syst. Tech. J.* **37**, 401 (1958).
44. A.S. Tager, A.I. Melnikov, G.P. Kobelkov, A.M. Tsebiev, *USSR author certificate No. 185965 and diploma on scientific discovery No. 24, priority of 27.X.1959.* [in Russian].
45. A.S. Tager, Perspectives of using of some semiconductor material for increasing of IMPATT diodes power, *Electronika SVCh* (12), 16-22 (1974) [in Russian].
46. S.M. Sze, *Physics of Semiconductor Devices* (Wiley, New York, 1981) pp. 98 and 522.
47. W. Muench, E. Pettenpaul, Saturated electron drift velocity in 6H silicon carbide, *J. Appl. Phys.* **48**(11), 4823-4825 (1977).
48. A.O. Konstantinov, Influence of temperature on impact ionization and avalanche breakdown in silicon carbide, *Sov. Phys. Semicond.* **23**(1), 31-34 (1989).
49. I. Mehdi, G.I. Haddad, R.K. Mains, Microwave and millimeter-wave power generation in silicon carbide avalanche devices, *J. Appl. Phys.* **64**(3), 1533-1540 (1988).
50. A.P. Dmitriev, A.O. Konstantinov, D.P. Litvinov, V.I. Sankin, Impact ionization and superlattice in 6H–SiC, *Sov. Phys. Semicond.* **17**(6), 686-689 (1983).
51. K.V. Vassilevski, Calculation of dynamic characteristics of silicon carbide IMPATT diode, *Sov. Phys. Semicond.* **26**(10), 994-999 (1992).

52. R.P. Joshi, S. Pathak and J.A. Macadoo, Hot-electron and thermal effects on the dynamic characteristics of single-transit SiC impact-ionisation avalanche transit-time diodes, *J. Appl. Phys.* **78,** 3492–3497 (1995).
53. R. Mickevicius and J.H. Zhao, Monte Carlo study of electron transport in SiC, *J. Appl. Phys.* **83,** 3161-3167 (1998).
54. E. Belotti, H. Nilsson, K. Brennan and P.P. Ruden, Ensemble Monte Carlo calculation of hole transport in bulk 3C-SiC, *J. Appl. Phys.* **85**, 3211–3217 (1999).
55. V. Gruzinskis, Y. Luo, J.H. Zhao, M. Weiner, M. Pan, P. Shiktorov and E. Starikcv, Monte Carlo simulation of 4H-SiC IMPATT diodes, *Mater. Sci. Forum* **338-342**, 1379-1382 (2000).
56. J.H. Zhao, V. Gruzinskis, Y. Luo, M. Weiner, M. Pan, P. Shiktorov and E. Starikov, Monte Carlo simulation of 4H-SiC IMPATT diodes, *Semicond. Sci. Technol.* **15**, 1093–1100 (2000).
57. B.K. Ridley, Soft-threshold lucky drift theory of impact ionization in semiconductors, *Semicond. Sci. Technol.* **2**, 116-122 (1987).
58. A.O. Konstantinov and N. Nordell, Temperature dependence of avalanche breakdown for epitaxial diodes in 4H silicon carbide, *Appl. Phys. Lett.* **73**, 1850–1852 (1998).
59. R. Raghunathan and B.J. Baliga, Temperature dependence of hole impact ionization coefficients in 4H and 6H-SiC, *Solid-State Electron.* **45**, 199-211 (1999).
60. M. Hjelm, H-E. Nilsson, A. Martinez, K.F. Brennan, E. Bellotti, Monte Carlo study of high-field carrier transport in 4H-SiC including band-to-band tunneling, *J. Appl. Phys.* **93**(2), 1099-1107 (2003).
61. I.A. Khan, J.A. Cooper, Measurements of high field electron transport in silicon carbide, *Mater. Sci. Forum* **264-268**, 509-512 (1998).
62. L. Yuan, M.R. Melloch, J.A. Cooper and K.J. Webb, Silicon carbide IMPATT oscillators for high-power microwave and millimeter-wave generation, *Proc. IEEE/ Cornell Conf. High Performance Devices,* 158-167 (2000).
63. L. Yuan, J.A. Cooper, M.R. Melloch, K.J. Webb, Experimental demonstration of a silicon carbide IMPATT oscillator, *IEEE Electron Device Letters* **22**(6), 266-268 (2001).
64. L. Yuan, J.A. Cooper Jr., Kevin J. Webb, M.R. Melloch, Demonstration of IMPATT diode oscillators in 4H-SiC, *Mater. Sci. Forum* **389-393**, 1359-1362 (2002).
65. *Medici Two-Dimensional Device Simulation Program User's Manual* (Technology Modeling Associates, Inc., Sunnyvale, 1996).
66. K.V. Vassilevski, A.V. Zorenko, K. Zekentes, X-band silicon carbide IMPATT oscillator, *Mat. Res. Soc. Symp. Proc.* **680E**, E10.11.1-E10.11.6 (2001).
67. K.V. Vassilevski, A.V. Zorenko, K. Zekentes, Experimental observation of microwave oscillations produced by pulsed silicon carbide IMPATT diode, *Electron. Lett.* **37**(7), 466-467 (2001).
68. K.V. Vassilevski, A.V. Zorenko, K. Zekentes, K. Tsagaraki, E. Bano, C. Banc, A.A, Lebedev, 4H-SiC IMPATT diode fabrication and testing, *Mater. Sci. Forum* **389-393**, 1353-1358 (2002).
69. K. Vassilevski, K. Zekentes, G. Constantinidis, I. Nikitina, Phase formation at rapid thermal annealing of Al/Ti/Ni ohmic contacts on 4H-SiC, *Mater. Sci. Eng., B* **80**(1-3), 370-373 (2001).
70. K. Vassilevski, K. Zekentes, A. Zorenko, L. Romanov, Isothermal I-V characteristics of 4H-SiC p-n diodes with low series differential resistivity at avalanche breakdown, *Mat. Res. Soc. Symp.* **622**, T1 8.1-T1.8.6 (2000).
71. K.V. Vassilevski, K. Zekentes, A.V. Zorenko, L.P. Romanov, Experimental determination of electron drift velocity in 4H-SiC p^+-n-n^+ avalanche diodes, *IEEE Electron Device Letters* **21**(10), 485-488 (2000).
72. K.V. Vassilevski, K. Zekentes, E. Bogdanova, A.V. Zorenko, Electrical characterization of 4H-SiC p^+-n-n^+ Avalanche Diodes, *Technical Program with Abstracts of 24th Workshop on Compound Semiconductor Devices and Integrated Circuits* (Agean Sea, Greece, 2000) VIII-19 - VIII-20.
73. K.V. Vassilevski, SiC IMPATT diode, PhD. thesis, A.F. Ioffe Physico-Technical Institute (2002) [in Russian].
74. S. Ono, M. Arai and C. Kimura, Demonstration of high-power X-band oscillation in p^+/n-/n^+ 4H-SiC IMPATT diodes with guard ring termination, *Mater. Sci. Forum* **483-485**, 981-984 (2005).
75. M. Arai, S. Ono and C. Kimura, IMPATT oscillation in SiC p^+-n^--n^+ diodes with a guard ring formed by vanadium ion implantation, *Electron. Lett.* **40**(16), (2004).

International Journal of High Speed Electronics and Systems
Vol. 15, No. 4 (2005) 931–996

SiC THYRISTORS

M. E. LEVINSHTEIN and S. L. RUMYANTSEV*

Solid State Electronics Division, The Ioffe Physical-Technical Institute of Russian Academy of Sciences, 194021, St. Petersburg, Russia

T. T. MNATSAKANOV

All-Russian Electrotechnical Institute, 111250, Moscow Russia

ANANT K. AGARWAL and JOHN W. PALMOUR

CREE Inc., 4600 Silicon Dr., Durham NC 27703, USA

Keywords: Silicon Carbide; thyristor; electron-hole scattering; transient processes; computer simulation.

1. Introduction

Thyristors made with Silicon Carbide (SiC) demonstrate great performance advantages over those made with Si or GaAs because of SiC's higher breakdown electric field, carrier saturation velocities and thermal conductivity than either Si or GaAs. A high breakdown electric field allows the design of SiC thyristors with thinner and higher doped base layers than identically rated Si or GaAs thyristors. Such devices show fast switching and low residual voltage drop at very high current densities. A high thermal conductivity should allow higher current density operation of SiC thyristor. The higher bandgap of SiC results in extremely high intrinsic temperature of about 1920 K (for 4H-SiC) for an extrinsic doping of 10^{16} cm^{-3}. Another advantage of high bandgap is negligible leakage currents up to 500-550K. Based on the above arguments SiC thyristors could find use in ultra high-voltage and high current applications with critical constrains on the size, weight, and elevated temperature characteristics of the power unit, such as motor controls for heavy electric vehicles and electric railways.

The first SiC dynistor was demonstrated in late 80-ies [1] . 6H-SiC [2] and 4H-SiC [3] thyristors have been demonstrated in early 90-ies. Presently, all main parameters of SiC thyristors have been investigated rather thoroughly.

*also with Department of Electrical, Computer, and Systems Engineering, CII 9017, Rensselaer Polytechnic Institute, Troy NY 12180-3590

The Chapter is organized as follows. In Section 2 we will discuss the turn-on processes in the thyristors. Section 3 is devoted to steady-state current-voltage characteristics. In Section 4 we will consider the peculiarities of the turn-off performances of silicon-carbide thyristors. Frequency properties of SiC thyristors will be discussed in Section 5. Section 6 is devoted to the peculiarities of the critical charge in silicon carbide thyristors. The main results will be summarized in the Conclusion.

2. Turn-on process in the thyristors

Turn-on processes in the thyristors depend appreciably on the width of blocking base, i.e. on the breakover voltage V_b. The switch-on processes in 4H-SiC thyristors with $V_b \leq 400$ V were investigated in Refs. [4,5,6].

2.1. *Turn-on process in low-voltage thyristor structures*

The device structure of low-voltage p^+np^-n 4H-SiC thyristor with breakover voltage $V_b \sim 400$ V is shown in Fig. 1.

The voltage blocking p^- layer was 4.5 μm thick and had an acceptor doping concentration of 2.8×10^{16} cm^{-3}. The n-base had a thickness of 0.55 μm and a doping of 4.5×10^{17} cm^{-3}. The operation area of the device was 3.6×10^{-4} cm^2. The devices were turned on by 3-200 ns gate pulse. The load resistance (R_L) connected in series with the thyristor was varied in the 1.1-50 Ω range to obtain the desired

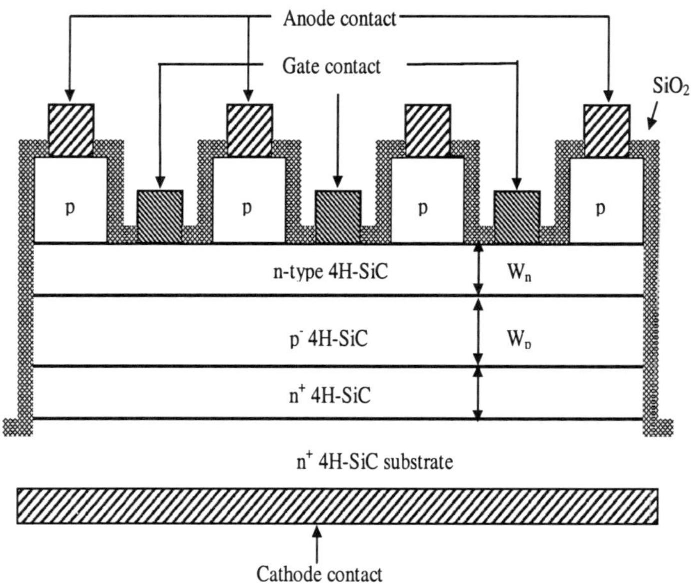

Fig. 1. Cross-section of low-voltage p^+np^-n 4H-SiC thyristor [6].

operating current density.

Over all range of initial bias V_0 and current density j up to $j \sim 10^4$ A/cm^2, the time dependence of the current I in turn-on process was described well by conventional formula:

$$I(t) = I_{\max}[1 - (\exp(-t/\tau_r)] \tag{1}$$

where τ_r is the current rise time constant.

However, unlike Si and GaAs thyristors, the τ_r in low-voltage SiC thyristors at 300 K did not depend on initial (blocking) bias V_0 (Fig. 2).

Si and GaAs thyristors (curves 2 and 3) with comparable breakover voltages V_b have been investigated in Ref. [7]. These devices had a voltage blocking n-drift layer thickness of 50-60 μm and the p-base layer thickness of 8-10 μm. The high voltage ($V_b \approx 1.5$ kV) Si thyristor had an n-drift layer thickness of approximately 500 μm and a p-base thickness of 40 μm [8]. As seen in Fig. 2, the rise time constants of Si and GaAs thyristors decrease monotonically with V_0 increase. On the opposite, the τ_r in SiC thyristors does not depend practically on V_0.

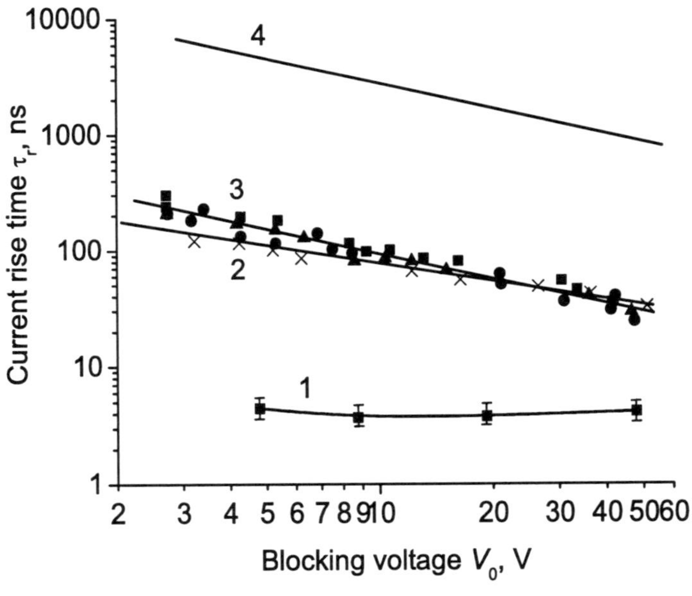

Fig. 2. The dependencies of the current rise time τ_r on the initial (blocking) voltage V_0 for various thyristor structures. 1 – low-voltage 4H-SiC thyristor; 2 - Si thyristor, 3 – GaAs thyristor, 4 - Si "thick-base "thyristor (W_n=500 μm). 300K [4]

It is usually assumed that for a small V_0 bias (about twice the built-in voltage of the $p-n$ junction), a thyristor switches on by the diffusion process [9,10]. (The minimum turn-on voltage is about 2 V for Si, 3 V for GaAs and 5 V for SiC thyristors). For the diffusion mechanism of the turn-on process, the rise time constant τ_r can be estimated by the simplified expression

$$\tau_r = (\Theta_1 \Theta_2)^{1/2} \tag{2}$$

where

$$\Theta_1 = \frac{W_1^2}{2D_2}; \qquad \Theta_2 = \frac{W_2^2}{2D_1} \tag{3}$$

The subscript "2" represents the thick blocking base. Subscript "1" represents the thin high doped base. The constant "D" is the diffusion coefficient of corresponding carriers, electrons or holes.

For the 400 V Si thyristor, assuming a hole diffusion constant of 10 cm^2/s for low doped thick n-base and electron diffusion constant of 15 cm^2/s for high doped p-base (where $N_a = 1 \times 10^{17}$ cm^{-3}), the τ_r is approximately 200 ns. For the GaAs thyristor, assuming a hole diffusion coefficient of 10 cm^2/s and electron diffusion coefficient of 50 cm^2/s for high doped base layer, the τ_r is approximately 100 ns. For the 1.5 kV Si thyristor, the τ_r estimated according to Eq. (2) is approximately 7 μs. A comparison of these estimates with the data in Fig. 2 shows a reasonable agreement.

For the SiC thyristor in question, assuming electron diffusion coefficient of 10 cm^2/s for low doped p-base and hole diffusion coefficient of 1.5 cm^2/s for highly doped n-base, the rise time constant is estimated to be $\sim$ 3 ns. Thus, the turn-on mechanism in low-voltage SiC thyristors is dominated by diffusion. Very small values of rise time constant are due to small thickness of the SiC thyristor base layers.

Figure 3 shows the time dependencies of the voltage across the load resistance $R_L = 50\ \Omega$ during turn-on process for different temperatures and for a forward bias $V_0 = 30$ V. Note that maximum value of V_L corresponds to the current $I_{max} \approx 0.5$ A and current density $j_{max} \approx 1400$ A/cm^2.

It can be seen from Fig. 3 that the turn-on process is strongly temperature dependent, especially at $T \leq 300$ K. The total turn-on time is $\sim$ 180 ns at 160 K and only $\sim$ 10 ns at 495 K. The inset in Fig. 3 shows the temperature dependence of current rise time constant τ_r. It is seen that τ_r decreases monotonically with increasing temperature from $\tau_r = 63$ ns at $T = 160$ K to $\tau_r = 1.9$ ns at $T = 495$ K. It should be noted that such a strong temperature dependence of τ_r has not been observed in either Si or in GaAs thyristors. As mentioned in Ref. [5], it is unlikely that this dependence can be explained by a fall in lifetime of minority carriers in SiC with decreasing temperature [11]. As is well known, even a very strong several fold decrease in minority carrier lifetime has barely any effect at reducing τ_r [12].

As mentioned above, the τ_r is practically bias independent at 300 K (Fig. 2). Just the same situation was observed in the high temperature region $T > 300$ K. In contrast, at low temperatures $T < 250$ K, the dependence of τ_r on the bias V_0 had a usual form: τ_r decreases monotonically with increasing bias. Decreasing τ_r with V_0 growth was observed also in high-voltage SiC thyristors at any temperature.

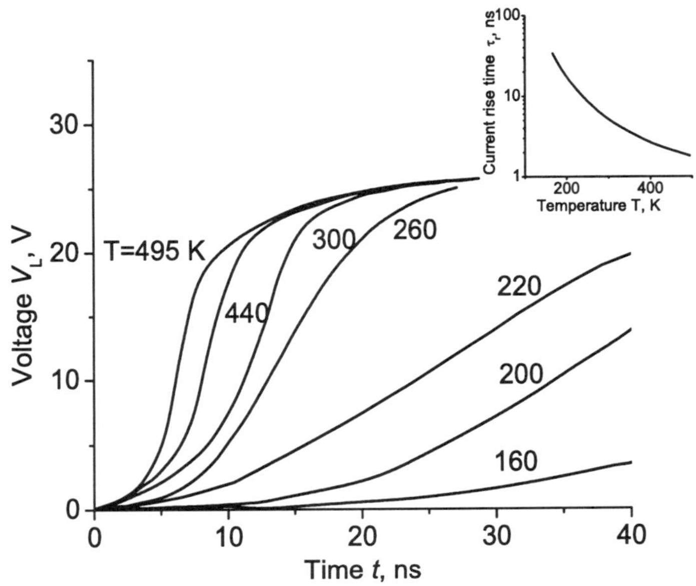

Fig. 3. Time dependences of voltage across the load resistor $R_L = 50\ \Omega$ during the turn-on process at different temperatures. $V_0 = 30$ V. Inset shows the temperature dependence of the time constant of current rise τ_r [5].

2.2. *Turn-on process in high-voltage thyristor structures*

Turn-on processes in 4H-SiC thyristors with a breakover voltage close to 700 V and pulse switch current density of 16 000 A/cm^2 were studied in Ref. [13]. The $p-$voltage blocking layer (p-base) had a thickness W_p of 11 μm. The n-base had a thickness W_n of 0.65 μm. The operation area of the device S was about 6×10^{-3} cm^2. The turn-on process was investigated in 50 Ω circuit with low parasitic inductance. The time resolution of the circuit was better than 0.5 ns.

Figure 4 shows the time dependences of V_L/V_0 at 300 K for different values of the bias voltages V_0 during the turn-on process. Here V_L is the voltage drop across the series load resistance $R_L = 50\ \Omega$. At the end of the turn-on process the steady state current density j_0 is equal approximately $j_0 \approx V_0/R_LS$. Hence, $j_0 \approx 1.5\times10^3$ A/cm^2 at $V_0 = 450$ V (curve 6 in Fig. 4) and $j_0 \approx 270$ A/cm^2 at $V_0 = 80$ V (curve 1, Fig. 4). It can be seen from Fig. 4) that current rise time τ_r decreases sharply with increasing bias at $V_0 \leq 200$ V. On the contrary, at $V_0 > 200$ V, both total time of the current rise and current rise time constant τ_r are practically independent of bias V_0. It should be noted that the $\tau_r(V_0)$ dependencies for these 700 V thyristors differ fundamentally from the $\tau_r(V_0)$ dependencies for low-voltage 400 V thyristors described in Section 2.1.

In low-voltage thyristors the τ_r is equal to $\sim 3-4$ ns and it does not depend on bias V_0 over the whole range of bias V_0 from $V_0 \approx 5$ V to $V_0 \approx 150$ V. As

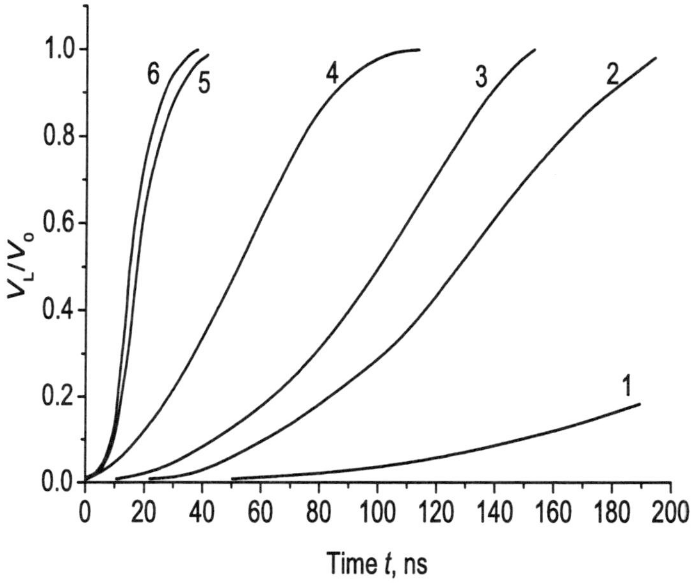

Fig. 4. Time dependencies of the voltage across the load resistor, V_L, at different voltages V_0. $T =$ 300 K. V_0 (V): 1 - 80, 2 - 100, 3 - 150, 4 - 180, 5 - 200, 6 – 450 (Ref. [13]).

mentioned, estimate of τ_r according to the diffusion model (formula (2)) shows a good agreement with experimental τ_r value for 400 V thyristors. On the contrary, strong dependence of τ_r on V_0 for 700 V thyristors demonstrates very clear the contribution of the field mechanism into the turn-on process [10].

Figure 5 presents the time dependencies of V_L for different temperatures at $V_0 = 200$ V. It can be seen that both total current rise time and τ_r decrease monotonically with increasing temperature. At $T = 500$ K $\tau_r = 1.2$ ns. This is the minimum value of τ_r observed in SiC thyristors. As mentioned above, the strong bias dependence of τ_r (Fig. 4) demonstrates that the turn-on mechanism in 700 V SiC thyristors is dominated by field process. Hence, one could expect that τ_r will increase with increasing temperature because the carrier mobility of carriers and the diffusion coefficient decrease monotonically with temperature growth over the whole temperature range 300 K - 500 K. However the opposite situation is observed experimentally (Fig. 5)).

The temperature dependencies of turn-on process in 2.6 kV 4H-SiC thyristors fabricated by CREE Inc. [14] were studied in Refs. [15,16]. A cross-sectional view of the thyristor structure is shown in Fig. 6.

Five epilayers were grown on 380 μm 8° off-axis 4H-SiC n-type substrates with resistivity of 0.02 Ω·cm. The blocking p^-(base) layer was 50 μm thick, doped to

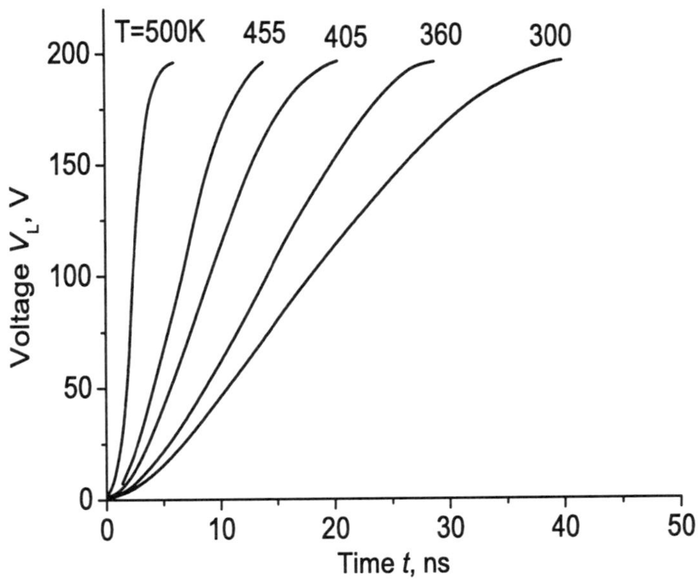

Fig. 5. Time dependencies of V_L for different temperatures at $V_0 = 200$ V. $R_L = 50\ \Omega$ [13].

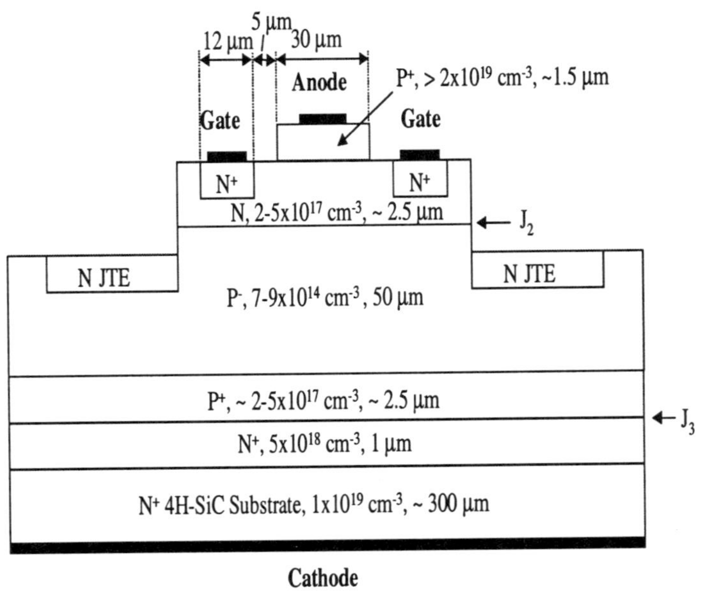

Fig. 6. Cross-sectional view of 2.6 kV thyristor structure [15].

around 7×10^{14} cm^{-3}. The proper injection efficiency of the p^{++}(anode)-n(base) junction was provided by very heavy doping of the p^{++}-layer to 1×10^{19} cm^{-3}. It is worth noting that, owing to the relatively large ionization energy of Al in SiC ($\sim$ 0.24 eV), only about 2 percent of Al atoms are ionized at room temperature. As a result, the concentration of holes in the p^{+}-emitter grows exponentially with temperature in the range from 300 to 450 K. The blocking junction was terminated with a junction termination extension formed by nitrogen implantation. The anode-gate configuration of the 1 mm $\times$1 mm device had an interdigitated geometry. There were 19 anode fingers, each 654 μm long, with the total anode area S of 3.7×10^{-3} cm^2.

Figure 7 shows the time dependencies of the current density during the turn-on process in a 2.6 kV SiC thyristor at different temperatures.

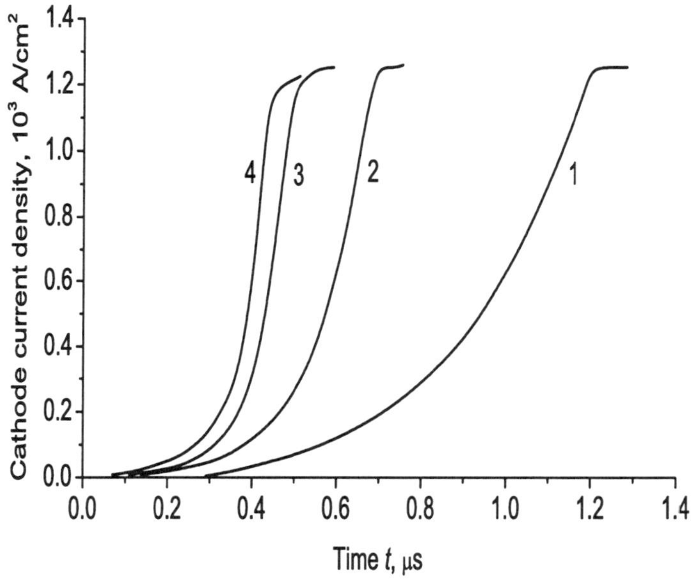

Fig. 7. Time dependencies of the current density during the turn-on of a 2.6 kV SiC thyristor at different temperatures. T (K): 1 – 293; 2 – 330; 3 – 379; 4 - 404. Cathode voltage V_0 = 200 V, load resistance $R_L = 50\Omega$ [15].

It can be seen that, just like in relatively low-voltage thyristors (Figs. 3 and 5), the turn-on process is strongly temperature dependent. The total turn-on time is a 1.2 μs at 293 K and only 0.4μ sec at 404 K. (It can also be seen that at $T > 380$ K the temperature dependence of the turn-on process tends to saturate).

The qualitative explanation of the fact that the turn-on time decreases with increasing temperature was proposed in Ref. [17]. It was supposed that the effect results from a very strong temperature dependence of the hole concentration in the p^{++}-emitter of a SiC thyristor. As mentioned, at room temperature, only 1–2 percent of Al atoms introduced into the p^{++}-emitter in a concentration of (1–2)$\times 10^{19}$cm^{-3} are ionized. With increasing temperature, the hole concentration in the p^{++}-emitter grows. Accordingly, the injection coefficient of the $p^{++} - n$ junction γ_1 increases, and, as a consequence, the common-base current gain of the $p^{++} - n - p^{-}$ transistor section of the thyristor becomes higher. However, more detailed interpretation of this effect in [17] is contradictory because it was based on the expression derived in [18], which is valid only for $\gamma_1 = \gamma_2 = 1$:

$$\tau_r = 2\left(\frac{\tau_1\tau_2}{\alpha_1 + \alpha_2 - 1}\right)^{\frac{1}{2}}, \tag{4}$$

where τ_1 and τ_2 are the minority carrier transit times, $\alpha_1 \equiv \gamma_1\alpha_{T1} = \alpha_{T1}$ and $\alpha_2 \equiv \gamma_2\alpha_{T2} = \alpha_{T2}$, ($\gamma_1 = \gamma_2 = 1$) are the gain coefficients in the common-base configuration of the top ($p^{++} - n - p$) and bottom ($n - p - n^{+}$) transistors, respectively (see Fig. 6). Here, γ_1 and γ_2 are the injection coefficients of the top $p^{++} - n$ and bottom $n^{+} - p$ emitter-base junctions, and α_{T1} and α_{T2} are the transport factors for the top and bottom transistors, respectively.

For silicon thyristors approximation $\gamma_1 = \gamma_2 = 1$ is quite reasonable for two reasons. First, the ionization energy of shallow donors and acceptors in Si is rather low (0.05–0.07 eV [19]). As a result, all these shallow levels, determining the equilibrium electron and hole concentrations in the bases and emitters of the structures, are fully ionized even at temperatures much lower than room temperature. Second, the working current density in Si thyristors (typically 50–100 A/cm^2) is too small for taking into account the decrease in the injection coefficients with increasing current density. Hence, Equation (4) can be applied to compare Si transistors with different α_T, or to analyze the case when the α_T values are temperature dependent. However, as mentioned above, it is the temperature dependence of γ_1 that is the reason for the decrease in τ_r in SiC thyristors. Hence, the use of Eq. (4) is incorrect in the given situation.

To obtain an expression for the current rise time constant τ_r, taking into account the deviation of the injection coefficients from unity, the charge control model can be used [20,21]. In the framework of this model, the time dependences of the hole charge in the n-base, $Q_1 = q\int\limits_{V_n} pd^3V$, and the electron charge in the p-base, $Q_2 = q\int\limits_{V_p} nd^3V$, are described by the following equations [10]:

$$\frac{dQ_1}{dt} = -\left[(1-\gamma_1)k_p + \frac{1}{\tau_p}\right]Q_1 + \gamma_1 k_n Q_2 + \gamma_1(I_{k0} + I_G - I_R)$$

$$\frac{dQ_2}{dt} = \gamma_2 k_p Q_1 - \left[(1-\gamma_2)k_n + \frac{1}{\tau_n}\right]Q_2 + \gamma_2 I_{k0} \tag{5}$$

where $k_p = \alpha_{T1}/(1-\alpha_{T1})\tau_p$; $k_n = \alpha_{T2}/(1-\alpha_{T2})\tau_n, \tau_p$ and τ_n are the minority carrier lifetimes in top $n-$ and bottom $p-$ bases respectively, I_{ko} is the current of reverse biased collector junction, I_G is the gate current, I_R — current of distributed leakage of top $p^{++} - n$ junction.

The total current I is given by

$$I = k_p Q_1 + k_n Q_2 + I_{k0} \tag{6}$$

In most cases $I_G \gg I_{k0}$ and $I_G \gg I_R$, and it is possible to neglect I_{k0} and I_R , compared with I_G.

The system (5) can be solved in the general form at an arbitrary time dependence of the gate current $I_G(t)$ [22]. However, in order to find the characteristic time τ_r, it is not necessary to analyze the appropriate cumbersome expressions. It suffices to note that at $I_G = const$ the time dependences of Q_1 and Q_2 can be expressed as

$$Q = A\exp(\lambda_1 t) + B\exp(\lambda_2 t) + C, \tag{7}$$

where the constants A and B are determined by initial conditions, C is a partial solution of the system (5), and λ_1 and λ_2 are the roots of the characteristic equation:

$$\lambda^2 - (a_1 + b_2)\lambda + a_1 b_2 - a_2 b_1 = 0, \tag{8}$$

where

$$a_1 = -\left[(1-\gamma_1)k_p + \frac{1}{\tau_p}\right], \quad b_1 = \gamma_1 k_n, \quad a_2 = \gamma_2 k_p, \quad b_2 = -\left[(1-\gamma_2)\,k_n + \frac{1}{\tau_n}\right]$$

are the coefficients in right-hand part of equations (5). It can be shown that well-known condition of the switch-on of a thyristor structure, $\alpha_1 + \alpha_2 > 1$, results in the negative sign of the quantity $(a_1 b_2 - b_1 a_2)$ since

$$a_1 b_2 - b_1 a_2 = \frac{1 - \alpha_1 - \alpha_2}{(1 - a_{T1})\,\tau_n\,(1 - a_{T2})\,\tau_p} < 0 \tag{9}$$

With $\tau_r = 1/\lambda_1$, where λ_1 is the positive root of the Eq. (5), we have

$$\tau_r = \frac{1}{2}\left\{\sqrt{[(1-\gamma_1\alpha_{T1})\,\tau_2 + (1-\gamma_2\alpha_{T2})\,\tau_1]^2 + 4\,(\gamma_1\alpha_{T1} + \gamma_2\alpha_{T2} - 1)\,\tau_1\tau_2)}\right.$$
$$\left. + (1-\gamma_1\alpha_{T1})\,\tau_2 + (1-\gamma_2\alpha_{T2})\,\tau_1\right\}(\gamma_1\alpha_{T1} + \gamma_2\alpha_{T2} - 1)^{-1} \tag{10}$$

where the minority carrier transit times τ_1 and τ_2 are related to the carrier lifetimes and transport factors α_{T1} and α_{T2} by the well-known expressions $\tau_1 = \tau_p(1 - \alpha_{T1}), \tau_2 = \tau_n(1 - \alpha_{T2})$ [10,20].

It can be seen from relation (10) that in the limit of low injection coefficients of the junctions, when $(\alpha_1 + \alpha_2) = (\gamma_1\alpha_{T1} + \gamma_2\alpha_{T2}) \Rightarrow 1 + \delta$(with $\delta \ll 1$), characteristic current rise time constant τ_r increases strongly with injection coefficient decrease. For this limiting case, we have from Eq. (10):

$$\tau_r = \frac{(1-\gamma_1\alpha_{T1})\,\tau_2 + (1-\gamma_2\alpha_{T2})\,\tau_1}{\gamma_1\alpha_{T1} + \gamma_2\alpha_{T2} - 1} \tag{11}$$

Taking $\gamma_1 = \gamma_2 = 1$ and $1 < \alpha_1 + \alpha_2 < 2$, we have from Eq. (10):

$$\tau_r = \frac{1}{2}\left\{\sqrt{[(1-\alpha_{T1})\,\tau_2 + (1-\alpha_{T2})\,\tau_1]^2 + 4\,(\alpha_{T1} + \alpha_{T2} - 1)\,\tau_1\tau_2}\right.$$

$$+ (1 - \alpha_{T1})\tau_2 + (1 - \alpha_{T2})\tau_1\} (\alpha_{T1} + \alpha_{T2} - 1)^{-1} \tag{12}$$

Equation (12) does not coincide with Bergman's formula (4). Expression (4) with accuracy of 2 can be obtained from the general expression (10) only in a nonrealistic limiting case $(\alpha_1 + \alpha_2) = (\gamma_1\alpha_{T1} + \gamma_2\alpha_{T2}) \Rightarrow 2 - \delta$ $(\delta << 1)$:

$$\tau_r = \sqrt{\frac{\tau_1\tau_2}{\alpha_1 + \alpha_2 - 1}} \tag{13}$$

At $\gamma_1 = \gamma_2 = 1$, this case should be associated with the condition $(\alpha_{T1} + \alpha_{T2}) \cong 2$, which never has been realized in real thyristor structures. As a rule, in the conventional thyristors, $(\alpha_{T1} + \alpha_{T2}) \cong 1.2 \div 1.5$ [10,21].

With the carrier lifetime in blocking p-base measured and the injection coefficient γ_2 known, a very rough estimate of the temperature dependence of τ_r can be found directly from equations (10) or (11). However, to compare correctly the experimental data and theoretical calculations, non-linear effects should be taken into account, such as Auger recombination and band-gap narrowing in the high-doped layers, dependences of the carrier mobilities in p^+- and n^+-layers on the doping densities, electron-hole scattering, etc. Such a comparison can be only done by means of a computer simulation.

The turn-on processes in 2.6 kV 4H-SiC thyristors were simulated using quasi one-dimensional computer program "INVESTIGATION" ("ISSLEDOVANIE") [23]. This program solves numerically a fundamental system of equations comprising two continuity equations, for electrons and holes, and a Poisson equation. The continuity equations are written in a form that takes into account all the main non-linear phenomena in bipolar devices: electron-hole scattering, semiconductor band gap narrowing, and Auger recombination.

Data on band-gap narrowing in 4H-SiC at high doping levels were taken from [24]. As bimolecular and Auger-recombination coefficients for 4H-SiC were adopted $B = 10^{-12}$ cm^{3}s^{-1} and $C = 7\times10^{-31}$cm^{6}s^{-1}, respectively [25].

The rate of monopolar recombination R was described by the well-known equation resulting from the Shockley-Read statistics (see, *e.g.* [26]):

$$R = \frac{np - n_i^2}{\tau_{p0}(n + n_1) + \tau_{n0}(p + p_1)}, \tag{14}$$

where p, n are the hole and electron concentrations; n_1, p_1, τ_{no} and τ_{po} are the standard parameters of the Shockley-Read statistics; and n_i is the concentration of intrinsic carriers. It was supposed that the lifetime in any layer of the thyristor structure is controlled by a deep level positioned at the midgap (see, e. g. [26]) which results in the condition $n_1 = p_1$.

The capture cross-sections of the level for electrons and holes are also supposed to be equal $(\tau_{no} = \tau_{po})$. At room temperature, $\tau_{no} = \tau_{po} = 300$ ns in the p-blocking base, which corresponds to an ambipolar carrier lifetime $\tau_b = 600$ ns according to experimental data for the thyristors under study. The decrease in the carrier lifetimes in heavily doped n$^+$- and p$^+$-layers was described by the semi-empirical formula proposed in [27]:

$$\tau_{n0,p0} = \frac{\tau_{n0,p0(base)}}{1 + (N/N_{n,p})^{\alpha_{n,p}}} \tag{15}$$

where $\tau_{n0(base)}$ is the τ_{n0} value in the diode base, N is the doping density in the n^+-layer, and N_n and α_n are characteristic parameters dependent on a material and a device fabrication technology. The magnitudes of N_n and α_n were taken to be 7×10^{17} cm^{-3} and 1, respectively. The parameters $\tau*_{n0} = \tau*_{p0}$ in the p^+-emitter were calculated using Eq. (15) with $\alpha = 1$. The characteristic parameter N_p served as the single fitting parameter.

The dependences of the carrier mobilities in the p^+- and n^+-layers on the doping densities were described using the experimental data reported in [28].

Figure 8 demonstrates the results obtained in calculating the current rise in the switch-on process at different temperatures.

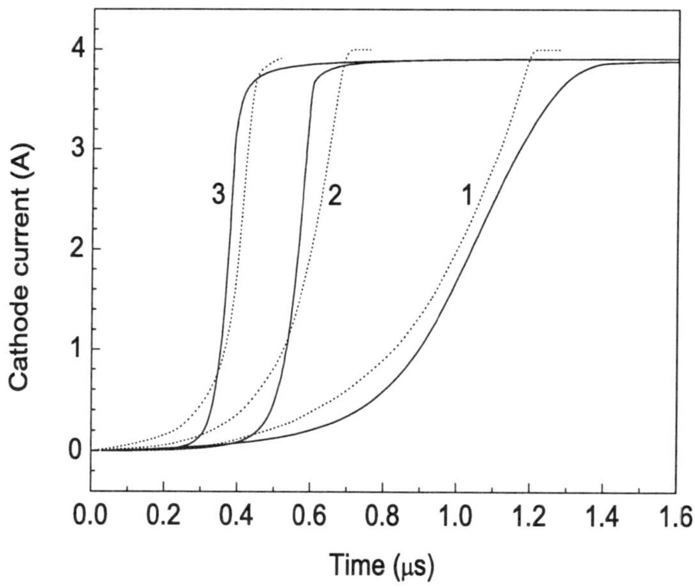

Fig. 8. Time dependences of the current rise in the switch-on process at different temperatures. Solid lines represent calculated results. Concentration of Al in the p^+-emitter 1×10^{19} cm^{-3}, $\tau*_{n0} = \tau*_{p0} = 0.8$ ns (see Eq. (15)). Dotted lines show the experimental data at T(K): (1) 293, (2) 330, and (3) 404 (ref. [15]).

It can be seen that there is a very reasonable agreement between the experimental and calculated data. Taking into account that many parameters, such as the temperature dependence of the Auger recombination, lifetimes and their temperature dependences in heavily doped layers, intensity of the electron-hole scattering, etc., can be estimated for SiC only very roughly, the agreement can be considered as very good.

Of course, not only injection coefficient of $p^{++}-n$ junction, but also many other important parameters of the thyristor are temperature dependent. However, as it was demonstrated in Ref. [15], the main contribution to the effect of temperature on the turn-on process in SiC thyristors comes from the temperature dependence of the hole concentration in the p^{++}-emitter.

As mentioned in Refs. [29,30], the current rise time constant τ_r can depend appreciably on the input signal. If the input signal corresponds to the low injection level in low doped blocking base, the initial stage of the turn-on process is very slow (the characteristic current rise time τ_r is very long). As soon as the current reaches the critical value, the turn-on process becomes steep (the τ_r is rather short).

A qualitatively similar behavior has been observed previously in Si-based thyristors (see, e. g. [31]). It has been shown that transition to a high injection level in the blocking base of a thyristor can speed up the current rise [31]. However, the results obtained in Ref. [31] can not be directly used in analyzing SiC-based devices. The point is that the injection coefficients of both emitter-base junctions were taken to be unity in Ref. [31]. This assumption, being justified for Si-based devices, is certainly incorrect for SiC-based structures (see, e. g. [32]). In addition, the system of transcendental equations obtained in [31] is too complicated to enable an explicit comparison of the current rise times at low τ_{rL} and high τ_{rH} injection levels.

Let us take as the starting point the Eq. (10) for the τ_r at arbitrary injection level, which can be rewritten as

$$\frac{1}{\tau_r} = -\frac{1}{2}\left[\frac{1-\gamma_1\alpha_{T1}}{(1-\alpha_{T1})\tau_p} + \frac{1-\gamma_2\alpha_{T2}}{(1-\alpha_{T2})\tau_n}\right] +$$
$$+\sqrt{\frac{1}{4}\left[\frac{1-\gamma_1\alpha_{T1}}{(1-\alpha_{T1})\tau_p} + \frac{1-\gamma_2\alpha_{T2}}{(1-\alpha_{T2})\tau_n}\right]^2 + \frac{\gamma_1\alpha_{T1}+\gamma_2\alpha_{T2}-1}{(1-\alpha_{T1})\tau_p(1-\alpha_{T2})\tau_n}}, \tag{16}$$

In the case of 4H-SiC-thyristors, Eq. (16) can be simplified by taking into account the following circumstances.

1) Owing to the rather low activation energy of donor impurities in 4H-SiC, the dopants must be completely ionized in the n$^+$-emitter layer. Hence, even at room temperature, the carrier concentration in the n^+-emitter is as high as (5-10)$\times 10^{18}$ cm^{-3} (see, for example, [33]). The acceptor dopant density in the blocking p-base lies within the range from 6×10^{14} to 10^{16} cm^{-3}. Thus, the injection coefficient of the n^+-p-junction (γ_2) can be taken equal to unity up to current densities of about 10^3 A/cm^2 [34]. Then Eq. (16) can be rewritten in the form:

$$\frac{1}{\tau_r} = -\frac{1}{2}\left[\frac{1-\gamma_1\alpha_{T1}}{(1-\alpha_{T1})\tau_p} + \frac{1}{\tau_n}\right] +$$
$$+\sqrt{\frac{1}{4}\left[\frac{1-\gamma_1\alpha_{T1}}{(1-\alpha_{T1})\tau_p}\right]^2 + \frac{1}{2}\frac{1-\gamma_1\alpha_{T1}}{(1-\alpha_{T1})\tau_p\tau_n} + \frac{1}{4\tau_n^2} + \frac{\gamma_1\alpha_{T1}+\alpha_{T2}-1}{(1-\alpha_{T1})\tau_p(1-\alpha_{T2})\tau_n}} \tag{17}$$

2) By contrast, the activation energy of the Al-acceptor impurity in 4H-SiC is rather high (about 0.24 eV). As a result, only 1–2 percent of Al atoms introduced into the p^+-emitter at a density of (1–2)$\times 10^{19}$ cm^{-3} are ionized at room temperature. So, the hole density in the p^+-emitter is relatively low and the γ_1 value is essentially less than unity. In this case, the transport factor of the thin n-base, α_{T1}, must be close to unity ($\alpha_{T1} \approx 1$) in order to provide the thyristor turn-on at room temperature.

It is also noteworthy that the minority carrier lifetime in the blocking p_o-base of high-voltage 4H-SiC thyristors, τ_n, is much longer than that in the thin base, τ_p ($\tau_n \gg \tau_p$).

Then it can be easily seen that the first term under the root sign in Eq. (17) has the highest magnitude, and the second term can be expanded in a power series. With minor terms in this expansion dropped, the following equation can be obtained for the current rise time τ_r:

$$\frac{1}{\tau_r} = \frac{\gamma_1 \alpha_{T1} + \alpha_{T2} - 1}{(1 - \alpha_{T2})\tau_n (1 - \gamma_1 \alpha_{T1})}. \tag{18}$$

Note that, in terms of the charge control model, Eq. (18) describes the current rise time for both the low and high injection levels in the blocking p_o-base. The only difference between these two cases is the transport factor, α_{T2}, which should be calculated in different ways for low and high injection levels [21].

At a low injection level, α_{T2} is given by

$$(\alpha_{T2})_L = \frac{1}{\cosh(W_p'/L_n)} \tag{19}$$

and at high injection level, by

$$(\alpha_{T2})_H = \frac{b}{b+1} + \frac{1}{b+1} \frac{1}{\cosh(W_p'/L_a)}. \tag{20}$$

Here W_p' is the width of the electrically neutral part of the base layer,

$$L_n = \sqrt{D_n (\tau_n)_L}$$

is the electron diffusion length at low injection level in the blocking base, and

$$L_a = \sqrt{[2b/(b+1)] D_p (\tau_n)_H}$$

is the ambipolar diffusion length, $b = \mu_n/\mu_p$ is the ratio of electron and hole mobilities. Then, the following expressions for τ_{rL} and τ_{rH} can be obtained from Eqs. (18) (19), and (20):

$$\left(\frac{1}{\tau_r}\right)_L = \frac{\gamma_1 \alpha_{T1} + (\alpha_{T2})_L - 1}{[1 - (\alpha_{T2})_L](\tau_n)_L (1 - \gamma_1 \alpha_{T1})}, \tag{21}$$

$$\left(\frac{1}{\tau_r}\right)_H = \frac{\gamma_1 \alpha_{T1} + (\alpha_{T2})_H - 1}{[1 - (\alpha_{T2})_H](\tau_n)_H (1 - \gamma_1 \alpha_{T1})}. \tag{22}$$

As regards the analysis of experimental data and verification of the adequacy of the analytical equations obtained, the most interesting parameter is the τ_{rL}/τ_{rH} ratio. From Eqs. (21) and ((22) we have:

$$\frac{(\tau_r)_L}{(\tau_r)_H} = \left[\frac{(\tau_n)_L}{(\tau_n)_H}\right] \cdot \left[\frac{[1-(\alpha_{T2})_L]}{[1-(\alpha_{T2})_H]}\right] \cdot \left[\frac{[\gamma_1\alpha_{T1}+(\alpha_{T2})_H-1]}{[\gamma_1\alpha_{T1}+(\alpha_{T2})_L-1]}\right] = F_1 \cdot F_2 \cdot F_3. \quad (23)$$

It is seen that the τ_{rL}/τ_{rH} ratio is governed by three factors.

The factor $F_1 = [(\tau_n)_L/(\tau_n)_H]$ is associated with a change in the carrier lifetime upon a low-to-high injection level transition. This factor is, as a rule, less than unity. If we assume that the recombination level lies at midgap and the electron and hole capture cross sections are equal to each other [35], then $F_1 = 0.5$.

The factor

$$F_2 = \left[\frac{[1-(\alpha_{T2})_L]}{[1-(\alpha_{T2})_H]}\right]$$

is essentially higher than unity because the transition from low-to-high injection level leads to a substantial increase in α_{T2} : from $(\alpha_{T2})_L$ to $(\alpha_{T2})_H \geq b/(b+1) \approx 0.9$. For the 4H-SiC thyristor structures in question, simple estimations give $F_2 \approx 7-8$. The factor

$$F_3 = \left[\frac{[\gamma_1\alpha_{T1}+(\alpha_{T2})_H-1]}{[\gamma_1\alpha_{T1}+(\alpha_{T2})_L-1]}\right]$$

accounts for the effect of the p^+n-junction injection coefficient. At room temperature, when $\gamma_1\alpha_{T1}+(\alpha_{T2})_L \approx 1+\delta$ (at $\delta \ll 1$) [15], the increase in α_{T2} from $(\alpha_{T2})_L$ to $(\alpha_{T2})_H$ allows the factor F_3 to reach a magnitude of about 10. So, the τ_{rL}/τ_{rH} ratio may be as high as $25-30$. At elevated temperatures, the γ_1 value increases [36], the contribution of the factor F_3 is reduced, and the $\tau_{rL}\tau_{rH}$ ratio decreases.

The results obtained are in qualitative agreement with the experimental data reported in Ref. [29]. However, detailed comparison is complicated because the current rise at low injection levels was not studied in Ref. [29]. To check the adequacy of the obtained analytical expressions, the computer program "INVESTIGATION" [23] was used for appropriate numerical calculations.

Figure 9 shows the time dependence of current for a 2.6-kV 4H-SiC gate turn-off (GTO) thyristors operating in the MOS-gate mode [29].

To turn the GTO off in this mode, the gate and anode are short-circuited using the low-resistance channel of a Si MOSFET. This shorts out the forward-biased gate-to-emitter p^+n-junction and the cathode current is by-passed through the MOSFET [29,36]. It was found in Ref. [29] that the condition $\Delta t_{off} > \Delta t^*_{off}$ should be satisfied for a thyristor to be turned-off completely (Δt_{off} is duration of the applied turn-off pulse, Δt^*_{off} is the minimum duration of the gate pulse which turns the thyristor off). If $\Delta t_{off} < \Delta t^*_{off}$, a spontaneous turn-on process takes place. In the calculations represented in Fig. 9, the duration of the turn-off pulse was taken to be $\Delta t_{off} = 0.80\ \mu s$. This value was extremely close to $\Delta t^*_{off} = 0.81 \mu s$. This means that the charge of non-equilibrium carriers in the base is very close to

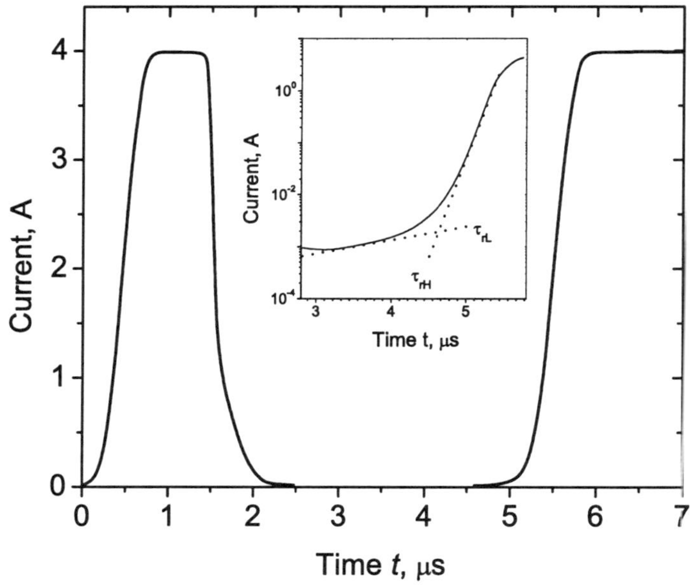

Fig. 9. Transient processes in a 4H-SiC GTO operating in the MOS-gate mode, calculated in terms of the computer model "INVESTIGATION". $T = 300$ K. At $t = 1.4$ μs, the turn-off pulse is applied to the Si MOSFET. The duration of the turn-off pulse $\Delta t_{off} = 0.80$ μs is very close to $\Delta t^*_{off} = 0.81$ μs. The insert shows that a turn-on process begins immediately after the end of the turn-off gate pulse. It can be seen that the rise time at low injection level is much longer than that at high injection level: $\tau_{rL}/\tau_{rH} = 16.5$ [29].

the "critical charge" of thyristor switch-on [37,38]. It can be seen in Fig. 9 that at $\Delta t_{off} = 0.80$ μs the thyristor structure turns-on spontaneously. However, there is a very long "delay" time between the end of the turn-off gate pulse and the beginning of the effective turn-on process.

The inset of Fig. 9 demonstrates clearly that, in fact, the turn-on process begins immediately after the end of the turn-off gate pulse. However, the turn-on time constant τ_{rL} at low currents (low injection level) is much longer than the turn-on time constant τ_{rH} at high injection level. The dotted lines in the insert show an exponential approximation to the current rise: $I = I_o \exp(t/\tau_r)$ for low and high injection levels. It can be seen that τ_{rH} is about 100 ns, and τ_{rL}, about 1.65 μs. Hence, the τ_{rL}/τ_{rH} ratio is ~ 16.5.

2.3. *Optical turn-on of thyristors*

Light-triggered semiconductor devices proposed by Auston [39] are being widely used for ultrafast switching of low-power structures and optical triggering of power semiconductor devices. Optical triggering of Si and GaAs thyristors has been used in ultrahigh-frequency operation with femto-Joule optical input [40], precise synchronization of power lasers for thermonuclear synthesis, electrical insulation of high-voltage rectifiers in power circuits, driving of high-power eximer lasers, and many other applications (see, e. g. [41,42,43,44,45,46,47,48]).

The optical turn-on processes in SiC thyristors were studied with 2.6 kV 4H-SiC GTOs described in previous Section [30,49]. The thyristor structures were switched-on by light pulses of an LGI-21 ultra-violet (UV) laser with wavelength of 337 nm, pulse duration of about 7 ns, and total pulse energy of about $1.5 \times 10^{-5} J$. The light intensity was attenuated with SZS-21 and SZS-23 calibrated UV glass filters. The structure was illuminated from the top (see Fig. 6). The diameter of the light spot was 3 mm.

Figure 10 illustrates the switch-on process under the action of UV pulse light at different illumination intensities.

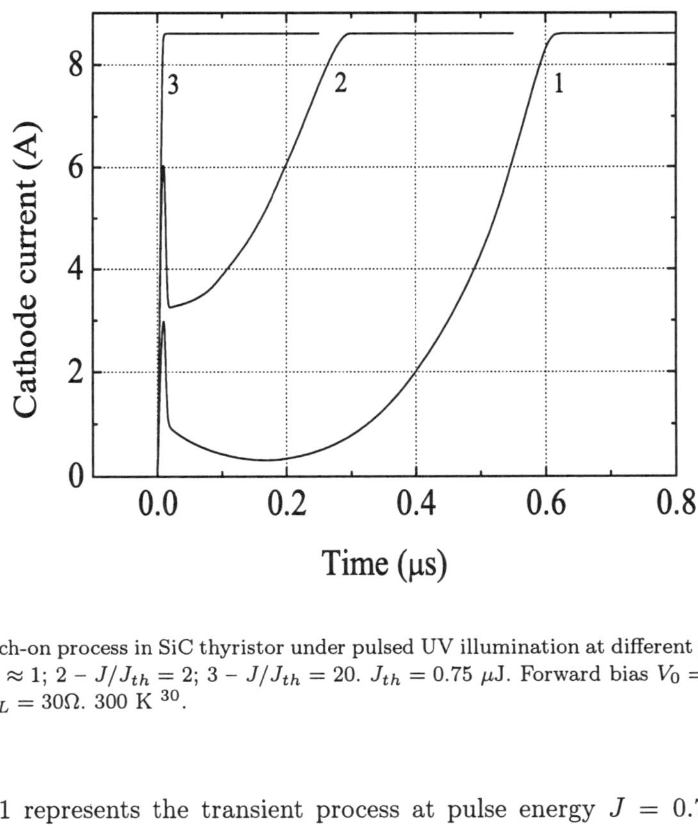

Fig. 10. Switch-on process in SiC thyristor under pulsed UV illumination at different pulse energies J. 1 – $J/J_{th} \approx 1$; 2 – $J/J_{th} = 2$; 3 – $J/J_{th} = 20$. $J_{th} = 0.75$ μJ. Forward bias $V_0 = 270$ V. Load resistance $R_L = 30\Omega$. 300 K [30].

Curve 1 represents the transient process at pulse energy $J = 0.75$ μJ. This energy is very close to threshold energy J_{th} required to switch the thyristor on. With room temperature carrier lifetime in the blocking base of the thyristor $\tau =$ 600 ns [29] and approximately the same thyristor turn-on time (about 600 ns at room temperature [15]), light pulse of duration 7 ns can be considered a "δ-shaped action". Immediately after the termination of this action, three main processes control the temporal behaviour of the thyristor [38]: (i) re-distribution of non-equilibrium carriers through their diffusion, (ii) carrier recombination in the base layers, and (iii) storage of non-equilibrium carriers in the bases as a result of the conventional positive thyristor feedback.

If the initial charge introduced into the thyristor by light exceeds the critical charge n_{cr} [37], the thyristor is switched on. In such a mode, the switch-on process is analogous to that in the conventional gate-controlled mode. As seen, the characteristic time of current rise is about 150 ns. Curve 3 in Fig. 10 illustrates another limiting case when the charge of non-equilibrium carriers, introduced into the thyristor structure by band-to-band illumination, is extremely high. In this case, the current reaches its quasi-steady-state magnitude of 8.5 A in about 10 ns. The characteristic current rise time is only 3 ns, *i. e.* 50 times shorter than that in the conventional gate-controlled mode. Curve 2 represents the intermediate case.

The "ultrafast switching" was observed experimentally with Si thyristors in Ref. [50] and analyzed in Ref. [51]. As shown in Ref. [50], for a thyristor to be switched on during a short laser pulse, the number of carriers created by light pulse in the thyristor bases must be equal to the number of carriers existing in the bases at a current I flowing through the thyristor in the on-state.

At a current I, the number N_I of carriers existing in the thin thyristor base can be roughly estimated [52] to be:

$$N_I \approx \frac{I\tau_p}{2q} \tag{24}$$

where τ_p is the hole lifetime in the thin n-base of the thyristor. The time constant τ_p can be found from the transport coefficient of the thin n-base, $\alpha_T \approx 0.7$ [16], with the use of the standard expression for α_T (see, e. g. [26]):

$$\alpha_T \approx 1 - \frac{W_n^2}{2L_p^2} \tag{25}$$

Here $W_n = 2.5$ μm is the width of the thin base, and $L_p = (D_p\tau_p)^{1/2}$ is the hole diffusion length in the n-base. At $I = 8.5$ A (see Fig. 10) and $D_p = 2$ cm^2/s [53], expression (24) gives $N_I \approx 1.4 \times 10^{12}$.

At photon energy $h\nu = 3.68$ eV, the absorption coefficient of SiC is rather high, $\sim 10^3$ cm^{-1} [53], and virtually all photons are absorbed in the thyristor bases (Fig. 6). The number of the electron-hole pairs created by laser light pulse can be estimated to be:

$$N_0 \sim \frac{J_0^s S}{h\nu} \tag{26}$$

where J_0^s is the pulse energy per unit area, and $S = 3.7\times10^{-3}$ cm^2 is the total anode area. At $J_0 =\sim 15$ μJ and light spot diameter of about 3 mm, $J_0^s \approx 2.1\times10^{-4}$ J/cm^2, and N_0 is 1.3×10^{12}. Taking into account the roughness of the input approximations, the agreement between the calculated values of N_I and N_0 should be considered excellent.

It is noteworthy that, at the same UV-light spot diameter and laser pulse energy, a thyristor structure with about 20 times larger area can be switched homogeneously with the same current density. At the same time, with light focussed into a spot of about 0.6 mm in diameter, it is possible to switch the thyristor on with a current of about 100 A.

3. Steady-state current-voltage characteristics

3.1. *Steady-state current-voltage characteristics of low-voltage thyristor structures*

The current-voltage characteristics of 6H-SiC thyristors with a forward breakover voltage close to 100 V were measured up to a current density $j \sim 5 \times 10^3$ A/cm^2 in Ref. [54]. For this current density the forward voltage drop V was $\sim$ 10 V, which is significantly larger than V values for identically rated GaAs and Si thyristors at the same current densities. This result made questionable practical use of SiC thyristors at high and super high current densities.

The steady-state current- voltage characteristics in 4H-SiC thyristors with a forward breakover voltage of about of 400V at very high current density (up to 7×10^4 A/cm^2) were investigated in Refs. [55,56]. The voltage drop across SiC thyristors has been found to be sufficiently less than the drop across 6H-SiC thyristor described in [54] and substantially less than the drop across identically rated Si and GaAs thyristors.

The cross-section of low-voltage p^+np^-n 4H-SiC thyristor is shown in Fig. 1. The parameters of the structures are described in Section 2.1.

Curve 1 in Fig. 11 shows the dependence of the current density j versus voltage drop V at 300 K.

It is seen that the curve consists of two characteristic parts. At small current densities $j < 300 - 400$ A/cm^2, the "on" state occupies only part of the thyristor's area [52,57]. In such a situation the higher the current I, the larger is the area of the thyristor occupied by the "on" state. The actual current density j_0 is not changed, and the differential resistance of the structure is close to zero. It is worth noting that the current density j_0 determined in such a manner, is in a good agreement with the j_0 value determined from optical measurements [57]. At $j > 300 - 400$ A/cm^2, the "on" state occupies the whole area of the thyristor structure and actual current density j is proportional to current I. Current-voltage characteristics have their conventional form: the differential resistance $R_d = \mathrm{d}V/\mathrm{d}j$ decreases monotonically with increasing bias. The R_d value falls from $R_d = 0.53 \times 10^{-3} \Omega$ cm^2 at $V = 3.1$ V to $R_d = 0.37 \times 10^{-3}$ Ωcm^2 at $V = 5.5$ V.

Curve 1' in the inset shows the current-voltage characteristic of the same 4H-SiC thyristor structure at superhigh current densities (up to 7×10^4 A/cm^2). It can be seen that in the current density range $10^4 - 7 \times 10^4$ A/cm^2 the current-voltage characteristic is a straight line. For this current density range $R_d = 0.31 \times 10^{-3}$ Ω cm^2. If one supposes that the contact resistance r_c makes the main contribution to the differential resistance R_d at high current density, the value $r_c = R_d/2 \approx 1.5 \times 10^{-4}$ Ω cm^2 seems to be quite reasonable [58].

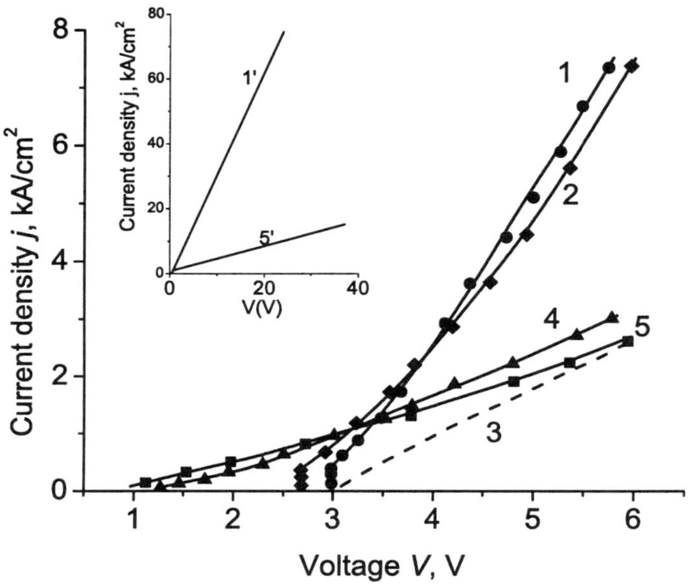

Fig. 11. Current-voltage characteristics of 4H-SiC (curves 1 and 2), GaAs (curve 4) and Si (curve 5) thyristors with breakover voltage $V_b = 300 - 400$ V. Curve 3 represents the room temperature current-voltage characteristic of 100 V 6H-SiC thyristor [54]. Curves 1,3,4,5 correspond to the temperature $T = 300$ K, curve 2 - $T = 600$ K. Inset shows the current-voltage characteristics of 4H-SiC (curve 1') and Si (curve 5') 400 V thyristors at super high current densities ($T = 300$ K) [55]

Curve 2 in Fig. 11 represents the current-voltage characteristic of the same structure at $T = 600$ K. As seen, at low current densities, the voltage drop across the structure decreases with increasing temperature owing to the reduction of the built-in potential. By contrast, at high current densities $j \geq 2500$ A/cm^2, the voltage drop at a given j grows with increasing temperature.

Such an "inversion" is also observed in Si diodes and thyristors, in which, however, the point of inversion corresponds to current densities of several tens of amperes per square centimeter, i.e. to j about two orders of magnitude lower than that in SiC diode structures. The voltage drop across the base at the inversion point does not exceed several tenths of a volt in Si-diodes. As seen in Fig. 11, the voltage drop across the base at the inversion point in SiC diodes is equal to several volts. If the temperature dependence of the hole lifetime in the base and the saturation current of the $p^+ - n$ junction, j_s, are known, the parameters of electron-hole scattering (EHS) can be extracted from isothermal $I - V$ characteristics as it will be discussed in Section 3.3

Dashed curve 3 in Fig. 11 represents the room temperature current-voltage characteristic of 6H-SiC thyristor with breakover voltage close to 100 V [54]. Large forward voltage drops are caused apparently by small minority carrier lifetime in the blocking base of the structure.

Curves 4 and 5 in Fig. 11 show the current-voltage characteristics of rated GaAs (curve 4) and Si (curve 5) thyristors with breakover voltage $V_b \sim 350-400$ V. It is seen that at $j > 2$ kA/cm^2 the voltage drop across the rated GaAs and Si thyristors is essentially larger than that across the SiC thyristor. It would be expected as to block the same voltage the rated Si and GaAs thyristors have more wide and lower doped bases.

3.2. *Steady-state current-voltage characteristics of high-voltage thyristors*

The current-voltage characteristics of 4H-SiC thyristors with a forward breakover voltage of 700 V were studied in Refs. [59,60]. Points in Fig. 12 represent experimental current voltage-characteristics of the thyristors measured at two temperatures [59]. As seen, the voltage drop in 700 V structures is substantially larger than that in 400 V thyristors at the same current densities j. Solid and dashed lines in Fig. 12 show current voltage characteristics of the thyristors calculated in the framework of simple analytical theory [61].

It is well known that the current-voltage characteristics of $n-p-n-p$ thyristors and n^+-p-n^+ diodes at high injection level in the base are described by the same system of equations [61].

The non-ideality of junctions is considered in the framework of the theory [61] by introducing saturation current densities j_{ps} and j_{ns}. Assuming that the hole diffusion length L_p in n^+emitter is smaller than the emitter width W, the estimate for saturation current densities j_{ps} can be expressed as

$$j_{ps} = q\left(\frac{D_p}{\tau_p}\right)^{1/2}\frac{n_i^2}{N_d} \tag{27}$$

The saturation leakage electron current density of $p-p^+$junction, j_{ns}, is determined similarly. The value of $j_{ps} = j_{ns} = j_s$ is an essential fitting parameter of the theory and characterizes the quality of the emitter junctions. In silicon diodes, j_s is usually in the range 10^{-14} - 10^{-12} A/cm^2 [61], in GaAs diodes j_s is within the range $10^{-18}-10^{-16}$ A/cm^2 [62] owing to the fact that the bandgap in GaAs ($E_g = 1.4$ eV) is essentially larger than in Si ($E_g = 1.1$ eV) The bandgap in 4H-SiC is about 3.3 eV. Thus, j_s in SiC diodes is expected to be essentially lower than in Si and GaAs ones.

The total voltage drop $V(j)$ at a given current density j can be calculated similarly to silicon and gallium arsenide n^+-p-n^+ structures as a sum of voltage drops across the emitter junctions, V_e, diode base, V_b, and contacts to the structure, V_c

$$V = V_e + V_b + V_c \tag{28}$$

where

$$V_e = \frac{kT}{q}\ln\frac{n_0 n_W}{n_i^2} = \frac{2kT}{q}\ln\frac{j}{j_0} \tag{29}$$

$$j_0 = \frac{qDn_i}{2L_a} \cdot \frac{b+1}{\sqrt{b}} \left[\left(1 + \sqrt{1 + 4jj_s \frac{1}{b+1} \left(\frac{L_a}{qDn_i} \right)^2} \right) \right.$$

$$\left. \times \left(1 + \sqrt{1 + 4jj_s \frac{b}{b+1} \left(\frac{L_a}{qDn_i} \right)^2} \right) \right]^{1/2} \tag{30}$$

Here $n_0(j)$ and $n_W(j)$ are the boundary electron concentrations in the p-base at $n^+ - p$ and $p - p^+$ junctions, respectively, $D = \frac{2}{b+1} D_n$ is the ambipolar diffusion coefficient, $L_a = \sqrt{D\tau}$ and τ - is charge carrier lifetime under high injection conditions.

The voltage drop across the base V_b is:

$$V_b \approx \frac{j}{q(\mu_n + \mu_p)} \int_0^W \frac{dx}{n(x)} \tag{31}$$

where

$$n(x) = \frac{n(0) \sinh \frac{W-x}{L_a} + n(W) \sinh \frac{x}{L_a}}{\sinh \frac{W}{L_a}} \tag{32}$$

$$n(0) = j \frac{1}{b+1} \cdot \frac{2L_a}{qD} \frac{1}{1 + \sqrt{1 + 4jj_s \frac{1}{b+1} \left(\frac{L_a}{qDn_i} \right)^2}} \tag{33}$$

$$n(W) = j \frac{b}{b+1} \cdot \frac{2L_a}{qD} \frac{1}{1 + \sqrt{1 + 4jj_s \frac{b}{b+1} \left(\frac{L_a}{qDn_i} \right)^2}} \tag{34}$$

Due to the very large SiC band gap, the leakage current j_s is as small as $\sim 10^{-45}$ A/cm^2 at 300 K. Such small value is very inconvenient in numerical calculations. Moreover, j_s is a drastically temperature dependent parameter: in the temperature range in question (300 K÷600 K) j_s varies from$\sim 10^{-45}$ to $\sim 10^{-17}$ A/cm^2. That is why it makes sense to change the Eqs. (30), (33), and (34) introducing the new parameter α:

$$\alpha = j_s \left(\frac{L_a}{qDn_i} \right)^2 \tag{35}$$

Within the temperature interval 300 K -600 K the α value varies 1.5–2-fold.

The voltage drop across the contacts of the structure, V_c, is:

$$V_c = 2R_c j \tag{36}$$

where R_c is the contact resistivity. The R_c value for the structures under discussion was measured in Ref. [13].

Considering the leakage current j_s as a fitting parameter of the theory, it is necessary to measure the carrier lifetime τ to calculate the current-voltage characteristics. To determine the τ in the blocking p-base of a thyristor, the technique proposed in Ref. [32] was used in Ref. [59]. The thyristor switch-off time (the recovery time of the blocking ability) should be measured as a function of the forward current I_{F0}, flowing across the thyristor before the switch-off. According to the theory, the switch-off time τ_{off} and effective carrier lifetime τ are related as

$$\tau_{off} = \tau \ln \left(\frac{I_{F0}}{I_0} \right) \tag{37}$$

where I_0 is the parameter of the theory. Hence, τ_{off} can be found from the slope of the $\tau_{off}(I_{F0})$ dependence.

The τ values in the thyristors in question were found to be equal to $\tau \approx 80$ ns at $T = 300$ K and $\tau \approx 280$ ns at $T = 500$ K [59]. This means that W/L ratio is equal to $W/L \approx 2.2$ at 300 K, and $W/L \approx 1.5$ at 500 K.

Knowing the lifetime, τ, and contact resistance, R_c, the current-voltage characteristics of the structures were calculated in Ref. [59] using the following 4H-SiC parameters: $\mu_n = 600 \cdot (T/300)^{-2.15}$ cm^2/Vs, $\mu_p = 75 \cdot (T/300)^{-2.15}$ cm^2/Vs, $D_n = kT\mu_n$, $D_p = kT\mu_p$, $E_g = 3.26 - 3.3 \times 10^{-4}(T - 300)$. The α value variations used as a fitting parameter of the theory.

As it is seen in Fig. 12, the calculations performed in the framework of simple theory [61] considering only the decrease of the emitter injection coefficient provide a very good agreement with the experimental results in a wide range of current densities and temperatures. Noteworthy is the very small variations of the fitting parameter α within a very wide temperature interval 300 K - 600 K.

The performed calculations make it possible to estimate the relative contribution of different components into the total voltage drop V. Curve 1 in Fig. 13 represents the current-voltage characteristic of 700 V thyristor structure at $T = 500$ K (dashed line in Fig. 12).

Curve 2 in Fig. 13 shows a calculated current-voltage characteristic of the same sample with $2R_c = 10^{-4}$ Ωcm^2 (the resistivity of good-quality contacts is equal to $R_c \approx 10^{-4}$ Ωcm^2 for the p-type, and $R_c \approx 10^{-5}$ Ω cm^2 for the n type SiC). It is seen that even with the same quality of junctions, the voltage drop can be reduced essentially. Curve 3 in Fig. 13 shows a current-voltage characteristic calculated for $2R_c = 10^{-4}$ Ωcm^2 and "ideal" junctions $\alpha = 0$).

High breakover voltage (2.6 kV) thyristors were reported in Refs. [14,15,16] (see Fig. 6). However, the data on steady-state current-voltage characteristics of 2.6 kV 4H-SiC thyristors have not been published in detail. The estimates of these characteristics can be made on the basis of the steady-state current- voltage characteristics of 5.5 – 6 kV 4H-SiC diodes of the same width of the base $W = 50$ μm [29,34,59]. Such situation is realized because not only the width of the bases but also the room temperature lifetime and even the temperature dependence of the lifetime coincide in these devices [63,64].

Figure 14 shows the temperature dependencies of carrier lifetime in the 5.5-kV 4H-SiC diode blocking base of n-type and in the 2.6 kV 4H-SiC thyristor blocking base of p-type [65,66].

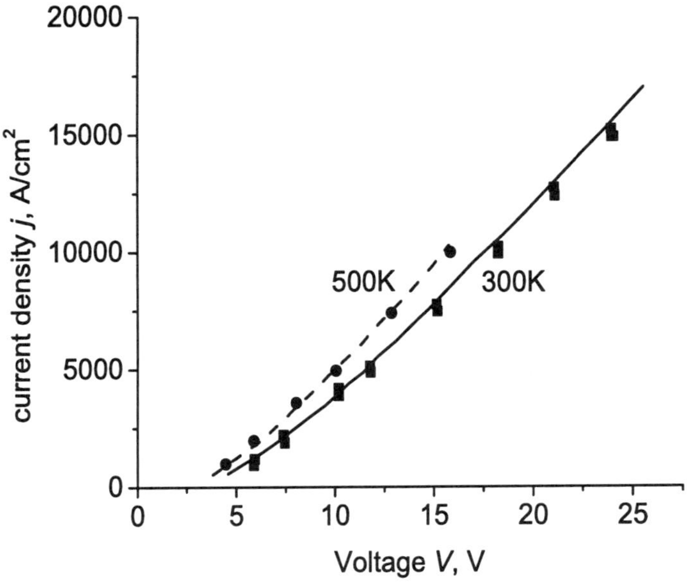

Fig. 12. Current-voltage characteristics of the thyristor structures with breakover voltage $V_b \sim 700$ V [59]. Points represent the experimental data. Solid and dashed lines show the results of relevant calculations:
$T = 300$ K; $\alpha = 7.0$ ($j_s = 1.1 \cdot 10^{-45}$ A/cm^2), $R_c = 4 \cdot 10^{-4}\Omega\cdot$ cm^2, $\tau_n = 80$ ns.
$T = 500$ K; $\alpha = 3.5$ ($j_s = 1.5 \cdot 10^{-23}$ A/cm^2), $R_c = 4 \cdot 10^{-4}\Omega\cdot$ cm^2, $\tau_n = 280$ ns.

As seen, these dependencies accord very closely exhibiting clearly that the just the same recombination level is responsible for the lifetime of non-equilibrium carriers in both devices [35].

Figure 15 demonstrates current-voltage characteristics of 5.5-kV 4H-SiC rectifier diodes for three temperatures [64]. (According to our measurements, these characteristics practically coincide with appropriate characteristics of 2.6-kV 4H-SiC thyristors). As seen, the voltage drop across the structure decreases with increasing temperature owing to the reduction of the built-in potential. As mentioned above, such temperature dependencies are characteristic also for Si diodes and thyristors at low current densities. However, in silicon devices, at current densities of about several tens of amperes per square centimeter, the voltage, on the opposite, increases with temperature growth due to mobility and diffusion coefficient reduce with temperature increase (so called "inversion").

As seen in Fig. 15, the negative temperature coefficient of the forward voltage drop is retained in silicon carbide devices even at current densities of about several hundreds of amperes per square centimeter.

Nevertheless, the "inversion" is observed also in SiC devices, however at much more higher current densities (Fig. 16). Points in Fig. 16 demonstrate isothermal $I-V$ characteristics of the same diodes as shown in Fig. 15, measured up to current

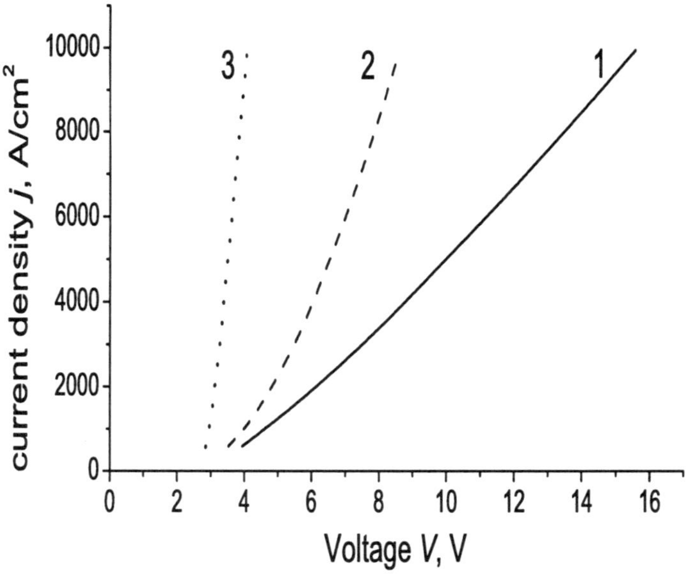

Fig. 13. Calculated current-voltage characteristics of 700 V thyristor. 1) Current-voltage characteristic reproduces dashed line in Fig. 12; 2) Calculation for the same set of the parameters, but with $2R_c = 10^{-4}\Omega\text{cm}^2$ (instead of $2R_c = 8 \cdot 10^{-4}\Omega\text{cm}^2$ as it was taken for the curve 1); 3) Calculation for $2R_c = 10^{-4}$ Ωcm^2 and " ideal" junctions ($\alpha = 0$) [59].

density of 10^4 A/cm^2 at three temperatures. As seen, at high current densities $j \geq 2500$ A/cm^2, the voltage drop at a given j grows with increasing temperature.

As mentioned above, at high current densities, the following non-linear processes: electron-hole scattering (EHS), band gap narrowing, and Auger-recombination, govern the formation of steady-state and transient characteristics of bipolar devices.

The parameters of band gap narrowing and Auger-recombination were studied for SiC in Refs. [25,67]. First theoretical estimates of EHS parameters in SiC were made in Ref. [68]. The detailed analysis of EHS in 4H-silicon carbide was performed in Ref. [69].

3.3. *Electron-hole scattering in silicon carbide*

As it has been shown in Ref. [69], if the temperature dependence of the hole lifetime in the diode base and the saturation current of the $p^+ - n$ junction, j_s, are known, EHS parameters can be extracted from current-voltage characteristics of the diodes at high current densities. The carrier lifetime τ in the diodes under study and its temperature dependence were measured in Ref. [64]. It was found that $\tau = 0.6$ μs at $T = 293$ K and grows steadily to $\tau = 3.8$ μs with the temperature increasing to $T = 553$ K (Fig. 14). The saturation currents of $p^+ - n$junction, j_s, for the diode structures in question were estimated in Refs. [34,59] ($j_s \approx 10^{-45}$ A/cm^2 at

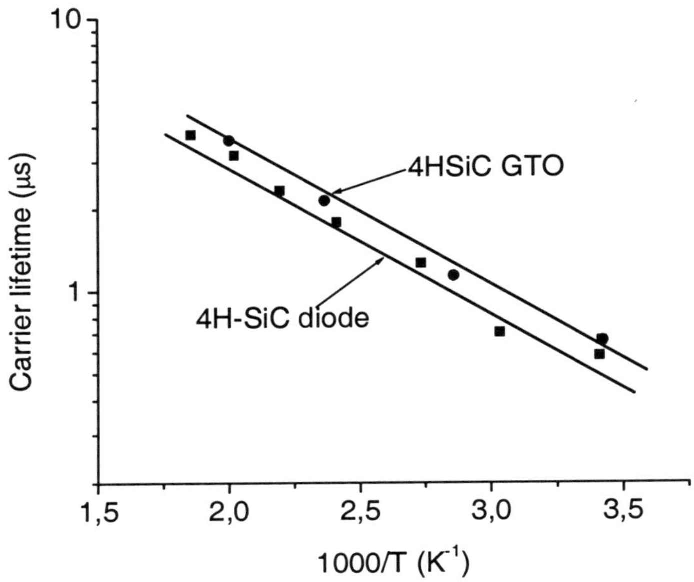

Fig. 14. Temperature dependencies of *hole* lifetime in the diode blocking base of n-type and of *electron* lifetime in the thyristor blocking base of p-type for 5.5-kV 4H-SiC rectifier diode and 2.6 kV thyristor, respectively [65,66].

$T = 293$ K).

The approach suggested in Ref. [70] was used in Ref. [69] to extract the EHS parameters from experimental results. It is worth noting that in the model developed in [70] the modern set of equations has been used to describe the transport phenomena in semiconductors [71,72]. The correctness of these equations was confirmed later in Ref. [73] (see also [74]).

Taking into account the EHS, total voltage drop V across the structure at a given current density j can be expressed as

$$V = V_{pn} + V_b + V_{eh} + jR_S + jR_C, \tag{38}$$

(compare with Eq. (28). Here V_{pn} is the voltage drop across the emitter-base junction; V_b is the ohmic voltage across the base n-layer due to all scattering mechanisms *except EHS*, i. e. scattering on phonons, impurities, dislocations, etc.; V_{eh} is the ohmic voltage drop across the base n-layer associated with EHS; R_S is the substrate resistance; and R_C is the total contact resistance.

An expression for V_{pn} for n-base diode can be written in the form

$$V_{pn} = \frac{kT}{q} \ln\left(\frac{p(0) \cdot p(W)}{n_i^2}\right), \tag{39}$$

(cf. Eq. (29)). Here $p(0)$ and $p(W)$ are the boundary concentrations of carriers at $x = 0$ ($p^+ - n$ junction) and $x = W$ (boundary between n-base and substrate),

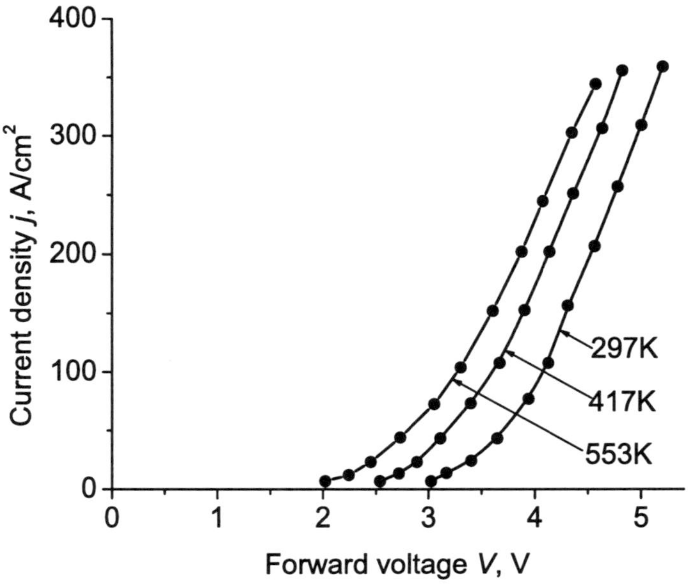

Fig. 15. Current-voltage characteristics of 5.5-kV 4H-SiC rectifier diodes for three temperatures. These characteristics represent with very good accuracy the appropriate characteristics of 2.6-kV 4H-SiC thyristors [64].

respectively.

The component of the carrier mobility μ_{np}, associated with EHS, is described by the following expression (see, e. g. [70]):

$$\mu_{np} = G\frac{p_0}{p}, \tag{40}$$

where G and p_0 are constants characterizing the EHS. Then the expression for V_{eh} in Eq. (41) has the form

$$V_{eh} = \frac{j}{q}\int_0^W \frac{dx}{p\mu_{np}} = \frac{jW}{qGp_0}. \tag{41}$$

When considering the expression for V_b, it is necessary to take into account that in the structures under study the characteristic ratio $W/L_a = 2.8$ at $T = 293$ K. With increasing temperature, L_a grows steadily owing to a rise in the lifetime (see Fig. 14), and $W/L_a = 1.8$ at $T = 553$ K. At such W/L_a values, a diffusion approximation of transport phenomena [75] should be used. In this approximation, the expression for V_b can be written as follows:

$$V_b = \frac{j}{q(\mu_n + \mu_p)}\int_0^W \frac{dx}{p(x)} = \frac{\pi}{2}\frac{jL_a \exp(W/2L_a)}{q(\mu_n + \mu_p)\sqrt{p(0)\cdot p(W)}}. \tag{42}$$

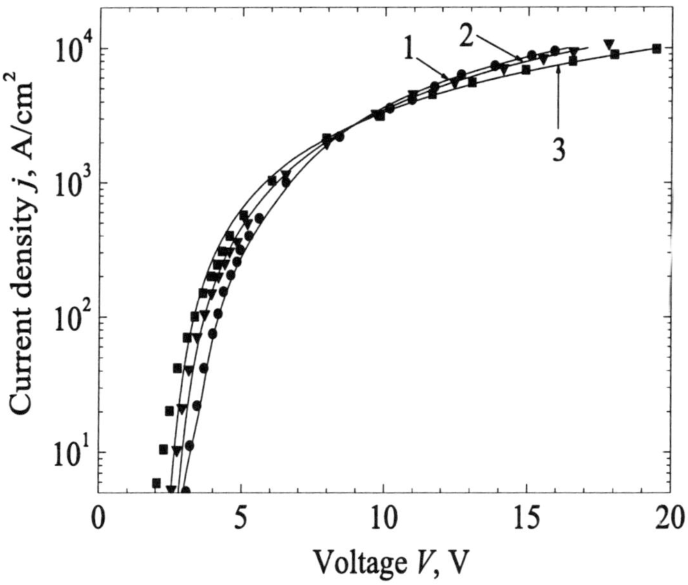

Fig. 16. Isothermal current-voltage characteristics of 5.5-kV 4H-SiC diodes. Points represent experimental data measured at different temperatures T (K): 1 – 293; 2 – 417; 3 – 553. Lines show $I - V$ characteristics calculated for the same temperatures, taking into account EHS [69].

It can be shown that at the established saturation currents of the $p^+ - n$junction ($j_s \approx 10^{-45}$ A/cm^2 at $T = 293$ K [34,59]) and current densities $j > q^2 D^2 n_i^2/(L_a^2 j_s) \approx$ 30 A/cm^2, the expressions for $p(0)$ and $p(W)$ are given by [70]

$$p(0) = n_i \sqrt{\frac{b}{b+1}\frac{j}{j_s}}, \quad p(W) = n_i \sqrt{\frac{1}{b+1}\frac{j}{j_s}}, \tag{43}$$

Using expressions (42) and (43), we can write Eq. (38) in the form

$$V = \frac{kT}{q} \ln\left(\frac{j}{j_0}\right) + A(T)\sqrt{j} + B(T)j, \tag{44}$$

where

$$j_0 = \frac{b+1}{\sqrt{b}} j_s, \quad A(T) = \frac{\pi}{2}\frac{kT}{q} \exp\left(\frac{W}{2L_a}\right) \frac{\sqrt{j_s}}{(qD_p(b+1)n_i/L_a)}$$

and

$$B(T) = R_S + R_C + \frac{W}{qGp_0}. \tag{45}$$

Equation (44) makes it possible to determine the values of the parameters j_0, A and B by means of fitting to experimental $I - V$ characteristics in a wide temperature region. Applying the least-squares method to achieve the best fit to the experimental data, we obtain the following values for the parameter B:

$B(293\text{ K}) = 8.3\times10^{-4}\ \Omega\cdot\text{cm}^2$; $B(417\text{ K}) = 9.8\times10^{-4}\ \Omega\cdot\text{cm}^2$; and $B(553\text{ K}) = 14.7\times10^{-4}\Omega\cdot\text{cm}^2$. In view of the fact that the EHS is caused by the Coulomb interaction, the temperature dependence of the constant Gp_0 in Eq. (40) was taken to be $(Gp_0)_T = (Gp_0)_{T_0}(T/T_0)^{1.5}$ ($T_0 = 293$ K). To take into account the temperature dependence of the substrate resistance, the standard expression $R_S = \left[qN_d^+\mu_n(N_d^+,T)\right]^{-1}$ was used (N_d^+ is the concentration of ionized donors in the substrate). The temperature and concentration dependencies of mobility $\mu_n(N_d^+,T)$ were calculated according to Ref. [76]. The temperature dependence of the contact resistance $R_C(T)$ was taken in the form of a simple power-behaved function $R_C(T) = R_C(T_0)(T/T_0)^\delta$. The magnitudes of $R_C(T_0)$ and δ are to be found from experimental data.

In terms of this approach, one can obtain from Eq. (44) a set of three equations which can be used to find $(Gp_0)_{T_0}$, $R_C(T_0)$ and δ:

$$B(T_i) = R_S(T_i) + R_C(T_0)\left(\frac{T_i}{T_0}\right)^\delta + \frac{W}{q(Gp_0)_{T_0}}\left(\frac{T_i}{T_0}\right)^{-1.5}, \tag{46}$$

where $T_i = 293$ K, 417 K, and 553 K.

Numerical solution of the set (46) gives the following values: $W/(qGp_0)_{T_0} = 5.4\times10^{-4}\ \Omega\text{cm}^2$, $R_C(T_0) = 2.1\times10^{-4}\ \Omega\text{cm}^2$, and $\delta = 2.25$.

At $T = 293$ K, the parameter Gp_0, which defines the contribution of EHS to the mobility (see Eq. (40)), is $Gp_0 = 5.8\times10^{19}V^{-1}\ \text{cm}^{-1}\text{s}^{-1}$; and the constant qGp_0 defining the contribution of EHS to the voltage drop across the structure (see (41)) is $qGp_0 = 9.3\Omega^{-1}\cdot\text{cm}^{-1}$.

The obtained estimate for the EHS parameters in SiC is approximately two times less than that in Si [77], four times less than that in Ge [78], and 60 times smaller than that in GaAs [79,80]. This means that the influence of EHS effects on the bipolar mobility and voltage drop across bipolar structures is 2, 4, and 60 times more effective than that in Si, Ge, and GaAs, respectively.

Computer simulation allows separate analysis of the contributions of different physical mechanisms responsible for the formation of $I-V$ characteristics at high current densities. To simulate isothermal $I-V$ characteristics, the computer program "INVESTIGATION" ("ISSLEDOVANIE") was used [23] (see Section 2.2.).

Curves in Fig. 16 represent $I-V$ characteristics calculated without any fitting parameters with the EHS constants estimated by means of above discussed semi-analytical approach. As seen, a very good agreement is observed at 293 K between experimental and calculated $I-V$ characteristics across a rather wide interval of current densities j from 5 to 10^4 A/cm^2. At elevated temperatures, good agreement between experimental and calculated $I-V$ characteristics is observed at current densities in the range from 50 to 10^4 A/cm^2. At low current densities, the calculated voltage drops at elevated temperatures are somewhat larger than experimental values because the unknown compensation ratio level in n-base has not been taken into account in the computer model.

Figure 17 makes it possible to estimate the relative contributions of EHS to the residual voltage drop at different temperatures. Curves 1 and 2 in Fig. 17 reproduce, on another scale, curves 1 and 3 in Fig. 16, respectively. Curves 1' and 2' demonstrate the results of calculations for which all parameters were taken to be just the same as those for curves 1 and 2, but EHS was disregarded.

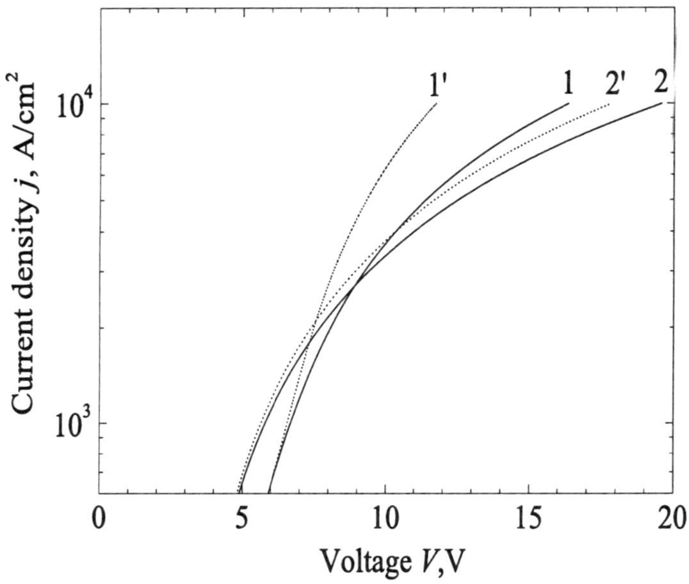

Fig. 17. $I-V$ characteristics calculated for two temperatures: 293 K (Curves 1 and 1') and 553 K (Curves 2 and 2'). Curves 1 and 2 are calculated taking into account EHS. Curves 1' and 2' are calculated with EHS disregarded and all other parameters being the same as for curves 1 and 2 (Ref.[69]).

First of all, it can be seen that the effect of EHS makes an essential contribution to the voltage drop only at $j \geq 10^3$ A/cm^2 even at the lowest temperature of 293 K and becomes increasingly pronounced with growing current density. Secondly, the higher the temperature, the weaker the effect of EHS on the residual voltage drop at a given current density. These results would be expected in view of the fact that EHS is caused by Coulomb interaction. At $T = 293$ K and $j = 10^4$ A/cm^2, EHS leads to an increase in the residual voltage drop from ~ 11.5 V to ~ 16.5 V. It should be noted that EHS has virtually no effect on the built-in potential V_{pn}, ($V_{pn} \approx 3$ V at $T = 293$ K). Hence, EHS causes an increase in the voltage drop across the diode base from ~ 8.5 V to ~ 13.5 V, i.e. by more than a factor of 1.5.

Consideration of the calculation accuracy performed in [69] shows that the Gp_0 value at room temperature can be estimated to be $Gp_0(293) = (8 \pm 4) \times 10^{19}$ V^{-1}cm^{-1}s^{-1}.

As seen, obtained results indicate that the efficiency of EHS in SiC is essentially higher than that in Si, Ge and GaAs. Computer modeling of the I-V characteristics of 4H-SiC diodes in a wide range of current densities and temperatures confirms the validity of the EHS parameters established by means of semi-analytical analysis and demonstrates very important contribution of EHS to the voltage drop across the diode and thyristor bases at high current densities.

4. Turn-off performances

There are three ways to turn the thyristor off. Any thyristor structure can be turned-off if cathode-anode bias reduces to the value at which the cathode-anode current is smaller than the holding current I_h. Gate turn-off thyristors (GTOs) can be turned-off by appropriate current pulse applying to the controlling gate electrode (conventional GTO mode). The third way to switch off a GTO is to short-circuit the gate and the anode (cathode) of the structure using low-resistance channel of field effect transistor (FET controlled mode). The second and third techniques provide the most effective, convenient and fast ways to turn the thyristor off.

4.1. *Conventional turn-off GTO mode*

Conventional gate turn-off operation in SiC thyristors was studied for the first time in low-voltage (100 V) 6H-SiC thyristors in Ref. [54]. The doping level in the n-type blocking base was approximately 3×10^{15} cm^{-3}, the operation area S was 3.5×10^{-4} cm^2. The turn-off time t_{off} of less than 100 ns was achieved, and very high frequency operation up to 600 kHz (at current density 30 A/cm^2) was realized.

The gate turn-off operation in 600-800 V 4H-SiC thyristors with p-type blocking base was studied in Refs. [81,82]. The t_{off} of less than 1 μs (at current density 800 A/cm^2) was observed in Ref. [81]. In addition to the voltage blocking capability, turn-off time, and holding current, the turn-off current gain K_G is one of the most important parameters of GTO's. (The K_G is determined as a ratio of the anode (cathode) current I to the gate current I_g which switches the thyristor off). The K_G was measured to be between three and seven in Ref. [81].

It has been demonstrated in Ref. [82] that 800 V 4H-SiC thyristor can be reliably switched on and turned off under the anode current density j_a of 5000 A/cm^2. The doping level in the p-type blocking base of 7 μm width was approximately 3×10^{15} cm^{-3}, the operation area S was as small as 3.6×10^{-5} cm^2. The anode current density of 5160 A/cm^2 could be turned off with the turn-off time of about 80 ns at room temperature. The turn-off performances of the same 800 V 4H-SiC thyristors were studied in Ref. [17] in the temperature range from 298 K to 513 K with switched anode current density j_a up to 10 000 A/cm^2. At $j_a \approx 8300$ A/cm^2, the total turn-off time t_{off} was equal to 175 ns at room temperature. The t_{off} magnitude increases with temperature growth from 175 ns at $T = 298$ K to $\sim$ 440 ns. at 513 K. The maximum temperature at which the GTO can be turned-off at $j_a \approx 10^4$ A/cm^2 was right 513 K.

As it was mentioned for the first time in Ref. [17], the time dependence of the cathode current I_C in the turn-off process consists of relatively fast and relatively slow parts. It was supposed that the slow part is caused by the electron-hole recombination in the blocking p-base. Hence, the dependences $I_C(t)$ can be used to estimate the lifetime of the non-equilibrium carriers (see Section 4.3). The electron lifetime in the blocking base of 800 V thyristors was estimated to be 5.2×10^{-8} s, 1.05×10^{-7} s, and 1.24×10^{-7} s at $T = 373$, 423, and 473 K, respectively.

The turn-off performances of high-voltage 2.6 kV 4H-SiC thyristors fabricated by CREE Inc. [14] were studied in Refs. [33,63,83]. Figure 18 illustrates the main transient processes in GTO's under study. At time $t = 0$, the negative pulse is applied to the gate, and after some delay, the thyristor is turned on (cf. Fig. 7). Cathode current in on-state, I_{Con}, is governed by the load resistance R_L and the cathode initial bias V_0. At $t = t_1$ ($t_1 = 2$ μs in Fig. 18), positive turn-off gate pulse is applied to the structure. After some delay ($t_2 - t_1$ in Fig. 18), the GTO is turned off.

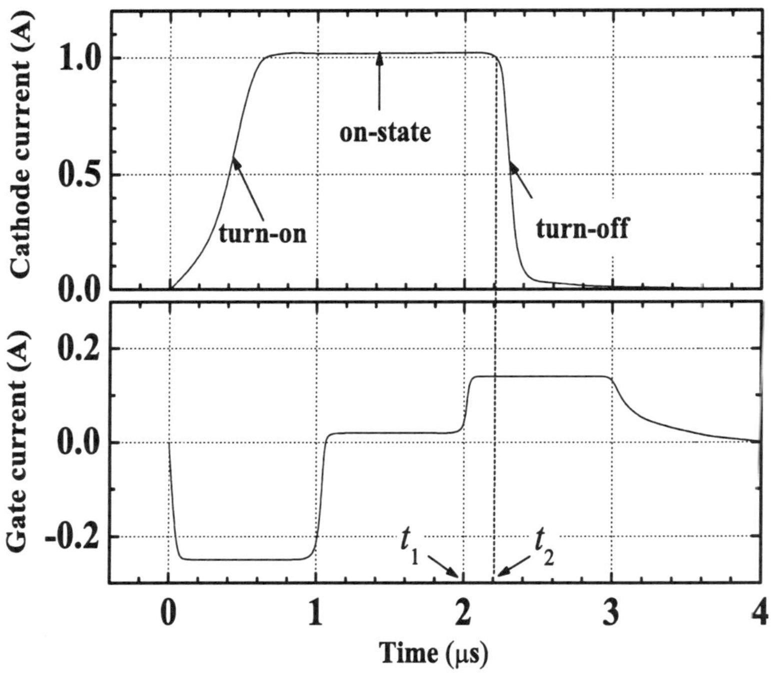

Fig. 18. Main stages of transient processes in 2.6-kV 4H-SiC GTO's. $T = 293$ K. Load resistance $R_L = 25$ Ω, $V_0 = 30$ V [63]

Turn-off time increases monotonically with temperature increase (see also [17]. At every temperature, there is a maximum value of I_{Con} which can be turned off by gate current. At room temperature, I_{Con} is equal to 3.3 A which corresponds to on-state current density $j_{Con} \approx 1000$ A/cm^2. The turn-off current gain K_G depends on temperature, gate pulse duration Δ_{off}, and current density in on-state. The K_G is at a maximum when Δt_{off} is large (quasi-static turn-off regime). At room temperature, the quasi-static value of K_G is equal to 6 at j_{Con} = 1000 A/cm^2. With j_{Con} decreasing, the quasi-static K_G increases, and it is equal to 12.5 at j_{Con} = 300 A/cm^2. The above value of K_G is the highest reported for SiC GTO's.

The amplitude of turned off gate current pulse, I_G increases with a decrease in Δt_{off}. Figure 19 demonstrates the dependence of I_G/I_{Gst} versus Δt_{off} at different temperatures. Here I_{Gst} is the turned off gate current in quasi-static regime. At given temperature T, the cathode current was chosen twice as high as the holding current. It is seen that at given I_G/I_{Gst} ratio, the turned off gate pulse duration increases with temperature growth. At T = 293 K (curve 1), I_G is practically equal to I_{Gst} at $\Delta t_{off} \approx 2.5\mu$s; at T = 502 K (curve 4), this time exceeds 10 μs.

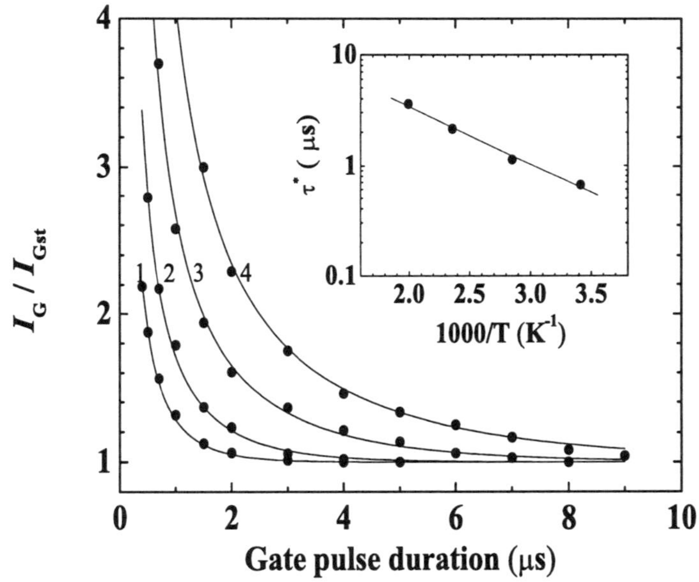

Fig. 19. Normalized turned-off gate currentI_G/I_{Gst} versus gate pulse duration at different temperatures. Points represent the experimental data. Solid lines are plotted according to Eq. (47) with $\tau*$ as a fitting parameter. 1 — 293 K, 2 — 351 K, 3 — 424 K, 4 — 502 K. R_L = 25 Ω. Inset shows the dependence of τ^* on $1000/T$ (ref. [63]).

In Ref. [84], the following semi-empirical formula was suggested for the dependence of I_G/I_{Gst} on Δt_{off}:

$$I_G/I_{Gst} = \frac{1}{1 - \exp(-\Delta t_{off}/\tau*)}, \tag{47}$$

where τ^* is a fitting parameter which can be considered as a rough estimate of the carrier lifetime in the blocking base. It is seen that equation (47) describes the experimental dependencies very well over the entire temperature range. Inset in Fig. 19 shows the temperature dependence of τ^*. As one can see, τ^* grows exponentially from 0.6 to 3.6 μs in the temperature interval 293–502 K. (Just the same dependence is shown in Fig. 14, upper curve).

Figure 20 shows the transient turn-off characteristics of GTO at different temperatures.

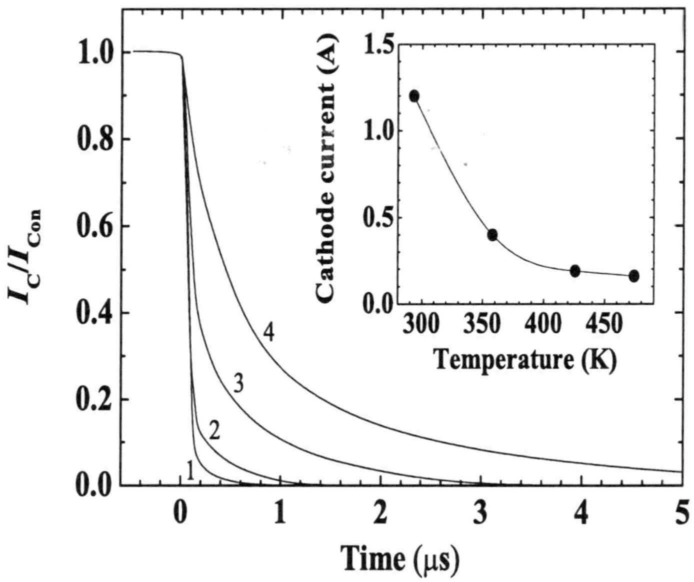

Fig. 20. Cathode current decay during the turn-off process at different temperatures. T(K): 1 – 293, 2 – 358, 3 – 426, 4 – 474. In all cases, very long quasi-static turn-off gate pulse of 100 mA is applied. Inset shows the temperature dependence of I_{Con}, which can be turned-off by a 100 mA gate pulse [63].

The $t = 0$ in Fig. 20 corresponds to $t = t_2$ in Fig. 18, i.e. only turn-off process itself is shown in Fig. 20. For each temperature, the cathode current $I_C(t)$ is normalized to the current I_{Con},, which can be turned off by 100 mA gate pulse. The inset shows the temperature dependence of I_{Con}. One can see that the turn-off time increases monotonically with increase in temperature. Furthermore, two steps can be observed in the turn-off process in the temperature interval 293 K $< T <$ 420 K (curves 1–3 in Fig. 20). At 293 K, cathode current falls down very sharp to $I_C \sim 0.1\ I_{Con}$. Then the current decay becomes more smoothly. At $T >$ 450 K, only the slow step is observed. It is worth to note that the duration of the slow step is very close to τ^* value at the corresponding temperature [17].

Figure 21 shows the temperature dependence of cathode holding current I_{Ch}.

The I_{Ch} decreases from 0.54 A to 0.052 A with an increase in temperature from 293 K to 502 K. One can suppose that two effects are responsible for I_{Ch}versus T dependence. First, the growth of the carrier lifetime in the blocking base (see Inset in Fig. 19) increases the common-base current gain of bottom $n-p-n$ transistor

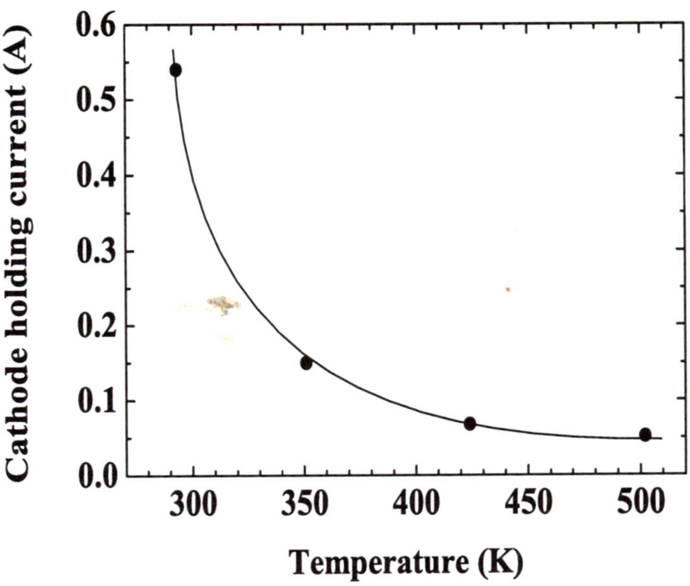

Fig. 21. Temperature dependence of the cathode holding current I_{Ch}.

(see Fig. 6). Second, the injection efficiency of $p^+ - n$ junction between the p$^+$-emitter and the n-base increases with temperature due to the enhanced thermal ionization of relatively deep Al acceptor level (see Section 2.2).

4.2. *FET controlled turn-off GTO mode*

As it can be seen from the results presented in previous Section, the maximum cathode current that can be turned off by a gate pulse is relatively small because of the leakage current and the breakdown of the low-voltage emitter-gate junction. Another way to switch off a 4H-SiC GTO has been proposed in Refs. [36,85]. To turn the GTO off, the gate and anode are short-circuited using the low-resistance channel of a 4H-SiC Junction Field-Effect Transistor (JFET). This shorts out the forward-biased gate-to-emitter $n-p^+$ junction and the cathode current is by-passed through JFET. Such a mode of turn-off operation was demonstrated in [36] at room temperature, the turn-off time being less than 100 ns.

FET controlled turn-off GTO mode of operation was investigated in detail in the temperature range from 293 to 496 K in Ref. [29]. A schematic of the experimental circuit is presented in Fig. 22.

The GTO anode was grounded during the measurements. The maximum on-state current of 6 A is governed by –300 V cathode bias and 50 Ω load resistance. To turn the GTO off in the MOS-gate mode, a Si n-channel MOSFET was used, as shown in Fig. 22.

Figure 23 shows temperature dependences of the maximum cathode current

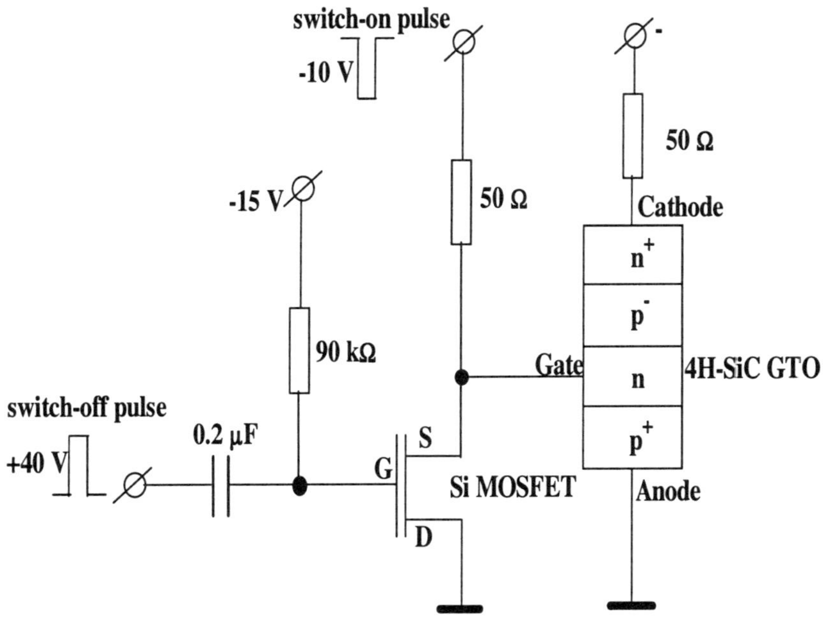

Fig. 22. Schematic of the experimental circuit for FET controlled turn-off GTO mode [29]

I_{cmax} that can be turned off in the conventional GTO mode (curve 1) and in the MOS-gate mode in question (curve 2).

Curve 3 in Fig. 23 represents the temperature dependence of the holding current I_h. It can be seen that in both modes I_{cmax} decreases monotonically with increasing temperature. In the MOS-gate mode, I_{cmax} decreases from 6 to 2.3 A in the temperature interval 397 - 496 K. Note that I_{cmax} achieved in the MOS-gate mode is much higher than that in the GTO mode. The room temperature I_{cmax} is estimated to be about 12 A (j_{cmax} = 3700 A/cm^2). This value is 3.6 times higher than that obtained at room temperature in the conventional GTO mode (see Section 4.1).

Figure 24 shows turn-on processes along with turn-off processes in the MOS-gate mode in the temperature range from 293 to 404 K, for an on-state cathode current of 4 A (j = 1250 A/cm^2). As seen in Fig. 24, there is some delay in the cathode current fall after applying the turn-off pulse (at t = 1.4 μs). The delay time τ_d grows with increasing temperature. At room temperature, τ_d is less than 30 ns, and τ_d is as long as 500 ns at 404 K.

Just like in the conventional turn-off regime, one can distinguish two parts in current-versus-time curves in the MOS-gate turn-off mode: "slow" and "fast" (cf. Fig. 20). The total turn-off time increases from 0.2 to 0.7 μs in the temperature interval 293–404 K.

As mentioned in Section 2.2, for a complete thyristor turn-off in the MOS-gate

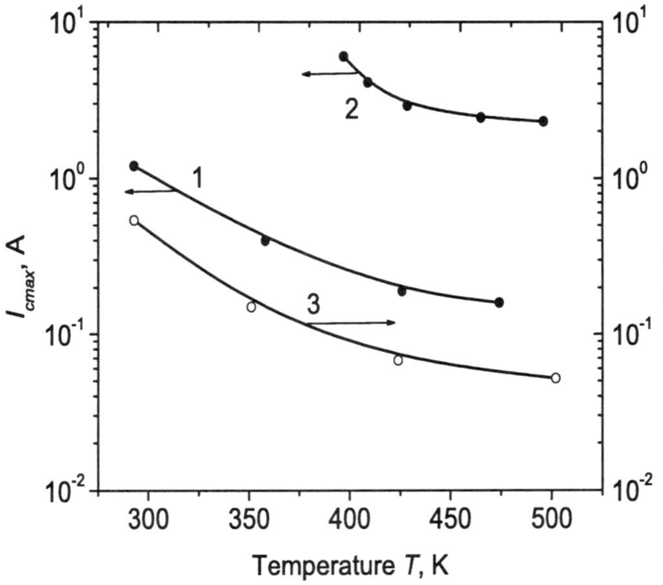

Fig. 23. Temperature dependencies of the maximum cathode current I_{cmax} that can be turned off in the conventional GTO mode of operation (curve 1) and in the MOS-gate mode in question (curve 2). Curve 3 shows the temperature dependence of the holding current I_h (Ref. [29]).

mode, the duration of a turn-off pulse applied to the MOSFET gate, Δt_{off}, should be long enough ($\Delta t_{off} \geq \Delta t^*_{off}$). Curve 1 in Fig. 25 shows the turn-off process with a long (quasi-static) turn-off pulse $\Delta t_{off} = 20$ μs). Curves 2 and 3 represent the "turn-off" processes at $\Delta t_{off} = 0.8$ μs and 0.2 μs, respectively. It can be seen that just after applying the gate pulse (at $t = 1.4$ μs), the thyristor is turned off. However, when the 0.8 μs gate pulse terminates (at $t = 1.4 + 0.8 = 2.2$ μs), a spontaneous turn-on process takes place (cf. Fig. 9). At $\Delta t_{off} = 0.2$ μs, the thyristor is switched on very fast after the termination of the turn-off pulse (curve 3 in Fig. 25). The character of the turn-on process in this case is similar to that observed when turning a thyristor on by a power light pulse (cf. Fig. 10).

The time Δt^*_{off} required to turn the thyristor off completely, weakly (logarithmically) increases with increasing cathode current I_c and increases rather fast (exponentially) as temperature is raised. The nature of these dependences is quite understandable and can be explained as follows. It is clear that Δt^*_{off} must be close to the time over which the concentration of excess carriers in the bases decreases to the critical charge density n_{cr} [37,57]. Under the well known assumption that the concentration of excess carriers is proportional to the current density [59], one can expect a logarithmic dependence $\Delta t^*_{off}(I_c)$. On the other hand, the increase in Δt^*_{off} with temperature is determined by two factors. First, the carrier lifetime in the structures under investigation increases exponentially with temperature (see

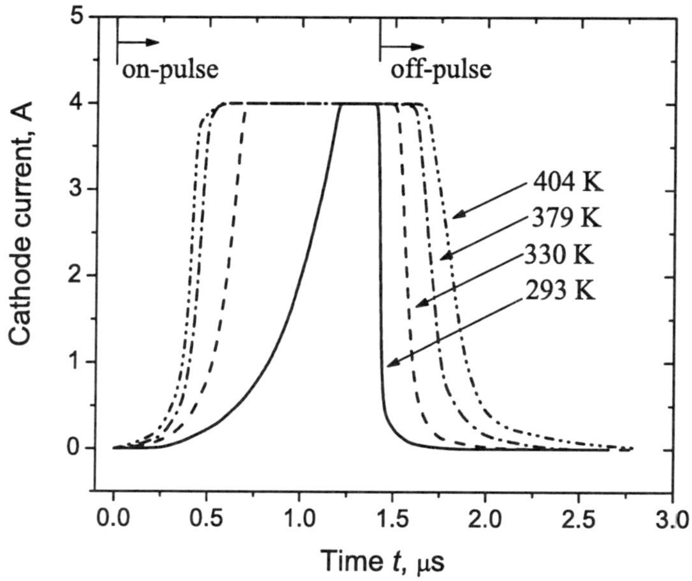

Fig. 24. Turn-on and turn-off time transient processes at different temperatures. Turn-on pulse is applied at $t = 0$. Turn-off pulse is applied to the MOSFET gate at $t = 1.4$ μs. $\Delta t_{off} = 20$ μs.

Fig. 14). Second, the critical charge density being approximately proportional to the holding current density [52] decreases with increasing temperature (see Fig. 21).

Figure 26 presents the temperature dependence of Δt^*_{off} (Arrhenius plot) plotted for $I_c = 4$ A. The activation energy of 0.12 eV, extracted from this dependence, is close to the activation energy of carrier lifetime (0.11 eV) obtained for these structures in Ref. [63] (see Fig. 14). This indicates that the temperature dependence of the carrier lifetime makes the main contribution to the temperature dependence of Δt^*_{off}.

The turn-off processes in the MOS-gate mode were simulated using a computer program "INVESTIGATION" [86,87]. According to experimental conditions, the following procedures were conducted in simulation process. The cathode bias V_0 was supposed to be applied to in-series connected GTO and load resistance R_L. The GTO was turned on with a +10 V, 1 μs gate pulse (the results of simulation of the turn-on were described in Section 2). After a steady state condition was reached, the forward-biased GTO's gate-to-anode junction was shorted out with a 1-Ω resistance R_{ch} for a controlled time Δt_{off}. The carrier (hole and electron) distributions, time dependences of the cathode current, voltage drop, and current flowing through R_{ch} were calculated.

Figure 27 shows time dependences of the GTO cathode current calculated with different Δt_{off}($V_0 = 200$ V, $R_L = 50$ Ω, and $T = 300$ K). The inset presents the corresponding experimental results [29] (cf. Fig. 9). Just as in Ref. [29], the gate-to-

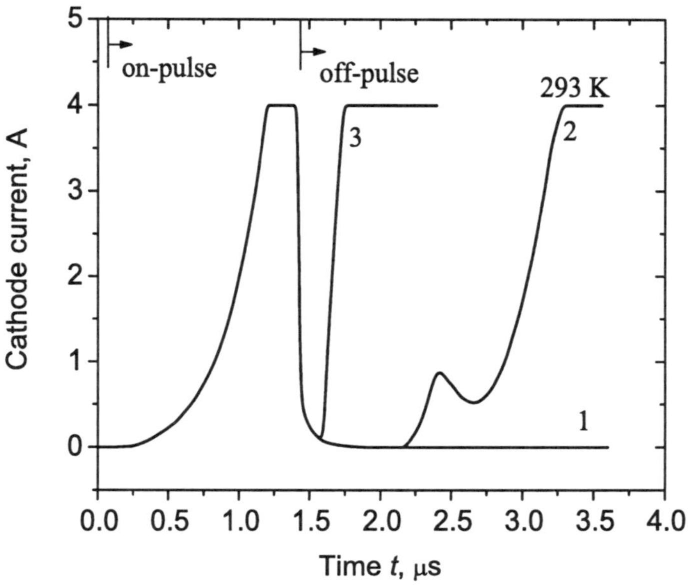

Fig. 25. Transient processes in the MOS-gate mode at different durations of the turn-off pulse, Δt_{off} ($T = 300$ K). Δt_{off} (μs): 1 — 20, 2 — 0.8, 3 — 0.2 (Ref. [29]).

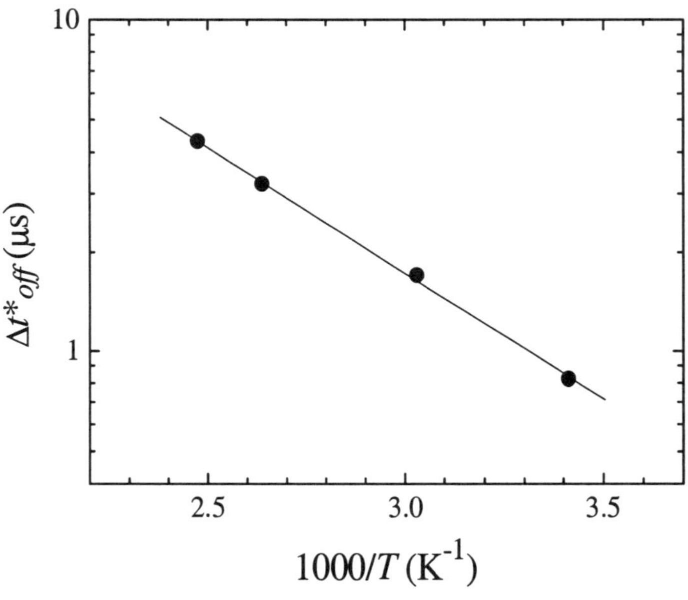

Fig. 26. Temperature dependence of Δt^*_{off}. (Cathode current $I_c = 4$ A)

anode junction is shorted out with resistance $R_{ch} = 1\ \Omega$ at $t = 1.4\ \mu s$. In the regime under study, the Δt^*_{off} value is 0.82 μs. As seen, a spontaneous turn-on occurs immediately after the turn-off pulse terminates if $\Delta t_{off} << \Delta t^*_{off}$ (Curve 1 in Fig. 27).

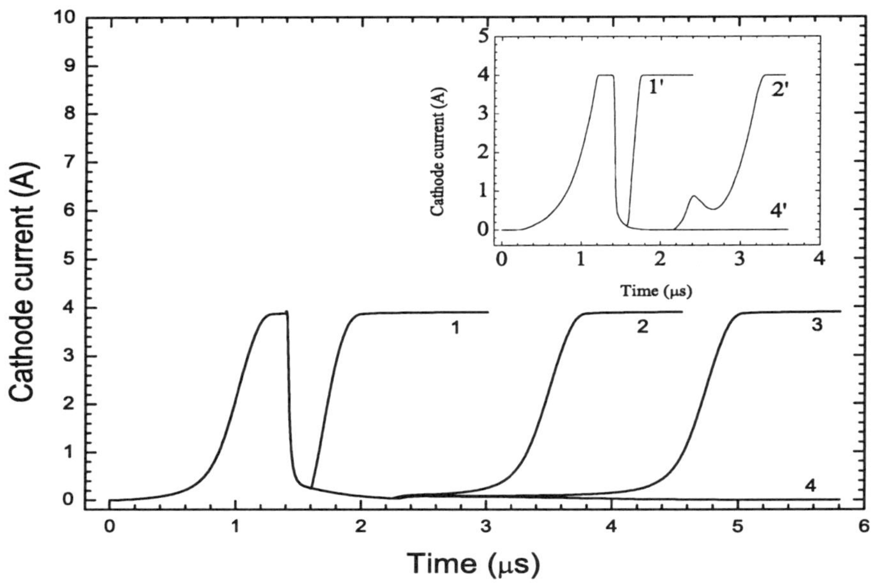

Fig. 27. Calculated time dependencies of current in the MOS-gate turn-off mode at different Δt_{off}. $T = 300$ K. $V_0 = 200$ V, $R_L = 50\Omega$. Just as in the experiment (see Inset), gate-to-anode junction is shorted out by the 1 Ω resistance at $t = 1.4 \mu s$. Δt_{off} (μs): 1 – 0.2, 2 – 0.8, 3 – 0.82, 4 – 0.83 ($\Delta t_{off} > \Delta t^*_{off}$). Inset shows measured time dependencies of the current [29]. $T = 300$ K. $V_0 = 200$ V, $R_L = 50\Omega$. Δt_{off} (μs): 1' - 0.2, 2' - 0.8, 4' – 0.9 $\Delta t_{off} > \Delta t^*_{off}$) (Ref. [86]).

With increasing Δt_{off} , some time delay appears between the instant of turn-off pulse termination and the stage of fast current rise in the spontaneous turn-on (Curves 2 and 3 in Fig. 27). At $\Delta t_{off} \approx \Delta t^*_{off}$, such a delay may be very long, i.e. essentially longer than both the turn-off pulse duration and the carrier lifetime in the blocking base (Curve 3 in Fig. 27).

The spontaneous turn-on actually indicates that, after the turn-off pulse terminates, the residual non-equilibrium carrier charge in GTO's bases remains higher than the critical charge density n_c [37,57]. By contrast, when $\Delta t_{off} > \Delta t^*_{off}$, the residual charge becomes lower than n_c, and the GTO does not turn on (Curve 4 in Fig. 27).

Comparison of the calculated current-time waveforms with the experimental ones (see inset of Fig. 27) shows good agreement between the simulation and the experiment. The only significant difference is that an additional small peak of current is seen in Curve 2' of the Inset. Presumably, the presence of this peak is due to two-dimensional effects, which can not be taken into account by our quasi one-dimensional computer program.

With increasing cathode bias V_0, the charge extracted from the base layer through the shunting resistance R_{ch} grows. On the other hand, the critical charge n_{cr} decreases with increasing V_0 [37,57]. As a result, these two effects virtually cancel each other out, and Δt^*_{off} depends only weakly on V_0 [86].

The Δt^*_{off} value strongly increases as temperature is raised. As mentioned in Ref. [29], the activation energy of 0.12 eV, extracted from Δt^*_{off} versus $(1000/T)$ plot, is close to the activation energy of the minority carrier lifetime (0.11 eV) obtained for analogous GTOs (cf. Figs. 14 and 26). This clearly indicates that it is the temperature dependence of carrier lifetime that controls the temperature dependence of Δt^*_{off}.

It is worth to note again, that at $\Delta t_{off} = \Delta t^*_{off}$, the residual nonequilibrium charge in GTO's bases is equal to the critical charge density n_{cr}. That is why a spontaneous thyristor turn-on occurs immediately after the turn-off pulse termination if $\Delta t_{off} < \Delta t^*_{off}$. However, the current rise is rather slow while the injection level in the thick blocking base remains low, and at this stage, the characteristic current rise time is essentially lower than that at high injection level (see Eqs. (21, 22) and Fig. 9). Transition from slow to fast stage occurs as soon as high injection level is reached in that part of the base, which lies near the n^+-emitter. Figure 28 illustrates this important statement.

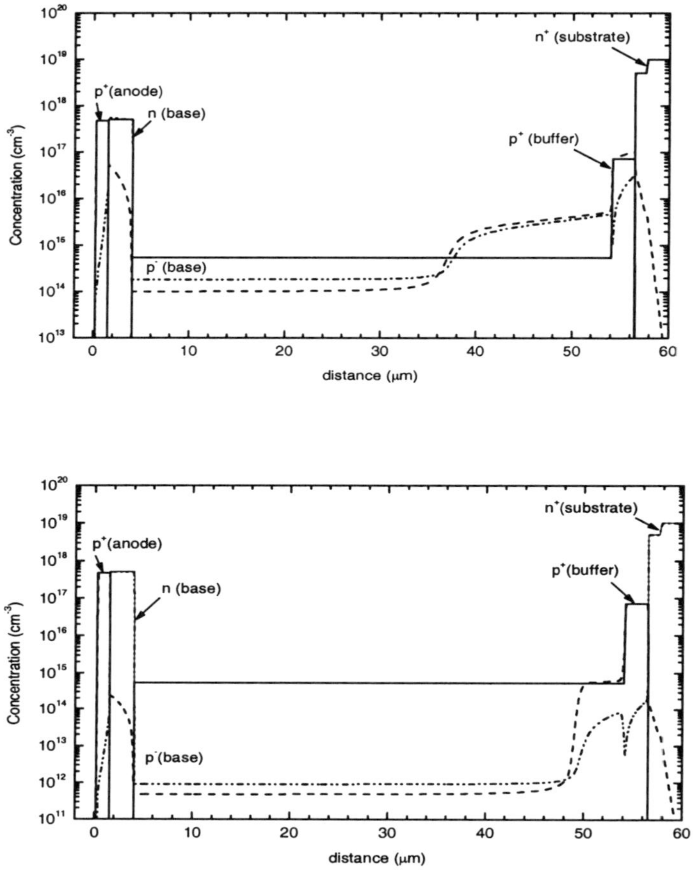

Fig. 28. Calculated hole (dashed line) and electron (dashed-dotted line) distributions across the thyristor structure during transient process (see Fig. 9) for two instants of time. The calculated distributions correspond to slow ($t = 2.5$ μs, upper figure) and fast current rise stages ($t = 5.2$ μs, bottom figure) (Ref. [86]).

5. Frequency properties

As mentioned in the Introduction, a high breakdown electric field allows the design of SiC thyristors with thinner and higher doped base layers than identically rated Si or GaAs structures. One can expect that the thinner base layers ensure much higher operating frequency of SiC thyristors. Indeed a very high operation frequency up to 600 kHz (at current density 30 A/cm^2) was realized in Ref. [54] for 6H-SiC thyristor. From practical point of view, however, the combination of high frequency and high current density is very important.

An operation frequency f_0 of 1000 kHz for 400 V 4H-SiC thyristors at a current density $j = 2700$ A/cm^2 and $f_0 = 500$ kHz at $j = 14000$ A/cm^2 were observed in Ref. [88]. For a 700 V SiC thyristor, f_0 was 1000 kHz at $j = 480$ A/cm^2 and f_0

was 500 kHz at $j = 750$ A/cm^2 [88]. The frequency properties of the thyristors were investigated in a circuit of simple linear modulator (Fig. 29).

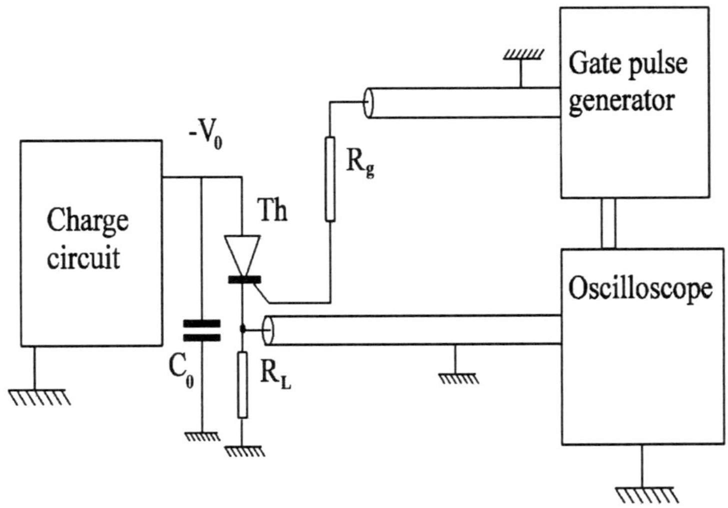

Fig. 29. Schematic of the experimental setup for high-frequency measurements [88]. $C_0 = 10$ nF, $R_L = 1\ \Omega$ (ultra low-inductive resistor). The total pulse duration is about 100 ns.

In this circuit the capacitor C_0 is charged from an external bias source. The thyristor Th is switched-on by a gate pulse, following with C_0 discharges through the thyristor and the series resistor R_L. The linear modulator is the simplest circuit that allows comparing correctly the limiting work frequencies of different types of the thyristors [89].

Experimental results obtained for 400 V and 700 V thyristors are compared in Fig. 30 with the theoretically estimated limiting operating frequency for 400 V thyristors and with the related data for Si thyristor KU221. One of the fastest Si industrial thyristors KU221 (made in Russia) has a blocking voltage of 400–600 V and a limiting operating frequency of about 30 kHz. This thyristor has the voltage blocking n-layer of 120 μm thick with donor concentration $N_d - N_a = 1.4 \cdot 10^{14}$ cm^{-3}. The effective pulse active area of the device with pulse duration of about 100 ns is equal to $S = 6.3 \cdot 10^{-3}$ cm^2.

The highest operation frequency achieved for 4H-SiC 400 V thyristor at current density $j \approx 2700$ A/cm^2 is about 1000 kHz. At $j \approx 14000$ A/cm^2 the highest operation frequency is 500 kHz. For rated Si thyristor, the highest operation frequencies are 30 kHz at $j = 1500$ A/cm^2 and 5 kHz at $j = 10^4$ A/cm^2. It is seen that at the same current density the operating frequency of SiC thyristors is approximately

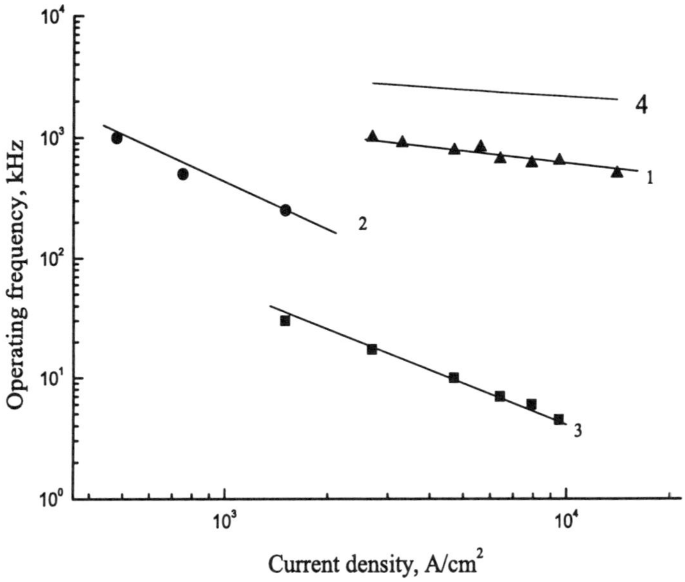

Fig. 30. Dependencies of limiting operating frequency on current density: 1- 400 V SiC thyristor, 2 – 700 V SiC thyristor, 3 - Si thyristor KU221 with blocking voltage of 400 –600 V. 4 - Theoretical estimate of limiting operating frequency [89].

two orders of magnitude higher than that for the identically rated Si thyristor.

Generally speaking, the limiting operating frequency of a thyristors depends on the operation circuit and on operation regime of [89]. In the simple circuit of linear modulator, the limiting operating frequency f_0 is mainly determined by switch-off time τ_{off}. As mentioned above, the switch-off time and the minority carrier lifetime in the voltage-blocking base, τ, are related by expression (37).

Figure 31 represents the dependence of τ_{off} on forward current density j_F for 400 V 4H-SiC thyristors [59].

The slope of this dependence gives the carrier lifetime in the voltage-blocking base, τ. The magnitude of forward current density j_F at $\tau_{off} = \tau$ defines the $j_0 \approx j_F/2.7$ value. As seen in Fig. 31, τ and j_0 are equal to 80 ns and 11.5 A/cm^2 respectively.

Estimating the limiting operating frequency f_0 as $f_0 \approx 1/\tau_{off}$, one can calculate the $f_0(j)$ dependence (Curve 4 in Fig. 30). As seen, the slopes of the theoretical and experimental dependencies virtually coincide. As expected, the theoretical estimate is several times higher than the experimentally achieved limiting frequency.

Experiments show that the main obstacle to increasing the frequency f_0 is the self-heating of the devices. Indeed, the lifetime τ increases with temperature growth (Fig. 14). At the same time, j_0 decreases with temperature increase. Both these effects contribute to growth of τ_{off} with temperature (Eq. (37)). Hence a thyristor

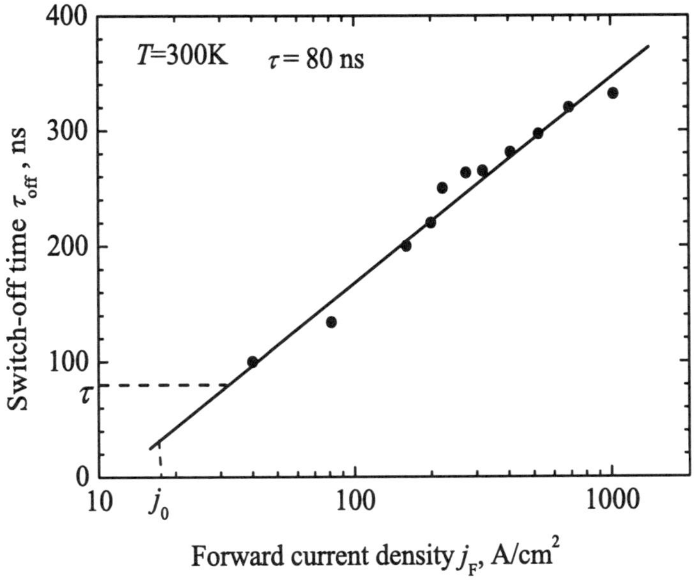

Fig. 31. Experimental dependence of switch-off time τ_{off} on forward current density for 400 V 4H-SiC thyristor at 300 K [59]. Dashed lines show carrier life-time in the blocking p-base, τ, and parameter of the theory j_0 (see Eq. (37)) [59].

structure with design optimal for $T = 300$ K, has too large τ_{off} and too small j_0 at elevated temperatures. From this follows that the effective way to increase the limiting operating frequency f_0 is to reduce τ at room temperature, so that the optimal τ values can be reached at elevated temperatures.

Frequency properties of high-voltage 2.6 kV thyristors were estimated in Ref. [29]. Figure 32 shows the temperature dependences of energy losses per one switching cycle for these thyristors at a current$I_c = 4$ A ($j_c = 1250$ A/cm^2).

It is seen that, in accordance with the temperature dependences of turn-on and turn-off times (Figs. 8, 9, 19, and 20), the switch-on losses fall and, by contrast, the switch-off losses grow with temperature. The minimum total losses of 70 μJ per cycle are observed at 350 K. Note that the static losses are about 48 W ($I_c = 4$ A and the residual voltage drop is 12 V, see Fig. 16). The upper switching frequency f_m, estimated as a frequency at which the switching losses are equal to the static ones, is $f_m \sim 680$ kHz.

6. Critical charge

The critical charge in a thyristor determines the maximum voltage ramp (dV/dt) of the thyristor [38], spread velocity of the "on" state [52], minimum gate control current density [90], holding current, and parameters of current filaments in gate-controlled thyristors [91]. The concept of the critical charge was put forward and developed in

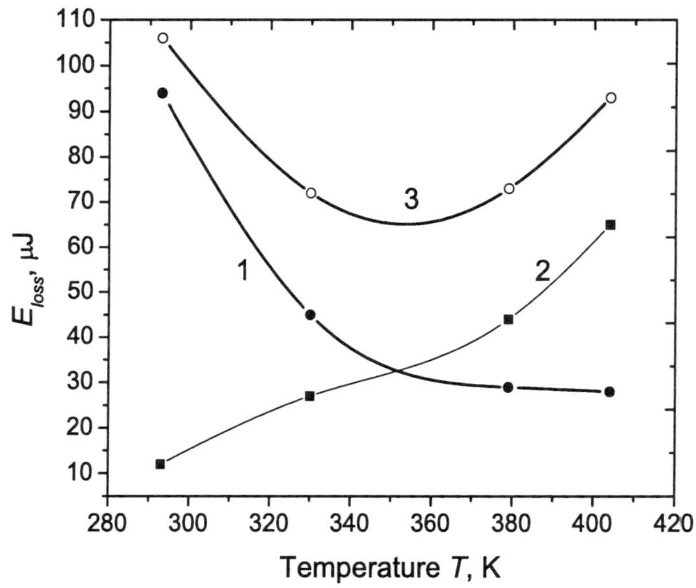

Fig. 32. Temperature dependencies of switching losses per one cycle for 2.6 kV 4H-SiC thyristors [29]. $I_c = 4$ A ($j = 1250$ A/cm^2). $T = 300$ K. 1 — switch-on losses, 2 — switch-off losses, 3 — total losses per cycle.

Refs. [37,38]. This concept unifies the numerous methods known to turn a thyristor on. These methods can be divided into two groups: pulse methods and dc methods.

The pulse methods to turn a thyristor on include dV/dt ramp, short gate pulse, short light pulses etc.(The term "short pulse" implies that the duration of the actuating pulse t_p is much less than the minority carrier lifetime τ_p in the thinner and higher doped base of the thyristor). The minimum critical charge Q_{cr}, required to turn on the thyristor is an intrinsic parameter describing the "sensitivity" of the thyristor. It is noteworthy that the current flowing through the thyristor cannot be used as a measure of this sensitivity for the pulse methods of the thyristors turn-on. For example, a thyristor evidently could not be turned on in the limiting case of δ-function current pulse, when the maximum current achieves an infinite value independent of the charge generated in the thyristor. By other words, even very high current will not turn the thyristor on if the charge generated in the bases is smaller than the critical charge.

The dc methods of thyristor turn-on include very slow (quasi-static) increasing of the anode voltage or gate current, varying temperature or light intensity at a constant anode and gate voltages etc. In such cases, the minimum current is a suitable parameter to describe the sensitivity of the thyristor. Depending upon the turn-on method, this could be either the gate current or the anode current. The charge introduced in the thyristor cannot be used as an intrinsic parameter in this case, because in the limiting case of very slow change in the actuating current, the charge tends to infinity.

It has been shown in Refs. [37,38] that the concept of the critical charge can be used for dc turn on methods also, but in this case, the critical charge represents the excess charge of the minority carriers and not the charge generated by instantaneous changes in operating conditions. For example, in the case of turn-on of the thyristor by gate current at constant anode voltage, the more is gate current, the more is the charge of minority carriers in the bases of the thyristor. Minimal minority carrier charge Q_1 which is required to turn-on the thyristor represents a suitable parameter to describe the sensitivity of the thyristor.

It has been shown [37,38] that Q_1 and Q_{cr} are of the same order of magnitude for a thyristor. References [37,38] establish the relation between these quantities, the dependences of Q_1 on the minority carriers distribution along the thyristor bases, their dependence on pulse duration etc.

The critical charge depends on the area where the thyristor is turned on. It may be the whole area of the thyristor, if the thyristor is turned on with the dV/dt effect, or a very small part of the whole area if the thyristor is turned on with a small beam of light (see for example Ref. [47]). Hence, critical charge per unit area is the more intrinsic parameter.

In some cases, however, the volume critical charge density should be used to describe adequately the sensitivity of the thyristor structure. It is well known, for example, that the turn-on of the thyristor by gate current results in an inhomogeneous distribution of excess minority charge along the thin higher doped base. The density of excess charge has a maximum near the gate electrode and decays monotonically along the base [92]. Hence, if the gate current increases slowly, the thyristor is turned-on at the point near the gate electrode where the minority carrier charge density has a maximum and it is equal to the critical charge density. With the critical charge Q_{cr} known, the volume critical charge density is given by

$$q_{cr} = \frac{Q_{cr}}{SW_1} \tag{48}$$

where W_1 is the width of the thinner and high doped base of the thyristor. Expression (48) contains only the thickness of the thin base and not the sum of the thicknesses of the thyristor bases. This is because for equal charge introduced to each of the bases, the efficiency of the minority charge is much more in the thin base of the thyristor [37,38]. The volume critical charge density is often expressed as

$$n_{cr} = q_{cr}/q \tag{49}$$

where q is the electron charge. In such a form, the critical charge density n_{cr} can be compared conveniently with the base doping densities and with the minority carrier concentrations in the bases. For silicon thyristors, the critical charge density falls in the range from 10^{16} cm^{-3} for power high voltage thyristors to 10^{13} cm^{-3} for special thyristors designed with very low dV/dt values and very low gate control current magnitudes [93,94]. The same range of n_{cr} values is typical for GaAs thyristors [95].

6.1. *Critical charge in low-voltage thyristor structures*

In low-voltage 4H-SiC thyristors with blocking voltage 300 – 400 V, the critical charge density was determined in Refs. [57,96]. To determine the critical charge density the well known "dV/dt technique" was used. First, a forward dc voltage V_0 was applied to the thyristor. Then, an additional pulse of amplitude ΔV_c (rise time < 1 ns) was applied to the cathode of the structure. The 50 Ω pulse generator with the rise time of 0.3 ns was used to form the dV/dt pulses. High frequency 50 Ω circuit with the low parasitic inductance value (< 1.2 nH) was used to apply the pulse to the anode of the thyristor. The minimum value of ΔV_c sufficient to turn on the thyristor was registered. The magnitude of the charge appearing in both bases of the thyristor Q_{cr} is equal to

$$Q_{cr} = \int_{V_0}^{V_0+\Delta V_c} C_{jc}(V)dV \tag{50}$$

where C_{jc} is the capacitance of the central (collector) $p-n$ junction of the thyristor The critical charge density n_{cr} was calculated according to Eqs (48) and (49). The $C_{jc}(V)$ dependence was measured at a frequency of 1 kHz in the range of forward voltage $0 < V_0 < V_b$ with an accuracy of 10^{-2} pF. It has been found that at $T > 550$ K and forward voltage $V_0 > 5$ V, the dV/dt switching in 4H-SiC thyristors is qualitatively analogous to the dV/dt effect in Si and GaAs thyristors. Curve (1) in Fig. 33 shows the dependence of the critical charge density n_{cr} versus forward voltage V_0 for 4H-SiC thyristor at 560 K. Curve 2 shows the $n_{cr}(V_0)$ dependence for GaAs thyristor with approximately the same breakover voltage V_b [97]. The critical charge Q_{cr} for these low-voltage SiC thyristors is rather small $Q_{cr} = 5.2$ pC at $V_0 = 5$ V and $Q_{cr} = 0.32$ pC for $V_0 = 100$ V.

However, due to small thickness of n-base ($W_n = 0.55$ μm) and very small thyristor area $S = 3.6 \times 10^{-4}$ cm^2 (see Section 2.1) the volume charge density is rather high: $n_{cr} = 1.5 \times 10^{15}$ cm^{-3} at $V_0 = 5$ V and $n_{cr} = 10^{14}$ cm^{-3} at $V_0 = 100$ V.

At $T < 450$ K the dV/dt effect has been found to be anomalous within the whole range of the V_0 values for a SiC thyristor. One can see this easily by comparing the $\Delta V(V_0)$ dependencies for SiC and Si thyristors (Fig. 34).

A conventional dV/dt effect is realized usually at an "infinitely long" current pulse. That is, the exponential increase of the anode current begins before the end of the dV/dt pulse as given by Eq (50). It is possible to turn-on the thyristor by a shorter pulse duration, however, that requires a correspondingly larger pulse amplitude ΔV. It can be seen from Fig. 34 that for a Si thyristor, a smaller pulse duration requires a larger ΔV magnitude over the whole range of dc voltage V_0. Such kind of $\Delta V(t)$ dependencies is typical for all Si and GaAs thyristors at any temperatures.

One can see in Fig. 34 that the opposite is true for SiC thyristor, however. At $V_0 < 20$ V, a smaller pulse duration requires a smaller ΔV magnitude to turn the thyristor on. To appreciate better this very unusual situation, let us assume, for example, that at $V_0 = 10$ V, a dV/dt pulse is applied to the thyristor. The

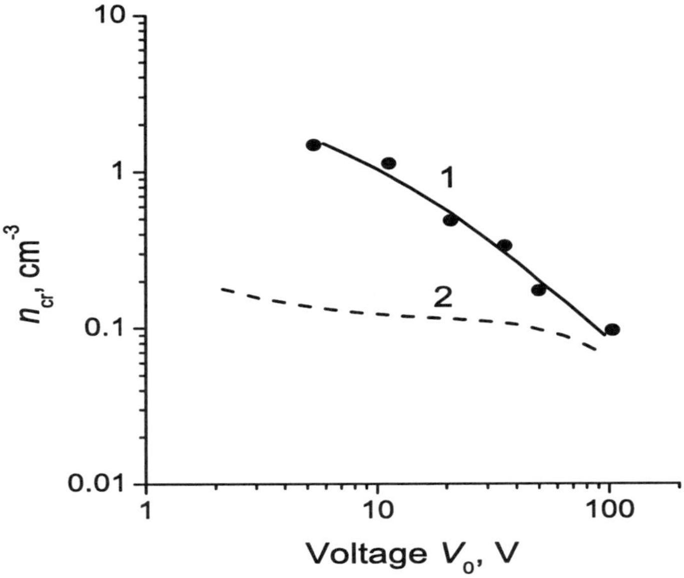

Fig. 33. Dependencies of the critical charge density n_{cr} versus forward voltage V_0 for 400 V 4H-SiC thyristor at 560 K (curve 1) and for rated GaAs thyristor (300 K) (Ref. [97]).

magnitude of the pulse ΔV is large enough to turn the thyristor on at $t = 20$ ns. However, if the pulse is not interrupted after 20 ns, the thyristor is not turned on. Hence the applied pulse impedes the turn on process. Only at $V_0 > 140$ V, the $\Delta V(t)$ dependence takes the conventional form.

This unusual behavior of dV/dt effect in low-voltage 4H-SiC thyristors was qualitatively explained in Refs. [57,96] by the very high concentration of deep levels close to central (blocking) $p-n$ junction of the thyristor structures. It is worth noting, however, that there is no clear physical picture of this phenomenon until now.

6.2. *Critical charge in high-voltage thyristor structures*

The critical charge in high-voltage 2.6 kV 4H-SiC thyristors was studied in Refs. [98,99]. Two techniques were used to determine the critical charge density in the high-voltage structures. The thyristors were switched-on by short ultraviolet (UV) light pulses (optical triggering) and by the conventional gate-switching.

Gate switch-on pulses with rise time of 50 ns were applied between the grounded anode and the gate pads at a frequency of 1 Hz to ensure isothermal conditions. The pulse duration was 30 μs, which much exceeded all the characteristic times of the thyristor structure: carrier lifetime in the blocking base, τ_n, (Fig. 14) and characteristic time constant of current rise, τ_r (Figs. 7-9).

The optical triggering of the thyristor structures was effected by light pulses of an LGI-21 UV laser with wavelength of 337 nm (photon energy $h\nu = 3.67$ eV). The

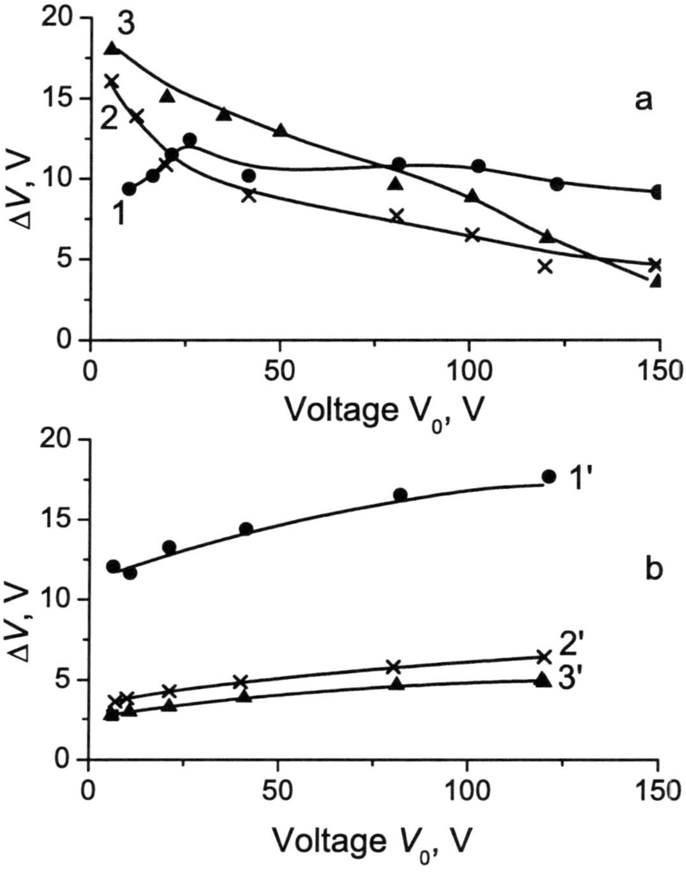

Fig. 34. Dependencies of the pulse amplitude ΔV_c turned-on the thyristor by dV/dt effect versus forward voltage V_0 at 300 K (Ref. [97]). (a) 4H-SiC thyristor. Pulse duration (ns): 1 — 20, 2 — 130, 3 — 400. (b) Commercial Si thyristor KU-103. Pulse duration (ns): 1' — 20, 2' — 130, 3' — 400.

pulse duration was 7 ns, and the total energy in the pulse was about 1.5×10^{-5} J. To attenuate the light intensity, calibrated UV glass filters SZS-21 and SZS-23 were used. The structure was illuminated from the top (see Fig. 6). The diameter of the light spot d_0 was equal to 3 mm.

One of the simplest ways to estimate the critical charge density was proposed in Ref. [37]. In the general case, the critical charge per unit area, Q_{cr}, can be represented as:

$$Q_{cr} = j_{R0} \times \tau_r \tag{51}$$

where j_{R0} is the effective leakage current density of the emitter $p^+ - n$ junction, and τ_r is the switch-on current rise time.

According to Ref. [38], the $(j_R)_0$ directly relates to the gate current pulse duration

t_g and gate current density j_g which can switch the thyristor on:

$$j_g = \frac{j_{R0}}{1 - \exp(-t_g/\tau_r)} \tag{52}$$

It follows from Eq. (52) that at $t_g \gg \tau_r$, the gate current density tends to its minimal value $j_{gmin} = j_{R_0}$, and equation (47) can be written as:

$$Q_{cr} = j_{gmin}\tau_r \tag{53}$$

where j_{gmin} is the minimum gate current density that can switch the thyristor on. Then, the critical charge density n_{cr} is given by:

$$n_{cr} \approx \frac{Q_{cr}}{qW_1} \approx \frac{j_{gmin}\tau_r}{qW_1}, \tag{54}$$

Figure 35 shows the temperature dependence of the minimum gate current i_{gmin} required to switch the thyristor on. The inset in Fig. 35 shows the temperature dependence of the current rise time τ_r for the thyristor in question, measured in Ref. [15]. It is seen that both i_{gmin} and τ_r decrease monotonically with increasing temperature across the entire temperature interval studied (300 – 500 K).

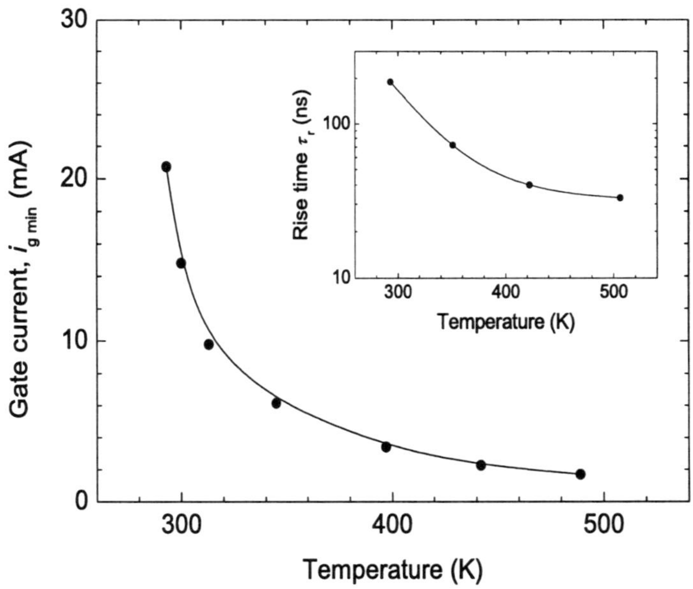

Fig. 35. Temperature dependence of the minimum gate current i_{gmin} required to switch the thyristor on. Cathode voltage $V_c = 300$ V, load resistance $R_L = 30\ \Omega$. Inset shows the temperature dependence of the current rise time τ_r (Ref. [15]).

It worth noting that rather long gate pulses should be used for the experimental results to be compared correctly with Eqs. (52) and (53). The minimum gate current that can switch the thyristor on strongly depends on the gate pulse duration t_g even at $t_g \geq \tau_n, \tau_r$. With $\tau_n = 600$ ns and $\tau_r = 200$ ns at room temperature, the minimum duration of the gate current pulse must be no less than approximately 10 μs.

At room temperature, with $i_{gmin} = 20.5$ mA (see Fig. 35) and $\tau_r = 200$ ns (see inset in Fig. 35), $j_{gmin} \approx i_{gmin}/S \approx 5.5$ A/cm^2, $Q_{cr} \approx 1.1 \times 10^{-6}$ C/cm^2, and $n_{cr} \approx 2.7 \times 10^{16}$ cm^{-3}.

Because both i_{gmin} and τ_r decrease with temperature growth, the critical charge density reduces monotonically with temperature increasing. At 500 K, with $i_{gmin} = 1.82$ mA (Fig. 35) and $\tau_r \approx 35$ ns (insert in Fig. 35) we have $j_{gmin} \approx i_{gmin}/S \approx 0.5$ A/cm^2, $Q_{cr} \approx 1.7 \times 10^{-8}$ C/cm^2, and $n_{cr} \approx 4.3 \times 10^{14}$ cm^{-3}.

Another independent way to estimate the critical charge density n_{cr} is the optical triggering of a thyristor [30].

Curve 1 in Fig. 10 represents the transient process at pulse energy $J = 0.75$ μJ. This J value is very close to minimum threshold energy J_{th} required to switch the thyristor on. At laser wavelength of 337 nm (photon energy $h\nu = 3.67$ eV), the absorption coefficient α of 4H-SiC is approximately 10^3 cm^{-1} [53]. This means that only a relatively small fraction of the photons penetrating into the thyristor structure is absorbed in the n-base of thickness $W_n = 2.5$ μm. This fraction can be estimated as:

$$N_n \approx N_0 \left[1 - \exp(-\alpha W_n)\right] \approx \alpha W_n N_0 \tag{55}$$

where N_0 is the total number of photons penetrating into the structure. With $\alpha \approx 10^3$ cm^{-1} and $W_n = 2.5$ μm, only about 25% of these photons are absorbed in the thin base of the thyristor.

The non-equilibrium carriers introduced in thyristor bases, both thin and thick, can switch the thyristor on. However, the efficiency of the carriers introduced into a thick base is much smaller than that of carriers generated in a thin base. With non-equilibrium carriers in both bases, the condition for the switch-on has the form [37]:

$$Q_1 + k_2 Q_2 \approx Q_{cr} \tag{56}$$

where Q_1 and Q_2 are the non-equilibrium charges (per unit area) introduced into the thin and blocking bases, respectively; $k_2 = (\Theta_1/\Theta_2)^{1/2}$ is the relative efficiency coefficient of the bases (thick blocking base with respect to the thin base), and Θ_1 and Θ_2 are the characteristic diffusion times for the thin and thick blocking bases, respectively (see Eq. (3)).

With $D_p \approx 2.5$ cm^2/s in the n-base with a doping level of (2–5)$\times 10^{17}$ cm^{-3} and $D_n \approx 20$ cm^2/s in low-doped p-base with a doping level of (7–9)$\times 10^{14}$ cm^{-3} (Fig. 6), $k_2 \approx 0.14$. Taking into account that $Q_2 \approx 3Q_1$, we have from Eq. (56) a simple expression for the critical charge Q_{cr}: $Q_{cr} \approx 1.42\, Q_1$.

The total number of photons penetrating into the structure per unit area, N_{0s}, at threshold light intensity J_{th} can be estimated as follows:

$$N_{0s} \sim \frac{(1-R)J_{th}}{2h\nu(\pi d_0^2/4)} \tag{57}$$

where $R \approx 0.25$ is the reflectance of 4H-SiC at $h\nu = 3.67$ eV [53], factor 1/2 in Eq. (57) accounts for the fact that half of the incident light is reflected from the top metallization.

With $J_{th} = 0.75$ μJ and light spot diameter d_0 of about 3 mm, N_{0s} is equal to 6.8×10^{12} cm^{-2}. The charge density introduced into the thin base of the thyristor, $n_1 \approx N_{0s}/4W_n$, and $n_{cr} \approx 1.42 n_1 \approx 9.5 \times 10^{15}$ cm^{-3}; $Q_{cr} \approx 0.38 \times 10^{-6}$ C/cm^2.

Taking into account that the techniques used here to find the critical charge density actually represent two opposite limiting cases in defining the critical charge, the agreement between the n_{cr} values obtained by these techniques can be considered quite reasonable [100].

To simulate pulse-gate-controlled switching and optical triggering, the "INVESTIGATION" software was used (see Sections 2.2, 3.2, and 4.2). Figure 36 presents the calculated temperature dependence of the minimum gate current i_{gmin} required to switch the thyristor on.

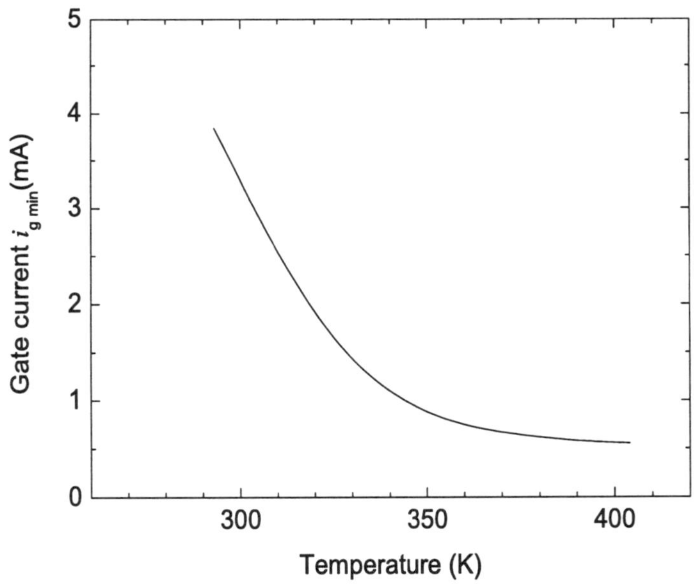

Fig. 36. Calculated temperature dependence of the minimum gate current i_{gmin} required to switch the thyristor on. The dependence is calculated in the absence of any shunt ($R_{sh} \rightarrow \infty$) (Ref. [98]).

The $i_{gmin}(T)$ dependence in Fig. 36 is calculated for the case of an unshunted anode-gate junction. It can be seen that the calculated temperature dependence is well consistent with the corresponding experimental dependence (Fig. 35), with, however, the calculated $i_{g\,min}$ values being about 5 times smaller than experimentally measured current densities.

It is well known that artificial shunting is used in Si thyristors to raise the critical charge, make higher the admissible dV/dt ramp, decrease the sensitivity of the thyristor structures to parasitic switching, etc. In SiC thyristors there are no artificial shunts; however, experiments show that the current-voltage ($I - V$) characteristic of the anode-gate junction differs very essentially from the "classical" one [35]. The simplest way to describe such effects in the framework of the standard theory of the thyristor is to introduce the ohmic (linear) leakage of the gate-anode junction [38,89]. However, calculations show that, to get a fivefold increase in i_{gmin}, unacceptably small shunting resistance (about 200 Ω) should be introduced into the calculations.

The temperature dependence of the current rise time τ_r for the thyristor in question was calculated in Ref. [15]. Within the width of the line, the calculated dependence coincides with the experimental curve (insert of Fig. 35). Hence, we can conclude that the calculated value of n_{cr} at room temperature for unshunted thyristor is five times less that experimental one and it is equal to $\sim 0.5 \times 10^{16}$ cm^{-3}.

Triggering of the thyristor by the optical pulse was simulated using the following procedure. At instant of time $t = 0$, a homogeneously distributed electron-hole "plasma" with a given concentration n was generated in both bases of the thyristors at a given cathode voltage V_c and with the series load resistance R_L. At small enough n, a pulse of current through the thyristor was observed; however, the thyristor was not switched on. If n is large enough, the thyristor was switched on. The minimum n_{min} value required to switch the thyristor on was calculated. After that, the n_{cr} value was found as $n_{cr} = 1.4 n_{min} \approx 0.68 \times 10^{16}$ cm^{-3} at room temperature for unshunted structure ($R_{sh} \to \infty$). At 400 K, the calculation gave $n_{cr} = 0.21 \times 10^{16}$ cm^{-3} ($Q_{cr} \approx 9 \times 10^{-8}$ C/cm^2). It is noteworthy that the room-temperature n_{cr} values calculated for these so different ways of thyristor switch-on are in very good agreement.

The results obtained show that n_{cr} established in this paper for high-voltage SiC thyristors is an order of magnitude larger than that in low-voltage SiC thyristors and of the same order as that in the high-voltage Si and GaAs power thyristors.

Experimental results and calculations for unshunted thyristor demonstrate a rather strong temperature dependence of n_{cr}. Such a result would be expected in view of above discussed properties of the $p^{++} - n$ anode-gate junction of silicon carbide structures (see Sections 2.2 and 4.1). As mentioned above, the concentration of holes in the p^{++}–emitter grows exponentially with temperature owing to thermal ionization of Al. Correspondingly, the injection coefficient of the $p^{++} - n$ junction grows with increasing temperature. As a result, i_{gmin}, τ_r, and $Q_{cr} = j_{gmin}\tau_r$ decrease with increasing temperature.

Comparison of the experimental data with the results of calculations clearly show that the experimentally observed minimum gate current i_{gmin}required to switch the thyristor on, is five times higher than the calculated value. It is impossible to achieve agreement between the calculated and experimental results with any reasonable shunting resistance of the anode-gate junction. This fact demonstrates evidently that Uvarov's theory of critical charge [37] should be modified for the case of SiC thyristors.

6.3. *Critical Charge Concept for 4H-SiC-based Thyristors*

The classical theory of the critical charge [37] assumes that γ_2, α_{T1}, and α_{T2} are constants, and only the effective injection coefficient γ_1 depends on the current density j. In Si thyristors, for which the classical theory [37] has been developed, such an approximation is valid in most cases. First of all, even in the highest-voltage Si-thyristors, the switch-on is achieved, as a rule, at low injection level in a low-doped blocking base. In this case, the transport factors α_{T1} and α_{T2} can be considered constant. In addition, the junction between the emitter and the thin highly doped base in Si thyristors is artificially shunted by the distributed resistance R_{sh}. If so, the dependence of γ_1 on j is extremely sharp near the switching point, and the $\gamma_2(j)$ dependence can be neglected.

At the same time, all SiC thyristors described until now have been fabricated without any artificial shunting. Moreover, as we mentioned above, the minimum gate current density j_{gmin} that can switch the thyristor on is rather high, and transition from low to high injection level can occur in the thick blocking base of the thyristor at these gate current densities. Hence, the condition of the thyristor switch-on can be satisfied not through an increase in γ_1, but owing to a rise in α_{T2} with increasing gate current.

The well-known condition of threshold stable state can be written as follows:

$$\gamma_1\alpha_{T1} + \gamma_2\alpha_{T2} = 1 - \frac{j_{k0}}{j}, \tag{58}$$

where j_{k0} is the collector leakage current density, γ_1 and γ_2 are the injection coefficients of the emitter junctions, α_{T1} and α_{T2} are the transport factors for the "component" transistors.

As shown in Ref. [21], the transition from low to high injection level in the low-doped base is accompanied by a change in the transport coefficient α_{T2} from $(\alpha_{T2})_L$ to $(\alpha_{T2})_H$ (see Eqs. (19) and (20)). Numerical estimates give for high-voltage SiC thyristor structure [16] $(\alpha_{T2})_L \approx (0.4 \div 0.5)$ and $\alpha_{T2})_H \approx (0.90 \div 0.95)$, which points to a very essential contribution of the current dependence of α_{T2} to the rise in the left part of Eq. (58) with increasing current density.

To study the problem analytically, the charge control model was used in Ref. [99]. In this technique, the equations describing the hole charge transient in the n-base, $Q_1 = q\int_0^{W_n} p(x)dx$, and the electron charge transient in the p-base, $Q_2 = q\int_0^{W_p} n(x)dx$, have the form:

$$\frac{dQ_1}{dt} = -\left[(1-\gamma_1)k_p + \frac{1}{\tau_p}\right]Q_1 + \gamma_1 k_n Q_2 + \gamma_1(j_{k0} - j_{R0}),$$

$$\frac{dQ_2}{dt} = \gamma_2 k_p Q_1 - \left[(1-\gamma_2)k_n + \frac{1}{\tau_n}\right]Q_2 + \gamma_2 j_{k0} \tag{59}$$

(see comments to Eqs. (5) for notation). The current density $j_{R0}$was defined in Ref. [37] as follows. It was assumed that the current flowing through the emitter p^+-n junction is a sum of the diffusion current and the excess current I_{R0}, which can be regarded as a recombination current, resistivity shunt current, or any additional "leakage" current.

Initial conditions for the set of equations (59) have the form:

$$\begin{aligned} Q_1|_{t=0} &= Q_{10}, \\ Q_2|_{t=0} &= Q_{20}, \end{aligned} \tag{60}$$

where Q_{10} and Q_{20} are the initial amounts of charge in the n- and p^0-bases, respectively.

Then the total current density can be expressed as:

$$j = k_p Q_1 + k_n Q_2 + j_{k0}. \tag{61}$$

Equations (59) can be simplified taking into account the specific features of SiC thyristors. First, the wide gap of SiC allows us to neglect the generation current j_{k0} in the right-hand part of Eqs. (59). Second, owing to the relatively low ionization energy of the donor levels in 4H-SiC, we can consider virtually all donors to be fully ionized at room temperature and above. As a result, at the typical doping levels of the n^+- emitter, $N_d^+ \approx (5-10) \times 10^{18}$ cm^{-3}, and the p^C-base, $N_a \approx (10^{14} \div 10^{15})$ cm^{-3} [15,16,29,63], the injection coefficient of the $n^+ - p^C$junction, γ_2, can be taken equal to unity ($\gamma_2 = 1$).

It is also worth noting that the transition from the low to high injection level in the blocking p^0-base may change the electron lifetime τ_n. In addition, this transition is accompanied by a monotonic increase in the transport factor α_{T2} and, as a result, by a monotonic increase in the parameter k_n. According to Eqs. (58), (59), it is the increase in α_{T2} that is the main factor governing the possibility of switching on a thyristor.

To avoid cumbersome analytical calculations, the τ_n value was changed in Ref. [99] for its effective value τ_{neff}, and the following monotonically growing approximation $k_n(Q_2)$for the parameter k_n was used:

$$\begin{aligned} k_n &= (k_n)_L \equiv \frac{(\alpha_{T2})_L}{(1-(\alpha_{T2})_L)\tau_{neff}}, \quad \text{at} \quad Q_2 \le Q_{2L}, \\ k_n &= k_{neff} - \frac{j_{Reff}}{Q_2}, \quad \text{at} \quad Q_{2L} \le Q_2 \le Q_{2H}, \\ k_n &= (k_n)_H \equiv \frac{(\alpha_{T2})_H}{(1-(\alpha_{T2})_H\tau_{neff}}, \quad \text{at} \quad Q_2 > Q_{2H} \end{aligned} \tag{62}$$

where

$$k_{neff} = \frac{(k_n)_H Q_{2H} - (k_n)_L Q_{2L}}{Q_{2H} - Q_{2L}},$$

$$j_{Reff} = \frac{((k_n)_H - (k_n)_L)\, Q_{2L} Q_{2H}}{Q_{2H} - Q_{2L}}, \quad Q_{2L} = \zeta q N_a W_p,$$

$$Q_{2H} = (1/\zeta)\, qN_aW_p,$$

ζ is a small parameter defining the injection level in the p^0-base ($\zeta \approx 0.1$). Note that k_{neff} is close to the $(k_n)_H$:

$$k_{neff} = \frac{(k_n)_H - (k_n)_L\zeta^2}{1-\zeta^2} \approx (k)_H + [(k_n)_H - (k_n)_L]\zeta^2 \approx (k_n)_H \tag{63}$$

Under the assumptions made, the set of equations (59) in the region of interest ($Q_{2L} \leq Q_2 \leq Q_{2H}$) has a form:

$$\frac{dQ_1}{dt} = -\left[(1-\gamma_1)k_p + \frac{1}{\tau_p}\right]Q_1 + \gamma_1 k_{neff}Q_2 - \gamma_1(j_{Reff} + j_{R0}),$$

$$\frac{dQ_2}{dt} = k_pQ_1 - \frac{1}{\tau_{neff}}Q_2 \tag{64}$$

In general form, the solution of the set (64) with the initial conditions (60) can be written as follows [101]:

$$Q_1 = \frac{\lambda_1 - b_2}{a_2}C_1\exp(\lambda_1 t) + \frac{\lambda_2 - b_2}{a_2}C_2\exp(\lambda_2 t) - \frac{b_2d_1}{b_2a_1 - b_1a_2},$$

$$Q_2 = C_1\exp(\lambda_1 t) + C_2\exp(\lambda_2 t) + \frac{a_2d_1}{b_2a_1 - b_1a_2}, \tag{65}$$

where constants C_1 and C_2 are defined by initial conditions:

$$C_1 = \frac{a_2}{\lambda_1 - \lambda_2}\left[Q_{10} - \frac{\lambda_2 - b_2}{a_2}Q_{20} + \frac{d_1}{\lambda_1}\right],$$

$$C_2 = -\frac{a_2}{\lambda_1 - \lambda_2}\left[Q_{10} - \frac{\lambda_1 - b_2}{a_2}Q_{20} + \frac{d_1}{\lambda_2}\right]. \tag{66}$$

The parameters a_1, b_1, a_2, and b_2 can be written as:

$$a_1 = -\left[(1-\gamma_1)k_p + \frac{1}{\tau_p}\right]; \qquad b_1 = \gamma_1 k_{neff};$$

$$a_2 = k_p, \quad b_2 = -\frac{1}{\tau_{neff}}, \quad d_1 = -\gamma_1(j_{Reff} + j_{R0}),$$

and λ_1 and λ_2 are the roots of the characteristic equation of the set (64), which has the following form:

$$\lambda^2 - (a_1 + b_2)\lambda + a_1b_2 - a_2b_1 = 0, \tag{67}$$

When the condition for the thyristor switch-on: $\gamma_1\alpha_{T1} + \gamma_2\alpha_{T2} > 1$ is satisfied, then $a_1b_2 - a_2b_1 < 0$ [15]. In this case the roots of the characteristic equation, λ_1 and λ_2, are real and have different signs. Let us suppose for the sake of definiteness that $\lambda_1 > 0$, and $\lambda_2 < 0$. Substituting the values of λ_1 and λ_2 in Eq.(65), and the resulting Q_1 and Q_2 values in Eq. (61), one can show that the condition defining the existence of the stable threshold state has the following form:

$$C_1 = 0 \tag{68}$$

Taking into account that $1/\lambda_1 \equiv \tau_r$ (here τ_r is thyristor switch-on time constant), we obtain from Eq. (68) the following condition for the existence of the stable threshold state

$$Q_{cr} = Q_{10} + \chi Q_{20}, \tag{69}$$

where $\chi = (b_2 - \lambda_2)/a_2$ is the efficiency coefficient of the charge stored in the blocking base [37], and the critical charge is given by the following expression:

$$Q_{cr} = \tau_r \gamma_1 (j_{R0} + j_{Reff}). \tag{70}$$

As mentioned above, the "classical" expression for the critical charge, obtained in Ref. [37] for Si thyristors has the form $(Q_{cr})_{Si} = \tau_r j_{R0}$ (see Eq. (51)). Comparing expressions (51) and (70), we can see, first of all, that the right-hand part of Eq. (70) contains a factor γ_1, which is the injection coefficient of the junction between p^+ emitter and thin highly doped base. In Uvarov's theory [37], the injection coefficients of the both emitter junctions are taken to be unity. In SiC thyristors, such an assumption cannot be justified for γ_1. The high ionization energy of the acceptor levels in the p^+ emitter makes the injection coefficient of the $p^+ - n$ junction, γ_1, at room temperature much lower.

At first glance, the character of the dependence of Q_{cr} on γ_1 looks paradoxical. Indeed, it is clear from the simple physical considerations that the larger γ_1, the smaller the critical charge.

The point is that in the approach developed here, τ_r is a function of γ_1. Using the expression for τ_r, (see Eqs. (21) and (22)) one can readily show that the product $\tau_r \gamma_1$ decreases with increasing γ_1, which results in the following inequality

$$\frac{dQ_{cr}}{d\gamma_1} = -\frac{Q_{cr}}{\gamma_1} \frac{\left[(\gamma_1 \alpha_{T1} + \alpha_{T2} - 1)^2 + \alpha_{T2}(1 - \alpha_{T2})\right]}{(\gamma_1 \alpha_{T1} + \alpha_{T2} - 1)(1 - \gamma_1 \alpha_{T1})} < 0. \tag{71}$$

Hence, we can conclude that Q_{cr} decreases with increasing γ_1 in conformity with simple physical considerations.

The second important difference between Eqs. (51) and (70) is the presence of the term j_{Reff} in Eq. (70). This term characterizes the contribution from the new mechanism of the critical charge formation, governed by a change in α_{T2} with increasing current density. The physical meaning of this term is very clear. If a thyristor structure cannot be switched on at low injection level, an additional charge of minority carriers, corresponding to transition from low to high injection level must be stored in the thyristor base. The j_{Reff} term characterizes just this mechanism of additional charge storage.

The appropriate simulations were performed in Ref. [99] using a quasi one-dimensional computer program "INVESTIGATION" (See Sections 2.2, 3.2, and 4.2). Figure 37 shows the hole and electron distributions calculated for 2.6-kV 4H-SiC thyristor (see Fig. 6) under the conditions that are in line with the experimental conditions of Ref. [98] (T = 300, cathode bias V_0 = 300 V). A critical gate current with density $j_{gmin} = 1.04$ A/cm^2 flows through the anode-gate circuit. (At $j_g > j_{gmin}$, the thyristor is switched on). It can be seen in Fig. 37 that there is an intermediate, rather than low, injection level in the quasi-neutral part of the blocking base. Hence, computer calculations demonstrate clearly that Uvarov's model [37], which has been developed for low injection level in the blocking base, cannot be applied adequately to describe the switch-on process in 4H-SiC high-voltage thyristors.

The most direct way to compare the predictions of classical Uvarov's model and the model in question is to calculate the density of critical charge Q_{cr} by means of Eqs. (51) and (70). Analytical estimate of $Q_{cr} = \tau_r \gamma_1 (j_{R0} + j_{Reff})$, based on Eqs. (62) and (70), gives a value $Q_{cr} \approx 0.6 \times 10^{-7}$ C/cm^2. Computer simulation allowed us to refine this estimate: according to our simulation, $Q_{cr} \approx 1.05 \times 10^{-7}$ C/cm^2.

At the same time, the critical charge density estimated in terms of Uvarov's model is two orders of magnitude smaller: $(\bar{Q}_{cr})_U = \tau_r j_{Rrec} \approx 2.0 \times 10^{-9}$ C/cm^2.

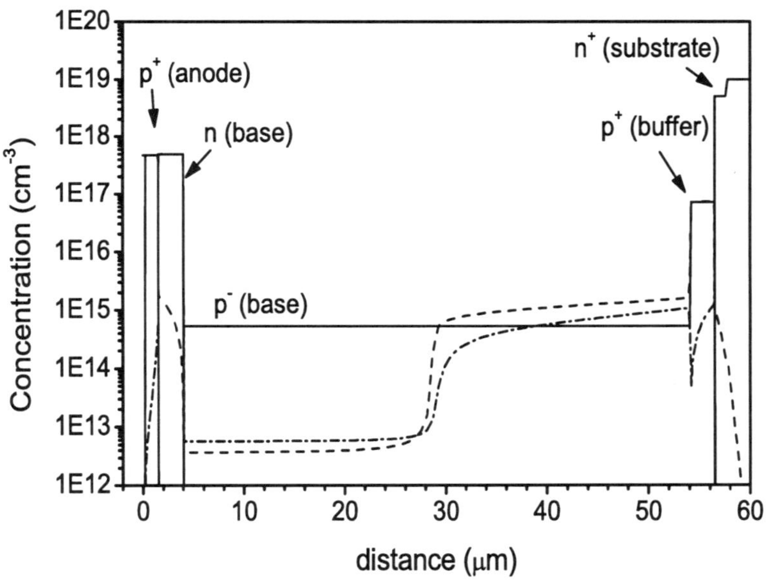

Fig. 37. Calculated hole (dashed line) and electron (dashed-doted line) distributions across thyristor structure at critical gate current density $j_{gmin} = 1.04$ A/cm^2. $T = 300$ K, $V = 300$ V (Ref. [99]).

The experimental Q_{cr} values established in Ref. [98] were $Q_{cr} \approx 1.1 \times 10^{-6}$ C/cm^2 for gate-controlled switching and $Q_{cr} \approx 3.8 \times 10^{-7}$ C/cm^2 for optical switching. Special mention should be made of the fact that, to process the experimental data, γ_1 was taken in Ref. [98] to be unity. The use of the real estimate of γ_1 for the structure in question makes better the agreement between the experimentally found Q_{cr} and the value calculated in terms of the modified model of the critical charge. Thus, it can be seen that the modified model provides reasonable agreement with experimental data. Some quantitative difference can be attributed to the presence of deep traps near the central $p-n$ junction, which cannot be taken into account in this version of the model [57]. The critical charge Q_{cr}predicted by "classical" Uvarov's theory is two orders of magnitude smaller than the experimental value.

7. Conclusion

Recent achievements in the silicon carbide technology resulted in the improvement of all main parameters of bipolar SiC devices.

The reduction in the "micropipe" density caused the increase of the operation area of SiC power devices [102].

SiC layers with the low level of compensation and donor concentration $(N_d - N_a) \sim 10^{14}$ cm^{-3} have been obtained. The $p^+ - n$ junctions on the base of these layers are able to block voltages higher than 20 kV [103].

High level of structural perfection and low concentration of the deep traps provide lifetime of the nonequilibrium carriers as high as several hundreds nanoseconds at room temperature and several microseconds at elevated temperatures in thick (~ 200 μm) 4H SiC layers. So high lifetimes of minority carriers cause the effective modulation of the low doped bases in high voltage bipolar devices at high forward current densities.

These achievements demonstrate clearly that all potential advantages of SiC devices declared for many decades have been already demonstrated experimentally. All main physical features of SiC thyristors such as turn-on and turn-off processes, steady-state current voltage characteristics, frequency properties, and critical charge peculiarities have been analyzed and successfully explained.

Acknowledgments

This work was supported by the Office of Naval Research MURI program under Contract No N00014–95–1–1302, monitored by Dr. J. Zolper. At All-Russia Electrotechnical Institute and Ioffe Physico-Technical Institute this work was supported by Russian Foundation for Basic Research.

References

1. V. A. Dmitriev, S. N Vainshtein, M. E. Levinshtein and V. E. Chelnokov. "First SiC dynistor" *Electron. Lett.*, **24**, 1031-1033 (1988).
2. J. A. Edmund, J. W. Palmour, C. H. Carter. "Junction devices in 6H-SiC" *Proc. Int. Semicond. Dev. Res. Symposium. ISDRS-91,* Charlottesville, p. 487, Dec. 4-6 (1991)
3. J. W. Palmour, S. T. Allen, D. G. Waltz. "4H-SiC power switching devices" in *Silicon Carbide and Related Materials, Inst. Phys. Conf. Series* **142**, 813 (1995)

4. M. E. Levinshtein, J. W. Palmour, S. L. Rumyantsev, and R. Singh "Turn-on process in 4H-SiC thyristors." in *Int. Phys. Conf. Ser.* No 155, Chapter 8, *Proceedings of 23rd Intern. Symposium on Compound Semiconductors* (ISCS-23), Sept.23-27, St. Petersburg, Russia, 601-603 (1996)
5. N. V. Dyakonova, M. E. Levinshtein, J. W. Palmour, S. L. Rumyantsev and R. Singh, "Temperature dependence of turn-on process in 4H-SiC thyristors" *Electron. Letters*, **33**, 914–915 (1997)
6. M. E. Levinshtein, J. W. Palmour, S. L. Rumyantsev and R. Singh, "Turn-on process of 4H-SiC thyristors" *IEEE Transactions on Electron Devices* **ED-44** (7), 1177-1179 (1997)
7. S .N. Vainshtein, Yu. V. Zhilyaev, M. E. Levinshtein *Sov. Phys. Techn. Lett.*, **31**, 788, (1986)
8. M. E. Levinshtein, S. V. Shenderei, *Sov. Phys. Semicond.*, **13**, 593, (1979)
9. An. A. Lebedev, A. I. Uvarov, V. E. Chelnokov, *Radiotechnika i Electronika*, **11**, 1458 (1966) (in Russian).
10. W. Gerlach, Thyristoren. Springer-Verlag, Heidelberg, 1981
11. A. M. Strel'chuk, 'Lifetimes and diffusion lengths of nonequilibrium charge carriers in SiC $p-n$ structures', *Semiconductors*, **29**, 614-623 (1995)
12. V. I. Brylevskii, V. P. Reshetin, V. B. Shuman, "Influence of γ-radiation on dynamic characteristics of thyristors", *Sov. Phys. Semicond.*, **11**, 822-824 (1977)
13. N.V. Dyakonova, M. E. Levinshtein, J.W. Palmour, S. L.Rumyantsev, R. Singh, "Turn-on Process in High Voltage 4H-SiC Thyristors" *Semicond. Sci. Technol.* **13** (2), 241-243 (1998)
14. A. K. Agarwal, S. H. Ryu, R Singh, J. W.Palmour "2600 V, 12 A, 4H-SiC, asymmetrical gate turn-off (GTO) thyristor development" *Abstracts of ICSCRM'99*, Research Triangle Park, NC, October 10-15, 1999, Abstract (474).
15. M. E. Levinshtein, T. T. Mnatsakanov, P. A. Ivanov, A. K. Agarwal, J. W. Palmour, S. L. Rumyantsev, A. G. Tandoev, S. N. Yurkov "Temperature dependence of turn-on processes in 4H-SiC thyristors" *Solid-State Electronics*, **45** (3), 453-459 (2001)
16. T. T. Mnatsakanov, M. E. Levinshtein, S. N. Yurkov, P. A. Ivanov, A. G. Tandoev, J. W. Palmour, and A. K. Agarwal "Current rise time constants in switch-on process of SiC thyristors" *Solid-State Electronics* **46** (4), 525-528 (2002)
17. L. Cao, B. Li, J. H. Zhao "Characterization of 4H-SiC gate turn-off thyristor" *Solid State Electronics* **44** (2), 347-352 (2000)
18. G. D. Bergman "The gate-triggered turn-on process in thyristors" *Solid State Electronics*; **8**, 757-765 (1965)
19. M. E. Levinshtein, S. L. Rumyantsev, and M. S. Shur, Editors: of Handbook Series of Semiconductor Parameters, v.1: Elementary Semiconductors and A3B5 Compounds, Si, Ge, C, GaAs,GaP,GaSb, InAs,InP,InSb. World Sci Publ Co, 1996.
20. F. E. Gentry, F.W. Gutzwiller, N. Holohyak, E. E. Von Zastrow Semiconductor controlled rectifiers. Englewood Clifs, N.J. Prentice-Hall, 1964.
21. A. Blicher Thyristor physics. Springer Verlag, New York-Heidelberg-Berlin, 1976.
22. E. Kamke Differential Gleichungen. Leipzig, 1959.
23. T. T. Mnatsakanov, I. L. Rostovtsev, N. I. Philatov, "Investigation of the effect of nonlinear physical phenomena on charge carrier transport in semiconductor devices" *Solid State Electron.*, **30**, 579-585 (1987)
24. U. Lindefelt "Doping-induced band edge displacements and band gap narrowing in 3C-, 4H-, 6H-SiC, and Si" *Journ. Appl. Phys.,* **84,** 2628-2637 (1998)
25. A. Galeskas, J. Linnros, B. Breitholtz "Time-resolved imaging of radiative recombination in 4H-SiC p-i-n diodes" *Appl. Phys. Lett.*, **74,** 3398-3400 (1999)

26. S. M. Sze "Physics of Semiconductor Devices". John Wiley & Sons, New York, 1981
27. P. T. Landsberg, G. S. Kousik "The Connection Between Carrier Lifetime And Doping Density In Non-Degenerate Semiconductors" *Journ. Appl. Phys.*, **56**, 1696-1700 (1984)
28. W. J. Schaffer, H. S. Kong, G. N. Negley, and J.W. Palmour "Conductivity anisotropy in epitaxial 6H and 4H SiC" in *Diamond, Silicon Carbide and Nitride Wide Bandgap Semiconductors*, ed. C. H. Carter, Jr, G. Gildenblat, S. Nakamura, R. Nemanich, MRS Symposium Proc., v. **339**, Pittsburg, 1994, p. 595.
29. P. A. Ivanov, M. E. Levinshtein, S. L. Rumyantsev , A. K. Agarwal, and J. W. Palmour "Turn-Off Operation of a MOS-Gate 2.6 kV 4H-SiC GTO Thyristor" *Solid-State Electron.*, **44** (11), 2155-2159 (2000)
30. M. E. Levinshtein, P.A. Ivanov, A.K. Agarwal and J.W. Palmour "Optical switch-on of silicon carbide thyristor" *Electron. Letters*, **38** (11), 592-593 (2002)
31. A. A. Lebedev, A. I Uvarov "On the theory of switch-on process in p-n-p-n structure" *Sov. Phys. Semicond.*, **1** (2), 211-216 (1967) (page numbers are given for the original Russian edition "Fizika i Tekhnika Poliprovodnikov").
32. V. A. Kuzmin "The thyristors of low and middle power" Moscow: Soviet Radio, 1971 (in Russian)
33. A. K. Agarwal, P. A. Ivanov, M. E. Levinshtein, J. W. Palmour, S. L. Rumyantsev, S. H. Ryu, M. S. Shur in: *Silicon Carbide and Related Materials ECSCRM2000. TRANS TECH PUBLICATIONS LTD,* Switzerland, Pensl G., Stephani D., Hundhausen M., editors, Materials Science Forum Vols **353-356,** 743-746 (2001)
34. N. V. Dyakonova, P. A. Ivanov, V. A. Kozlov, M. E. Levinshtein, J. W. Palmour, S. L. Rumyantsev, R. Singh "Steady-state and transient forward current-voltage characteristics of 4H-silicon carbide 5.5 kV diodes at high and superhigh current densities" *IEEE Transactions on Electron Devices* **46** (10), 2188-2194 (1999)
35. P. A. Ivanov, M. E. Levinshtein, J.W. Palmour, S. L. Rumyantsev, R. Singh "Classical" current-voltage characteristics of 4H-SiC p^+n junction diodes" *Semicond. Sci. Technol.* **15**(9), 908-910 (2000)
36. A. K. Agarwal, S. Seshadri, M. MacMillan, S. S. Mani, J. Casady, P. Sanger, P. Shah "4H-SiC $p-n$ diodes and gate turn-off thyristors for high-power, high-temperature applications" *Solid State Electronics*; **44**, 303-308 (2000)
37. A. I. Uvarov "Critical turn-on charge of a thyristor" in: Ryvkin S. M., Shmartsev Yu. V., editors. *Physics of p-n junctions and semiconductor devices.* New York-London: Consultants Bureau, 1971, 170-179
38. A. I. Uvarov "Conditions for turning on a thyristor by short gate current pulses" in: Ryvkin S. M., Shmartsev Yu. V., editors. *Physics of p-n junctions and semiconductor devices.* New York-London: Consultants Bureau, p. 216-223 1971
39. D. H. Auston. "Picosecond optoelectronic switching and gating in silicon." *Appl. Phys. Lett.*, **26**, 101-103 (1975)
40. G. Bickel, P. Heremans, M. Kuijk, R. Vounkx, and G. Gorghs. "Receiver with optical thyristors operating at 155 Mbit/s with 3 femto-Joule optical inputs." *Appl. Phys. Lett.*, **70**, 750-752 (1997)
41. V. Korobov and and V. Mitin. "Reverse recovery of a GaAs optoelectronic thyristor." *Journ. Appl. Phys.*, **79**, 1143-1150 (1996)
42. Y. Tashiro, K. Kasahara, N. Hamao, M. Sugimoto, and T. Yanase. "High speed response in optoelectronic gated thyristor." *Japan. Journ. Appl. Phys. Letters*, **26**, *L*1014-*L*1016 (1987)
43. J. H. Hur, R.Hadizad, S. G. Hummel, K. M. Dzurko, P. D. Dapkus, H. R. Fetterman, and M. A. Gundersen. "GaAs based opto-thyristor for pulsed power applications."

IEEE Trans. Electron Dev., **ED-37**, 2520-2525 (1990)
44. D. Giorgi, P. K. L. Yu, J. R. Long, V. D. Lew, T. Nouvapanih, and O. S. F. Zuker. "Light-activated silicon diode switches" *Journ. Appl. Phys.*, **63**, 930-933 (1988)
45. O. S . F. Zuker, J. R. Long, V. L. Smith, D. J. Page, and P. L. Hower. "Experimental demonstration of high power fast-rise-time switching in silicon junction semiconductors" *Appl. Phys. Lett.*, **29**, 262-264 (1978)
46. V. M. Aleksandrov, O. I. Buzhinskii, I. V. Grehov, M. E. Levinshtein, A. I. Moshkunov, and V. G. Sergeev. "Many-channel nanosecond commutator for excitation of copper vapor by a transverse discharge" *Kvantovaya Elektronika (Quantum Electronics)*, **8**, 191-193 (1983) (papers are given for original Russian edition)
47. V. A. K. Temple. "Controlled turn-on thyristor." *IEEE Trans. Electron Dev.*, **ED-30**, 816-824 (1983)
48. J. M. Proud and S.L. Norman "High-frequency waveform generation using optoelectronic switching in silicon." *IEEE Trans. on MTT*, **MTT-26**, 137-140 (1978)
49. M. E. Levinshtein, P. A. Ivanov, A. K. Agarwal, and J.W. Palmour "Optical Switch-on of Silicon Carbide Thyristor" *Proc. of 4-th European Conference on Silicon Carbide and Related Materials ECSCRM2002*, September 01-05, 2002, Linkoping, Sweden, Materials Science Forum, Vols 433-436, 851-854 (2003)
50. I. V. Grekhov, M. E. Levinshtein, and V. G. Sergeev. "Homogeneous optical switching of large-area semiconductor structures." *Sov. Phys. Semicond.*, **8**, 431-434 (1974)
51. I. V. Grekhov, M. E. Levinshtein, and V. G. Sergeev. "Some new possibilities of fast switching of large-area $p-n-p-n$ structures." *Sov. Phys. Semicond.*, **10**, 206-208 (1976)
52. M. I. Dyakonov and M. E. Levinshtein. "Theory of propagation of the turned-on state in a thyristor." *Sov. Phys. Semicond.*, **12**, 426-433 (1978)
53. "Properties of Advanced Semiconductor Materials: GaN, AIN, InN, BN, SiC, SiGe" ed. by M.E. Levinshtein, S.L Rumyantsev, and M. S. Shur. John Wiley & Sons, 2001
54. K. Xie, J. H. Zhao, J. R. Flemish, T. Burke, W. R. Buchvald, G. Lorenzo, and H. Singh "A high-current and high-temperature 6H-SiC thyristors" *IEEE Trans. Electron Dev. Lett*, **17**, 142-144 (1996)
55. M. E. Levinshtein, J. W. Palmour, S. L. Rumyantsev, and G.S. Singh "Steady state current-voltage characteristics of 4H-SiC thyristors at high and superhigh current densities." *Semicond. Sci. Technol.* **12**, 1498-1499 (1997)
56. M. E. Levinshtein, J. W. Palmour, S. L. Rumyantsev, and R. Singh "Steady state current-voltage characteristics of 4H-SiC thyristors" *Electrochem. Society, Inc. Meeting Abstracts,* v.97-1, Spring meeting - Montreal, Quebec, Canada, May 4-9, 488 (1997)
57. M. E. Levinshtein, J. W. Palmour, S. L. Rumyanetsev, and R. Singh "The critical charge density of 4H-SiC thyristors", *IEEE Trans. Electron Dev.*, **ED-45** (1), 307-312 (1998)
58. M. M. Anikin, P. A.Ivanov, A. A. Lebedev, S. N. Pytko, A. M.Strel'chuk, and A. L. Syrkin, in *"Semiconductor interfaces and microstructures"*, ed. by Z. C. Feng, World Scientific, Singapore-New Jersey-London, 280-311 (1992)
59. M. E. Levinshtein, J. W. Palmour, S. L. Rumyantsev, and R. Singh, "Forward current-voltage characteristics of silicon carbide thyristors and diodes at high current densities" *Semicond. Sci. Technol.***13**, 1006-1010 (1998)
60. M. E. Levinshtein, J. W. Palmour, S. L. Rumyantsev, and G.S. Singh, "Forward current-voltage characteristics of silicon carbide thyristors and diodes at high current densities" *Abstracts of 2nd European Conference on Silicon Carbide and Related Materials*, Montpellier, France, Sept. 02-06, 231-232 (1998)

61. A. Herlet and R. Raithel, " The forward characteristic of silicon power rectifiers at high current densities" *Solid State Electron.*, **11**, 717-742 (1968)
62. L. A. Delimova, Yu. V. Zhilyaev, V. Yu. Kachorovskii, M. E. Levinshtein, V. V. Rossin "Forward current-voltage characteristics of gallium arsenide power diodes at high current densities" *Solid State Electron.*, **31**, 1101-1104 (1988)
63. A. K. Agarwal, P. A. Ivanov, M. E. Levinshtein, J. W. Palmour, S. L. Rumyantsev, and Sei-Hyung Ryu "Turn-off performance of a 2.6 kV 4H-SiC asymmetrical GTO thyristor" *Semicond. Sci. Technol.* **16**(4), 260-262 (2001)
64. P. A. Ivanov, M. E. Levinshtein, K. G. Irvin, O. Kordina, J. W. Palmour, S.L. Rumyantsev and R. Singh "High hole lifetime (3.8 ms) in 4H-SiC diodes with 5.5 kV blocking voltage " *Electron. Lett.*, **35** (15), 1382-1383 (1999)
65. A. K. Agarwal, P. A. Ivanov, M. E. Levinshtein, J. W. Palmour, S. L. Rumyantsev, and M. S. Shur "Turn-off Performance of a 2.6 kV 4H-SiC Asymmetrical GTO Thyristor" *Proceedings of 3rd European Conference on Silicon Carbide and Related Materials ECSCRM2000* September 03 - 07, Klaster Banz, Germany, Materials Science Forum, vols. 353 - 356 (2000), pp. 743-746 (2001)
66. M. E. Levinshtein, P. A. Ivanov, T. T. Mnatsakanov, J. W. Palmour, A. K. Agarwal, and R. Singh "Properties of modern bipolar SiC devices fabricated by CREE Inc" *Abstracts of IV International Seminar on Silicon Carbide and Related Materials* (ISSCRM-2002), May 30-31, 2002, Novgorod the Great, Russia, pp. 83 – 86 (2002)
67. U. Lindefelt "Doping–induced band-edge displacement and bandgap narrowing in 3C-, 4H-, 6H-, and Si" *J. Appl Phys.* **84** (5), 2628-2637 (1998)
68. E. Velmre and A. Udal, *Physica Scripta,* **79**, 193 (1999)
69. T. T. Mnatsakanov, M. E. Levinshtein, P. A. Ivanov, J. W. Palmour, A. G. Tandoev, and S. N. Yurkov "Parameters of electron-hole scattering in silicon carbide" *J. Appl. Phys.* **93** (2), 1095-1098 (2003)
70. T. T. Mnatsakanov, L. I. Pomortseva, V. B. Shuman "Investigation of the effect of Auger recombination on charge carrier transport and static characteristics of silicon multilayer structures" *Solid-State Electronics,* **38** (11), 1871-1880 (1997)
71. T. T. Mnatsakanov, I. L. Rostovtsev, N. I. Philatov "Einstein relation in semiconductors under strong electron-hole scattering conditions" *Sov. Phys. Semicond.,* **18** (7), 807-809 (1984)
72. T. T.Mnatsakanov "Transport equations and Einstein relation in a high density plasma of solids" *phys. status solidi* (*b*), **143** (1), 225-234 (1987)
73. D. E. Kane, R. M. Swanson "Effect of electron-hole scattering on current flow in semiconductors" *J. Appl. Phys.,* **72** (10), 5294-5304 (1992)
74. M. E. Levinshtein, T. T. Mnatsakanov "On the transport equations in popular commercial device simulators *IEEE Trans. on Electron Devices* **49** (4), 702 -703 (2002)
75. T. T. Mnatsakanov, D. Schroder, A. Schlogl "Effect of high injection level phenomena on the feasibility of diffusive approximation in semiconductor device modeling" *Solid-State Electronics* **42** (1), 153-163 (1998)
76. T. T.Mnatsakanov, L. I. Pomortseva, S. N. Yurkov "Semi-empirical model of carrier mobility in SiC for analyzing its dependence on temperature and doping level" *Semiconductors,* **35** (4), 394-397 (2001)
77. J. R. Krausse "Die Abhangigkeit der Tragerbeweglichkeit in Silizium von der Konzentrazion der freuen Ladungstrager". *Solid-State Electronics* **15** (11), 1377-1381 (1972)
78. L. V. Davies "Electron-hole scattering at high injection levels in Germanium", *Nature,* **194** (4830), 762-763 (1962)
79. R. A. Hopfel, J. Shah, P. A. Wolff, A. C. Gossard "Negative absolute mobility of holes in n-doped GaAs quantum wells" *Appl. Phys. Lett.*, **49** (9), 572-574 (1986)

80. B. N. Gresserov, T. T. Mnatsakanov "Influence of of electron-hole scattering on charge carrier transport in GaAs multilayer structures" *Sov. Phys. Semicond.* **24** (9), 1031-1032 (1990)
81. A. K. Agarwal, J. B. Casady, L. B. Rowland, S. Seshadri, R. R. Siergiey, W. P. Valek, C. D. Brandt "700-V asymmetrical 4H-SiC gate turn-off thyristors (GTO's)" *IEEE Electron. Dev. Lett.*, **18** (10), 518-520 (1997)
82. B. Li, L. Cao, J. H. Zhao "High-current density 800-V 4H-SiC gate turn-off thyristors" *IEEE Electron. Dev. Lett.*, **20** (5), 219-222 (1999)
83. A. K. Agarwal, P.A. Ivanov, M.E. Levinshtein, J.W. Palmour, and S. L. Rumyantsev "Dynamic Performance of 2.6 kV 4H-SiC Asymmetrical GTO Thyristors", International Conference on Silicon Carbide and Related Materials ICSCRM 2001, October 28-November 02, 2001, Ibaraki, Japan, pp. 660-661 (2001)
84. I. L. Kaganov and E. S. Malahov *Reports of Scientific-Technical Conference of All-Union Electrotechnical Research Institute,* (1966) Moscow (in Russian)
85. J. B. Casady, A. K. Agarwal, S. Seshadri, R. R. Siergiej, L. B. Rowland, M. F. MacMillan, D. C. Sheridian, P. A. Sanger, C. D. Drandt "4H-SiC power devices for use in power electronic motor control" *Solid-State Electronics* **42** (11), 2165-2176 (1998)
86. M. E. Levinshtein, T. T. Mnatsakanov, S .N. Yurkov, P.A. Ivanov, A. G. Tandoev, A.K. Agarwal and J.W. Palmour "Turn-off operation of a 2.6 kV 4H-SiC gate turn-off thyristor in MOS-gate mode" *Solid State Electronics,* **46** (10), 1953-1957 (2002)
87. M. E. Levinshtein, T. T. Mnatsakanov, S. N. Yurkov, P. A. Ivanov, A. G. Tandoev, A. K. Agarwal, and J.W.Palmour "Turn-off operation of a 2.6 kV 4H_SiC gate turn-off thyristor in MOS-gate mode" The 26-th Workshop on Compound Semiconductor Devices and Integrated Circuits held in Europe (WOCSDICE), May 21 - May 25, 2002, Chernogolovka, Russia, pp.123-124
88. M.E. Levinshtein, J.W. Palmour, S.L. Rumyantsev, and R. Singh, "Frequency properties of 4H-SiC thyristors at high current density" *Semicond. Sci. Technol.* **14** (2), 207-209 (1999)
89. W. Gerlach "Thyristoren", Springer , Heidelberg 1981
90. M. I. Dyakonov and M. E. Levinshtein "Theory of propagation of the turned-on state in a thyristor in the presence of a gate current" *Sov. Phys. Semicond.*, **12,** 992-997 (1978)
91. M. I. Dyakonov and M. E. Levinshtein "Parameters of current filaments in a gate-current-controlled thyristor and its turn-off gain. *Sov. Phys. Semicond.*, **14,** 283-286 (1980)
92. J. R. Hauser "The effects of distributed base potential on emitter-current injection density and effective base resistance for stripe transistor geometries" *IEEE Trans. on Electron Devices* **11** (5), 238-243 (1964)
93. M. Suzuki, N. Sowaki, K. Iwata, T. Nishinaga "Current distributions at the lateral spreading of electron-hole plasma in a thyristor" *IEEE Trans. on Electron Devices,* **29**, 1222-1225 (1982)
94. H. Dodson, R. L. Longini "Probed determination of turn-on spread of large area thyristor" *IEEE Trans. on Electron Devices,* **13**, 478 485 (1966)
95. S. N. Vainshtein, Yu. V. Zhilyaev, M. E. Levinshtein "Propagation of turned-on state in gallium arsenide thyristor" *Sov. Phys. Semicond.*, **21**, 77-80 (1987)
96. N.V. Dyakonova, M.E.Levinshtein, J.W.Palmour, S.L.Rumyantsev, and R.Singh "Main parameters of turn-on process in 4H-SiC thyristors" *4-th Intern. Symp. on Semicond. Device Research. (ISDRS-97),* Charlottesville, USA, Dec. 10-13, 367-370 (1997)

97. S. N. Vainshtein, I. I. Diakonu, Yu. V. Zhilyaev, M. E. Levinshtein "Fundamental switching parameters of gallium arsenide thyristors" *Sov. Phys. Techn. Lett.*, **28**, 359 (1983)
98. M. E. Levinshtein, P. A. Ivanov, T. T. Mnatsakanov, S.N. Yurkov, A. K. Agarwal, and J. W. Palmour "The Critical Charge Density in High Voltage 4H-SiC Thyristors" *Solid State Electron.*, **47**, 699-704 (2003)
99. T. T. Mnatsakanov, S.N. Yurkov, M. E. Levinshtein, A. G. Tandoev, A. K. Agarwal and J. W. Palmour "The critical charge concept for 4H-SiC-based thyristors", *Solid State Electron.*, **47**, 1581-1587 (2003)
100. V. A.Kuzmin, Editor: "Calculations of power semiconductor devices" Moscow, Energiya, 1980 (in Russian)
101. E. Kamke "Differential Gleichungen" Leipzig; 1959
102. R. Singh, D.C. Capell, K.G. Irvine, J.T. Richmond, J.W. Palmour "7.4 kV, 330 A (pulsed), single chip, high temperature 4H-SiC *pin* rectifier " *Electron. Lett.*, **38** (26), 1738-1740 (2002)
103. Y. Sugawara, D. Takayama, K. Asano, R. Singh, J. Palmour and T.Hayashi "12-19kV 4H-SiC pin Diodes with Low Power Loss" *Proceedings of 2001 International Symposium on Power Semiconductor Devices & Ics*, Osaka, Japan (June 4-7, 2001), 27 – 30 (2001)
104. Tanaka Y., Nishizawa S., Fukuda K., Arai K,. Ohno T., Oyanagi N., Suzuki T., Yatsuo T. "Fabrication of 1.4-kV mesa-type p ¿ – n diodes with avalanche breakdown and without forward degradation on high-quality 6H-SiC substrate" *Appl. Phys. Lett.*, **83** (2), 377-379 (2003)

International Journal of High Speed Electronics and Systems
Vol. 15, No. 4 (2005) 997–1033

World Scientific
www.worldscientific.com

SILICON CARBIDE STATIC INDUCTION TRANSISTORS

Gregory C. DeSalvo

Compound Semiconductor Research Group
Science and Technology Center
Northrop Grumman Corporation – Electronic Systems

1. Introduction and History of the Static Induction Transistor

The static induction transistor (SIT) is a three terminal, vertical current flow, semiconductor device that most closely resembles the solid state version of the triode vacuum tube.[1–4] The SIT is also a voltage controlled, majority carrier device, and so it is one version of a field effect transistor (FET), where the voltage on the gate modulates the current flow between the drain and source regions. However, it has a few distinct differences. Most FETs are planar devices and have only a single gate to control current flow in the channel (where all terminals are in the same plane, and current flows horizontally parallel to this plane). However, the SIT is a vertical device, with two gate contacts on opposing sides of its channel. Thus, while planar FETs only exhibit current–voltage (I–V) characteristics similar to pentode vacuum tubes with saturating current levels, the vertical current flow and extra gate control of the SIT allows for an additional I–V regime which exhibits non-saturating current characteristics similar to triode vacuum tubes.[5–8] That is, under high drain bias conditions (V_{DS} = 10's to 1000's V), the SIT exhibits current–voltage characteristics similar to vacuum tube triodes. Therefore, the SIT does not behave exactly as other FET devices, and so it should not be considered as just a "vertical FET".

The static induction transistor was invented by Watanabe and Nishizawa in 1950 as disclosed in a Japanese patent, and was first demonstrated experimentally in 1972 using Si.[20–27] This first implementation of the SIT showed that this device follows a space charge injection model when in triode-like mode.[4,28,29] In 1952, Schockley proposed the "analog transistor" whose current conduction is based on space charge limited current, analogous to vacuum tube triode operation.[22,30] The general characteristics of the analog transistor are similar to static induction current of the SIT under triode-like operation. Although the first SITs were made using Si,[2,3,10–16] and a few SIT designs have used GaAs,[31–37] the most active material system for SIT devices over the past 10 years has been SiC. This is because the static induction transistor is well suited for high frequency and high power applications, due to (a) its short channel length with low doping to minimize the RC time constant (short transit time) for high frequency response, (b) its vertical structure allowing for high voltage operation, and (c) its high density of gate and source fingers allowing for high current operation. The static induction transistor is also considered a low noise device, with wide dynamic range, and provides excellent low harmonic and intermodulation distortion (IMD) when compared to the bipolar junction transistor (BJT) and the planar FET.[28,38,39] Furthermore, the reliability, noise, and radiation hardness of the SIT are superior to the MOSFET.[3]

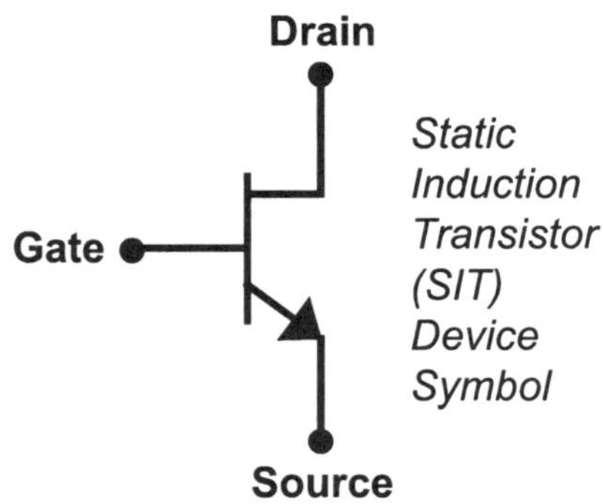

Figure 1. The device symbol for the SIT combines aspects of both the FET and the BJT, due to its early history. The first Si-based SITs used a p/n junction between the gate and source, with current flowing in a vertical direction as for early bipolar junction transistors (BJT). However, the SIT is a voltage controlled, majority carrier device, and behaves as a field effect transistor (FET), as represented by the gate to drain portion of the symbol.[1,3,9–19]

Using SiC material, with its wide energy band gap, very high electric field breakdown strength, large electron saturation velocity, good electron mobility, and excellent thermal conductivity, SiC becomes the ideal semiconductor choice for the static induction transistor to provide very high power at high frequencies. The ability of SiC to operate at a higher junction temperature, combined with its high thermal conductivity, allows the SiC static induction transistor to dissipate higher power than conventional Si transistors.[40,41] SiC SITs have been operated at greater than 225°C without experiencing any deleterious effects. The SiC static induction transistor has demonstrated the highest pulsed power density of any solid state transistor and it is the most advanced microwave power device in SiC.[42] Pulsed SiC SITs are advantageous for state-of-the-art RF transmitter applications in radar and electronic warfare due to their improved system performance, providing lower weight, smaller volume, and lower operating cost.[42]

The first SiC static induction transistor was demonstrated in 1993, which used 6H-SiC material and Schottky barrier sidewall gates. These first SiC SITs produced 37 W of pulsed output power with 60% power added efficiency[a] (PAE) at 175 MHz.[4,43] As SiC

---

[a] Power added efficiency is defined as $PAE = \left(\frac{P_{out}^{RF} - P_{in}^{RF}}{I_{ds}^{DC(Q)} \bullet V_{ds}^{DC(Q)}}\right) \bullet 100\%$, where the numerator represents the difference between the output and input RF power of the device, divided by the DC input power (I × V)

material quality and device processing techniques improved, so did the SiC SIT device performance. The first 4H-SiC SIT was fabricated in 1995, and demonstrated 225 W output power at 600 MHz, with 47% PAE and 8.7 dB gain.[29,44] This resulted in a power density of 1.35 W/mm source periphery, which was significantly higher than 0.76 W/mm at 225 MHz and 0.52 W/mm at 900 MHz achieved with Si. These results are significantly better than Si devices in terms of output power density, breakdown voltage, and frequency response,[29] demonstrating the superior high power capability obtained with an immature SiC technology.[44]

The first practical demonstration of a SiC SIT as a pulsed high power amplifier at radar frequencies occurred in 1996, providing more than 450 W output power at 600 MHz with good linearity, and giving over 1000 W when combined in a multi-transistor module.[45,46] To achieve high power at even higher frequencies, the ion implanted gate SiC SIT was first demonstrated in 2002 and showed 220 W output power (3.38 W/mm) with 52% drain efficiency and 10.4 dB gain at 1.3 GHz, and 156 W output power (2.40 W/mm) and 45% drain efficiency and 8.1 dB gain at 3 GHz.[47] Recent published results have shown that the SiC SIT can produce over 2 kW of pulsed output RF power with 56% PAE and 7.4 dB gain at 450 MHz, 900 W at 1.3 GHz with 65% drain efficiency and 11 dB gain, and 240 W at 3.1 GHz with 29% drain efficiency and 6.5 dB gain.[42] For CW operation, over 100 W has been demonstrated at 750 MHz with 59% PAE and 8.7 dB gain, requiring only air cooling.[48]

2. Static Induction Transistor Device Structures

The static induction transistor is a majority carrier device, and since the electron mobility of the semiconductor is much higher than the hole mobility (~10×), the SIT uses an n-type channel region where electron current flow is modulated. The static induction transistor is a vertical structure device, comprised of multiple channels beneath finger-like protrusions (or pillars) on the semiconductor surface, with the source contact residing on the top of these pillars. Gate finger contacts surround each source pillar, and a blanket drain contact resides under the substrate.[4,49] This shown pictorially in Figure 2. Electrons flow from the source contact at the top of the wafer, moving vertically down towards the drain contact at the bottom of the wafer. This electron current flow is modulated by the gate contacts on both sides of the channel, directly below the source. For SiC SITs, a Schottky barrier gate or an ion implanted gate contact is used to form the

---

biased at the quiescent point (Q) of the load line. Drain efficiency is defined as $DE = \left(\dfrac{P_{out}^{RF}}{I_{ds}^{DC(Q)} \bullet V_{ds}^{DC(Q)}} \right) \bullet 100\%$, which does not subtract out the RF input power, and provides higher efficiency numbers. Power added efficiency is a better figure of merit because it incorporates the transistor power gain. As the power gain ($P_{out}^{RF} / P_{in}^{RF}$) increases, the difference between power added efficiency and drain efficiency becomes small.

rectifying gate region to deplete (and thus control) the electron flow in the channel region. In contrast with the planar FET, moving the drain contact from the topside to the backside allows for a higher density of current carrying channel regions, and thus the SIT provides the highest power density achievable for any transistor configuration. Furthermore, the self shielding and field plate effects built into this vertical structure permit high voltage operation, depending on the type of device edge termination used.

2.1. *SIT device structure layout*

The corresponding symbolic dimensions associated with the detailed SIT structure cross section are shown in Figure 2, and will be referred to repeatedly throughout this chapter. Although this device schematic is an example of an ion implanted gate SIT, the Schottky barrier SIT structure operates in the similar manner as will be described. Regardless of whether the SiC SIT is a Schottky barrier SIT (metal–semiconductor junction) or an ion implanted SIT (p/n junction), both SIT types form a rectifying contact which depletes the channel region and controls current flow in the same manner as described here.

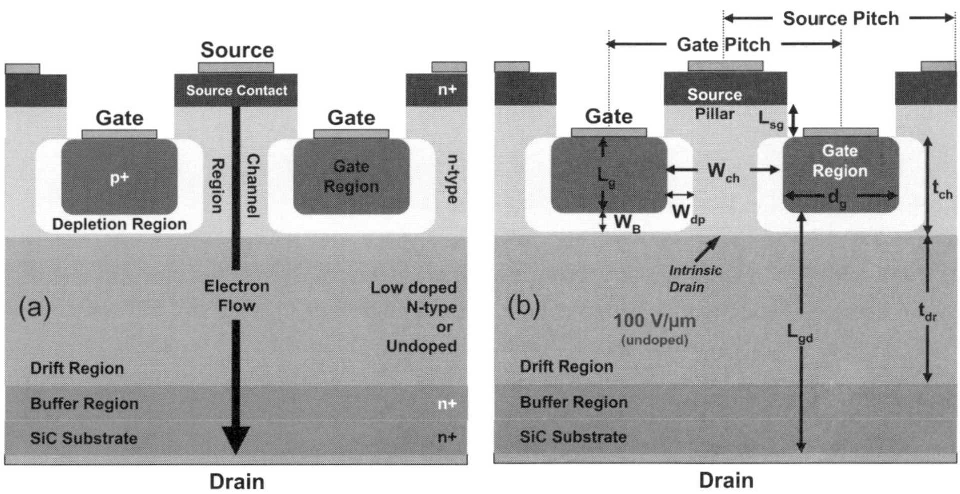

Figure 2. Cross sectional schematic of the static induction transistor structure, showing the (a) device layer component names with the (b) device layer dimensions labeled. An ion implanted gate SIT device structure is shown here, but the Schottky barrier gate SIT structure is very similar (cf. Figure 5). The width of the built-in depletion region is given by W_{dp}, which occurs when no gate bias is present ($V_{GS} = 0$). All other dimensions are fixed and defined either by the growth or the device fabrication process.

The SIT device layout consists of n-type source contact, channel, drift, and buffer layers on an n+ substrate, all consisting of SiC. Most of the time, all of these layers are grown epitaxially, but in some instances the n+ source contact layer is formed instead by ion implantation of nitrogen into the top surface of the channel region.[50] Source fingers (or pillars) are formed lithographically by removing the n+ source contact layer, exposing the channel region. Gates are formed on both sides of the source pillar either by (a) Schottky metallization along the source pillar sidewalls, forming a metal–semiconductor rectifying contact [cf. Figure 5], or by (b) ion implanting a p+ region into the channel layer, forming a p/n junction between the gate p+ region and the n-channel layer. For the ion implanted gate SIT, ohmic metal contacts are then formed on the p+ gate region, usually at the same time as the source metal ohmic contacts are being formed.[47] Further processing steps are conducted to connect a large number of source pillars together in parallel, and also to connect all of the gate fingers in a similar fashion. A cross section of the topside of an ion implanted gate SiC SIT is shown in Figure 3.

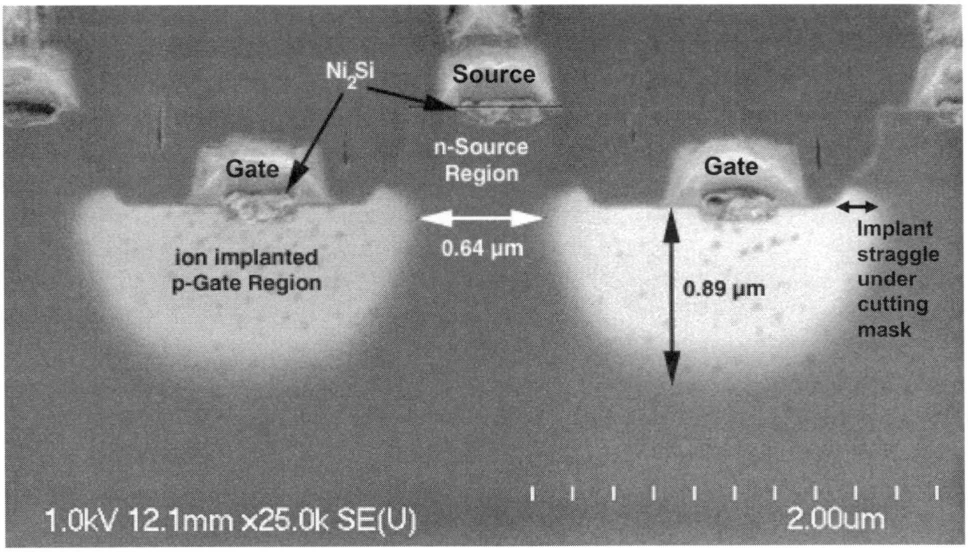

Figure 3. Cross sectional scanning electron microscope (SEM) image of an ion implanted gate SiC static induction transistor. Due to the nature of scanning electron microscopy, the heavily doped p+ gate regions are clearly seen as bright regions on both sides of the channel.

Finally, the wafer backside is metallized, forming an ohmic drain contact to the n+ substrate. For SiC, ohmic contacts are normally formed using Ni metallization, along with a high temperature annealing step for form an ohmic nickel silicide. The completed SiC SIT device contains a large number of gate fingers connected together in parallel and forming a single topside contact pad, and also in a similar manner, the source pillars are connected in parallel and forming a single topside contact pad. This completed device is usually referred to as a device "cell", which are grouped into individual die, consisting of several cells. Presently, the cell size (device active area), and thus the current carrying capacity, is limited by the quality of the SiC substrate material. As the device area becomes large (~0.5 cm^2), the probability of the device lying within a substrate micropipe defect increases dramatically, killing the device and reducing cell yield. Therefore, device areas within the mm^2 size are used, and grouped together into neighboring die. Several die are then mounted into a transistor package, and wire bonded together to operate as a single transistor, as shown graphically in Figure 4. Within the transistor package, tuning is performed using internal capacitors and wire bond inductance to match both the input and output impedance of the combined devices for a given frequency range. In this manner, a large number of devices can be combined to form a transistor package capable of producing kW levels of RF power.

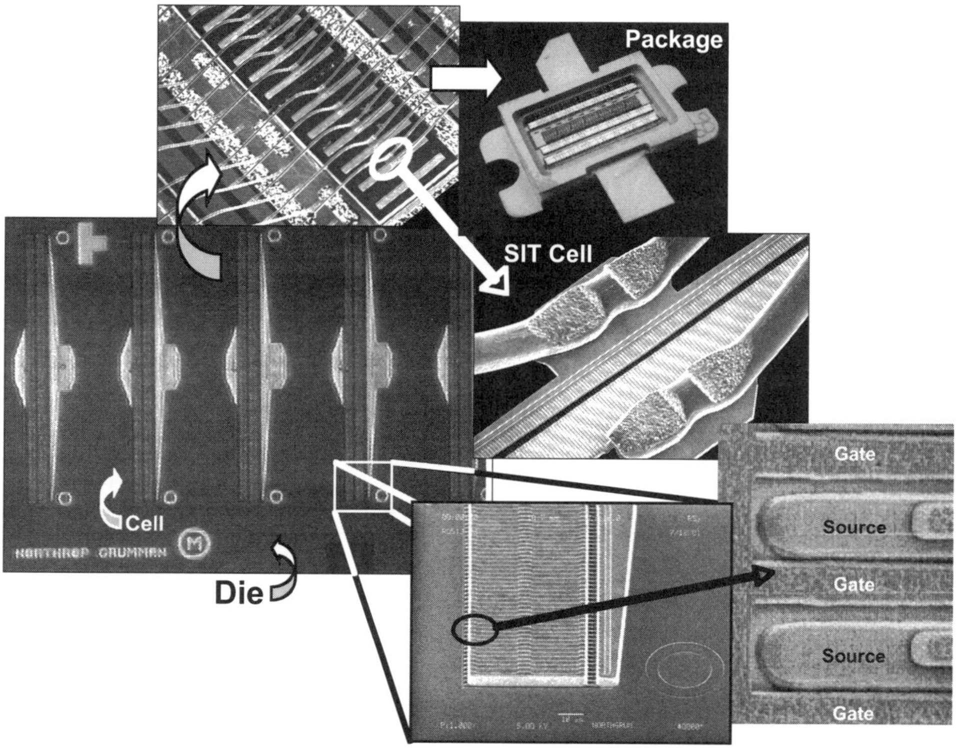

Figure 4. Numerous gate and source fingers are combined together to form a device cell, and several neighboring cells are grouped on the wafer into die. Several die can be separated from the wafer and mounted and wire bonded together into a transistor package, capable of producing high RF output power.

2.2. *Optimization of SiC SIT device structure "properties"*

Each of the device layers is optimized in terms of doping density and thickness depending on the application needs, in order to provide the highest performance needed for the SIT. However, the general device structure optimization includes:

Source Contact Layer: In general, the source contact layer is reasonably thin and doped as heavily n-type as possible, to minimize contact resistance and ohmic losses with the source metal contact. However, this layer must also be thick enough to permit ohmic nickel silicide formation of the source contact without penetrating into the channel layer below (~0.3 μm). This layer can be formed by ion implantation or epitaxially grown in the same manner as the lower layers. Using ion implantation to form the source contact layer allows for capacitance-voltage (C–V) measurement of the channel region's critical doping density before the heavily doped n+ source contact is made. However, this method requires an additional ion implantation process step. On the other hand, epitaxially growing the n+ source contact layer as the last step in the SIT device layer growth requires only one additional layer to the growth process, but does not allow for measurement of the critical channel doping density until most of the processing has been completed.

Channel Layer: The channel layer is usually lightly doped, in an effort to achieve a compromise between doping as high as possible to minimize channel resistance for large current flow, yet doping lightly enough to allow for sufficient channel depletion.[51] Using a light channel doping density allows for higher electron mobility, due to lower scattering by ionized donors, which results in higher frequency response.[50] The same compromise is made for the channel thickness (t_{ch}), where the channel is thin enough to minimize resistance losses for large current flow, while still being thick enough to provide sufficient blocking voltage. As part of controlling current flow, the turn-off voltage (V_{to}) is dependent on both the channel doping and thickness.[52]

Drift Layer: The drift region usually comprises an undoped or lightly doped n-type layer that is grown very thick relative to the channel and source contact layers (several microns). The drift layer thickness primarily determines the breakdown voltage required for device operation, and needs to be thick enough to withstand and block the high gate-drain voltages during normal SIT operation.[4] The maximum blocking voltage (i.e., the breakdown voltage) is a function of the doping concentration and the thickness of this drift region, as well as the critical electric field strength (E_c) of the semiconductor.[4] For SiC, E_c equals 3×10^6 V/cm, providing one of the highest breakdown values among all semiconductors. For practical purposes in SiC SITs, one conservatively assumes a blocking voltage capability of about 100 V for each micron of low doped or undoped drift layer thickness (i.e., 100 V/μm).

Buffer Layer: Finally, the buffer layer is used by all epitaxy growth techniques as an interface layer between the substrate material and the uppermost epitaxial layers (drift, channel, and source contact), and is heavily doped n-type to provide a minimum resistance loss between the drift region and the n+ substrate and drain metal contact.

Separation of Channel and Drift Layer: In the early days of Si-based SIT fabrication, the channel layer doping density was not different from the drift region, as both regions were considered to be the same layer. The objective of having the channel and drift layers as a single, lightly doped layer was to maximize the gate control of the channel region without regard of the channel thickness, and to turn-off the current flow with a minimum gate voltage. However, this leads to a high channel resistance which increases the "on-state" resistance of the device and lowers overall current capacity.[53]

Optimization of the combined channel/drift layer doping density to both minimize resistance and maximize blocking voltage is too problematic. Therefore, by separating the channel doping from the drift doping, providing for two independent and separate regions, one can optimize the drift region for blocking voltage separately while maximizing current capacity through the channel. That is, the channel doping is normally increased to provide for higher current capacity, while the drift region is lightly doped or undoped to maximize the blocking voltage.

As just described, the carrier concentration and depth profile of the SIT is very demanding of the epitaxial growth process, requiring extremes in both doping and thickness.[4,29] That is, the drift region requires controlled, undoped or low n-type doping ($\sim 10^{15}$ cm^{-3}) that can be several microns thick (> 5 μm), while the source contact requires very high doping concentrations ($>10^{19}$ cm^{-3}) in a very thin (0.2 μm) n+ layer.[4,29] Accurate control of the epitaxy doping level and layer thicknesses is important in achieving the designed turn-off voltage at the correct gate and drain biases, without degrading the on-state performance of the SIT.[54]

The SIT is a structurally sensitive device, and its electrical parameters strongly depend on the device structure and its material parameters.[55] Table 1 shows a comparison of the important electrical parameters that are controlled by the material parameters of the device structure.

Table 1. Comparison of important SIT electrical parameters which are primarily influenced by the corresponding device structure material variables. Separation of the channel and drift layers allows for independent control of most of the electrical parameters.

Static Induction Transistor Material Parameter:

	Channel			Drift		Source
Electrical Parameter:	Width (W_{ch})	Thickness (t_{ch})	Doping (N_{ch})	Thickness (t_{dr})	Doping (N_{dr})	Height (L_{sg})
Current Capacity (I_{dss} or I_{max})	√	√	√			
Transconductance (g_m)	√	√	√			
On-State Resistance (R_{on})	√	√	√			
Turn-Off Voltage (V_{to})	√	√	√			
Voltage Gain (μ)	√	√	√			
Blocking Voltage	√	√	√	√	√	
Breakdown Voltage (BV_{GD})				√	√	
Breakdown Voltage (BV_{GS})						√
Gate-Drain Capacitance (C_{GD})			√	√	√	
Gate-Source Capacitance (C_{GS})			√			√

2.3. *Schottky barrier gate and ion implanted gate SiC static induction transistors*

A comparison between the Schottky barrier gate SIT, the ion implanted gate SIT, and the planar metal semiconductor field effect transistor (MESFET) for SiC is shown in Figure 5. In the Schottky barrier gate, trenches are etched to define the channel region and Schottky metal gate contacts are formed in the bottom and along the side walls of the trench.[56] Thus, a Schottky barrier, rectifying metal-semiconductor junction is formed along the channel sidewalls, and most closely approximates a vertical MESFET structure. For the ion implanted gate SIT, p+ gate regions are implanted on either side of the source pillar, providing a p/n junction along the side of the channel. Since the p+ gate region forms the rectifying junction, then only planar, ohmic metal contacts to the p+ gate region are needed, and deposition along the SiC channel sidewall is avoided. The ion implanted gate SIT most closely approximates a vertical junction field effect transistor (JFET).

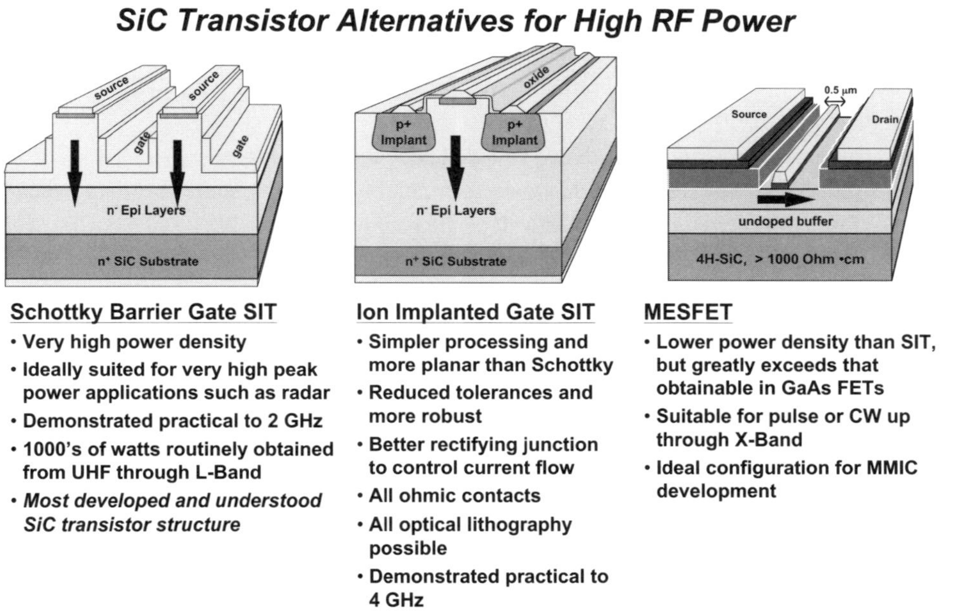

Figure 5. The SiC transistor alternatives for high frequency power include the (a) Schottky barrier gate static induction transistor, the (b) ion implanted gate static induction transistor, and the (c) planar metal-semiconductor field effect transistor. The SIT devices provide the highest power density among all transistor designs, including the MESFET. However, being a vertical device, the SIT is not easily adaptable for monolithic microwave integrated circuits (MMIC).

The first SiC SIT was made using the Schottky barrier gate structure, because forming p+ dopant regions in SiC has been difficult. Although p+ gate regions have been formed by diffusion for Si SIT structures, this is extremely difficult for SiC, which requires extremely high diffusion temperatures and has very low dopant diffusitivities.[57] Therefore, the most practical method to provide a p+ region in SiC to make a p/n junction SIT is to use ion implantation of Al or B to form the p+ gate region into the lightly doped channel layer (Al is preferred over B due to its higher electrical activation).

With the advent of controlled ion implantation into heated SiC substrates (up to 1000°C) and the development of precise high temperature masking (annealing cap) processes to protect the SiC surface during activation of the ion implanted species, ion implanted gate SITs are now being fabricated in SiC.[58] Ion implanted junctions have many positive features for high power operation, and are good candidates for high yield, high performance, rugged microwave power transistors. Ion implanted gates are annealed at up to 1750°C to repair the SiC lattice damage and electrically activate the p-type dopant, and are therefore rugged and reliable under high temperature and high electric field conditions.[59] Ion implanted gate SiC SITs are resistant to electrostatic discharge (tested up to 16 kV without device failure or change in operation),[60] unaffected by voltage overdrive in the gate and drain circuits, and have reduced surface topology when compares to Schottky barrier gate SITs [cf. Figure 5].

In general, the improvements obtained by using an ion implanted gate SIT structure in SiC include:[47,61,62]

- higher voltage operation due to rounded edge terminations,
- reduced source resistance due to uniform, highly doped n+ source regions (when formed by ion implantation as well as the gate regions)
- normally-off operation is possible due to the higher built-in voltage of p/n junctions,
- robustness of p/n junctions at elevated junction temperatures during device operation,
- immunity to electrostatic discharge due to the non-destructive breakdown properties of p/n junctions,
- reversible junction breakdown (i.e., non-destructive),
- improved radiation hardness, and
- significant yield improvements expected from more planar ion implanted device processing.

Using ion implantation, with its known scattering ability in the horizontal direction that undercuts masking during patterned implantation (i.e., straggle), a reduced gate-to-gate spacing can be achieved. That is, a gate-to-gate spacing smaller than the mask dimension used to define the channel width can be obtained without requiring finer line width lithography. The advantages of using this straggle to form narrower channels include:[17,47,61]

- Narrower channel widths will provide higher blocking voltage (i.e., higher voltage gain), and the superior rectifying p/n junction will maximize the voltage gain.
- The use of straggle will allow the source region to overlap the gate region, such that strict alignment accuracy is not required when the source region is formed.
- The minimum channel width will occur below the surface of the SiC wafer, such that on-state voltage drops can be reduced in the channel region since the width of the channel in the vicinity of the surface is larger than in conventional 1-2 μm gate structures. Using this ion implanted gate structure, on-state voltages have

been reduced by almost tenfold less than conventional SIT structures, without decreasing the blocking gain.

- Shrinking the gate-to-gate spacing and the device pitch reduces the parasitic capacitances, increasing channel current, transconductance, and voltage gain, which allow for higher frequency operation.

Furthermore, the high built-in p/n junction voltage (~3 V, proportional to the energy band gap of SiC equal to 3.2 eV) offers the potential for the fabrication of a normally-off microwave power transistor that would simplify efficient transmitter circuitry.[42]

2.4. *Variations in the gate structure of the static induction transistor*

There are three main variations to the gate structure of the SIT, which are known as the (1) buried gate structure, (2) the planar surface gate structure, and the (3) recessed gate structure. These are compared schematically in Figure 6. The buried gate structure embodies the original concept of the SIT, and of all the SIT structures most closely resembles the vacuum tube triode. Many Si-based SITs have been fabricated using the buried gate structure.[1–3,63–67] The buried gate design usually requires a second epitaxy regrowth step, as the use of deep ion implantation will usually damage the channel and source contact regions above the buried ion implanted gates. This requirement of a second regrowth makes the buried gate SIT difficult for commercial SIT production when compared to the other SIT designs. In general, the buried gate construction gives a lower channel resistance, minimizing the voltage loss across the channel.[2] However, there are problems associated with parasitic resistance of the buried gate structure, since the gate resistance is due solely to the p+ layer (i.e., no gate metal present along the gate fingers).[67] So, the surface and recessed gate structures have better switching properties than the buried gate SIT, due to lower gate resistance.[12,58]

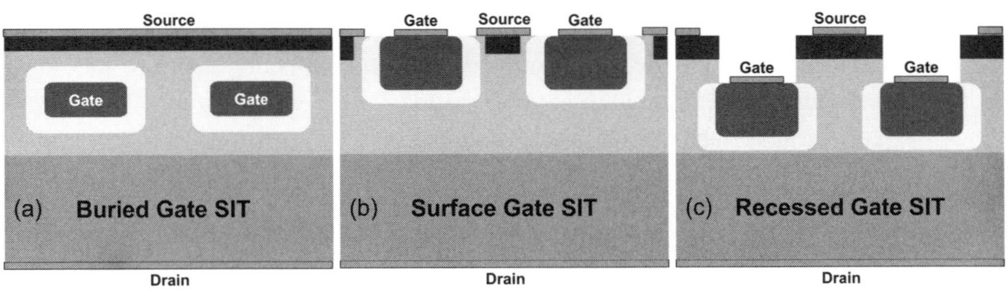

Figure 6. Cross section schematic comparing the three main variations to the gate structure of static induction transistors, including the (a) buried gate SIT, the (b) planar surface gate SIT, and the (c) recessed gate SIT. Gate metal contact to the buried gate SIT only occurs at the ends of the gate fingers, by etching down to the gate fingers and forming a busbar across all fingers in a single device cell.

For high frequency operation, the surface gate and recessed gate SIT structures are considered superior to a buried gate SIT structure because:[68]

- the very short distance between source and gate (L_{sg}) in the surface and recessed gate structures result in a small series channel resistance, and with a shorter carrier transit time which results in higher frequency operation,
- the gate resistance of the surface and recessed gate structures is much smaller than for the buried gate structure, as gate metal (and not p-type SiC material) is used along the entire gate finger length, and
- the surface and recessed gate structures do not have the depletion region surrounding the entire gate (as for the buried gate structure), which results in lower capacitance between the source and gate, and allows for higher frequency operation.

In comparison between the surface gate and the recessed gate structures, the gate resistance and capacitance is lower using the recessed gate SIT structure, as the gate region is located along the sidewalls of the channel. Furthermore, the recessed gate structure exhibits a higher frequency performance because the gate-source capacitance and the gate-drain capacitance can be reduced, and the transconductance and gate-source breakdown voltage can be increased.[68] Therefore, work has focused on the surface and recessed gate SiC SIT structures owing to their lower gate resistance. Since the surface and recessed gate SIT structures are particularly suited for high frequency and high power applications,[65] there have not been any buried gate SiC SIT structures fabricated to date yet.

2.5. *Advantages of a vertical FET structure*

By removing the drain contact from the front surface and moving in onto the backside, several enhancements over planar FETs are achieved. Moving the drain contact to the backside allows for:

- a higher density packing of source and gate fingers, providing for very high current and power density,
- placement of gate regions on both sides of the conducting channel, providing more active gate control of the channel region,
- a very large distance between the drain and both the gate and the source, which enables very high voltage operation without causing device breakdown, and
- the absence of "gate lag" or RF dispersion owing to the removal of surface state traps between the gate and drain contacts.

With a planar FET, the gate and drain are very near each other [cf. Figure 5], such that the voltage between the gate and the drain that can be supported without causing device breakdown is usually low. Normal FET operation occurs for a positive drain bias ($V_{DS} > 0$) and negative gate bias ($V_{GS} \leq 0$), such that the voltage between the gate and the drain will always be the largest between any contacts (i.e., $V_{DG} = V_{DS} - V_{GS}$). Due to both (a) close proximity between the gate and drain electrodes in a planar FET, and (b) the existence of surface interface traps always present in semiconductors, the gate-drain breakdown voltage becomes the critical voltage beyond which the FET cannot operate, limiting the output power obtainable[b]. By moving the drain contact from the topside over to the wafer backside, the semiconductor surface interface problem is removed, and the distance between the gate and the drain is significantly increased. This allows for much higher gate-drain breakdown voltages, and thus also allows for higher operating voltages and higher output power. As an example, the normal operating voltage allowable for an S-band SiC MESFET is ~50 V V_{DS}, with gate-drain breakdown voltages approaching 100 V (V_{DG}), whereas for L-band and S-band SiC SITs, the normal operating voltages can be ~100 V V_{DS}, with gate-drain breakdown voltages greater than 1000 V being possible. In fact, the breakdown voltage now becomes a modeling parameter, related to the doping concentration and thickness of the drift region (t_{dr} and N_{dr}). For low doped drift regions, the rule of thumb for breakdown in SiC SITs is 100 V/μm of drift region thickness.

2.6. *Normally-on versus normally-off SIT design*

Although most SITs are designed to be normally-on devices, where the channel is "open" and conducting significant current ($I_{DS} > 0$) when no gate voltage is present ($V_{GS} = 0$), the SIT can be designed to be normally-off, where the channel is "turned-off" and prevents current flow even though there is no bias on the gate (i.e., $I_{DS} = 0$ at $V_{GS} = 0$). A normally-off SIT design can be accomplished by reducing the gate-to-gate spacing and minimizing the channel region width (i.e., smaller W_{ch}) until the built-in depletion regions from both gates come together and pinch-off the channel without requiring any gate bias ($V_{GS} = 0$).[47,64,69] Other methods include increasing the depletion region, by either reducing the channel doping density, and/or increasing the gate region doping density (if possible), such that the depletion regions will move further into the channel for a given reverse bias condition.[64] Regardless of how this is accomplished, a normally-off SIT device will occur when $W_{ch} \leq 2\ W_{dp}$. Since the ion implanted gate SIT provides a superior p/n junction relative to the Schottky barrier, a normally-off SIT device is usually obtained using an ion implanted gate structure, with reduced channel widths than is used for normally-on SITs. Furthermore, by utilizing the "straggle" that naturally occurs when performing patterned ion implantation, such that the implantation penetrates slightly underneath the mask, smaller gate-to-gate spacings can be achieved that are smaller that the actual mask dimension used to define the channel width during ion implantation. This is shown pictorially in Figure 3.

[b] However, the gate and source contacts remain close together, such that the gate-source breakdown voltage is still an important consideration, depending on how high a gate-source voltage is needed to provide the required blocking voltage. This is directly relates to the voltage gain defined later in this chapter.

3. Current-Voltage Characteristics of the Static Induction Transistor

From the description of the SIT device structure, it can be seen that a rectifying junction is formed between the gate and the channel region, whether this occurs due to a metal-semiconductor Schottky barrier, or a p/n semiconductor junction. In this manner, a depletion region will naturally form in the channel layer, since the doping concentration of the channel layer is much less than the gate region (either from the Schottky metal contact or the p+ doped SiC region). So, a built-in depletion will be formed into the channel region at zero gate bias ($V_{gs} = 0$), as shown graphically in Figure 2. In this manner, current flow through the channel region occurs when there is a positive bias on the drain relative to the source ($V_{DS} > 0$), which can be regulated by placing a negative bias on the gate relative to the source ($V_{GS} < 0$), causing the depletion region to widen and even turn-off the current completely when the depletion regions from both sides of the channel touch. The negative gate voltage where the current in the channel becomes negligible is referred to as the "turn-off voltage" (V_{to}). This is the basic operation of all field effect transistors. The SIT, however, being a vertical FET device, has some unique properties that make it different from planar FETs.

The static induction transistor, being a vertical FET with active gate control on both sides of the channel, can operate in four different current-voltage (I–V) modes, depending on the operating current and voltage magnitude, and device structure geometry. The four I–V operating modes are known as:

- pentode-like mode,
- triode-like mode,
- mixed mode, and
- bipolar mode.

The SIT I–V characteristics are strongly dependent on device structure. The channel width (W_{ch}), channel thickness (t_{ch}), the ratio of (W_{ch}/t_{ch}), and the channel doping (N_{ch}) are all essential parameters in determining the I–V characteristics of the SIT.[54] Except for the special case of the mixed mode, these current-voltage modes can be obtained from any SiC SIT, depending on the magnitude of the I–V region the device is operating in. It should be noted that the "pentode-like" and the "triode-like" designations should only be given relative to a *linear* I–V plot representation.[52]

3.1. *Pentode-like mode*

Under low voltage (V_{DS}) and high current (I_{DS}) operation, the SIT exhibits the FET family I–V curves normally seen in planar semiconductor FETs [Figure 7], which resemble those of the vacuum tube pentode.[7,8] That is, for a fixed gate bias, the current increases linearly at very low drain biases and then quickly reaches a saturated level with respect to increased drain bias (beyond the knee voltage or pinch-off voltage). As for planar FET devices, the unit of measure to quantify this gain control between the current flow (I_{DS}) and the gate bias (V_{GS}) is called the "transconductance gain" (g_m), and is defined as:

$$g_m = \left. \frac{\partial I_{DS}}{\partial V_{GS}} \right|_{V_{DS}=\text{constant}} \tag{1}$$

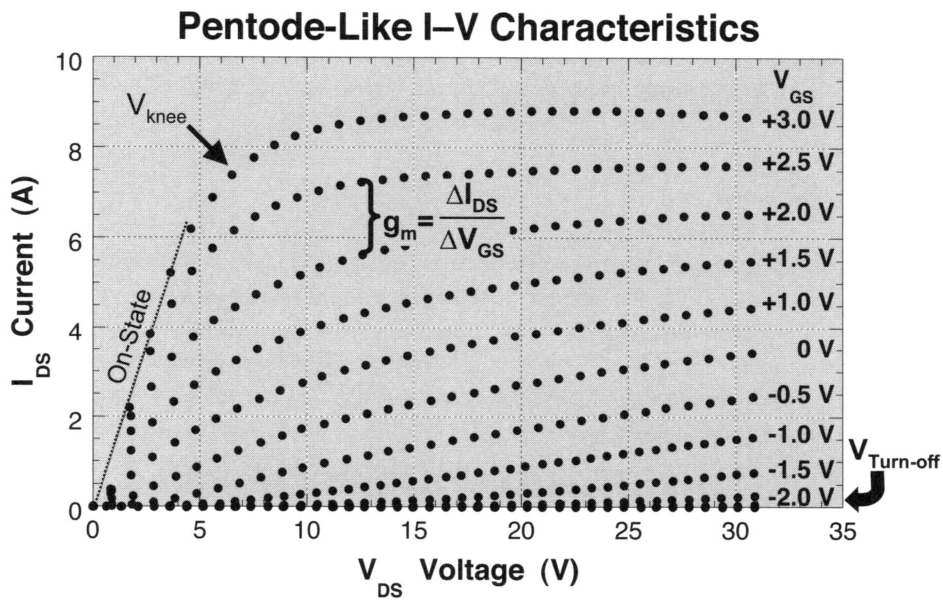

Figure 7. Measured current–voltage characteristics of a SiC static induction transistor which resemble the I–V curves of the vacuum tube pentode, and is normally seen for planar field effect transistors. This pentode-like mode of operation occurs in the range of low drain voltages and high currents. The identical SiC SIT transistor was used to provide both the triode-like mode and pentode-like mode of operation shown here (cf. Figure 8).

In this pentode-like mode of operation, the SIT acts as a normal FET and behaves as a voltage controlled current source. Even though the SIT is constructed similar to the vacuum tube triode, which exhibits triode-like I–V characteristics at low I_{DS} and high V_{DS}, under high drain current and low drain bias the SIT behaves as a normal FET and displays pentode-like I–V characteristics. The nature of current saturation beyond channel pinch-off is somewhat complicated, as the limiting factor for current saturation may be due to the electron saturation velocity rather than the effect of electron injection into the depleted channel region (V_{GD} negative).[70] When comparing the semiconductor SIT with the vacuum tube triode, the pentode-like operation will be more readily observed in the SIT due to the lower electron saturation velocity (~10^7 cm/sec) that occurs in all solid state semiconductor materials relative to vacuum tubes, providing a saturation limit at high current levels.[71] That is, the electron saturation velocity in a vacuum tube approaches the speed of light (~10^{10} cm/sec) in vacuum, allowing for three orders of magnitude higher electron velocity than in solid state semiconductors, and potentially higher current saturation levels.

Closer examination of the pentode-like I–V characteristics show that the SIT is not operating in pure pentode-like mode, since the on-state resistance is not constant for low drain voltages (i.e., the slope of the I–V curve before saturation occurs), but improves significantly as the channel is made more conducting by applying a gate bias in the positive direction ($V_{GS} > 0$). However, the current does saturate as the drain bias is increased, approximating the vacuum tube pentode-like characteristics. This change in on-state resistance is an important consideration when operating the SIT as a power switch.

3.2. *Triode-like mode*

From the pentode-like regime, if the gate is then negatively biased significantly past the turn-off voltage ($V_{GS} \gg V_{to}$), the current flow between the drain and the source becomes negligible and is essentially zero ($I_{DS} \approx 0$). However, if the drain voltage (V_{DS}) is further increased to a much higher level beyond those normally seen in pentode-like mode, the SIT changes from pentode-like characteristics to a triode-like behavior after the forward blocking voltage is overcome.[5,6,72]

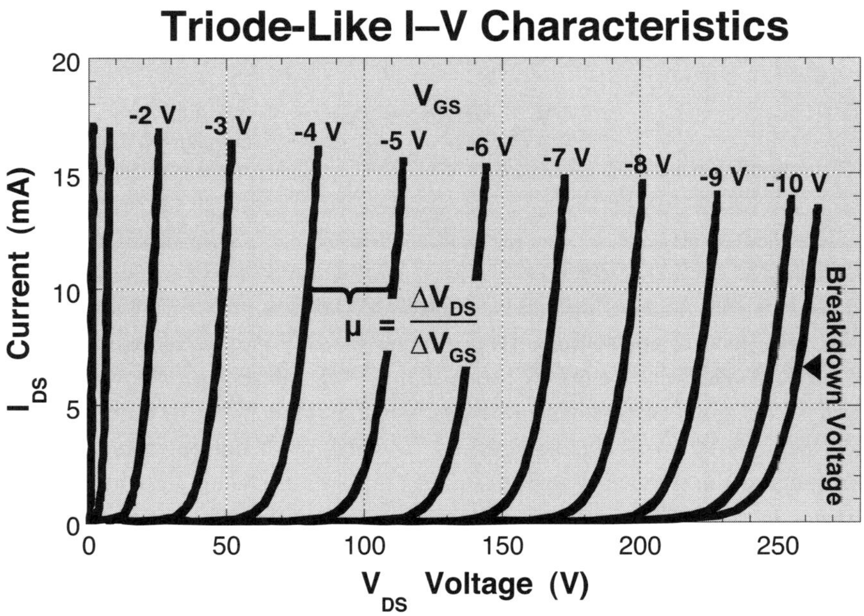

Figure 8. Measured current–voltage characteristics of a SiC static induction transistor which resemble the I–V curves of the vacuum tube triode. This mode of operation only occurs in the semiconductor SIT or the vacuum tube triode, and is a unique feature of the SIT among semiconductor transistors. This triode-like mode of operation occurs in the range of high drain voltages and low currents. The identical SiC SIT transistor was used to provide both the pentode-like mode and triode-like mode of operation shown here (cf. Figure 7).

That is, for a fixed, large negative gate bias (V_{GS} »V_{to}), as the drain bias is increased significantly, the current will no longer be blocked by the reverse biased gate and current will begin to flow in the channel again. As the drain bias is further increased, the current will increase exponentially.[65] The drain bias (V_{DS}) at which the current begins to flow again is defined as the "blocking voltage"(BkV_{DS}), and varies depending on the value of gate bias (V_{GS}). The drain bias at which current begins to flow regardless of how high a negative the gate bias is, is referred to as the "breakdown voltage" (BV_{DG}). This breakdown occurs across the gate-drain contacts in the drift layer, and represents the avalanche breakdown in the SiC material. These I–V characteristics are similar to how a triode vacuum tube behaves, and is referred to as the "triode-like" mode, as shown in Figure 8. The SIT exhibits triode-like I–V characteristics under high voltage (V_{DS}) and low current (I_{DS}) operation, where the channel is depleted of carriers, and current flow is non-saturating, and instead increases exponentially with higher drain bias.

3.2.1. *Drain induced barrier lowering (static induction)*

Since this is a special characteristic of the static induction transistor, further explanation is presented. Under triode-like operation, the SIT channel is fully depleted of carriers, due to the large negative bias on the gate ($V_{GS} > V_{to}$). This channel depletion is caused by the depletion regions spreading from both gates and meeting in the center of the channel, blocking current flow. For small drain voltages, the depletion regions create a potential barrier to current flow in the channel, which initially peaks at the center of the channel, both in terms of the channel width (W_{ch}) and the channel thickness (t_{ch}). The peak of this potential barrier is known as the "saddle point", and represents a two dimensional potential distribution in the device,[65,71] that prevents electron flow from source to drain[c]. A three dimensional representation of this potential barrier is shown graphically in Figure 9. As the drain voltage is increased ($V_{DS} > 0$), perpendicular to the gate voltage inducing the channel depletion, the potential barrier to current flow in the center of the channel is lowered, and the saddle point begins to move towards the source contact. This effect is known as "drain induced barrier lowering" (DIBL), and is also called "static induction".[4,73] As the drain bias is raised, the potential field from the drain-source direction penetrates into the gate depletion regions in the channel, which is directed perpendicular to the drain-source field.[71] With further increases in drain voltage, the saddle point potential continues to be lowered, and moves closer to the source contact. Finally, a point is reached at some high drain voltage where the potential barrier to current flow is almost zero, and the saddle point has reached the source contact at the edge of the channel region. Although in theory, the saddle point goes to zero,[71] in reality the potential barrier does not completely vanish at the source, due to the presence of

---

[c] Since the potential distribution is due to two voltages in perpendicular directions (biasing from V_{GS} and V_{DS}), the shape of the two dimensional potential barrier resembles a "saddle", with the single peak in the distribution referred to as the "saddle point". Also, since electrons are negatively charged, the potential barrier to electron flow is actually negative, and the peak potential barrier, or "saddle point", is in reality a potential minimum [cf. Figure 9].

mobile electronic space charge in the channel, but is easily overcome by the high drain bias.[52] After this point, as the drain voltage is increased further, electrons are injected over this minimal potential barrier and current begins to flow in the channel. After passing the barrier at the source, the electrons travel over most of the channel thickness at saturated drift velocity.[52] This realm of current flow is also referred to as space charge limited current.

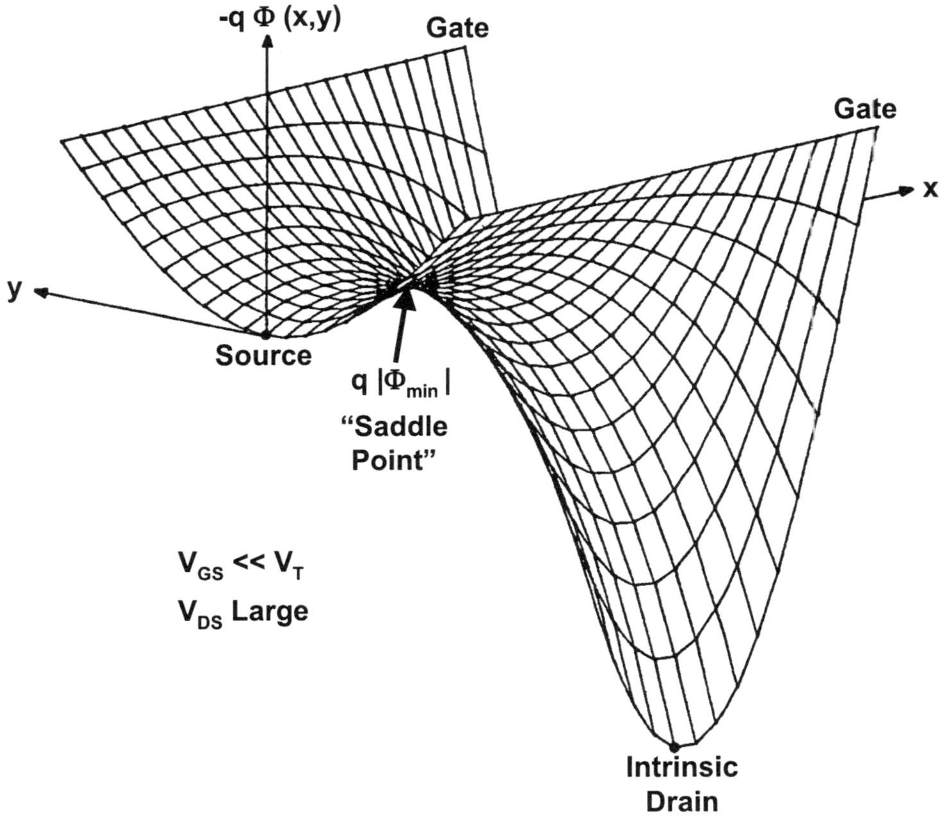

Figure 9. Three dimensional representation of the electron potential under large reverse gate bias, showing the potential barrier present between the gates and within the channel region. The potential surface resembles a "saddle", and the peak of the barrier is referred to as the "saddle point". The intrinsic drain is defined as the potential at the interface between the drift and channel regions (cf. Figure 2 where the channel layer (t_{ch}) and drift layers (t_{dr}) meet).[52] [Figure reprinted from Reference 52 with permission from Elsevier].

The drain voltage (V_{DS}) at which current now begins to flow in the channel is referred to as the "blocking voltage" for a given gate voltage. That is, any drain bias lower than this blocking voltage will not result in current flow, such that a particular negative gate bias is able to block this amount of drain voltage before current conduction occurs. The current flow in the channel beyond the blocking voltage then varies exponentially with increasing drain bias, and does not saturate as occurs for the pentode-like mode.[65] This triode-like current flow is not as large as for pentode-like mode, and it should be noted that this current flows through a channel still depleted of carriers. That is, the free carriers flow with an electron density less than the doping concentration of the channel layer. Modeling has shown that this triode-like current is confined to a very narrow stripe at the center of the channel, equidistant between the gate regions.[52,73] This creates a very small "effective" channel for current flow, the width of which is proportional to the magnitude of the drain current.[52] As the drain current increases beyond the doping density of the channel layer, the minimum saddle point at the source contact (in the center of the channel) becomes slightly flattened, and the potential barrier becomes surmountable over a wider region at the center of the channel, extending towards the gate regions.[52] So, changes in the current flow occur within this center neutral stripe, and grow wider as the drain current magnitude (electron concentration) attempts to increase beyond the doping concentration of the channel layer. In this manner, the width of the "effective" channel for current flow is adjusted such that the electron concentration never goes beyond the doping density of the channel layer.[52]

In a similar manner with pentode-like I–V characteristics, each of these triode-like I–V curves will vary depending on the gate voltage chosen. Higher blocking voltages can be obtained by increasing the gate voltage more negatively, up to the limit of the breakdown voltage between the gate and drain regions of the device itself. Higher breakdown voltages can be designed by increasing the drift layer thickness. Therefore, under triode-like operation, current flow changes exponentially depending on the magnitude of both the gate voltage (V_{GS}) and the drain voltage (V_{DS}),[74] and the current does not saturate at these low current levels.[75] Also, the SIT can hold off (or "block") very large voltages (V_{DS}) across the drain–source region and prevent current flow (I_{DS}) with a reasonably small gate voltage (V_{GS}). The unit of measure to quantify this gain control between the voltage across the device (V_{DS}) and the gate bias (V_{GS}) is called the "voltage gain" or "blocking voltage gain", but is also sometimes referred to as "voltage amplification".[73] The voltage gain (μ) is defined as:

$$\mu = \left.\frac{\partial V_{DS}}{\partial V_{GS}}\right|_{I_{DS}=\text{constant}} \tag{2}$$

One can also define the "static blocking voltage gain", "forward voltage blocking gain", or "DC blocking gain" (G) at any given point by using $G = V_{DS}/V_{GS}$.[52,65,73]

3.2.2. *Blocking voltage and breakdown voltage in the static induction transistor*

The channel width (or gate-to-gate spacing) plays a very important role in defining the blocking voltage gain,[58,65] and this is the main reason why SIT devices require tight control of the channel width in the fabrication process.[73] Narrower channels increase the blocking voltage capability for the SIT, but also decrease the current carrying capability, which results in lower output power and a higher on-state resistance.[71]

Since the modulation of the potential barrier to current flow in triode-like mode is controlled by both the gate and drain voltages, the geometric shape of the channel (W_{ch} and t_{ch}) and its doping level (N_{ch}) strongly influence the triode-like characteristics of blocking voltage, both G and μ.[4,65] The intrinsic (i.e., inside the channel region only) electrostatic gain (μ_0) characterizes the relative importance of the channel geometry in controlling the electrostatic potential and blocking voltage of the channel. It is the solid state equivalent of the triode gain factor (μ_A) used in vacuum tube literature, and it is given by:[76]

$$\mu_0 = \exp\left(\frac{\pi t_{ch}}{W_{ch}}\right) \tag{3}$$

This is related to the intrinsic blocking voltage gain (G*), which at high values of negative gate bias ($V_{GS} >> V_{to}$) approaches a limit value of $G^* \rightarrow (\mu_0 - 1)^2 / 2\mu_0$.[52] When the ratio of the channel thickness is greater than the channel width ($t_{ch}/W_{ch} > 1$), then the electrostatic gain is much larger than one ($\mu_0 >> 1$), and the blocking voltage gain approaches $G^* \sim \mu_0/2$, or the blocking voltage gain can be expressed in terms of the channel thickness and width as:

$$G^* \approx \frac{\mu_0}{2} = \frac{1}{2}\exp\left(\frac{\pi t_{ch}}{W_{ch}}\right) \quad \text{for} \left(\frac{t_{ch}}{W_{ch}} > 1\right) \text{and} \left(V_{GS} >> V_{to}\right) \tag{4}$$

This shows that, as a general trend, the SIT blocking voltage at a fixed gate voltage depends exponentially on the ratio of the channel thickness (t_{ch}) to the channel width (W_{ch})[d]. Therefore, high forward voltage blocking gains can be obtained in SITs having thick enough channels (i.e., large enough t_{ch}/W_{ch} aspect ratios).[52] Since a significant part of the drain-to-source voltage drops across the drift region, the overall forward voltage blocking gain (G) of the device will be sensibly in excess of G*, but it will still be proportionally related to the (t_{ch}/W_{ch}) channel ratio.[52]

So, by properly adjusting the channel thickness to width ratio, as well as the thickness and doping of the drift region, blocking voltages over 2000 V can be achieved in SiC SITs, as long as:[59]

- the drain-gate breakdown voltage (BV_{GD}) is larger than the blocking voltage,
- the reverse breakdown (avalanche) voltage of the gate-source junction is higher than the negative gate bias (V_{GS}) needed to block the high drain-source forward voltage, and
- the resistive edge terminations surrounding the device and spreading the electric field out does not breakdown before the gate-drain breakdown voltage is reached.

---

[d] Technically speaking, it is not the channel thickness (t_{ch}) that should be used in the equation of the intrinsic blocking voltage gain (G*), but instead the thickness of the gate region (L_g) plus the built-in depletion width (W_B) at the bottom the gate. However, in many SIT designs, $t_{ch} \approx L_g + W_B$, as shown graphically in Figure 2.

The breakdown voltage relates to the voltage handling limit of the device structure itself, and is almost exclusively dependent on the drift layer thickness (t_{dr}) and doping concentration (N_{dr}).[71] The breakdown voltage BV_{GD} between gate and drain decreases with drift doping level (N_{dr}), while it increases with drift thickness (t_{dr}),[77] which unfortunately, lowers the frequency characteristics of the SIT.[77] Another method to raise the breakdown voltage (BV_{GD}) and to decrease the gate-drain capacitance (C_{GD}) is to separate the doping concentration between the drift and channel regions, such that the drift region has a lower doping concentration than that of channel region.[77] In addition, by providing a step structure between the gate and source regions (i.e., source pillar with height [L_{SG}]), the breakdown voltage between the gate and source can be raised, while also decreasing the gate-source capacitance (C_{GS}).[77] Both of these innovations have been incorporated into the modern SiC SIT fabrication process.

3.3. *Mixed mode*

Mixed mode operation refers to a specialized static induction transistor device structure, where the current–voltage characteristics exhibit neither the pentode-like mode nor the triode-like mode, but a combination of both modes (i.e., "mixed mode"), as shown graphically in Figure 10. To obtain mixed mode operation for a SIT, three key device parameters must be properly chosen. These device parameters include the (1) channel width (W_{ch}), (2) channel thickness (t_{ch}), and (3) the zero gate bias, built-in depletion width of one side of the gate region (W_{dp} when $V_{GS} = 0$ V).

Mixed mode operation for a SIT occurs when two ratios of these three device parameters fall within a narrow range. Basically, mixed mode operation will occur when the ratio of the channel width to the single sided depletion width lies between 2 and 3, and when ratio of the channel thickness to the channel width is less than 1.5, which can be described mathematically as:[54,71]

$$\text{Mixed Mode Operation When:} \quad 2 \le \frac{W_{ch}}{W_{dp}} \le 3 \quad \text{and} \quad \frac{t_{ch}}{W_{ch}} < \frac{3}{2}$$

$$\text{Or:} \quad 2W_{dp} \le W_{ch} \le 3W_{dp} \quad \text{and} \quad t_{ch} < \frac{3}{2} W_{ch} . \tag{5}$$

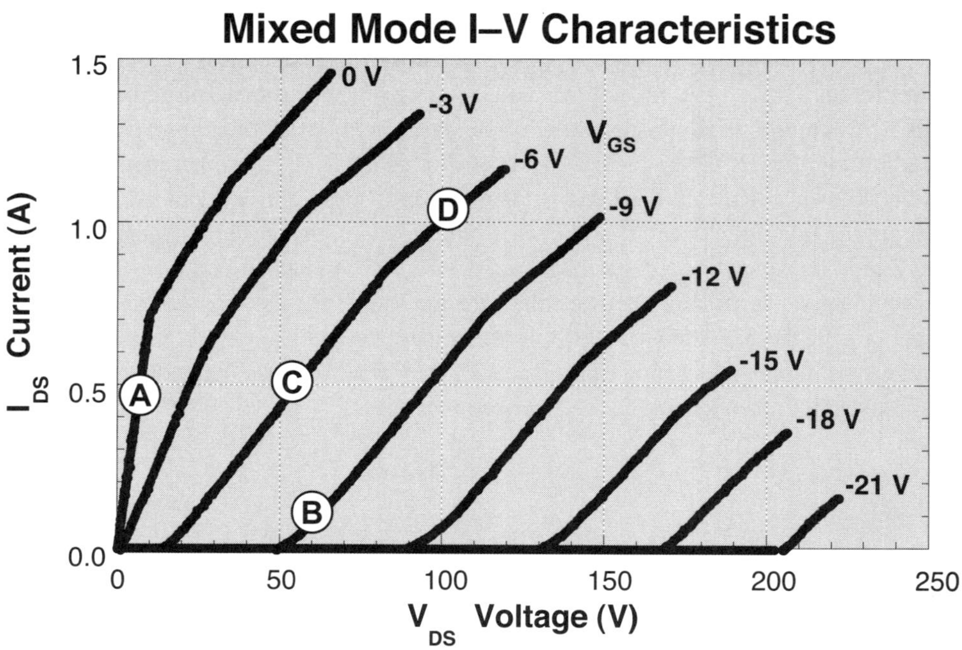

Figure 10. Measured current–voltage characteristics of a SiC static induction transistor demonstrating mixed mode operation, which is a "mix" between pentode-like and triode-like I–V characteristics. This type of operation occurs only for a specific SIT geometry that falls within a narrow range of parameters depending on channel width (W_{ch}), channel thickness (t_{ch}), and the built-in depletion width (W_{dp}). The four distinct regions of the mixed mode I–V curve are identified as (A) linear "ohmic" region, (B) thermionic emission region, (C) space charge limited current region, and (D) space charge limited current region curtailed by electron saturation velocity.

These conditions are shown graphically in Figure 11, and also show where purely triode-like and pentode-like regions can occur for SIT structures outside of this range. When the channel width is less than or equal to twice the built-in depletion width, the device is considered normally-off (where $I_{DS} = 0$ when $V_{GS} = 0$). Normally-off devices provide very high blocking voltages, and current conduction mainly occurs due to the triode-like mode. As the channel width becomes wider, the SIT device changes from a normally-off device to a normally-on device (where $I_{DS} > 0$ when $V_{GS} = 0$). However, if the channel becomes too wide, a large gate bias will be required to turn-off the channel, and the device will have poor blocking voltage characteristics and exhibit pentode-like characteristics. Therefore, the mixed mode SIT operation must lie in between the normally-off (triode-like) condition and the "wide open" normally-on (pentode-like) condition. Furthermore, if the channel thickness becomes much larger than the channel width ($t_{ch} > 1.5\ W_{ch}$), the field induced by the drain voltage will be strongly screened by the gate depletion in the channel, and will not be strong enough to lower and overcome the potential barrier in the channel.[71] So, when the channel thickness becomes larger than 1.5× the channel width, the drain voltage no longer has any appreciable control over the drain current and the SIT operates more as the pentode-like mode. So, both device structure conditions must be met to obtain mixed mode operation.

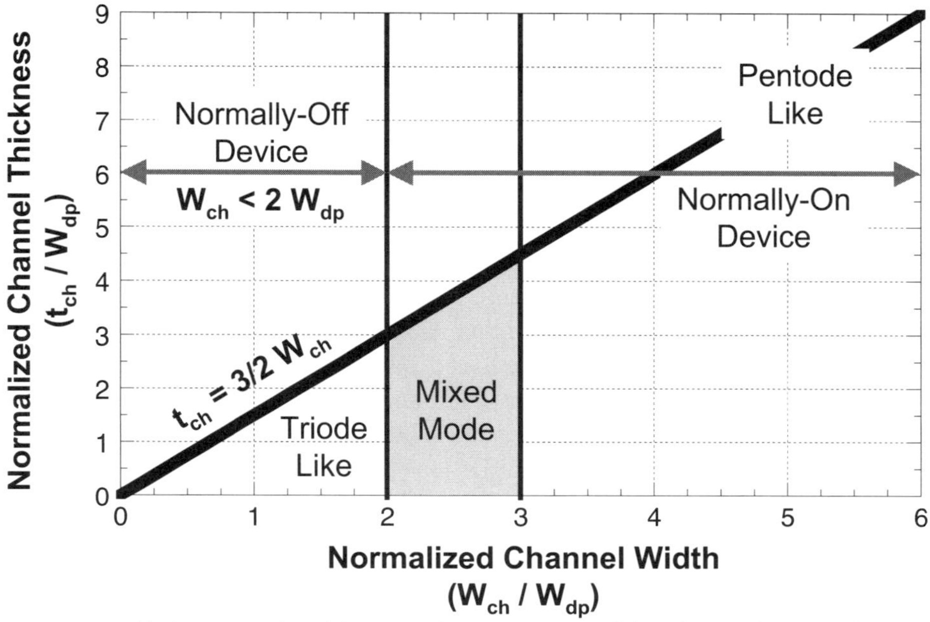

Figure 11. Graphical representation of the geometric SIT structure conditions showing the range where mixed mode I–V operation can take place as a function of the channel width (W_{ch}) and thickness (t_{ch}) relative to the built-in depletion region (W_{dp}).[71] Mixed mode operation exists within a narrow range of SIT device geometry.

Mixed mode is sometimes preferred over triode-like mode because of higher transconductance and higher R_{DS}, both of which allow for improved frequency response (f_{max}).[78] The mixed mode characteristics are considered more desirable in comparison to purely triode-like characteristics, because the depleted "dead" region (i.e., the region in triode mode when no current flows) is reduced and the merits of low output resistance and low distortion from triode-like characteristics are maintained.[71] In order to maximize channel current and minimize the effect of knee voltage on device efficiency, mixed mode SIT operation is usually employed. For mixed mode operation, enough doping is added to the channel to provide large enough currents at medium gate bias, while at higher gate bias the device reverts to triode-like SIT operation to achieve large blocking voltages.[29] The mixed mode I–V characteristics of the SIT, when applied to low resistance load, direct drive circuits, are more desirable because of their low on-state resistance and high AC power efficiency.[55]

3.4. Bipolar mode

Finally, the bipolar mode of operation can occur in a p/n gate junction static induction transistor at very low drain biases (i.e., a few Volts), when the p/n junction between the gate and the source is forward biased ($V_{GS} > 0$) sufficiently so that the junction is turned on ($V_{GS} > V_{Diode\ Turn\text{-}On}$), allowing gate current to flow into the channel region.[1,28] In this mode of operation, the SIT is no longer a voltage controlled, majority carrier device, but is now a minority carrier, current controlled current source. In this bipolar mode, as for standard bipolar junction transistors (BJT), a current gain (β) between the drain and gate currents can be measured as:[1]

$$\beta = \left.\frac{\partial I_{DS}}{\partial I_{GS}}\right|_{V_{GS} = \text{constant}} \tag{6}$$

Although the SIT is a fast switching device, it has a relatively large on-state resistance, due in part to the thickness of the channel, which is needed to provide a high blocking voltage capability.[9] This on-state resistance can be lowered by forward biasing the gate-source junction and operating the SIT in the bipolar mode, as this solution introduces some minority carriers into the channel and lowers the channel resistance due to conductivity modulation.[9] However, this injection of minority carriers into the channel also increases the gate-source capacitance and lowers the frequency response of the SIT.[9] In general, the bipolar mode is worthwhile in reducing the on-state resistance of the channel when switching relatively low voltages (<1000 V), as the drift region can be kept thin to block this amount of voltage. However, for blocking much larger voltages, the drift region must be made thicker, causing a large portion of the on-state resistance to exist in the low doped drift region. Since operating in the bipolar mode primarily reduces the on-state resistance in the channel, the bipolar mode is not very beneficial when switching large voltages, where the on-state resistance is dominated by the drift region.

Furthermore, the current gain (β) of the SIT decreases inversely with the square of the channel thickness ($\beta \propto t_{ch}^{-2}$), which dramatically effects the blocking voltage capability of the SIT.[10] Even though the current gain is improved by reducing the channel thickness, this in turn reduces the current carrying capability of the SIT,[10] while increasing the blocking voltage. Thus, the current gain of the bipolar mode SIT decreases with high current density, contrary to what is desired.[79] However, Si bipolar SITs have been made with high current gains and faster switching speeds that Si BJTs, and can be easily driven in parallel.[14] The Si bipolar SIT has been found suitable for applications with low breakdown voltages and large currents, and is used in motor inverters and large current commutators.[11] Although this bipolar mode has been explored for Si based SIT devices,[80–84] it has not been employed in SiC SITs, due to the fact that the hole mobility in SiC is dramatically worse (≤ 100 cm^2/V sec) than the electron mobility (≥ 1000 cm^2/V sec) as compared to Si or GaAs semiconductors. Therefore, use of the bipolar mode for SiC is not presently advantageous.

4. Static Induction Transistor Applications

Static induction transistors have been used for many applications requiring high power at radio frequencies (RF). Practical applications for Si-based SITs, include (a) linear mode audio, VHF/UHF, and microwave amplifiers,[2,34] AM/FM radio transmitters,[2,11,66] high voltage power supplies,[11,15,17,31,34,85,86] ultrasonic generators,[2,11] motor controllers,[34] linear power amplifiers,[2,85,86] high frequency inverters,[3,15,18,19,87,88] DC-DC converters,[89,90] cycloconverters,[91] image sensors,[73,92] dynamic random access memory (DRAM),[93] logic circuits,[94] and induction heating.[2,11,15,17,18,19,34,54,88]

In SiC, the static induction transistor has been used for high power amplification and transmitter applications from the UHF to microwave frequencies (300 MHz – 3 GHz) and for high power switching in power conditioning systems. Presently, there are several ongoing development and production efforts with SiC SITs for commercial and military applications at both the industry and university level.

4.1. *High RF pulsed power amplification with SiC static induction transistors*

The characteristic features of the static induction transistor which offer the ability to produce high power at high frequencies include:[68]

- The short channel length (t_{ch}) and the low doping concentration of the channel region result in high electron mobility across the channel with a small RC time constant.
- The potential barrier of the saddle point between the gates is controlled by static induction (DIBL) and not by the gate (base) resistance as for a BJT, so there is no frequency limitation due to gate (base) resistance.
- The negative temperature coefficient of the majority carrier current offers a thermally stable operation for large current operation.
- The relatively long distance between gate and drain allows for a high breakdown voltage, with small gate-drain capacitance.
- The vertical structure allows for the fabrication of a multiple channel structure with very high density to achieve high current levels, and provides for self shielding which enables high gate-drain breakdown voltages.
- The excellent characteristics of the SIT such as low noise and low distortion are exhibited even at high frequency.

The high power density achieved with SiC SITs lend themselves to very high pulsed power applications such as television and radar transmitters across the VHF, UHF, L-band, and S-band frequency range.[42] The SiC microwave power SIT provides a pulsed transmitter technology that outperforms conventional solid state transmitters, and is capable of replacing all present vacuum tube based transmitters in pulsed radar systems up through S-band operation.[42] SiC SIT technology offers many military system advantages, including higher power and higher temperature operation with higher efficiency, allowing for significantly reduced cooling and resulting in lower overall system cost and weight.[42]

In order to attain very high power levels (~kW) at microwave frequencies, numerous SIT device cells must be combined in parallel to raise the current level and increase the power triangle. The positive temperature coefficient of the channel resistance forces current equalization across the channel area, and therefore permits easy paralleling of large numbers of device cells.[1,3] That is, as the temperature rises due to I^2R heating in the device, the electrical resistance of the SiC material also increases. This, in turn, reduces the current flow across the hot spot, forcing the current to flow to cooler, less resistive regions of the SIT device. Thus, many SITs can be paralleled without requiring extensive device matching. However, one needs to choose devices with similar turn-off voltage characteristics, such that all devices will turn-off at the same voltage. Parallel combining of SiC SIT device cells will provide increased microwave power, until one of the following fundamental limits is reached:[42]

- The physical size of the parallel cells must not exceed approximately one-tenth (1/10) the wavelength of signal operation. With sufficiently large die, phase variations along the gate bus cause transistor cells to operate out of phase, reducing combining efficiency.
- Paralleling to increase circuit current reduces input and output impedances, reaching a limit when the microwave circuit can no longer be matched to a 50 Ω system. (In practice, impedance levels of 10 Ω and below exhibit severe matching difficulties.) SiC devices have inherently higher impedances than other semiconductors, but are not immune to this limitation.[42] Although the parallel capacitance from the cells can be compensated by the parallel inductance from the bond wires, this is possible only within a narrow frequency range, and limits the operating bandwidth. This compensation becomes harder to do for higher frequencies.[68,95]

Packaging technology for microwave power levels above 1 kW and greater have been demonstrated.[42] A 51 cm periphery package at 1.3 GHz shows 900 W of output power, 65% efficiency, and 11 dB of associated gain.[42] Pulsed power from SiC SITs at S-band (3.1 GHz) is still substantial, with 240 W from a single 17 cell package, 29% drain efficiency, and 6.5 dB associated gain.[42] In the radio frequency range below 1 GHz, published data has shown that SiC SITs have produced over 2 kW of pulsed output power with excellent power added efficiency (PAE) up to 78%.[48] In the GHz frequency range, SiC SITs have repeatedly demonstrated pulsed output power between 400 W–700 W at L-band (1.67 W/mm power density), 80 W–300 W at S-band, and 47 W at 4 GHz, achieving power densities exceeding 60 kW/cm^2.[50,96]

By using e-beam lithography to define the channel width to sub-micron dimensions (<1 μm), higher operating frequencies can be achieved.[97,98] Also, by using the air bridge technique to join the multiple channel source pillars together, the parasitic capacitance of the field dielectric can be reduced by a factor of four, resulting in higher gain and higher frequency response.[99] In this manner, high power SiC SIT operation at S-band can be accomplished, demonstrating 15 dB small signal gain at 3 GHz with an extraordinary high power density equal to 300 kW/cm^2.[99] Similar SiC SITs had cut-off frequencies (f_t) as high as 7 GHz, and produced 2.2 W/mm power density at S-band.[97,98] By using controlled straggle from ion implanted gates to reduce the channel width further, output power density levels of 3.38 W/mm have been achieved at 1.3 GHz, and 2.40 W/mm at 3 GHz.[47] This corresponds to 156 W output power (108 kW/cm^2) with 45% PAE and 8 dB gain at 3 GHz for ion implanted gate SiC SITs.[47]

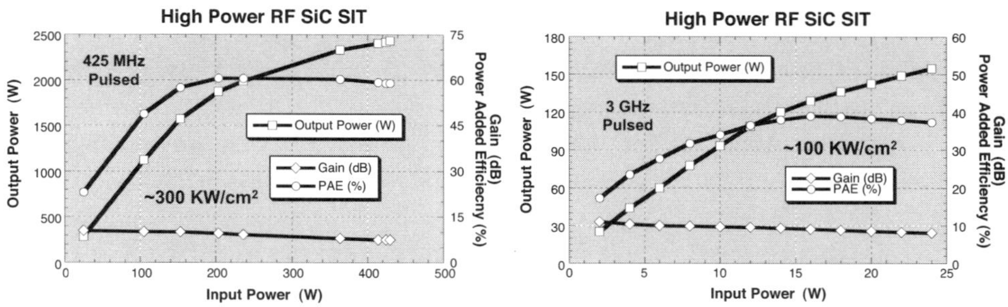

Figure 12. SiC static induction transistors have demonstrated output power levels above 2 kW at 425 MHz to over 150 W at 3 GHz, and are ideally suited for pulsed power applications such as radar and television transmitters.

Further improvements in package and circuit design have led to higher efficiency at these same power levels, providing greater than 70% PAE.[100,101] Control of wafer quality and chemical vapor deposition (CVD) growth uniformity of the SIT device layers has reached the levels required for pilot production, and processing yields are currently in excess of 50%.[100] Taking advantage of these high power levels, SiC static induction transistors have been used as solid state replacements for vacuum tubes in pulsed RF power applications such as radar modules and high definition television (HDTV) terrestrial transmitters.[41,100,101]

SiC SITs have been demonstrated for UHF television transmitters with good signal fidelity at the 2000 W peak envelope power (PEP) level.[102] In 1996, an over-the-air broadcast of HDTV using SiC SIT solid state transmitters was demonstrated, providing 200 W average (1 kW peak) output power out of the transmitter with better than –30 dB third order products.[102] When lower linearity requirements were allowed, the SiC SIT based transmitter produced 400 W average and 2 kW peak power. This demonstration showed that SiC SITs are reliable and economically viable alternatives to traditional high power vacuum tube transmitters in UHF HDTV service.[41] The SiC SIT has a very high impedance and can operate over the entire UHF television band without appreciable reduction in gain or power output.[41] In addition, the higher operating voltage and power dissipating capability of SiC allows the static induction transistor to deliver higher linear output power than Si-based devices.[41]

4.2. *High RF CW power amplification with SiC static induction transistors*

Commercial applications such as wireless base stations and military communication require high RF continuous wave (CW) power in the kilowatt to megawatt range. To date, these high RF power requirements can only be accomplished by incorporating traveling wave tubes (TWT) or inductive output tubes (IOT). However, recent advances in the output power capability of SiC power devices for radar systems now suggest that high power SiC static induction transistors (SIT) can be used to develop a 10 kW CW solid state RF power amplifier. From analysis of existing high power RF amplifiers using SiC SITs, development of a transistor cell that can produce 10 W CW output power would provide the necessary building block for assembly of a 10 kW power amplifier. As a first demonstration, ten SiC static induction transistor cells were combined and packaged together as described previously [cf. Figure 4]. Initial RF measurements have produced 100 W CW output power at 750 MHz, with 59% power added efficiency and 8.7 dB gain, as shown graphically in Figure 13. This corresponds to a CW power density of 8.23 W/cm gate periphery (36.5 kW/cm^2 active area) and 10.7 W/cell, meeting the initial SIT device cell building block requirements needed to develop a 10 kW RF solid state driver amplifier.[48]

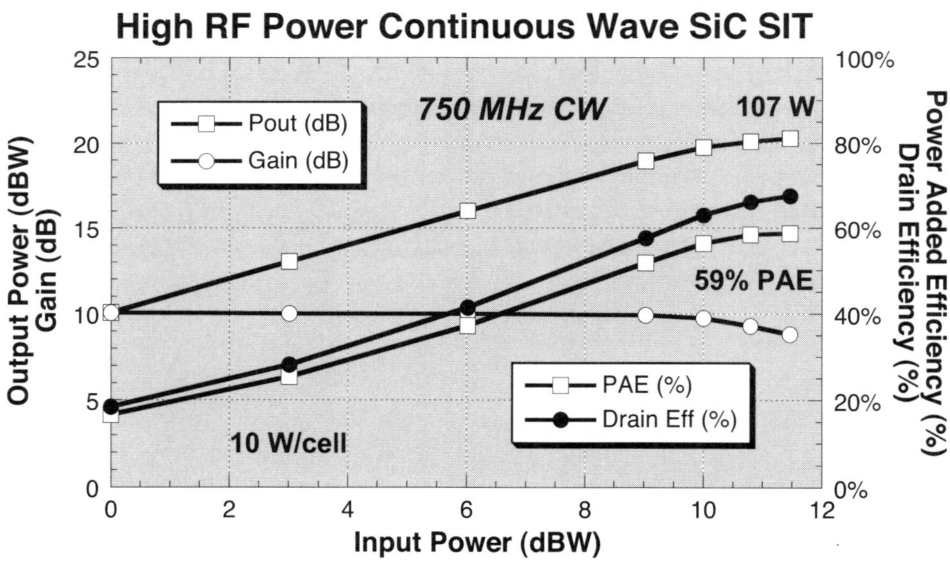

Figure 13. RF measurements at 750 MHz continuous wave (CW) operation for a 10 cell ion implanted gate SiC SIT have demonstrated up to 107 W output power with 59% power added efficiency (PAE) and 8.7 dB gain. These measurements were performed without any active cooling, and the measured operating junction temperature was approximately 200°C, with no signs of degradation in performance.[48]

4.3. *Power switching with SiC static induction transistors*

Another important application for the SiC static induction transistor is the power switch used in power conditioning equipment, providing a solid state electronic means to convert AC to DC, DC to DC, or invert DC to AC. In general, the cost of the power semiconductor devices in power conditioning equipment usually does not exceed 30% of the system cost, but the total equipment cost is still highly influenced by the performance of these power devices.[1] That is, the power switching efficiency, switching frequency, and the maximum operating temperature of the semiconductor devices determines the cooling requirements needed for the power conditioning equipment, which becomes a significant portion of the total equipment cost and weight.[1] Lowering the cooling requirement for power conditioning equipment is often more important than the switching efficiency, as operating at higher temperatures reduces the heat sinking size which contributes to lower overall size and cost of the power conditioning equipment.[1,103] Furthermore, higher frequency switching allows for smaller magnetics and improved performance of the total power electronics system.[1]

Previous researchers have developed several figures of merit to calculate the best semiconductor material for power switching.[104–107] A comparison between SiC and conventional Si and GaAs semiconductors is presented in Table 2, which shows that SiC provides superior material properties for power switching applications for all figures of merit (i.e., the "Keyes" figure of merit [FM_K],[104] the "Johnson" figure of merit [FM_J],[105] and the "Baliga" figure of merit [FM_B][106]). The "Baliga" figure of merit (FM_B) shows that SiC is better in power switching applications where efficient power control is important, by a factor of 100× over Si and 20× over GaAs.[43]

Table 2. Power Semiconductor Device Figure of Merit Comparisons of Semiconductor Material Properties for High Frequency Applications.[104–109]

	Material Property	**Units**	**Si**	**GaAs**	**4H-SiC**
E_g	Energy Band Gap	(eV)	1.1	1.4	2.9
T	Max Operating Temperature	(°C)	150	200	500
κ	Thermal Conductivity	(W/cm K)	1.5	0.5	3.5
E_c	Electric Breakdown Field	(10^6 V/cm)	0.3	0.4	3.0
μ_n	Electron Mobility	(cm^2/V sec)	1500	8500	1150
v_{sat}	Electron Saturation Velocity	(10^7 cm/sec)	1.0	2.1	2.1
ε_r	Relative Dielectric Constant		11.8	12.8	9.7
FM_K	Figure of Merit † $FM_K = \kappa\sqrt{c \cdot v_{sat}/4\pi\varepsilon_r}$	Ref. 104	1	0.5	3.7
FM_J	Figure of Merit † $FM_J = E_c v_{sat}/2\pi$	Ref. 105	1	2.8	21
FM_B	Figure of Merit † $FM_B = \mu_n E_b^3 \varepsilon_r$	Ref. 106	1	15	630

† Figure of Merit Relative to Si

The intrinsic material properties of SiC with regards to high electric field breakdown, high operating temperature (up to 500°C),[43] and high thermal conductivity are ideally suited for power switching. The static induction transistor design has many advantages for power switching, including its ability to operate at high frequencies, to provide for very high blocking voltages, to have a low on-state resistance, and to demonstrate good thermal stability.[79] Therefore, the combination of using the static induction transistor design with SiC is well suited for power switching applications.

Since power conditioning normally involves switching very high currents and voltages, both the pentode-like and triode-like I–V characteristics of the SIT are employed [cf. Figure 7 and Figure 8]. In power switching, the only two states used are the (1) high current output with low voltage loss ("switch closed" or in the "on-state") and the zero current at high voltage ("switch open" or in the "off-state"). This is demonstrated for a SiC SIT used as a power switch in Figure 14. The vertical nature of the SIT allows for very high blocking voltages when the SiC SIT is in the off-state, with very low leakage currents due to the wide energy band gap material of SiC. For the on-state condition, there will be some voltage drop across the SIT due to the non-vertical slope (i.e., $R_{on} > 0$) of the I–V curve when the device is turned on, resulting in a static power loss. This is caused by the resistance in the conducting channel and drift regions during high current flow.[12] Therefore, special care must be given to designing the SIT structure to minimize the on-state resistance for power switching application.

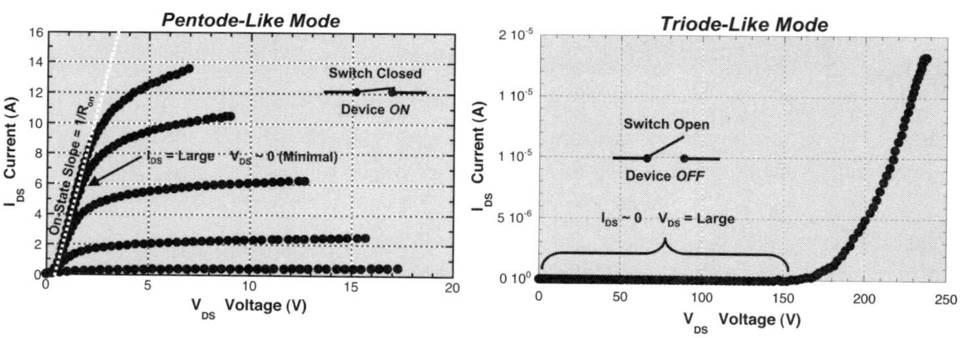

Figure 14. Measured I–V characteristics of a SiC SIT power switch which operates between the (a) high current pentode-like mode when the switch is closed (ON) to allow large amounts of current to flow with minimal voltage drop ($V_{DS} \leq 2$ V), and the (b) triode-like mode when the switch is open (OFF) to prevent current flow and block the high voltage present ($I_{DS} \sim 0$).

The forward voltage drop in a SIT is determined by the on-resistance of the device, which is given by $R_{on} = R_{sub} + R_{ch} + R_{dr} + R_{cont}$, which includes the series resistances from the substrate (R_{sub}), channel (R_{ch}), drift (R_{dr}), and the combined drain and source contact resistance (R_{cont}).[72,110] The most important resistance components occur in the channel and drift region, and are given by:[72,110]

$$R_{ch} = \frac{1}{2q\mu_n N_{ch} Z}\left[\frac{t_{ch} + W_{dp}}{W_B - W_{dp}}\right]\ln\left(\frac{d_g - 2W_{dp}}{W_g - 2W_B}\right)$$
$$R_{dr} = \frac{1}{q\mu_n N_{dr} Z}\left[\frac{t_{dr} - W_{dp}}{W_B - \left(d_g - 2W_{dp}\right)}\right]\ln\left(\frac{W_B}{d_g - 2W_{dp}}\right) \qquad (7)$$

where W_B is the depletion width at the bottom of the gate region next to the low doped drift layer. As expected, the trends show that lower resistance is possible by reducing channel and drift layer thickness (t_{ch} and t_{dr}), increasing doping concentration (N_{ch} and N_{dr}), or by shrinking the built-in depletion width (W_{dp}) [since $t_{dr} >> t_{ch}$]. However, each of these have consequences that would reduce the turn-off voltage, blocking voltage, breakdown voltage, and switching speed. Therefore, a trade-off between the switching performance and device layer characteristics will be required to obtain an optimum solution.[17]

For SiC, the most dramatic drop in on-state resistance occurs when the gate is forward biased in the positive direction from 0 V to +2.5 V, up to the point where the p/n junction of the gate-source diode would turn-on. Since the p/n junction turn-on voltage for SiC is approximately 3 V, the drop in the on-state voltage caused by the gate forward bias is not due to conductivity modulation from minority carrier injection, but is caused by narrowing of the depletion layer in the channel region and a further lowering of the potential barrier between the gate and the source.[17] Forward biasing the gate beyond the turn-on voltage will only increase the gate-source current and will not significantly reduce the on-resistance any further, such that forward biasing the gate beyond turn-on is not desirable.[17]

5. Summary

The SiC static induction transistor has only existed since 1993, and has already demonstrated very high output power levels (from 100's W to >2 kW) across a very wide frequency range (175 MHz up to 4 GHz) with reasonable gain and efficiency under both pulsed and CW operating conditions. To date, the SiC SIT has demonstrated the highest pulsed power density of any solid state semiconductor transistor and it is the most advanced microwave power device in SiC. The SiC SIT has demonstrated sufficient process yield and device reliability to warrant pilot production operation and is ready for insertion into military and commercial systems which require high power at RF and microwave frequencies. The superior electric field breakdown and excellent thermal conductivity of SiC, coupled with its large electron saturation velocity and high temperature operation capability, make the SiC SIT ideally suited for state-of-the-art RF power applications in harsh environments. Recent work in power switching show that the SiC SIT can provide the capability to switch large voltages (>1000 V) and large currents (>20 A) efficiently at high switching frequencies (kHz range) while requiring minimal cooling. Further work in this area should provide significant improvement in both switching voltage and current levels by an order of magnitude. The SiC SIT has the potential to play a significant role in the development of future RF high power transmission and high power conditioning systems, providing for robust operation at elevated temperatures.

6. Acknowledgements

The author wishes to thank R. Chris Clarke, Stephen Van Campen, Eric Stewart, Ty McNutt, and Victor Veliadis for the many helpful technical discussions and assistance in preparing this chapter.

7. References

1. B. Bose, Evaluation of modern power semiconductor devices and future trend of converters, IEEE Industry Appl. Soc. Annual Mtg, 790 (1989).
2. B. Bose, *Recent advances in power electronics*, IEEE Trans. Power Electronics, **7**, 2 (1992).
3. B. Bose, *Evaluation of modern power semiconductor devices and future trends of converters*, IEEE Trans. Industry Appl., **28**, 403 (1982).
4. R. Clarke, R. Siergiej, A. Agarwal, C. Brandt, A. Burk, A. Morse, and P. Orphanos, *30 W VHF 6H-SiC power static induction transistor*, Proc. High Speed Semicon. Dev. and Circuits (Cornell Conf.), 47 (1995).
5. K. Spangenberg, *Vacuum Tubes*, McGraw–Hill, New York, NY (1948), pp. 201–237.
6. E. Chaffee, *Theory of Thermionic VacuumTubes*, McGraw–Hill, New York, NY (1933), pp. 144–191.
7. K. Spangenberg, *Vacuum Tubes*, McGraw–Hill, New York, NY (1948), pp. 266–297.
8. E. Chaffee, *Theory of Thermionic VacuumTubes*, McGraw–Hill, New York, NY (1933), pp. 588–621.
9. M. Grecki and A. Napieralski, *Two dimensional analysis of SIT (static induction transistor) during switching process*, 5th European Conf. Power Electron. and Appl., 95 (1993).
10. H. Takagi, H. Kawauchi, T. Yoshida, S. Miura, H. Ito, and H. Tadano, *High voltage bipolar mode SIT with low power loss*, 5th Intl. Symp. Power Semicon. Dev. and ICs, 276 (1993).
11. M. Tatsuta, E. Yamanaka, and J. Nishizawa, *High frequency, high power static induction transistor*, IEEE Industry Appl. Soc. Annual Mtg., 1321 (1993).
12. A. Napieralski and M. Grecki, *Transient temperature evaluation during switching process in static induction transistor*, Power Electronics and Variable Speed Drive Conf., 537 (Oct. 1994).
13. H. Ogiwara and M. Nakaoka, *Zero current soft switching mode parallel load resonant tank high frequency inverter using normally-off bipolar mode SIT*, Power Electron. Spec. Conf. (PESC), 95 (1994).
14. K. Yamamoto, T. Sugai, and K. Tanaka, *A compact, high efficiency, high power DC/DC converter*, Telecommunications Energy Conf. (INTELEC), 251 (1993).
15. A. Okuno, M. Hayashi, H. Kawano, H. Yasutsune, E. Hiraki, and M. Nakaoka, Phase-lock loop operated load-resonance inverter using static induction power transistors and its practical characteristic evaluations, Intl. Conf. Industrial Automation and Control, 1 (1995).

16. H. Ogiwara, S. Haq, and M. Nakaoka, High efficiency zero current soft switching mode high frequency inverter using normally-off bipolar mode SIT for induction heating, Intl. Conf. Power Electronics, Drives, Energy Syst, for Industrial Growth, 801 (1996).
17. T. Iwasaki, T. Oono, K. Asano, Y. Suguwara, and T. Yatsuo, Electrical characteristics of a novel gate structure 4H-SiC power static induction transistor, IEEE Intl. Symp. Power Semicon. Dev. and ICs, 149 (1997).
18. A. Okuno, H. Kawano, J. Sun, M. Kurokawa, A. Kojina, and M. Nakaoka, Feasible development of soft-switched SIT inverter with load adaptive frequency tracking control scheme for induction heating, IEEE Trans. Industry Appl., **34**, 713 (1998).
19. A. Okuno, S. Shirakawa, and M. Nakaoka, Latest developments of voltage-fed resonant high frequency inverter with load resonant frequency tracking scheme for induction heating, Power Electronics and Variable Speed Drives Conf., 152 (1998).
20. Y. Watanabe and J. Nishizawa, Semiconductor Devices Including High Resistive Layer, Japanese Patent 205,068 Dec 1950.
21. J. Nishizawa, T. Terasaki, and J. Shibata, Field effect transistor versus analog transistor (static induction transistor), IEEE Trans. Electron Devices, vol 22, pp.185-197 Apr 1975.
22. K. Ng, Complete guide to semiconductor devices, J. Wiley and Sons, pp. 225–233 (2002).
23. J. Nishizawa, A low impedance field effect transistor, Intl. Electron Dev. Mtg. (IEDM), 144 (1972).
24. J. Nishizawa, T. Terasaki, and J. Shibata, Field effect transistor versus analog transistor (static induction transistor), IEEE Trans. Electron Dev., **22**, 185 (1975).
25. H. Ogawa, A. Abe, T. Nakajima, and T. Kajiwara, Direct observation of the depletion layer in a multi-channel vertical JFET (MJFET), Electron. Lett., **11**, 223 (1975).
26. K. Yamaguchi, T. Toyabe, and H. Kodera, Two dimensional analysis of triode-like operation of junction gate FETs, IEEE Trans. Electron Dev., **22**, 1047 (1975).
27. Y. Mochida, J, Nishizawa, and T. Ohmi, Characteristics of static induction transistor: Effects of series resistance, IEEE Trans. Electron Dev., **25**, 761 (1978).
28. T. Tamamushi, T. Akiyama, and J. Nishizawa, Current–Voltage Characteristics and Noise Performance of a Static Induction Transistor for Video Frequency, IEEE Trans. Electron Dev., **35**, 1130 (1988).
29. R. Siergiej, R. Clarke, A. Agarwal, C. Brandt, A. Burk, A. Morse, and P. Orphanos, *High power 4H-SiC static induction transistors,* Intl. Electron Dev. Mtg. (IEDM), 353 (1995).
30. W. Shockley, *Transistor electronics: Imperfections, unipolar and analog transistors*, Proc. IRE, **40**, 1289 (1952).
31. P. Hadizad, J. Hur, S. Hummel, M. Gundersen, and H. Fetterman, *High speed static induction transistor for pulsed power applications*, IEEE 19th Intl. Power Modulator Symp., 343 (1990).
32. M. Mori and T. Yatsuo, *A high voltage GaAs power static induction transistor*, Proc. 19th Conf. Solid State Dev. And Matl., 279 (1987).
33. P. Hadizad, J. Hur, M. Gundersen, and H. Fetterman, *Design of an opening and closing GaAs static induction transistor for pulsed power applications*, Proc. 7th IEEE Pulsed Power Conf., (June 1989).
34. P. Hadizad, J. Hur, H. Zhao, K. Kaviani, M. Gundersen, and H. Fetterman, *A high voltage optoelectronic GaAs static induction transistor*, IEEE Electron Dev. Lett., **14**, 190 (1983).
35. J. Nishizawa, P. Plotka, and T. Kurabayashi, *Light emission from tunneling mode GaAs static induction transistor*, IEE Proc. Circuits Dev. Syst., **146**, 27 (1999).
36. J. Nishizawa, P. Plotka, and T. Kurabayashi, *Ballistic and tunneling GaAs static induction transistors: nano-devices for THz electronics*, IEEE Trans Electron Dev., **49**, 1102 (2002).
37. X. Gu, C. Myles, A. Kuthi, Q. Shui, and M. Gundersen, *GaAs and Si FET-type switches for repetitive pulsed power applications*, IEEE 25th Intl. Power Modulator Symp., 437 (2002).

38. J. Nishizawa, K. Motoya, and A. Itoh, *The 2.45 GHz 36 W CW Si recessed gate type SIT with high gain and high voltage operation,* IEEE Trans. Electron Dev., **47**, 482 (2000).
39. T. Mizoguchi, I. Takayanagi, E. Shimizu, H. Nakajima, S. Hashimoto, S. Yokoyama, J. Nishizawa, and M. Imai, *A 250 k pixel SIT image sensor operating in its high sensitivity mode*, IEEE Trans. Electron Dev., **38**, 1021 (1991).
40. C. Brandt, R. Hopkins, A. Agarwal, J. Casady, R. Siergiej, S. Seshadri, G. Augustine, L. Rowland, and J. Perkins, *SiC temperature tolerant power electronics for dual use systems applications*, Proc. Intersoc. Energy Conv. Engr. Conf. (IECEC), 705 (1997).
41. C. Davis, J. Hawkins, and C. Einolf, *Solid state DTV transmitters*, IEEE Trans. Broadcasting, **43**, 252 (1997).
42. R. Clarke and J. Palmour, *SiC microwave power technologies*, Proc. IEEE, **90**, 987 (2002).
43. M. Driver, R. Hopkins, C. Brandt, D. Barrett, A. Burk, R. Clarke, G. Eldridge, H. Hobgood, J. McHugh, P. McMullin, R. Siergiej, and S. Sriram, *Advances in SiC device processing and substrate fabrication for high power microwave and high temperature electronics*, GaAs IC Symp., 19 (1993).
44. C. Weitzel, J. Palmour, C. Carter, K. Moore, K. Nordquist, S. Allen, C. Thero, and M. Bhatnagar, *Silicon Carbide high power devices*, IEEE Trans. Electron Dev., **43**, 1732 (1996).
45. A. Morse, P. Esker, R. Clarke, C. Brandt, R. Siergiej, and A. Agarwal, *Application of high power SiC transistors at radar frequencies*, IEEE Intl. Microwave Symp. (MTT-S), 677 (1996).
46. R. Clarke, A. Agarwal, R. Siergiej, C. Brandt, and A. Morse, *The mixed mode 4H-SiC SIT as an S-band microwave power transistor*, IEEE 54th Dev. Res. Conf., 62 (1996).
47. T. Knight, R. Clarke, R. Barron, J. Ostop, B. Morick, J. Gigante, W. Malkowski, A. Morse, G. DeSalvo, and K. Petrosky, *A SiC self-aligned and ion implanted static induction transistor (SAI-SIT) for 150 Watt S-band operation*, IEEE 60th Dev. Res. Conf., 179 (2002).
48. P. Esker, G. DeSalvo, T. Flint, J. Ostop, E. Stewart, T. Knight, K. Petrosky, S. Van Campen, R. Clarke, and G. Bates, *Ion Implanted SiC Static Induction Transistor with 107 W Output Power and 59% Power Added Efficiency under CW Operation at 750 MHz,* IEEE Proc. Lester Eastman Conf. High Performance Dev., 52 (2004).
49. J. Fuerherm, Y. Zeng, and M. White, *A study of interface charges on the operation of 4H-SiC static induction transistors*, Intl. Semicon. Dev. Res. Symp., 134 (2003).
50. R. Bojko, R. Siergiej, G. Eldridge, L. Chen, A. Morse, J. Ostop, P. Esker, R. Barron, R. Clarke, and C. Brandt, *Recent progress in 4H-SiC static induction transistors for high frequency power generation*, IEEE Dev. Res. Conf., 96 (1998).
51. K. Yano, C. Kim, M. Kimura, A. Tanaka, S. Motoyama, and T. Sukagawa, I*nfluence of majority carrier accumulation in the SIT with a high purity channel on voltage amplification factor*, IEEE Trans. Electron Dev., **40**, 2092 (1993).
52. C. Bulucea and A. Rusu, *A first order theory of the static induction transistor*, Solid State Electron., **30**, 1227 (1987).
53. K. Yano, C. Kim, M. Kimura, A. Tanaka, S. Motoyama, and T. Sukegawa, *Effects of shielded-gate structure on on-resistance of the SIT with a high purity channel region*, IEEE Trans. Electron Dev., **39**, 1257 (1992).
54. S. Ortolland and C. Johnson, *4H-SiC SIT device for heating applications*, IEE Colloq. Adv. Semicon. Dev., 2/1 (1999).
55. S. Li, R. Liu, and S. Liu, *Control on the characteristics of the static induction transistor (SIT)*, 4th Intl. Conf. Solid State and Integrated Circuit Technol., 473 (1995).
56. C. Weitzel, *Wide Bandgap semiconductor power electronics*, Intl. Electron Dev. Mtg. (IEDM), 51 (1998).
57. G. Harris, ed., *Properties of Silicon Carbide*, INSPEC, London, UK (1995), pp. 153-156.

58. H. Onose, T. Yatsuo, A. Watanabe, T. Yokota, T. Ishikawa, I. Sanpei, T. Someya, and Y. Kobayashi, *Design consideration for 2 kV SiC SIT*, Proc. Intl. Symp. Power Semicon. Dev. and ICs, 179 (2001).
59. H. Onose, A. Watanabe, T. Someya, and Y. Kobayashi, *Switching properties of 2 kV SiC SIT*, Proc. 14th Intl. Symp. Power Semicon. Dev. and ICs, 189 (2002).
60. R. Clarke, *private communication* (2003).
61. A. Agarwal, L. Chen, G. Eldridge, R. Siergiej, and R. Clarke, *Ion implanted static induction transistors in 4H-SiC*, IEEE 56th Dev. Res. Conf., 94 (1998).
62. V. Sankin and P. Shkrebiy, *4H and 6H-SiC vertical static induction transistor with p-n junction as a gate*, Intl. Semicon. Dev. Res. Symp., 579 (2001).
63. M. Tatsuta, E. Yamanaka, H. Wada, T. Tamamushi, and J. Nishizawa, *1 MHz, 1 kW series resonant converter for ultrasonic transducer using the high frequency power SITs*, Power Electron. Spec. Conf. (PESC), 1236 (1988).
64. M. Hanes, J. Bartko, J. Hwang, and P. Rai-Choudbury, *Radiation hard static induction transistors*, IEEE Trans. Nuclear Sci., **35**, 1475 (1988).
65. A. Strollo and P. Spirito, *A self consistent model for the SIT DC characteristics*, IEEE Trans. Electron Dev., **38**, 1943 (1991).
66. T. Yamaguchi, H. Naka, Y. Nasu, K. Hayeiwa, and T. Miyazaki, *Development of high power medium wave radio transmitter with SIT*, Intl. Broadcasting Convention (IBC), 29 (1992).
67. K. Nuttall and W. Chen, *An optically activated SIT for high voltage applications*, 5th European Conf. Power Electronics and Appl., 283 (1993).
68. J. Nishizawa and K. Yamamoto, *High frequency high power static induction transistor*, IEEE Trans. Electron Dev., **25**, 314 (1978).
69. S. Aoki, N. Tsukiyama, T. Irie, H. Tadano, and J. Nishizawa, *Low loss high gain 300 V–200 A Class normally-off SIT module for DC motor control*, Power Electron. Spec. Conf. (PESC) 1988, 703 (1988).
70. B. Streetman, *Solid State Electronic Devices*, 2nd ed., Prentice–Hall, Englewood Cliffs, NJ (1980), pp. 286–293.
71. K. Yamaguchi and H. Kodera, *Optimum design of triode-like JFETs by two dimensional computer simulation*, IEEE Trans Electron Dev., **24**, 1061 (1977).
72. R. Singh and B. Baliga, *Cryogenic operation od static induction (SIT/SITh) devices*, Proc. Intl. Symp. Power Semicon. Dev. and ICs, 267 (1995).
73. A. Strollo and P. Spirito, *A two dimensional analytical model for the output I–V characteristics of the static induction transistor (SIT)*, Proc. Intl. Symp. Power Semicon. Dev. and ICs, 196 (1990).
74. J. Nishizawa, T. Tamamushi, M. Ushirozawa, and J. Kim, *Spectral response of an SIT image sensor with an improved structure*, IEEE Electron Dev. Lett., **11**, 143 (1990).
75. K. Yano, C. Kim, M. Kimura, A. Tanaka, S. Motoyama, and T. Sukegawa, *Improvement in on-resistance of a high purity channel SIT by a shielded gate structure*, Proc. Intl. Symp. Power Semicon. Dev. and ICs, 226 (1992).
76. P. Gray and C. Searle, *Electronic Principles – Physics, Models, and Circuits*, Wiley, New York, NY (1969).
77. Y. Kajiwara, Y. Yukimoto, and K. Shirahata, *High power microwave static induction transistor*, Intl Microwave Symp. (MTT-S), 281 (1977).
78. C. Weitzel, *Wide bandgap semiconductor RF power devices*, IEEE Intl. Symp. Compound Semicon., 421 (1997).
79. H. Tadano, M. Ishiko, S. Kawaji, and Y. Taga, *Low loss static induction devices (transistors and thyristors)*, Proc. 20th Intl. Conf. Microelectronics (MIEL), 353 (1995).
80. C. Kim, M. Kimura, K. Yano, A. Tanaka, and T. Sukegawa, *Effects of the doping profile on current characteristiocs in BSITs*, IEEE Electron Dev. Lett., **13**, 95 (1992).

81. J. Stork and J. Plummer, *Small geometry depleted base bipolar transistors (BSIT)–VLSI devices*, IEEE Trans. Electron Dev., **28**, 1354 (1981).
82. J. Nishizawa, T. Ohmi, and H. Chen, *Analysis of static characteristics of a BSIT*, IEEE Trans. Electron Dev., **29**, 1233 (1982).
83. K. Yano, M. Mitsui, H. Moroshima, J. Morita, M. Kasuga, and A. Shimizu, *Rectifier characteristics based on bipolar mode SIT operation*, IEEE Electron Dev. Lett., **15**, 321 (1994).
84. K. Yano, M. Kasuga, A. Shimizu, M. Mitsui, H. Moroshima, and J. Morita, *A novel rectifier based on bipolar mode SIT structure*, Proc. Intl. Symp. Power Semicon. Dev. and ICs, 273 (1995).
85. Y. Uesugi, T. Imai, K. Kawada, and S. Takamura, *Fundamental and third harmonic operation of SIT inverter and its applications to RF thermal plasma generation*, Proc. Power Conv. Conf. (PCC), 1473, (2002).
86. Y. Uesugi, T. Imai, K. Kawada, and S. Takamura, *Third harmonic operation of SIT inverter RF source for ICRF heating in the divertor plasma simulator NAGDIS-II*, Digest of Pulsed Power Plasma Sci. (PPPS) 1677 (2001).
87. H. Ogiwara and M. Nakaoka, *Resonant High Frequency Inverters using Static Induction Transistors,* 3rd Intl. Conf. Power Electron. and Drive Syst. 29 (1988).
88. H. Ogiwara, L. Garnage, and M. Nakaoka, *Quasi-resonant soft switching PWM voltage-fed high frequency inverter using SIT for induction heating applications*, IEE Proc. Electr. Power Appl., **148**, 385 (2001).
89. Y. Hatanaka, M. Yamagami, and M. Nakaoka, *ZCS-ZVS Characteristics of high frequency inverter and DC-DC converter*, 13th Intl. Telecomm. Energy Conf. (INTELEC), 488 (1991).
90. Y. Hatanaka, M. Yamagami, T. Kubo, M. Osakabe, and M. Nakaoka, *High frequency inverter and DC-DC converter with ZCS-ZVS mode*, Intl. Conf. Industrial Electronics, Control, and Instrumentation (IECON), 237 (1991).
91. S. Miyazawa and F. Nakamura, *New cycloconvertor with static induction transistors*, IEE Proc. **136**, Pt. B, 293 (1989).
92. J. Nakamura, Y. Gomi, M. Uno, and H. Hayashi, *Nondestructive readout mode static induction transistor (SIT) photo sensors*, IEEE Trans. Electron Dev., **40**, 334 (1993).
93. J. Nishizawa, T. Tamamushi, Y. Mochida, and T. Nonaka, *High speed and high density static induction transistor memory*, IEEE J. Solid State Circuits, **13**, 622 (1978).
94. Y. Horiba, K. Murakami, S. Ohsaki, K. Kijima, T. Nakano, and H. Oka, *Static induction transistor logic compatible with 1 GHz ECL circuits*, IEEE Intl. Solid State Circuits Conf. (ISSCC), 64 (1980).
95. R. Clarke, C. Brandt, S. Sriram, R. Siergiej, A. Morse, A. Agarwal, L. Chen, V. Balakrishna, and A. Burk, *Recent advances in high temperature, high frequency SiC devices*, High Temp. Electronic Matl. Dev. and Sensors Conf., 18 (1998).
96. R. Siergiej, R. Bojko, R. Clarke, and W. Curtice, *4H-SiC continuous wave SITs*, IEEE 57th Dev. Res. Conf., 98 (1999).
97. J. Henning, A. Przadka, M. Melloch, and J. Cooper, *Design and demonstration of C-band static induction transistors in 4H silicon carbide,* IEEE 57th Dev. Res. Conf., 48 (1999).
98. J. Henning, A. Przadka, M. Melloch, and J. Cooper, *A novel self-aligned fabrication process for microwave static induction transistors in SiC*, IEEE Trans. Electron Dev. Lett., **21**, 578 (2000).
99. R. Siergiej, A. Morse, P. Esker, T. Smith, R. Bojko, L. Rowland, and R. Clarke, *High gain 4H-SiC static induction transistors using novel sub-micron air bridging*, IEEE Dev. Res. Conf. (DRC), 136 (1997).
100. R. Clarke, A. Morse, P. Esker, and W. Curtice, *A 16 W, 40% efficient, continuous wave, 4H-SiC, L-band SIT*, IEEE Conf. High Performance Dev. (Cornell Conf.), 141 (2000).

101. R. Clarke, *Silicon carbide devices for power applications*, IEEE Proc. Bipolar/BiCMOS Circuits Technol. Mtg. (BCTM), 124 (2001).
102. A. Morse, P. Esker, S. Sriram, J. Hawkins, L. Chen, J. Ostop, T. Smith, C. Davis, R. Barron, R. Clarke, R. Siergiej, and C. Brandt, *Recent application of SiC to high power microwave*, IEEE Intl. Microwave Symp. (MTT-S), 53 (1997).
103. J. Wang and B. Williams, *Evaluation of high voltage 4H-SiC switching devices*, IEEE Trans. Electron Dev., **46**, 589 (1999).
104. R.W. Keyes, *Figure of merit for semiconductors for high speed switches*, Proc IEEE, **60**, 225-232 (1972).
105. E.O. Johnson, *Physical limitations of frequency and power parameters of transistors*, RCA Review, **26**, 163-177 (1963).
106. B. Baliga, *Power semiconductor device figure of merit for high frequency applications*, IEEE Electron Dev. Lett., **10**, 455 (1989).
107. S. Delage, D. Floriot, and C. Brylinski, *Solid state RF power amplifiers: status and perspective*, IEEE 10th Intl. Symp. Electron Dev. Microwave and Optoelectronic Appl., 136 (2002).
108. K. Shenai, R. Scott, and B. Baliga, *Optimum semiconductors for high power electronics*, IEEE Trans. Electron Dev., **36**, 1811 (1989).
109. R. Trew, J. Yan, and P. Mock, *The potential of diamond and SiC electronic devices for microwave and millimeter wave power applications*, Proc. IEEE, **79**, 598 (1991).
110. B.J. Baliga, *Modern Power Devices*, Krieger Malabar, FL,1992.

International Journal of High Speed Electronics and Systems
Vol. 15, No. 4 (2005) 1035–1038

World Scientific
www.worldscientific.com

AUTHOR INDEX
Volume 15 (2005)